Barracks, Forts and Ramparts:

Regeneration Challenges for
Portsmouth Harbour's
Defence Heritage

Celia Clark PhD with Martin Marks OBE

Barracks, Forts and Ramparts:
Regeneration Challenges for Portsmouth Harbour's Defence Heritage

Celia Clark PhD with Martin Marks OBE

All rights reserved. No part of this publication may be reproduced, stored in any retrieval system or transmitted in any form or by any means, electronic, mechanical, photocopying, recording or otherwise, without the prior written permission of the copyright holder for which application should be addressed in the first instance to the publishers. The views expressed herein are those of the authors and do not necessarily reflect the opinion or policy of Tricorn Books or the employing organisation, unless specifically stated. No liability shall be attached to the authors, the copyright holder or the publishers for loss or damage of any nature suffered as a result of the reliance on the reproduction of any of the contents of this publication or any errors or omissions in the contents.

In all cases every effort has been made to contact the original owners of the imagery used in this book - where not credited

ISBN 9781912821648

Published 2020 by Tricorn Books
Aspex 42 The Vulcan Building
Gunwharf Quays Portsmouth PO1 3BF
www.tricornbooks.co.uk
Printed & bound in the UK by W&G Baird

Barracks, Forts and Ramparts:

Regeneration Challenges for
Portsmouth Harbour's
Defence Heritage

Acknowledgements

Deane Clark and Margaret Marks were very patient about our obsession over several years to document changes in Portsmouth Harbour. This book is dedicated to them, and to the memory of Ray Riley: prolific author, mentor and friend and to Greg Ashworth: heritage expert, author of *War and the City* and friend. Chris Donnithorne generously contributed his expert historical knowledge by editing several chapters. Dan Bernard was responsible for the book's handsome design.

We give voice to many people's passion for Portsmouth Harbour. We are very grateful to all those who contributed to it – in interviews, publications and by their achievements in the remaking of these military spaces into civilian places. There are many more than those listed below.

Gail Baird Tricorn Books
Dr. Michael Bateman University of Portsmouth
John Bell Former resident Eastney Barracks
John Bingemen Society for Nautical Research
Eric Birbeck Haslar Heritage Group for Haslar Hospital
Pam Braddock Former English Heritage curator at Fort Brockhurst
Darren Bridgman Peter Ashley Activity Centre
David Brock Historic England
Paul Brown Naval Dockyards Society
Sue Bruley University of Portsmouth
Deane Clark Historic Buildings Architect Hampshire County Council
Myrtle Clark-Bremer Southsea Rock Gardens
Jonathan Coad English Heritage Dockyard Historian
Dr. Ann Coats Naval Dockyards Society
Wayne Cocroft Historic England
Deborah Croker JPImedia.co.uk
Dr. David Dennison Council for British Archaeology
Christopher Donnithorne Historian Blockhouse/HMS Dolphin
Caroline Dinenage MP Haslar Hospital resident
Jeff Downing Portsmouth City Council
Rosemary Dunne Fort Brockhurst, Eastney Barracks
Cllr. Peter Edgar Hampshire County Council
Dave Edmundson Fort Cumberland
Martin England Solent Forts
Mike Farquharson-Roberts Retired Surgeon Rear Admiral, former MDG(N)
Stephen Fisher LCT 7074, Photographer
Dave Fricker British Armed Forces Historical Small Craft Group
Joanne Godfree former Librarian Portsmouth Grammar School
Peter Goodship Portsmouth Naval Base Property Trust
Debbbie Gore Gosport Borough Council
Steve and Julia Hender Square Tower
Cheryl Hewitt Former resident Eastney Barracks
Richard Hibbert CBE RN
Robert Hind The News
Terry Hinkley Heritage Walks and Talks, Gosport Society
Richard Holdsworth Chatham Historic Dockyard Trust

Richard Holme Naval Dockyards Society
Rob Hoole HMS Vernon
Nigel Hosier Royal Armouries
Jim Humberstone Hampshire Buildings Preservation Trust
Bob Hunt Portsdown-tunnels.org.uk
William C. Hutchin, FCCI Video Production Equipment
Warwick Jacobs Hovercraft Museum founder
Mark Jones University of Portsmouth
Rob Kennedy HMS Vernon
Brian Kerr Fort Cumberland Historic England
Pete Lambert-Maude HMS Vernon
Michael Laird University of Portsmouth
Michelle Lees Gosport Heritage Action Zones
Danny Lovell HMS Vernon
Jennifer Macey Historic Environment Record Portsmouth City Council
Poo Madge Sailor
Archie Malley Portsmouth Royal Dockyard Historical Trust
Phil McGrath Royal Armouries Fort Nelson
Brian Mansbridge Lee Residents' Association
Lee Matthews Arty's Royal Clarence Yard
Andy Mason Chartered surveyor
Dennis Miles Portsmouth Royal Dockyard Historical Trust
David Moore Palmerston Forts Society
Mick Morris Architect
Louis Murray Gosport Society
Stephen Payne Naval Dockyards Society
Jacky Percival Gosport Aviation Society
Giles Pritchard Architect
Sarah Quail Former Portsmouth City Records Officer
John Roberts, Former volunteer at Explosion Museum
Sarah Roberts Designer
Vaughan Roberts Historic England Archive
Shaun Roster Photographer
John Sadden Portsmouth Grammar School Archivist
Geoffrey Salvetti Palmerston Forts Society
Dr. Roger Saunders Veterans Village
Dave Sherren Map Librarian University of Portsmouth
Captain Nick Stanley RN HMS Vernon
Dr. John Stedman Records Manager Portsmouth City Council
James Thomson Fort Purbrook
Howard Thompson Captains Row Old Portsmouth
Richard Trist Former Chief Executive Portsmouth City Council
Phil Turner RTPI Hampshire Buildings Preservation Trust
Mark Waldron Editor The News
Stephen Weeks Film Director Portsmouth Grammar Schooboy
Duncan Williams Palmerston Forts Society
Dennis Wills HMS Vernon

Contents

Chapter 1   Introduction   9

Chapter 2   Shaped by the Nation's Defence   29

Chapter 3   The Dockyard downgraded; defence sites close   47

Chapter 4   'Beware of Little Expenses: A Small Leak Will Sink a Big Ship'   69

Chapter 5   The conservation tide comes in: local responses   85

Chapter 6   Heritage tourism to the rescue? Portsmouth Historic Dockyard: 105

Chapter 7   Portsmouth Historic Dockyard: buildings, ships, boats – and one that got away   34

Chapter 8   Make Love Not War: Museums and galleries in military spaces   163

Chapter 9   A tale of two hospitals: the Royal Naval Hospital Haslar and Queen Alexandra Hospital Portsdown Hill   192

Chapter 10   Ordnance Yards: HMS Vernon/Gunwharf and Priddy's Hard   218

Chapter 11   Early flight to a civilian airfield   268

Chapter 12   What can you do with the harbour's legacy of military lines, batteries and forts?   284

Chapter 13   Forts, batteries and military lines: Portsdown Hill to the forts in the sea   309

Chapter 14   Education: University and schools in former defence buildings   338

Chapter 15   Research – cutting edge developments in two centuries – and into a third   352

Chapter 16   Defence sites repurposed as housing   364

Chapter 17   Field of fire to common, redoubt to rose garden; bastion and military lines to bosky walks   398

Chapter 18   The bigger picture; local lessons   426

Appendix   439

Index   447

Barracks, Forts and Ramparts

Regeneration Challenges for Portsmouth Harbour's Defence Heritage

Chapter 1

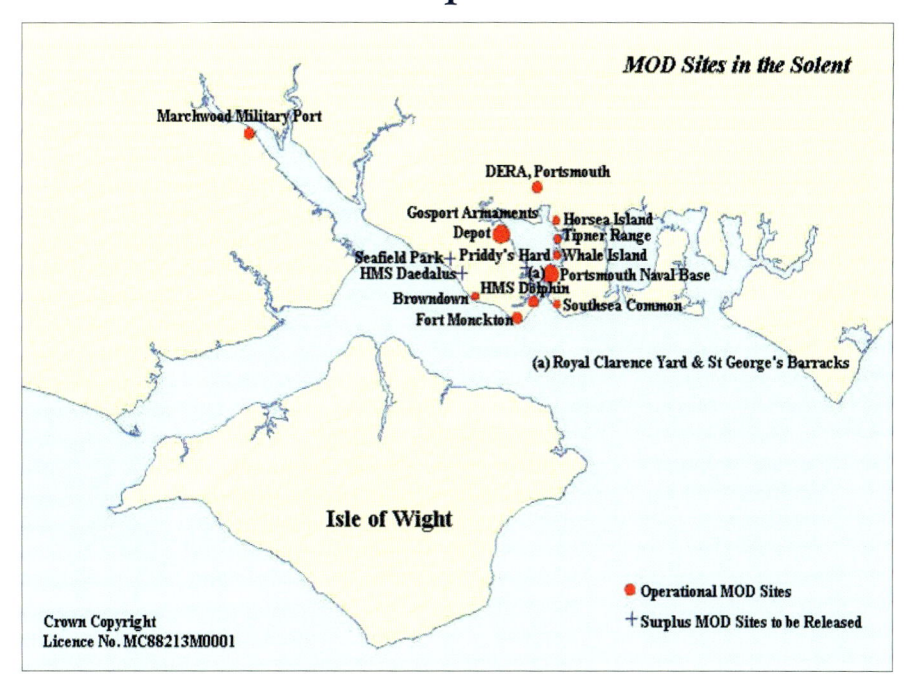

Map of MOD Sites in the Solent

"Without the Royal Dockyards there could have been no Royal Navy…The English royal dockyards, victualling yards and hospitals formed what are arguably the largest industrial centres in Britain before the Industrial Revolution, while their economic impact was out of all proportion to their size'

The Royal Dockyards 1690-1850 *Architecture and Engineering Works of the Sailing Navy*

Jonathan Coad 1989

Introduction

Portsmouth Harbour has one of the densest concentrations of specialised defence establishments in the country, within the context of South Hampshire on England's south coast. Its extensive area, narrow entrance from Spithead within the shelter of the Isle of Wight and proximity to our rivals on the high seas made it ideal for the development of the dockyard. For many centuries fleets and armies sailed from the country's premier naval port to fight in turn the French, Spanish, Dutch, Americans, Germans and Russians, and to supply and garrison the global British empire.

During wars, for every uniformed person in action, there were thousands of men and women and a substantial portion of the nation's material and financial resources supplying and equipping them. A complex system of military, naval and air force support facilities developed over the centuries around Portsmouth harbour: gunwharf, victualling and ordnance yards, hospitals, barracks and airfields, defended by successive rings of fortifications.

In the nineteenth century the government feared the threat of a French invasion, with the dockyard as a particular target, so the 'ring fortress' of twelve forts was constructed around the harbour. Until after WWII many local defence installations were guarded by a substantial army presence.

Inspired by living in Portsmouth and Gosport for nearly fifty years, Celia Clark and Martin Marks live within this military framework to our lives. We have observed and taken part in the transformation of these physical survivors of the harbour's rich military and naval past and their moving on to new, civilian futures. Over the past century, a whole alphabet of military structures and sites across the world, including many local examples, have found new uses, or are in the process of doing so: abattoirs, airship sheds, arsenals and armouries, bakeries, basins, balloon sheds, barracks, batteries, boathouses and breweries, casemates, caponiers, chapels and cooperages, depots, drill halls, drydocks, factories, forts, fortresses and foundries, gunpowder works and gun wharfs, magazines, mould lofts and monuments, pillboxes and pontoons, radar stations, redoubts, rigging houses and roperies, sawmills, scrieve boards, slaughterhouses, slips, smitheries and storehouses, turrets and towers, victualling yards, wardrooms, watchtowers, wind tunnels, workshops and zoos; listed by SAVE Britain's Heritage in 1993.

Even within a comparatively small compass around the harbour, the new uses include industrial and residential estates, museums, art gallery, art cinema and multi-purpose auditorium, artists' studios, film locations, shopping centre, squash court, leisure facilities, traditional boatbuilding and restoration school, marina, conservation laboratory, archaeology and marine research centre, schools, riding stables, luxury hotels, university facilities, student hostel, rifle range, veterans centre, distillery, band practice room, boxing gym, ship recycling yard, squash courts, storage for historic military vehicles, nature trails, open air arena, climbing wall, fitness trail, rose garden, public open space… and this conversion process is still ongoing.

As participants and observers, Celia Clark and Martin Marks decided to document this complicated transition with first-hand accounts by people who played key parts in it, and to offer a snapshot of what has been achieved by the end of the second decade of the twenty-first century. As the wheel of time turns, it takes with it the living memory of how these extraordinary transformations from military to civilian life were achieved. We focus on more recent events - from the late nineteenth century releases of defence land to more detailed accounts from the mid-twentieth century onwards. Of course, it's a challenge to document a moving picture. In such a broad canvas and long sweep of time there will be omissions and mistakes – which we fully acknowledge and are happy to have corrected. We are grateful for permission to use the many illustrations from various sources, which are acknowledged. Other photographs are by Celia Clark.

Understanding each site's history and how it came to be how it is now is an essential stage is determining appropriate futures for it. We do not offer detailed histories using documentary sources, but short accounts of these diverse places and how they came to be preserved and new uses found for them. We celebrate the extraordinary creativity that gives these places new and unusual life – as they continue to evolve.

As we explore our home territory it is changing fast, as more defence sites are released, and ambitious plans are proposed to transform them. In 2020 there is a dramatic plan to transform the Tipner peninsula in the centre of the harbour into a huge £1bn new development that involves reclaiming land around the edges to create an entirely new district of the city. Development of Fort Blockhouse and Fraser Battery is currently being discussed. Gosport's Heritage Action Zone is breaking new ground in cooperation between the Defence Infrastructure Organisation - the MOD's property arm, with Historic England, the local authorities and the local civic society to come up with positive reuses of challenging sites in Gosport's diverse military legacy.

As yet there is little research into the conversion of airfields, depots, barracks, dockyards, training grounds and fortifications into civilian uses in the UK, which leaves significant gaps in our knowledge about the effect of base closures and the prospects for civilian uses. Different countries have contrasting systems of disposing of redundant government land – from free transfer to meet community needs (the United States) to sale to the highest bidder (the UK and Germany). These differences influence how the land is subsequently used.

Celia Clark and Samer Bagaeen's book *Sustainable Regeneration of Former Military Sites* published by Routledge in 2016 was the first to explore the complexity of this transition in many different countries. It identifies the factors that contribute to sustainable regeneration of these very special places. The challenges include poorly maintained historic buildings, contamination, poor transport links, loss of skilled employment…. Positive and sustainable reuse may be defined as the creation of new long lasting economic, social and cultural activity which benefits ex-defence communities, employment that replaces the income and work of the soldiers, sailors, airmen and civilian staff who were once employed in defence facilities, as well as the adaptive reuse or reconfiguration of the spaces and surviving structures - plus the addition of new buildings to house sustainable activities. How the sweeping restructuring of UK planning proposed in the 2020 White Paper will affect this widespread but largely undocumented transition process and its outcomes is as yet unclear.

We focus on the experience of reuse of defence sites around Portsmouth Harbour - for the excellent reason that all the challenges to restoration and reuse of our defence legacy are concentrated and demonstrated in this one small area on the south coast of England. Together they offer a microcosm of how similar sites around the world can successfully be brought back to new sustainable civilian life.

Within these dense and varied case studies contained in this small area, how much local benefit have formerly defence-dependent communities gained when they were redeveloped? We examine the conflicts that arise in the transformation process, what the actors' motivations are and whose view of these special sites' future prevails. One underlying theme of the book is who has the power to change these spaces into something new, to transform them into their extraordinary variety of new uses? There appears to be a democratic deficit in the extent to which local people can influence government agencies such as the Environment Agency and the MOD's Defence Infrastructure Organisation or local government quangos such as PUSH (Partnership for Urban South Hampshire) and the East Solent Coastal Partnership which is responsible for designing the area's sea defences. This book is a contribution to the debate about how much local people can influence the new land uses that develop on redundant sites no longer needed for national defence. We identify good practice and the many factors that have a bearing on what happens after the military leave.

Specialist interest groups who may find it useful include geographers, planners and surveyors and other built environment professionals working on sustainable conversion and adaptation of these complex structures, local authorities responding to proposed disposals of defence sites, as well as people with interests in recent local and post-defence history - including those who remember what happened - and those who participated in the transformation process.

These include a farsighted politician initiating the exponential growth of defence heritage tourism; a visionary architect anticipating the millennium; military architecture experts researching and restoring Victorian forts; divers determined to celebrate achievements underwater over the centuries in a new museum; developers rebuilding historic defence sites with high-return new uses for their shareholders' profit; a local voluntary society with a vision to transform a depressed and inward-looking town with a renewed emphasis on history and built heritage; and a determined local authority conservation officer researching and identifying a site's special history in military architecture and technology to ensure that the key structures survive its redevelopment.

Physical clues to the harbour's role in national defence

Once your eye's attuned, echoes of the dominance of national defence are still evident in the harbour's townscapes. Rachel Woodward's *Military Geographies* recognises the "tenacity in urban forms and lives of a military inheritance into a civilian present, and the possibilities or otherwise of conversion of post-military landscapes."

The evidence of the harbour communities' martial past is there - if you look for it. St. Jude's church spire in Southsea is extra tall, because the Admiralty gave its architect Thomas Ellis Owen a £1000 grant so it could serve as a seamark for the Swashway channel into the harbour. Less benign traces are visible too. A star shaped crater in a barracks wall speaks eloquently of terrifying bombing raids – as does the roofless ruin of the Garrison Church in Old Portsmouth. The strange zigzag shape of Portsmouth's Terraces on the way to the seafront is an echo of the town's vanished star-shaped defences. Gosport's surviving ramparts around the town, barracks, naval hospital and victualling yard have shaped subsequent development. Fragments such as the hefty iron links in Capstan Square at the harbour entrance are evidence of the chain once hoisted on pontoons to close off the harbour entrance to enemy ships.

Dragons' teeth Ferry Road Eastney

Anti-tank traps at Fraser Battery Eastney 2019

Concrete blocks – 'tank traps' or 'dragons' teeth' – line the beach at Eastney in front of Fraser Battery and on the road to Hayling Ferry and near Fort Brockhurst, all designed to deter the Germans from landing in WWII. Sloping concrete spine walls on the southern shore of Horsea Island and on Hayling Island's Langstone Harbour channel are evidence of the Mulberry building yards – where concrete pontoons were constructed and towed across the Channel to create an artificial harbour for the D-Day landings.

Broken Mulberry Harbour section Langstone Harbour 2014

The shallow Langstone Harbour to the east of Portsea Island was where a replica city was picked out in lights to deceive German bombers in WWII. The brightly painted buoys, anchors and chains that line the road into Port Solent may well have originated in the dockyard. According to a plaque there the headless Mark 8 torpedo now lying in front of the restaurants was developed on Horsea Island in 1928; an improved version was used during the Falklands campaign in 1982.

Observation Point or gun emplacement Portsdown Hill west of Fort Purbrook   Mick Morris

Earlier military land releases

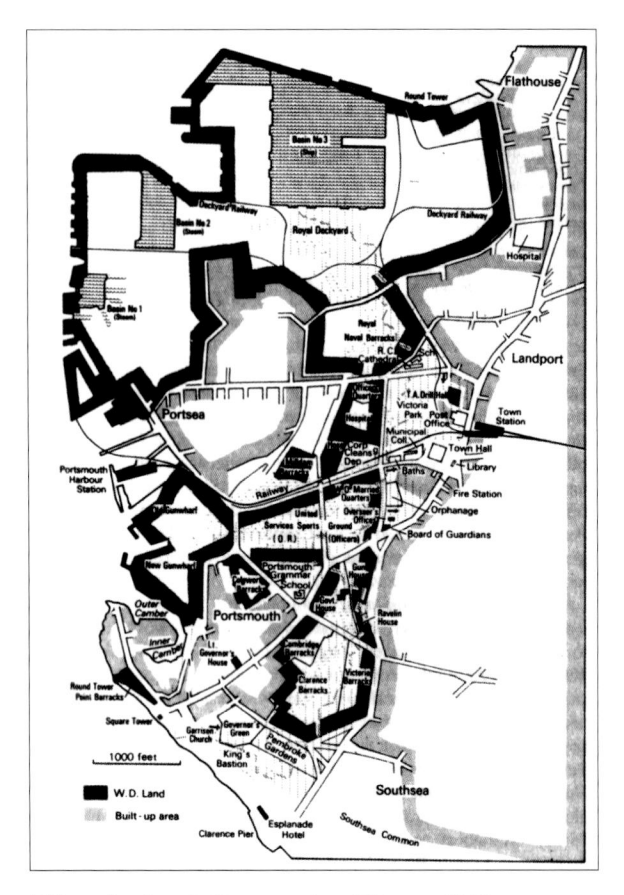

Military land use in Portsmouth and Portsea, 1860
The Geography of Defence Routledge 2016 p. 56

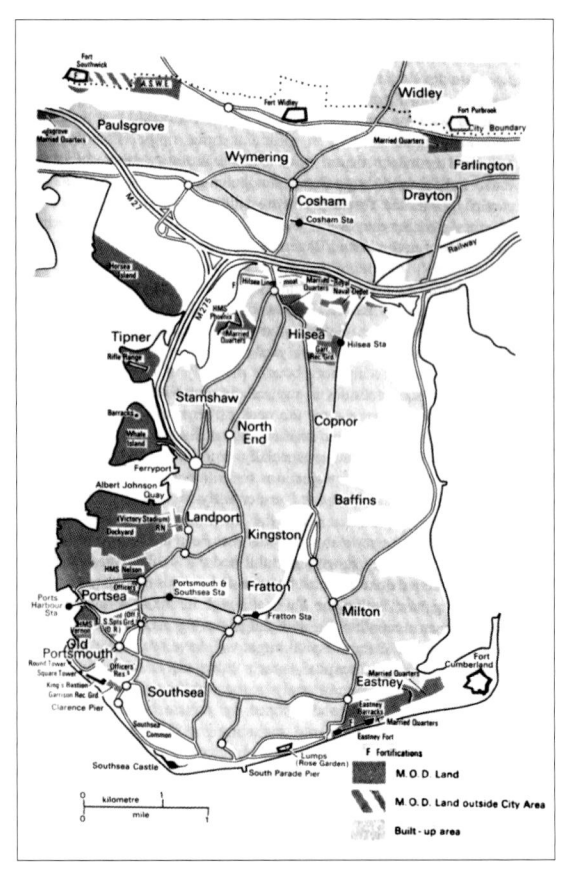

Military land use in Portsmouth and Portsea, 1986
The Geography of Defence Routledge 2016 p. 56

Our second chapter outlines how the earlier defences of Portsmouth and Gosport were created and how some disappeared in the nineteenth and twentieth centuries as the towns expanded, the dockyard contracted, and the army garrison closed. This process was profoundly important to the harbour communities' future, yet it was unreported and not studied until a hundred years after the first releases began.

The Geography of Defence was commissioned for the meeting of Institute of British Geographers at Portsmouth Polytechnic in 1986 and published by Croom Helm in 1987. Edited by Michael Bateman and Ray Riley, invited articles examined for the first time the geographical impact of national defence, with Portsmouth as a central case study. Ray Riley's chapter: 'Military and Naval Land Use as a Determinant of Urban Development – the Case of Portsmouth' is a masterly analysis of the successive layers added to defend Portsmouth harbour – and also of their subsequent release into civilian uses from the late nineteenth century onwards. Together with Ray Riley's many Portsmouth papers on the dockyard and local archaeology it is the source and inspiration for this book. It's not surprising that as a seminal work, Routledge republished *The Geography of Defence* in 2015.

Why are local defence facilities no longer needed to defend the country?
Who influences what happens next?

Why and how were these establishments, so often important in national history declared redundant from their defence roles? In response to geo-political and economic change, defence cuts and developments in military technology, more and more military land and facilities are becoming surplus to countries' defence. Since 1979 the UK government has sold some two million hectares of defence, health and transport land to private owners – or about 10% of the entire British land mass.[1] In planning civilian futures for a significant subset: defence land, there are considerable difficulties, explored in this book.

Gosport Borough Council and Portsmouth City Council produced this map to show the extensive areas the MOD occupied or released in 2017 for the third seminar of professionals, developers and academics involved in the transformation of defences sites to civilian futures.

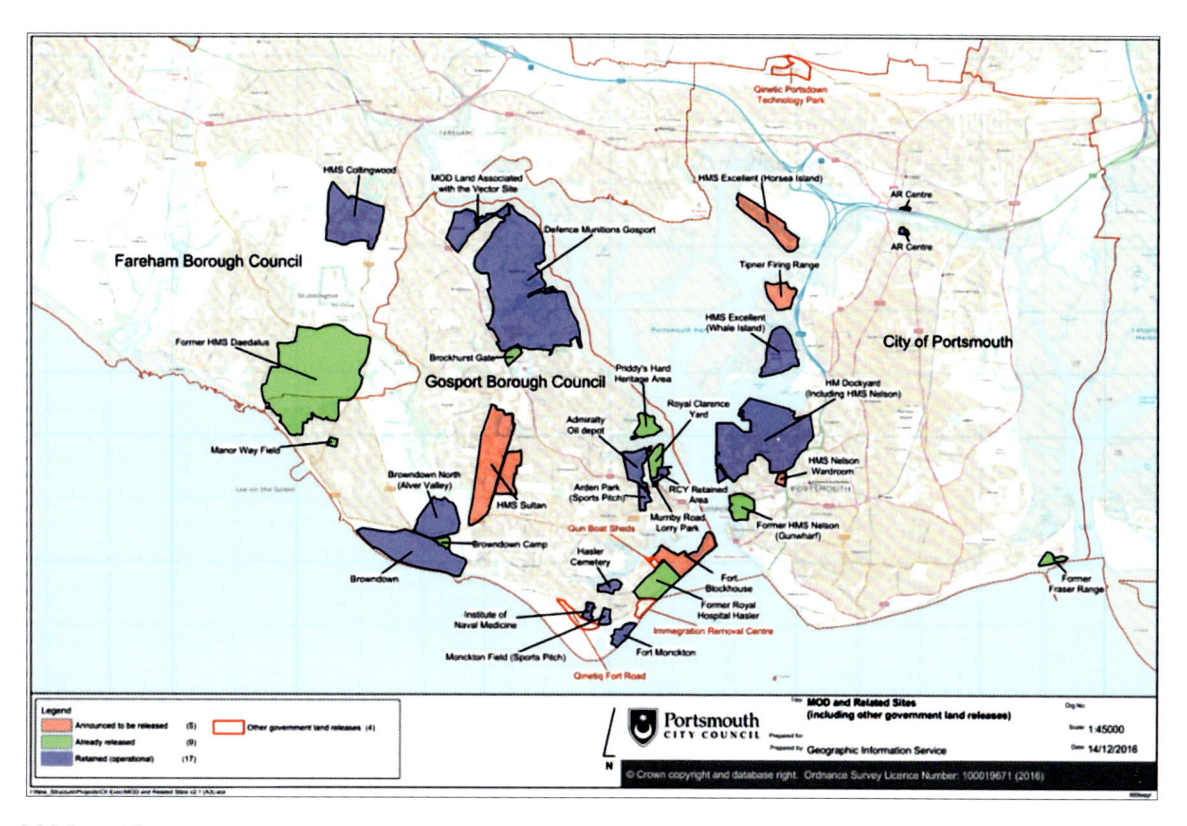

MOD and Related Sites (including other government land releases   2017 Portsmouth City Council

We explore the UK's current system for the disposal of state property set out in the Treasury's 1996 rules: sale at maximum value within three years of closure. Government departments disposing of historic areas are recommended to take into account that maximisation of receipts should not be the overriding aim in cases involving the disposal of historic buildings. Sites should be considered as a whole to preserve settings, using other methods than sale on the open market or competitive tender where these will increase the chances of securing appropriate ownership and use. Early consultation with all interested parties to help to overcome any difficult or controversial issues is recommended. Clearly, from our local experience, this does not always happen.

Changes of mind by the MOD also raise difficulties for local authority planners and conservationists. In January 2016 the MOD announced that they were selling twelve military sites, including HMS Nelson wardroom in Queen Street Portsea in order to generate £500m to reinvest in defence and to provide land for 15.000 new homes as part of the government's initiative to build 160,00 new homes by 2020. This was the first part of an MOD plan to reduce the defence estate by a third.[2] There were hopes that the wardroom's sale and conversion into a hotel might finance the restoration of the Old Academy in the dockyard which remains in a seriously decayed state. Three years later, the wardroom is still unused and in naval hands, with no sign of it being put up for sale. Is this because it too needs serious money spending on it? The Ministry of Justice, too is selling Haslar Barracks, dating from 1802, last used as a detention centre for refugees.

The effect of closure, especially of historic military sites, now defined as 'defence heritage', has been a local issue for the last forty or fifty years. We examine the effect of the removal of the Crown Exemption from civilian legislation in 2006 - which in theory means that the Ministry of Defence must maintain its historic estate in good repair with regular inspections and funding, finding appropriate new uses where necessary. Leaving buildings empty and unused is never a good idea. Historic England publishes a Historic Buildings At Risk Register every year; in our area the greatest number are always those belonging to the MOD. The list's purpose is to stimulate owners into repairing their neglected buildings, but the MOD tends to say that they are not funded to keep buildings they no longer need in good repair – their budget priority being defending the country. The decay of important dockyard buildings currently left empty and unmaintained and other defence heritage buildings on the At Risk register in Gosport is of considerable concern to many, including the Naval Dockyards Society and Hampshire Buildings Preservation Trust. What is not acknowledged is that the cost of eventual repair and reuse escalates during years of neglect and inaction.

As long ago as 1991 in its annual Conservation series, Hampshire County Council Historic Buildings Bureau focused on 'A Strategy for our Defence Heritage'. SAVE Britain's Heritage, a national building conservation lobby, helped to raise Its survival and reuse in their pioneering study: *Deserted Bastions: Historic Naval and Military Architecture* published in 1993. *Beauty or the Bulldozer,* SAVE's proposals for Peninsula Barracks Winchester in 1994 made conservation-led recommendations that were largely implemented but this is by no means always the case.

Economic, social and environmental interests converge in decisions about the transformation of the former defence sites around the harbour. An extraordinary range of people and groups have risen to the challenge of what to do with these specialised places and took action, from schoolboys, specialist developers, local civic groups, Prince Charles, to charitable trusts and campaigning heritage bodies. Stephen Weeks was a sixteen-year-old boy who petitioned the Minister of Housing to stop the demolition of 56 listed historic buildings by Gosport Borough Council between 1947 and 1965 in order to build blocks of council flats. The Portsmouth Society, a local civic group, criticized plans for a new naval headquarters on Whale Island as 'an ugly great lump' - which caused outrage in the Admiralty, but led to a complete change of design. Volunteers are crucial to many of the museums around the harbour, including the Diving and Hovercraft museums and the Historic Dockyard.

In 1983 the Portsmouth Naval Base Property Trust (PNBPT) took on what became the dockyard Heritage Area which includes significant historic buildings as well as the *Mary Rose, HMS Victory, HMS Warrior 1870* and *M33,* all of which have their own trusts as tenants of PNBPT, along with the National Museum of the Royal Navy. In the trust's 20 Year Report published in 2006 the enormous debt of gratitude owed to those who have given and continue to give their time freely in helping to achieve its ambition in spelt out. "The value to the Trust of the professional time and skill of its volunteers cannot be overestimated". PNBPT also acquired Priddy's Hard in Gosport across the harbour in 2009. A new channel was dredged for a ferry connecting it and the Explosion Museum to the historic dockyard.

The Prince's Trust sponsored an Enquiry by Design workshop on the future of the Royal Naval Hospital at Haslar in 2009. SAVE Britain's Heritage supported the Portsmouth Society's campaign to save the dockyard's Boathouse 4, which was threatened with demolition. QinetiQ, an ex-government research agency continues to test ship hull design research in Haslar's pioneering ship testing tanks. Gosport's local authority conservation officer anticipated Ministry of Defence disposals and worked with their property arm, Defence Estates, later the Defence Infrastructure Organisation (DIO) to prepare agreed conservation plans for both Haslar Hospital and Fort Blockhouse.

The process of conversion is still continuing. In 2015/6 three historic defence sites in Portsmouth were restored and converted to new uses. Boathouse 4 was restored and new facilities added for the International Boatbuilding Training College. The casemates in Old Portsmouth, which had belonged to Portsmouth City Council since the early 1960s, were converted in 2018 to the Hot Walls Studios and the lively Canteen café, with a balcony to the Round Tower suspended over the sea. The Cell Block forming the seaward wall of Portsmouth dockyard was renovated as the Cell Block Studios for starter business spun off from the university. In 2016 Portsmouth city council purchased Tipner Range and reclaimed land north of Horsea Island for residential and business development.

Defence disposals and their regeneration are so important to the southern region that there have been three local seminars on the subject: in Oxford in 1996, Winchester in 2002 and Burseldon Brickworks in 2017, to examine what was happening to the many sites coming onto the disposal list - and to identify good practice by the MOD, local planning authorities and developers. Since 2017 the Hampshire Building Preservation Trust has gathered evidence from local authorities across the county on defence site disposals and their subsequent civilian transformations - in order to raise the issue with the Defence Select Committee and to enter a dialogue with the MOD's Defence Infrastructure Organisation which is responsible for its property, but so far - without success.

Defence works in the Portsmouth harbour landscape and economy

Hilsea Lines' earthen linear hill defending the northern shore of Portsea Island from the mainland is planted with trees covering the brick casemates, once bristling with heavy armament. High red forts punctuate the skyline of Portsdown Hill, their guns facing northward against any invading French troops. Two of the four round sea forts planted on the seabed at Spithead between Portsmouth and the Isle of Wight are now luxury event venues. Hampshire County Council restored the moat, ditches and sharp slopes of the southernmost Gosport rampart and the overgrown entrance gate to Bastion No. 1 in the 1970s and again as part of the Heritage Action Zone. Although the remains of Portsmouth's vanished landward ramparts briefly reappeared in late 2018, they were reburied under the university's new sports centre.

Sometimes changes of use follow each other quite rapidly. Connaught Drill Hall in Stanhope Road was converted into two nightclubs: Liquid Envy in 2007. It now has a gym as well as Pryzm, the largest nightclub in Portsmouth. The red brick Renaissance Clarence Barracks in Old Portsmouth was converted into Portsmouth City Museum and Art Gallery in 1972-3, but the next-door NAAFI that housed the City Records Office from the early 1970s is once again empty. A few years ago it began to deteriorate, its backbone cracked over the underlying city ramparts, and it is likely to be demolished. In contrast, some sites show a continuity of use. The Explosion Museum of Naval Firepower developed from the collection at the former Royal Naval Armaments Depot, Priddy's Hard.

Residential conversion often seems the obvious next use for redundant barracks. We explore how Eastney Barracks and St. George Barracks East and West in Gosport

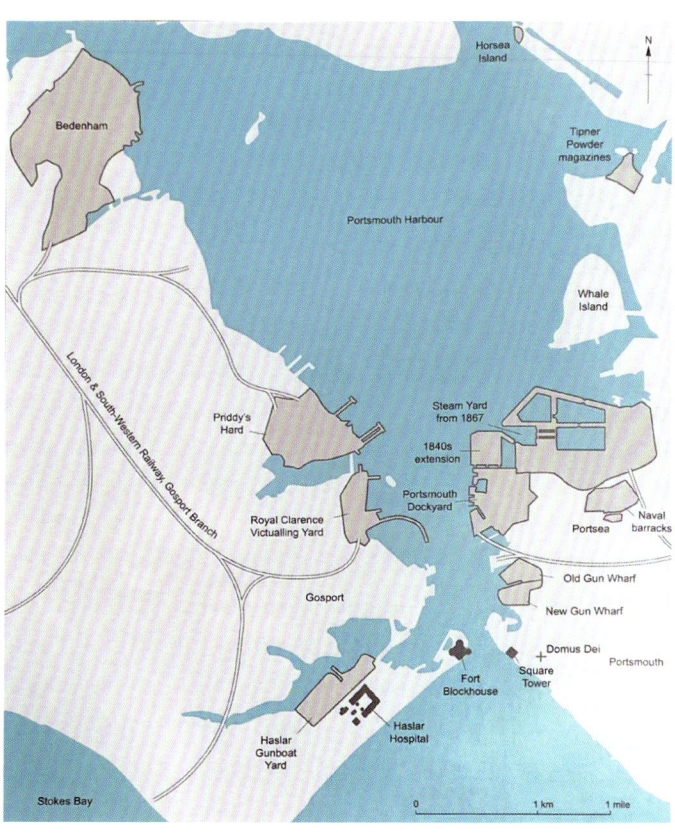

Naval installations at Portsmouth   Jonathan Coad 1989
Historic England Archive

were converted into private housing estates. Other sites too, such as the Royal Clarence Victualling Yard are now residential.

The Millennium walkway on the Portsmouth and Gosport shores of the harbour entrance links many former Ministry of Defence sites together. Defence heritage tourism is a huge sector of local employment. In 2015 Portsmouth welcomed 9.2 million visitors to the city's attractions. In 2014 they spent about £463.5m – an important part of the city economy. The prime attraction is the Historic Dockyard. Its evolution from part of the active naval base to the major tourist attraction on the south coast of England is explored in two chapters: the establishment of the Portsmouth Naval Base Property Trust and the National Museum of the Royal Navy and individual buildings, basins and dry docks, historic ships and boats.

How do people feel about local defence heritage?

This book owes much to local historians, whether focused on documentary history, naval boats and ships, particular sites – HMS Vernon, Priddy's Hard, Fort Blockhouse – or historic buildings and planning. Here are two contrasting responses expressing local pride in our defence legacy, nearly two hundred years apart.

Dr Henry Slight described Gunwharf as he saw it in 1820 in eloquent terms:

And now the Gun-wharf's ample space
Detains, each stately pile to trace –
Stupendous buildings, which might vie
With proudest boast of Italy!
For though no statue meets the eye –
Yet sacred to utility,
High walls and sculptured panels grand,
And domes and towers, on haven strand –
Large stores, where once the billows rude
And vessels, fraught with treasures, rode.
The iron bridge, revolving round
Thrown o'er the foaming stream profound;
The dreadful implements of war –
The mansions neat - the gay parterre –
And mighty stores for gun and ball
Compose this spacious Arsenal.

Dr. Henry Slight 1820

According to a 2015 poll commissioned by the Heritage Fund to coincide with 20 years of National Lottery investment into the UK's heritage, 81% of people in Portsmouth feel proud to live there because of the city's heritage - and say it makes it a better place to live.[3] The Fund has given over £6 billion – including £58 million to 67 projects in Portsmouth alone. Most of this investment was into the city's naval and maritime heritage: the historic dockyard, the *Mary Rose* and Fort Nelson. Gosport has also benefited from Heritage Lottery Grants.

Key findings of the poll were that heritage plays a powerful role in bringing people together and helps to improve perceptions of quality of life, and that the benefits of heritage are seen as both transactional and emotional, encouraging local pride and fostering social cohesion. The sample was not large and did not include Gosport, but 353 Portsmouth people said that heritage was important for the city and made it a place people were likely to visit. 79% of them said that heritage benefits them personally with cultural and recreational opportunities, especially for families, but "More needs to be done to engage people on a local level in their heritage and spread investment across the city. While only 4% of respondents surveyed hadn't heard of the Royal Naval Museum and Mary Rose Museum, a significant 60% were unaware of the Gosport Milestones Walking and Cycling Trail, and 52% didn't know about the Arthur Conan Doyle Collection at Portsmouth Library."

"A stakeholder taking part in the research said: 'Funding for heritage has permitted Portsmouth's big attractions to develop and succeed. Without the investment in heritage, Portsmouth wouldn't have the attractive product it has today. It wouldn't be able to compete on a national and international level.'"

Two mass events demonstrate the harbour communities' strong attachment to their historic defence establishments and to commemorating their role in national history. Gosport people's eloquent campaign against the government's proposal to close the historic Haslar Naval Hospital in 2002 attracted over 20,000 demonstrators in 1999. Public meetings, vigils and deputations ensured that the hospital remained open until 2009. Our Tale of Two Hospitals chapter documents the campaign in detail.

D-Day 75 Commemoration Southsea 2019
Deane Clark

Portsmouth Harbour's role in national and international history was highlighted in the D-Day commemorations in Portsmouth and Gosport on the 40th, 50th and 75th anniversaries, when a diminishing number of veterans, American presidents, the royal family and heads of state attended ceremonies on Southsea Common and Stokes Bay. The 75th event in 2019, which was attended by thousands of people, was the largest of all. The Ministry of Defence said that more than 4,000 personnel were involved – one of the biggest mobilisations of the UK forces in recent years, not for defence, but commemoration. After the reception with veterans the world leaders met to discuss NATO and world security. Many veterans were then ferried by cruise ship across to France, accompanied by a flotilla of naval vessels.[4] People's memories and research identified the many places where preparations for the greatest invasion force ever mounted were based.[5]

Stokes Bay D-Day 75 Re-enactment June 2019

Impressively, in 2019 Gosport Borough Council planned to rebuild the vanished ramparts around the town with a public walkway on top. These once stood 20 to 30 feet high, extending from the surviving No. 1 Battery to the ramparts around Royal Clarence Yard. "This is possibly the most innovative and exciting part of the Local Plan. The Council is to be commended for the ambition of its vision, which combines the protection and conservation of both the historic structures and the natural environment with the plans to improve community and visitor access to some of the most important heritage assets in the Town and along the Waterfront." [6]

Gosport's military history stretches back more than 800 years. In the 21st century it has five military sites, five cadet units and 800 service homes. 13% of the population are members of the armed forces community. In 2020 the Borough Council's support for them was recognised by a gold award in the MOD's Employer Recognition Scheme under which it offers extra leave for reservists to attend training, flexible working for staff whose partners are in the armed forces and help for members of the armed forces community to apply for jobs and training.

How deeply the armed services ethos still runs in the local psyche still surfaces regularly: whether it's the enthusiasts who booked out all the 2019 Heritage Open Day tickets for HMS Excellent, the Hotwalls History Tour, Fort Cumberland, Fort Blockhouse, Royal Haslar Hospital, Shore Leave Haslar in the Memorial Garden at the hospital, Royal Clarence Yard, Priddy's Hard, Fort Brockhurst… to which we might add the thousands who have attended the D-Day commemorations over the years. Many people protested when the Royal Marines Museum's public galleries were closed in April 2017. They also signed the petition against moving the Yomper statue as part of the Royal Marines Museum's planned move into the dockyard's Boathouse 6. The weight of local opinion led to the decision to leave the Yomper where it is.[7] The anniversary of the Battle of Jutland in 2016 stirred up memories of lost relatives a hundred years before and well as inspiring an exhibition in Boathouse 5. In 2019 Princess Anne attended the commemoration on Whale Island of the 834 men lost in the sinking of the *Royal Oak* at Scapa Flow on 14 October 1939.[8] Portsmouth received a government grant to commemorate VE Day 75 in May 2020. Union jacks, bunting and pavement parties were welcome reliefs to the lockdown. But in the evening, because of coronavirus, rather than the wild celebrations at the end of WWII, hundreds of us congregated on Southsea ramparts to watch searchlights mounted on Southsea Castle probe the night sky over the city, standing in sombre silence – remembering the many thousands dead and dying then and now all over the world. Sad too, are the decommissionings of ships with close links to the city.

Ark Royal Decommissioning Parade, Guildhall Square 2011

Locals enjoy the ceremonial events put on by the navy and the Royal Marines. Thousands of us attend the moving Remembrance Day ceremonies each November with the Royal Marines band providing the music in Portsmouth's Guildhall Square. Their School of Music is based in a converted prison in the dockyard. Locals flock to the Field Gun competition on Whale Island and Armed Forces Day each June. Edward King's evocative paintings of Portsmouth's war damaged fabric were shown at a popular exhibition of his work in Portsmouth City Museum in 2016-7. Richard Eurich's vivid depiction of the devastating bombing raid on Portsmouth dockyard loaned from the Tate Gallery was the focus of sold-out lectures about Portsmouth in the Blitz in January and March 2017.

The extent of people's attachment to local place and association with their lives is not often recognised by planners, who distance themselves from the sounds and smells of real places to the abstract plane of maps. "A low response to the proposals" (Gunwharf) may be all they hear in response to a redevelopment application, but an alert ear to the ground will pick up all sorts of voices on these specialised sites which were once humming with human life - in wartime, peace, or times of scientific or technical innovation.

Family history – a Sapper (Royal Engineers) grandfather who was stationed in Fort Gilkicker, a neighbour who remembered growing up in Fort Brockhurst - a paradise for children – lines of washing strung across the parade ground, or a little girl imagining that a Gosport fort is a fairy castle, with water lilies growing in the moat… all show what a hold our military surroundings have on us. Sadder memories well up too. When she gave a lecture about women's work in Portsmouth Dockyard in WWI Celia Clark was moved by a poignant photograph of a woman wearing her badge "On War Service" dressed for her work in the dockyard. In 1914 her husband had just been killed and her baby son had died of bronchitis. She had never told her son about her war work.

Retired people evoke their working lives in barracks, mess, dock, battery, hangar, laboratory, workshop, gymnasium, chapel, sickbay or stables. We describe how in 1982 in response to the historic downgrading of Portsmouth Royal Dockyard to a Fleet Maintenance and Repair Base, a Times obituary mourned the end of centuries of dockyard craft tradition. Celia Clark arranged this at the request of four shipwrights and a boilermaker who founded the Portsmouth Royal Dockyard Historical Society - while working eighteen hours a day in preparation for defence of the Falklands Islands. These craftsmen collected tools, equipment, and documents to preserve the dockyard's material history. Celia was their group's founding secretary for two years. Using a tape recorder from Portsmouth City Museum she recorded as wide a range of specialised dockyard trades as she could. Some of these recordings are transcribed; they can be listened to in Portsmouth History Centre.

Fort Cumberland Guard
Re-enactment 2005

All kinds of specialist interests converge on old defence sites. The ecologist is enthralled by natural reclamation of forts by plants and animals (Fort Gomer). A restorer of a Saxon necklace or a Roman sword works in a high-tech laboratory inside a former motor transit shed (Fort Cumberland). A web designer spins her webs in a cell once used to lock up drunken sailors (Cell Block Studios). John Green, the artist and retired dockyard rigger inspired by his long experience as a skilled craftsman paints powerful images of his colleagues' working lives. Turban-headed gravestones commemorate the Turkish sailors who died of cholera in Clayhall cemetery in Gosport.

Other defence sites further afield offer other visions. Homeless people imagine living in those empty barrack rooms in the ready-made village (Attercliffe Fort Dover). A visionary planned a sustainable community on a former airfield, which gave back as well as taking from the earth, while Daniel Scharf wanted a museum dedicated to the dangerous times of the Cold War (Upper Heyford US Air Force Base, Oxfordshire).

Back in Portsmouth craftspeople and artists set up workshops in the empty casemates in Old Portsmouth (Clay Station/Portsmouth Makers; Hot Walls Studios). Kites fly, barbecues are lit and hymns sung on Southsea Common, once an assembling ground for armies preparing for invasion of the continent. A pony owner taught disabled children how to ride in the empty military playing field so they can enjoy physical contact with another creature in motion (Priddy's Hard). Fortifications freaks enjoy exploring dank tunnels, torch and old creased plans in hand, peering out through bramble obscured gun slits (Palmerston Forts Society). Adventurous children lead expeditions and battles, building dens or pushing over walls, away from adults' censorious view and fisherman dream peacefully of that monster pike in the moat (Fort Rowner HMS Sultan). Teenagers gather alongside their piled-up bikes by a weedy moat (No. 1 Bastion Gosport). Boys on motorbikes launch themselves down the eroded earth cover of military lines (Hilsea Lines). Lovers' intensity's cocooned in bowers of flint and brick (Fort Widley). Leaf cutting insects make busy curving streets over tumbled bricks, stone heaps and twisted rails. Owls swoop over ramparts, moles undermine smooth parade grounds, spiders spin silk universes across crumbling casemates ….

The human voices, rarely heard by officialdom, also argue with each other: 'old buildings are an investment of energy and money that should be recycled' (Gunwharf)'; 'that rare rosy weevil will be destroyed if you clear the earth mantle from the ramparts' which the building conservationists are convinced is causing damp (Fort Gilkicker); destructive boys could be given constructive things to do in a fort's wide spaces - car repair or boat building or go-carting - or: 'they must be kept out by higher fences, vigilant security guards and Closed Circuit Television' (Fort Cumberland). Even when their military life has ebbed away, these sites don't stand still, they evolve.

These voices represent the complexity of the interests and hopes for the future that converge onto redundant military sites. All over the world, formerly defence dominated communities are seeking new economic, social and physical regeneration in these previously closed areas. Local authorities aim to find new jobs to replace lost skilled employment, and to seize the opportunity to meet the needs of the homeless for affordable homes. There may be opportunities for developing leisure activities for different generations including outdoor development for young people or services for the disabled. Historical research and fieldwork can usefully inform the choices of new land uses. Sites that have long remained empty are a focus for voluntary military heritage groups. There are sometimes competing demands from specialists interested in building and ecology/natural conservation. The need for public and private open space may compete with developers' desire for profit.

Experience gained in the redevelopment of ex-defence sites around Portsmouth Harbour shows how difficult it is to respond to the challenge of long-lasting sustainable development while raising the finance to pay for redevelopment of places with a complicated history and a complex physical legacy. There is no collected body of experience of successful and sustainable reuse to draw upon, so as already mentioned, this book, along with its international counterpart: *Sustainable Regeneration of Former Military Sites* (Routledge 2016) is a contribution to redress this absence.

When there's talk of defence cuts leading to closure of dockyards it's instructive to compare how local newspapers in Plymouth and Portsmouth react. In the 1990s the Portsmouth *News* ran a doughty campaign opposing the proposal, linked to the university economics department, which assessed the likely collateral damage to local firms including suppliers on defence contracts. In research for her PhD about decision-making when defence sites are sold and redeveloped Celia Clark did not find the same reaction in Plymouth – where South Yard was disposed of. How local people feel about these changes in places so long dominated by the priorities of national defence may or may not be reflected in what the local papers or tv stations are saying, or in pub, street and supermarket conversations.

But perhaps as memories of our military and naval past fade, as we enjoy the roses in Lumps Fort in their scented and colourful glory, we may have forgotten that it was once one of the many forts defending the approach to Portsmouth Harbour and the dockyard from attack by the French. Exploring the Japanese garden inside it, we may not know that what links Maizuru to Portsmouth are common traditions – as naval bases and as centres of brick production. The view from the top of Lumps Fort ramparts offers the view of a line of stumps stretching into the distance towards the Isle of Wight – the WWII barrier across the Solent with small gaps for ships and submarines. The Cockleshell Heroes who launched the Commando raid on Bordeaux harbour in 1942 lived at 9 Spencer Road Southsea while they trained at Lumps Fort, Hilsea Lido and Fort Cumberland are commemorated in a blue plaque put up by the city council in 2012.

Dark legacies - economic effects of defence sites' security closures

What was not anticipated in a major tourist attraction – and was then a highly unusual eventuality in a popular venue – although unfortunately more common since – were the 1980s closures of the dockyard Heritage Area during the height of the Irish Republican Army bombing campaign and the barring of all tourists during Red Alerts: the MOD's highest rate of threat. This demonstrated the dockyard's continuing strategic importance to the country's defence. On 10th March 1980 following the Netheravon attacks by the IRA/INLA in Salisbury, the Naval Base and tourist attractions were closed, though no duration is given in the report.[9] On 10th November 1984 the Naval Base was expected to remain closed throughout the weekend with *HMS Victory* and the *Mary Rose* "off limits ... amid fears of a new IRA bombing campaign" and the discovery the day before of a suspect device at the Royal Naval Armaments Depot, Elson.[10] On 16th April 1986, the Naval Base, *HMS Victory* and *Mary Rose* were again closed to the public and all visitors due to potential terrorist reprisals after America's bombing raid on Libya. The report says that " . . . the attractions could remain closed until late August . . ."[11] Loss of revenue to the dockyard attractions during those periods must have been significant.

During the Cold War Celia Clark remembers the nuclear submarines coming and going from HMS Dolphin. She heard about the siren that was to sound from the dockyard in the event of a nuclear attack – or accident - but it was not at all clear what we were supposed to do if we heard it.

The threat of nuclear war – or accident – may have receded, but when nuclear submarines were stationed at HMS Dolphin this leaflet issued by Gosport, Portsmouth and Hampshire County Council instructed people who lived within quarter of a mile in a leaflet to "Go In, Stay in, Tune in" - to the local radio. "In the unlikely event of a radiation emergency at Portsmouth naval base... The design of nuclear-powered submarines and ships means the likelihood of a radiation accident is very remote. There is absolutely no risk of an 'atom bomb'-type explosion. If an accident happened, it probably wouldn't affect people more than half a kilometre (500 yards) away from the vessel... You live in this zone."

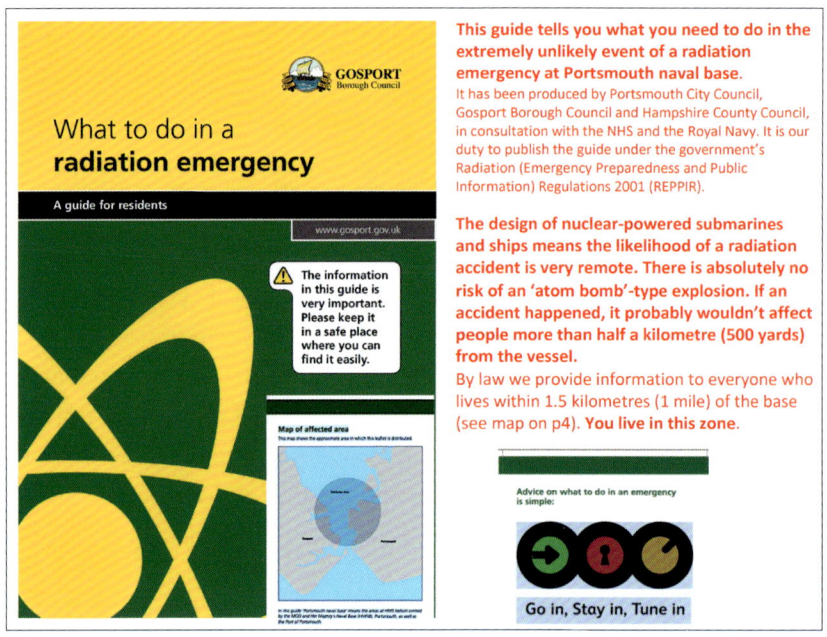

Leaflet Gosport Borough Council

Chris Donnithorne of HMS Dolphin said "It always struck me as ironic that when serving in a nuclear submarine I was not allowed to wear a watch with luminous hands because this source of radiation would far exceed anything from the reactor, and make measurements difficult."

Physical reminders of the harbour's strategic past are not always benign. Deep contamination is one such malign legacy. As we explore in our open space chapter, the notorious 'Glory Hole' on the southern shore of Eastney Lake was where decades of dockyard waste – some of it toxic - were dumped,[12,13] thus neatly sidestepping the responsibility. According to the council's statistics as much as 10% of Portsmouth has been reclaimed from tips; the military has operated in several areas. Contamination has been found at Langstone Marina, Milton Common and the Moneyfield/Longmeadow allotments.[14]

The Palmerston Forts Society was asked by Portsmouth Naval Base Property Trust to survey the C18th ramparts around Priddy's Hard in Gosport, but they found so much contamination by explosives and other materials they declined to do so.[15]

Unexploded ordnance is another hazard. The harbour's military and naval past suddenly came into sharp focus again on November 16 2016 when a dredging barge preparing the deep channel required for the huge new aircraft carriers: *HMS Queen Elizabeth II* and *Prince of Wales* (65,000-tonnes) uncovered an unexploded 500 lb German WWII bomb. A 500-metre cordon was thrown around the bomb, which meant that the historic dockyard, the most popular tourist attraction on the south coast had to be completely evacuated, along with the shoppers in Gunwharf Quays – as well as the residents of Gunwharf and Old Portsmouth – while the bomb was towed out to sea twelve hours after it was found, for a controlled detonation off the east of the Isle of Wight. A month later on Christmas Day 2016 the historic district of the city of Augsburg was also evacuated – on discovery of a 1.8 ton unexploded Allied bomb. More German bombs have been found in Portsmouth Harbour. On 22 February 2017 another 500 lb bomb was found. Rail, ferry and bus services were all suspended, while "divers from the Royal Navy's Portsmouth- based Southern Diving Unit 2 towed the bomb away from the harbour, lowered it to the seabed, and planted explosive charges for a controlled detonation of the device. Shortly after 11:00am, it was destroyed in a plume of smoke and spray. The entrance to Portsmouth Harbour was closed", stopping cross channel ferry services "until around 7:30am as a precaution while the bomb disposal team assessed the swiftest and safest way of removing the device. Some transport services were temporarily suspended, including Gosport Ferry and Wightlink services, and Gunwharf Quays was closed."

Perhaps these legacies are inevitable in a site dedicated to war. Also inevitable are rising sea levels and storm surges. Portsmouth is the only island city in the country threatened with major flooding. Large built up areas are below sea level. Historic England defines about 60% of the Heritage at Risk in Portsmouth's vicinity as coastal defence sites. These include Long Curtain Moat, Southsea Castle and Fort Cumberland all requiring new sea defences to protect them from the sea.

Flooding on Southsea Common by Storm Roger landward of Portsmouth War Memorial March 2020

But nothing could have prepared us for the enormous effects of Covid-19 in 2020 - and the consequent lockdown and physical distancing: of people, places and losses of employment. In an extraordinarily short space of time everything in our lives was abruptly reduced. A "Do not Enter Portsmouth" sign went up on the M275, but of course essential shipping into the ferryport continued to supply the country with food. But for some people blessed with creativity, welling local focus and energy was intensified by confinement to our home areas. Finishing the book during the three months of lockdown, it's a hymn to our home harbour. The Victory Gate to the dockyard remained closed, with a notice from the National Museum of the Royal Navy that only "Base pass holders" may enter: "Please knock for entry." Team Portsmouth 'Working Worldwide' whose partners are the Royal Navy, the Ministry of Defence and BAE Systems posted an adjacent notice: "Coronavirus Social Distancing. For everyone's safety, you must remain 2 metres apart, even if you're not working". How this was being achieved in the confined spaces of the two aircraft carriers moored alongside was not clarified. The loss of tourist revenue to the Naval Base Property Trust, historic ships and museums was off the scale compared with the comparatively short closures described above.

Sources

Printed sources are listed at the end of each chapter. Earlier publications which document the history of defence sites around the harbour include David Lloyd's *Buildings of Portsmouth and Its Environs (1974)* and his Hampshire volume in the Buildings of England series with Nikolaus Pevsner (1967), which was revised and greatly expanded by Charles O'Brien for Yale University Press in 2018 as *Hampshire: South*. It provides authoritative architectural descriptions of buildings new as well as old. As already mentioned *The Geography of Defence* edited by Michael Bateman (1987 Croom Helm; reprinted 2016) by pioneering geographers at Portsmouth Polytechnic was the first book to analyse how defence – and reuse of redundant defence sites – has shaped defence dominated towns such as Portsmouth. Three recent books are invaluable. As well as *Hampshire South*, Paul Brown's *The Portsmouth Dockyard Story* published by the History Press also in 2018 brings the dockyard's history right up to date. Brett Christophers' 2018 book *The New Enclosure. The Appropriation of Public Land in Neoliberal Britain* analysed the absence of public benefit – and even serious losses to the state and to local communities from defence land disposals. The Palmerston Forts Society's detailed publications are a primary source, particularly for the chapters about the forts. Planning records, local newspapers and personal recollections – but also fieldwork – exploring the area as it evolves – are both essential. As historian Bettany Hughes says: "I cannot write about the past unless I go where history happened..." All history is local because it happens in particular locations – and what makes Portsmouth harbour unusual is that those local events are sometimes of national and indeed international significance.

The authors

What are our motivations for writing the book? Celia Clark and Martin Marks met in a restaurant in a converted Victorian fort in Sliema, Malta, and realised they shared a passion for defence heritage. Celia has served as secretary, chair and president of the Portsmouth Society since 1973. The Society celebrated its hometown in *Maritime City Portsmouth 1945-2005* (Sutton Publishing 2005), to which she contributed chapters on building conservation and education buildings. As Education Officer of the Civic Trust from 1989 to 1991 she was involved in shaping the built environment aspect of the first national curriculum. She taught the history of architecture and building conservation to Portsmouth university undergraduates for nineteen years, some of whom worked on the post-fire restoration of Uppark and Windsor Castle. Celia's MSc. dissertation and PhD thesis compared dockyard and other service sites' experience of closure and changing functions. She is a founder member of the Hampshire Buildings Preservation Trust, the Wymering Manor Trust and the Naval Dockyards Society. She has published many articles and conference papers about the conversion of former defence establishments to new uses. Her book *Sustainable Regeneration of Former Military Sites* edited with Samer Bagaeen (Routledge 2016) explored post-defence redevelopment in different parts of the world. This book is its locally focused successor. On-going research focused on civilian reinvention around Portsmouth Harbour may be usefully compared with what's happening in other formerly defence-dependent areas.

Commander Martin Marks OBE served for 36 years in the Royal Navy as a mechanical engineer. He spent time down in the South Atlantic during the Falklands War and later managed the Royal Naval Hospital Haslar for five years, first as Executive Officer and later as the Commanding Officer of the Support Unit when the hospital became a tri-service establishment. His OBE was awarded for "Services to military medicine". After a second career as CEO of a sheltered housing association and a military charity, he has since become involved in operating a 100-year old steam gunboat for the National Museum of the Royal Navy and is a committee member for The Diving Museum, an all-volunteer and award winning organisation housed in a Grade II* gun battery in Gosport. He has lived in that area for nearly 40 years and was chair of the 1100-member residents' association for five years.

Structure of the book

As you walk round Portsmouth and Gosport or cross the harbour on the ferry the physical reminders are there, but this book aims to help us to be more aware of our defence-dominated local history - to recognise them as parts of a complete military and naval system which shaped urban development around the harbour for many centuries – but which are entering new, quite different leases of life: the justification for this book. It's structured around thematic chapters which explore how ex-defence sites released on both sides of the harbour since the mid-20th century have been redeveloped, either according to the type – forts, ordnance yards, barracks and hospitals, or to the new uses: education, museums and galleries, residential development and open space. The book ends with a summation of research at international level, identifying further directions for research useful for the many communities facing similar challenges.

As TS Eliot says:

We shall not cease from exploration,
And the end of all our exploring
Will be to arrive where we started,
And know the place for the first time.

Little Gidding The Four Quartets

Sources

Michael Bateman and Ray Riley 1987 *The Geography of Defence* Croom Helm

Paul Brown 2018 *The Portsmouth Dockyard Story* - The History Press

Hampshire Mills Group *"Chesapeake" Hampshire.* https://www.hampshiremills.org/History-%20%20Chesapeake%20Mill.htm

Council for the Protection of Rural England 1996 Design Competition for USAF base Upper Heyford Oxfordshire in the Old Fire Station Oxford

David W Lloyd 1974 *Buildings of Portsmouth and Its Environs A survey of the dockyard, defences, homes, churches, commercial, civic and public buildings* City of Portsmouth

Bettany Hughes 2017 'My working day. If you're going to inhabit someone else's world, the very least you can do is spend a little time in it". *The Guardian Review* Saturday 11 February p. 5

Charles O'Brien et al Editor 2018 *Hampshire South* The Buildings of England Yale University Press

Nikolaus Pevsner and David Lloyd 1967 *Hampshire and the Isle of Wight* Penguin Books

Daniel Scharf 2013 'Upper Heyford US Air base and Greenham Common' Unpublished chapter for book: *Sustainable Regeneration of Former Defence Sites* Routledge 2016

The Economic Impact of Tourism Portsmouth 2014 Tourism South East Research Unit Eastleigh Hampshire

https://www.nmrn.org.uk/news-events/nmrn-blog/closure-public-galleries-royal-marines-museum-heralds-next-step-museum-plan accessed 19Feb 2017

https://www.theguardian.com/uk-news/2016/nov/16/unexploded-wwii-german-bomb-found-portsmouth-harbour retrieved 24 January 2017

http://www.wbaltv.com/article/54000-evacuated-on-christmas-day-after-wwii-bomb-is-discovered/8536925 retrieved 24 January 2017

https://www.groundsure.com/resources/newsenvironmental-contamination- and-warfare/ accessed 13 July 2020

Portsmouth Naval Base Property Trust 2006 *20 Year Review 1986-2006*

Rachel Woodward 2004 *Military Geographies* Wiley-Blackwell ISBN: 978-1-405-12777-6 p. 43

Brett Christophers 2018 *The New Enclosure The Appropriation of Public Land in Neoliberal Britain* Verso 2018 p. 276; p. 278

Celia Clark 2014 'Women at Work in Portsmouth Dockyard 1914-19' British Dockyards in the First World War Transactions of the Naval Dockyards Society Vol. 12 pp.1-33

References

1. Brett Christophers *The New Enclosure The Appropriation of Public Land in Neoliberal Britain* Verso 2018 p.2, 104

2. https://www.forces.net/news/tri-service/mods-ps500-million-property-sale

3. Britain Thinks https://www.heritagefund.org.uk/news/portsmouths-heritage- makes-us-happier

4. Steven Morris and Ben Quinn 'Heroism and sacrifice of those who lost their lives will never be forgotten' *The Guardian* 6 June 2019 pp.6-7

5. https://theddaystory.com/discover/researching-local-d-day- connections/portsmouth-and-d-day/ accessed 18 June 2019

6. Gosport Waterfront and Town Centre SPD Draft Consultation: Gosport Society Response. 28 September 2017 p.10

7. *The News* July 2017

8. https://www.gosportsociety.co.uk/SDP%20final.pdf

9. http://www.scapaflowwrecks.com/wrecks/royal-oak/index.php 9 *Portsmouth Evening News*, 10th March 1980 p.1

10. *Portsmouth Evening News* 10th November 1984 p.1 11 *Portsmouth Evening News* 16th April 1986, p.1

11. https://hansard.parliament.uk/commons/1984-12-19/debates/5e85aadc- 5191-470d-b197- 90126b9745d5/GloryHolePortsmouth(Dumping)

12. *The News* 2014 Better to be safe than sorry over Portsmouth's poisoned land' 1 May

13. https://atom.lib.strath.ac.uk/oeda-shewell-1986-naval-married-quarters-site-glory-hole- eastney-portsmouth-contaminated-land-aspects

14. Geoffrey Salvetti Interview with Celia Clark 11 August 2019; http://teamlocals.co.uk/royal- navy-detonate-wwii-bomb-found-portsmouth-harbour/ accessed 24 February 2017

Chapter 2
Shaped by the Nation's Defence

Portsmouth is one of the most mapped and most often mapped places in Britain.[1] The harbour's military shaping is apparent as you explore it, particularly around its watery edges. This chapter is a brief outline of how the army, navy and airforce developed the many defence installations around the harbour, and what has happened to them in subsequent centuries.

Once they were no longer needed to defend the country, how does the Ministry of Defence dispose of them? The next chapter also describes how the surviving legacy of fortifications and dockyard and support facilities were revalued from the mid twentieth century onwards, and in many cases are now protected from demolition by new laws and regulations.

History as Geography

Both the history and future of the towns around the harbour are rooted in Portsmouth Harbour's geography. Since the Middle Ages the protected harbour in its strategic location on the north shore of the Solent between the mainland and the Isle of Wight has been vital to the nation's defence, in several roles: to house its premier naval dockyard, where the navy's ships were built and maintained, to house the army garrison to defend the dockyard, and as a base from which to launch invasion forces.

The town and later city of Portsmouth and the borough of Gosport developed to house the workers in the many army and navy facilities and the dockyard and their families. For hundreds of years - until late in the twentieth century - both communities' existence and economy were almost solely dependent on decisions by central government, which invested in, developed or ran down and closed the many defence establishments - as national policy and geo-politics dictated. Even as late as the 1970s Gosport was still the largest garrison town in the country.

Reflecting their importance as frontier towns, Berwick upon Tweed in the sixteenth and Portsmouth in the seventeenth century had the most extensive defences in England, their elaborate design resembling those in Europe rather than simpler British town walls. Key defences and facilities: the Round and Square Towers, Gunwharf, the dockyard, Whale Island, Tipner Battery and Horsea Island are still prominent on Portsmouth Harbour's western edge, while the surviving parts of Gosport's town defences also stand out in the townscape.

The nineteenth century ring of sea and land forts were commissioned by Lord Palmerston to defend Spithead and a landward attack on the dockyard. Those on the top of Portsdown Hill punctuate the northern skyline above Hilsea Lines' defensive earthwork along Portscreek, while those in the sea stand out like cakes or plugs at Spithead in the approach to the harbour.

We now explore the fortifications and defence facilities around the harbour, and discuss that happened to them in subsequent centuries when they became obsolete. Our Make Art Not War chapter describes museums and galleries in military spaces, while others describe the many barracks around the harbour and what's happened to them, and our penultimate chapter the new open spaces created where once only the military had access.

Portchester Castle

The best preserved Roman fort north of the Alps,[2] the 4th century walls of Portchester Castle still stand to their original height, dominating the head of Portsmouth Harbour. The fortress is square; each wall is 600 foot long. The keep and bailey walls are protected by a moat. The impressive keep was probably built before 1100. Its height was raised by two storeys to ninety feet later in the century, when the royal palace buildings were constructed. William Ponte de L'Arche founded the priory of Augustinian canons in about 1128 inside the Roman walls; St. Mary's Church was built soon after in about 1130, now serving as the parish church.

From 1665 the castle was used to house foreign prisoners-of-war, most notably during the wars with France between 1793 and 1815. In 1632 Charles I sold the castle to a local landowner, Sir William Uvedale, whose descendants, the Thistlethwayte family, still own it today. In 1926 the Thistlethwaites decided to place the castle in the guardianship of the Ministry of Works as an ancient monument. In the 1920s and 1930s the Ministry cleared vegetation, repaired the walls and excavated the castle's moats. Many of the workmen were unemployed Welsh miners. In 1984 the site came into the care of English Heritage.[3]

Southsea Castle

In 1538 the alliance of France and Spain were a serious threat to England when Henry VIII was in conflict with the Pope over the autonomy of the English church. To protect our shores he built a series of forts or castles from Kent to Dorset, including Southsea Castle on the southernmost part of Portsmouth in 1544-5. Its role was to defend the Solent and the eastern approach to Portsmouth Harbour. Unlike the earlier rounded forms its central keep is square, within a low-walled enclosure with two rectangular gun platforms to the east and west and two angled bastions to the front and rear, surrounded by a deep dry moat. It is an early English example of the popular continental *trace italienne* fortification. The 1778 Cowdray engraving of the Battle of the Solent in 1545 shows Henry VIII visiting the castle, with the French and English fleets approaching and the sinking of the Mary Rose. Made long after the event, it is not regarded as a reliable picture of the local fortifications. Despite several serious fires, the castle remained in service and saw brief action at the start of the English Civil War in 1642 when it was stormed by Parliamentary forces.

Concerns about a potential Dutch attack in the 1680s prompted Charles II to commission his Dutch-born master engineer, Sir Bernard de Gomme, to carry out a major scheme to improve Portsmouth's defences. In about the 1660s a glacis or outer earthwork was added to the castle. A new castle gate approached by a drawbridge across the dry moat was crowned with Charles's coat of arms and the date: 1683 to celebrate these alterations. The outlines of the original gunports are visible inside the keep.[4]

There was a major explosion in 1759 that left it as a ruin. One consequence of this was a major rebuild of Blockhouse because Southsea Castle was out of action. In 1785 the government took possession of Southsea Common, the marshland surrounding the castle to ensure it remained an open field of fire for the fort's guns. However in 1803 Major General Whitlocke reported in a letter to Col. Clinton that 'The Fortifications are in tolerable repair with the exception of South Sea Castle... its dilapidation at this time is the most unfortunate'. Convicts were deployed to fill in and flatten out the marshes from 1831 onwards. The castle was substantially reconstructed in 1814 at the end of the Napoleonic Wars, when its angular seaward bastion was made into a curve, forming a large gun platform facing the sea. The outside was refaced in Purbeck stone. The keep was filled with massive local red brickwork in two wide tunnel vaults, creating two new floors under a strong roof platform to take more guns. A spiral staircase dating from 1814 accesses the roof. The northern part of the castle was reconstructed, preserving the Tudor diagonal salient but with the walls rebuilt further out to provide more space for barrack accommodation – turning the castle into a compact fortress manned by artillerymen. The arms of Charles II were reset in the wall. De Gomme's glacis was reconstructed with a new covered way for patrolling soldiers on the vertical inner face or counterscarp gallery flanking the moat, reached by the caponier, a triangular stone covered way across the bed of the moat. The black and white painted lighthouse commissioned by the Admiralty was constructed on the western gun platform in 1828, rising 34 feet (10 m) above its base and crowned with a fox weathervane. The lighthouse keeper was housed in the castle.

Fears of a French invasion in the 1850s and 1860s led to expansion of the fortifications with additional gun batteries added to the castle to the east and west. Between 1863 and 1869 new underground magazines were constructed on either side to form part of a 17-acre (6.9 ha) complex enclosed by a defensive wall. In 1886 the site was well armed, with 25 rifled muzzle-loading guns; a visiting Russian military engineer praised the design.

At the start of the 20th century Southsea Castle formed part of the Fortress Portsmouth plan for defending the Solent. In the interwar years some fortifications were decommissioned, but the castle saw service again in the Second World War when it was involved in Operation Grasp, the seizure of French naval vessels in Portsmouth harbour. In 1960 its military use ended. It was sold to Portsmouth City Council for £35,000, and after restoration was open to the public.

Town and Port Defences: Portsmouth

Portsmouth dockyard's first dry dock dates from 1495. From the 15th century onwards it was recognised that Portsmouth Harbour needed to be defended on both sides. Two towers were built at the narrowest point on either side of the harbour's entrance to protect it. A chain barrier, first mentioned in about 1420, was raised across the harbour mouth by capstans in yards on the north side of the towers - hence Capstan Square in Old Portsmouth adjacent to the Round Tower. The Blockhouse tower (1417, 1492-4) was built to defend the Gosport end of the chain. The Round Tower was constructed 1418-1426, rebuilt in stone in 1495 and rebuilt again in the time of Queen Elizabeth with six gun-ports, later externally blocked, probably by de Gomme. Another storey was added in Napoleonic times when the interior was reconstructed: a massive circular brick vault supported on a central pillar with casemates formed behind the surviving gun-ports: work comparable to the contemporary Martello towers. The top was reconstructed in about 1847-50, adapted for gun positions. To protect the western approaches of the Solent Hurst Castle was built from 1541, and another round tower, Calshot Castle between 1539 and 1540.

Round Tower and WL Wyllie's house

The Square Tower on the Portsmouth side was added in 1494 to function as a residence for the Governor. Traces of Tudor fireplaces are visible inside. It has been used in many other ways since, including as a gunpowder store at the end of the Siege of Portsmouth during the Civil War. Gunpowder was a dangerous substance to store near where people lived. A plan by Lempriere dated 1715 shows the 'New Magasine' in the Camber, while his second plan of 1716 labels the Square Tower as the Old Magazine. In response to local protests, the storage of gunpowder was relocated to Priddy's Hard in 1779. The tower was used a meat store for the navy Victualling Board until 1850. Substantial wooden warehouse flooring to support barrels of salted meat was inserted in the 1770s at the same time as the vaulting. The remnant of the cellars, left open when the building was conserved and remodelled by Portsmouth city council in the 1980s, fills and empties with the tide. The wooden floor on the seaward side was removed during this remodelling to make a grand vaulted two-storey space.

The gilded bust of Charles I (a replica of the original of 1635 by Hubrecht Le Sueur) above a weathered coat of arms faces the High Street, with part of a longer description commemorating Charles I's safe landing here from France and Spain in 1623. The original bust is in the city museum. In 1823 the Board of Admiralty installed a semaphore on the roof, the first in a signalling chain on hills to transmit messages from Spithead to the Admiralty in London and back. The dockyard had its own signal station on the Semaphore Tower. This signalling system was made redundant by the arrival of the electric telegraph in Portsmouth. In 1847-50 the Square Tower was included in the remodelled defences with gun emplacements on the roof and reconstructed.

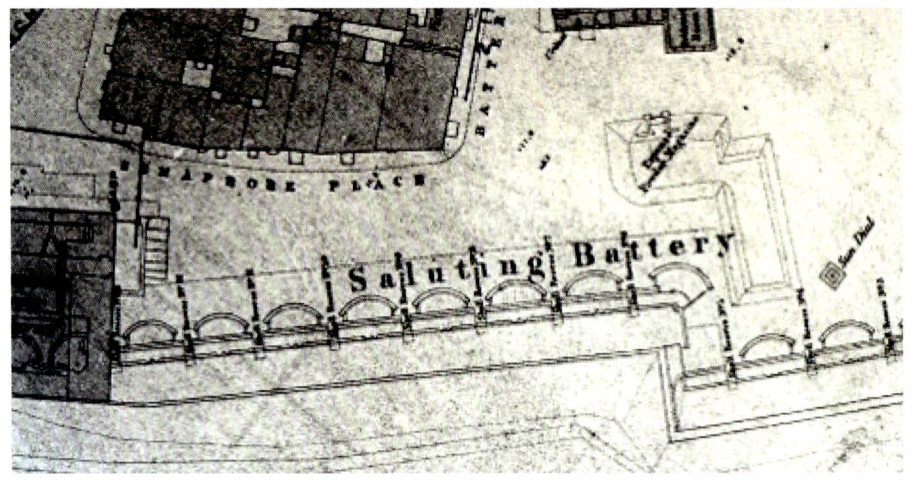

Saluting Battery and Platform 1861 http://historyinportsmouth.co.uk/opp/fortifications/ south.htm

Today the Square Tower looks significantly different from its original late fifteenth century appearance as it was refaced in 1827, concealing the original gun ports. The Saluting Platform which adjoins the Square Tower dates from 1568, the only surviving structure from the Tudor defences of Portsmouth. When its outer wall collapsed in a storm in 1987, the packed oyster shells of its inner filling were exposed.

King's Bastion Spur Redoubt excavated prior to restoration in 1975   Deane Clark

The Domus Dei, a medieval hospital, was converted as a new residence for the Governor of Portsmouth after the Reformation, and it was in the governor's chapel that Charles II was married to Catherine of Braganza in 1662.

There was another battery to the north of the Round Tower, roughly where the Still & West sits. Constructed in 1681 with heavy guns mounted, it was initially called Leake's battery after the Master Gunner of England, and subsequently known as Point Battery.

While only the lowest parts of de Gomme's landward defences in Portsmouth survive, the dramatic seaward stretch includes a length of curtain wall, part of the 16th century defences known as the Long Curtain, an earthen rampart with an inclined inner face, flat top and stone outward face rising almost vertically from the moat. Above this is a brick parapet bordering a walkway along the outward side of the rampart. De Gomme also designed the substantial King's Bastion and the associated Spur Redoubt: the bastion is four-sided, with two short flanks and two longer diagonal sides coming to an apex, faced in stone with a low brick parapet wall. The top where the mid-day gun was once fired is banked and grassed. The curtain wall and bastion were reused during the Second World War: traces of the artillery concrete bases survive. The triangular Spur Redoubt (1680) was designed to strengthen the defences to seaward. Access to it was through a brick lined sally port under the curtain wall and a light wooden bridge across the moat. Lord Nelson passed through this sally port in 1805 on his last departure from England.

Following periods of dereliction and rebuilding the upper parts of the Spur Redoubt were removed and the lower levels were covered in 1934 when the public promenade along the seafront was created. In the 1970s a wooden bridge designed by Deane Clark echoing the old design was built across the moat, reusing timbers from the Hayling Island railway bridge. This was replaced in 2004/5 by a curving steel structure. The concrete covering the Spur Redoubt was removed in the 1980s and its apex crowned with a triangular sculpture. The bridge over the exposed remains was constructed from New Forest timber blown down in the great storm of 1987. Another storm in 2014 punched a long breach in the outer wall to the moat, a portent of climate change; in 2020 the old wall was encased within a new much higher one. This engineering work destroyed the beautiful seaward Portland stone paving at sea level, shaped by wave action over the centuries.

De Gomme also augmented the defences controlling access to the harbour. The battery has two parts: Eighteen Gun Battery parallel with the shore and the shorter Flanking Battery at right angles, connecting to the Round Tower. These were the last stages of De Gomme's late seventeenth century fortifications, later reconstructed from 1847 to 1850 when the casemates were deepened and a second tier of three added to the top of the Flanking Battery. The Common Sally Port to allow civilian access from Broad Street to the foreshore where merchandise was landed pierced the eastern end of the main battery, doglegged through the fortifications. This was blocked in the 1847-50 reconstruction when the casemates were deepened and a second tier of gunports was added to the Flanking Battery. The Sally Port was replaced by the modern broad shallow archway. The Eighteen Gun Battery has 12 tunnel-vaulted casemates, each with a gunport facing seaward, now blocked.

The late C17 remains of the stone abutment to King James' Gate, which was demolished in about 1870 and re-erected in Burnaby Road supports a flight of steps to the popular walkway along the ramparts created in the 1960s. Arthur Corney identified these steps as a remnant of the Elizabethan demi-bastion that preceded the King James Gate. They are visible in the print of Catherine of Braganza landing in Portsmouth in the museum collection. The late eighteenth century Old Sallyport formerly had a stepped landing stage known as the Kings Stairs. This second Sally Port was used via the King's Stairs as the point of entry for the crews and passengers of the many vessels visiting the town every year, as well as the point of departure for important military and political figures in the country's history.[5] The inscription reads: "From this place naval heroes innumerable have embarked to fight their Country's battles. Near this spot Catherine of Braganza landed in state May 14 1662 previous to her marriage to Charles II at the Domus Dei a week later." The two-storey Point Battery, at right angles to Eighteen Gun battery provided flanking fire via narrow gunslits to protect Point Beach. It is a Grade I listed building and a scheduled monument. A staircase offers access to the public walkway and through the battery to the Round Tower. In the 1890s, along with other updates, two electric searchlights were installed by the Submarine Mining Engineers to illuminate the harbour entrance.

Gosport

In about 1539 a further Gosport fortification, Haselworth was built, probably to deter people from using the anchorage. It fell into ruin and the Gilkicker leading mark was built there in 1643. This and a stand of trees close to Alverstoke church were used together to guide ships sailing into Portsmouth. When the trees were cut down by the bishop of Winchester they were replaced by the Kickergill mark which stood in Clayhall Road until 1965. To make way for the building of Fort Monckton, the Gilkicker mark was demolished in 1779 and replaced with something smaller mounted on the new fort. Another triangular white stone seamark constructed in 1735, a scheduled monument, is still visible on the summit of Ashey Down to the southeast of Ryde.

In 1642 early in the Civil War, a significant action took place that demonstrated the importance of fortifying Gosport to the defence of Portsmouth. This was the first time that this was fully understood. Parliamentary forces who already had control of the nary, fired shots from a gun platform somewhere near the present Ferry Gardens, so damaging the nave and tower of St. Thomas's Church used as a lookout by Royalist defenders with the bells for communication that it remained a ruin until it was rebuilt after the Restoration. After an assault on Southsea Castle the Governor of Portsmouth surrendered to Parliamentary forces on 3 September 1642.

Defending and supplying the Dockyard

At the same time as the outer fortifications, the naval and dockyard defences and facilities were remodelled in the seventeenth century in order to protect British trade and expanding sea routes against rivals France and Spain, and later in the century, the Netherlands. Warships were able to access the English Channel more quickly from Portsmouth and Plymouth – where the dockyards were considerably expanded - than from Deptford (1513) and Woolwich (1512) on the Thames and Chatham (1567) on the Medway.

In response to threats from France and Holland, in 1665 King Charles II issued formal instructions for to strengthen the defences around Portsmouth and Gosport. Dutchman Sir Bernard de Gomme, Charles II's Surveyor General and Chief Engineer was appointed. Designer of the magnificent Plymouth Citadel, the fort at Tilbury as well as fortifications in Dunkirk, Italy and France and Hull, Sheerness and Harwich, De Gomme completely remodelled Portsmouth's Elizabethan fortifications, obliterating the earlier fourteenth century earthern fortifications and Henry VIII's ramparts.

He reshaped the existing bastions, modifying the system of moats and adding a glacis and a ravelin, a detached triangular outwork near the Landport Gate in 1669: Town Mount. A series of bastions were constructed, raised in height by excavation of a moat 20 metres wide and 4m deep flushed by the tide.

No. 1 Bastion before restoration. Gosport Borough Council

De Gomme's Gosport's Lines consisted of a rampart for the men and guns, a firing step for musketeers, a parapet to fire over and a scarp dropping into a wet moat. Near the bottom of the scarp was a level space to prevent debris from the rampart from falling into the moat. Beyond the moat was a level area known as a covered way for the first line of defence and a gentle slope beyond: the glacis over which the guns on the ramparts could rake approaching enemies with fire. The purpose of the glacis was to mask the walls of the fortification so an enemy could not fire directly at them. These works were improved in the mid 18th Century at which time the southern sections were moved and left as neglected mounds. Between 1797 and 1803 the southern sector of the Gosport Lines were reconstructed. Following the building of Haslar Gate in 1800 the present Bastion was constructed in 1802-1803 using convict labour.

This elaborate design was restored by Hampshire County Council and as part of the Heritage Action Zone and is visible today. The zigzag trace continued to where the Ritz Cinema stood, crossing Walpole Road to the corner of Ordnance Road near the site of the old Drill Hall. A bastion enclosed what is now the southern part of St. George Barracks Parade Ground and a demi-bastion in the Watercraft area, bringing the Lines back to the harbour just short of the present Royal Clarence Yard. De Gomme also designed Fort Charles where Camper & Nicholson's main building now stands, and Fort James on Burrow or Rat Island and the beginnings of Fort Blockhouse. A stone marking the main gate to the fort in the Queen Anne rebuild with a date of 1708 is still in situ. The development of Blockhouse is described in our first fort chapter: a detailed history of the harbour's many forts and military lines.

The Royal Dockyards, 1690-1850: Architecture and Engineering Works of the Sailing Navy J B Coad Scholar Press/The Royal Commission on Historic Monuments (England) 1989
Historic England Archive

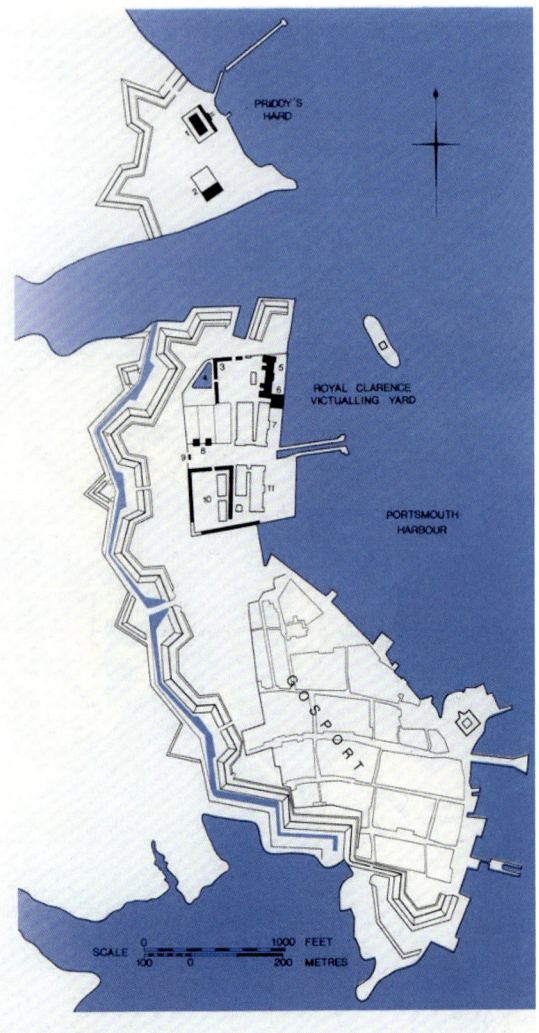

Figure 10. Gosport: Naval Establishments: Principal Buildings c. 1850

(Buildings in solid black still survive)

Priddy's Hard
1. Powder Magazine and Shifting House
2. Ordnance Storekeeper's House

Royal Clarence Victualling Yard
3. Stockyard and Cattle Pens
4. Reservoir
5. Bakery
6. Granary/Mill
7. Storehose
8. Victualling Officers' Houses
9. Main Gateway
10. Cooperate Green
11. Brewery

In 1678 defences were started to protect Gosport and the western side of Portsmouth harbour. Modifications were made to its enclosing walls. A smaller fort, James Fort, was constructed within Portsmouth Harbour to act as a final line of defence. Between 1745 and 1756 Col. Desmaretz remodelled the harbour's defences further. In 1779 the Gosport Lines' gates, platforms and palisades had to be repaired, and later the rampart and moat near Weevil Lane were improved. In the period 1797 to 1803 the Lines reconstructed in the southern sector. Following the building of Haslar Gate in 1800 the present Bastion was constructed in 1802-1803 using convict labour. Not all Gosport's defences survive. Charles Fort was 27 foot high with a 60 foot square tower, built in 1678-79. The armaments were removed in 1742. In ruins by 1828, the site is now completely built over. St James Fort was built in 1678 - 79 on Burrow Island. It was in ruins by 1800 and now only part of the stonework remains above ground. At Priddy's Hard most of the rampart survives and the moat, though nearly all filled in, is still clearly visible.

Seaward wall of Fort Blockhouse

By the eighteenth century Portsmouth dockyard was one of the largest industrial complexes in the world. The introduction of steam power in the early nineteenth century to power dockyard machinery – first in Block Mills and later complexes and to power ships led to considerable expansion: three basins, new locks and docks and machine shops. The development of the dockyard is described in detail in our dockyard chapters.

Support facilities developed around the harbour to house, supply and maintain the army and navy: barracks, gunwharf, ordnance manufacture and supply, victualling, barracks, and the earliest British naval hospital were developed over the centuries. In the late nineteenth century ship testing tanks were constructed at Haslar. Their history and post-defence experience is explored in later chapters.[6]

Jonathan Coad's map of the two walled towns in 1860 and Riley's map of military land use in the same year[7] show Portsmouth's 300 yard wide ramparts and outworks, together with the 375-450 yard width of Portsea's. Nineteenth century topographers and railway guides lauded the size and complexity of these impressive structures, whose area was actually greater than that of the towns they surrounded. In addition, in 1813 the Crown purchased about 364 hectares to the south of Hilsea Lines and to their north on the mainland – half as large again as the combined area of Portsmouth, Portsea, their fortifications, the 1860 dockyard and Southsea Common. Apart from extensions to the Hilsea defences, most of this continued in agricultural use.

J B Coad *The Royal Dockyards, 1690-1850: Architecture and Engineering Works of the Sailing Navy* Scholar Press/The Royal Commission on Historic Monuments (England) 1989 Historic England Archive

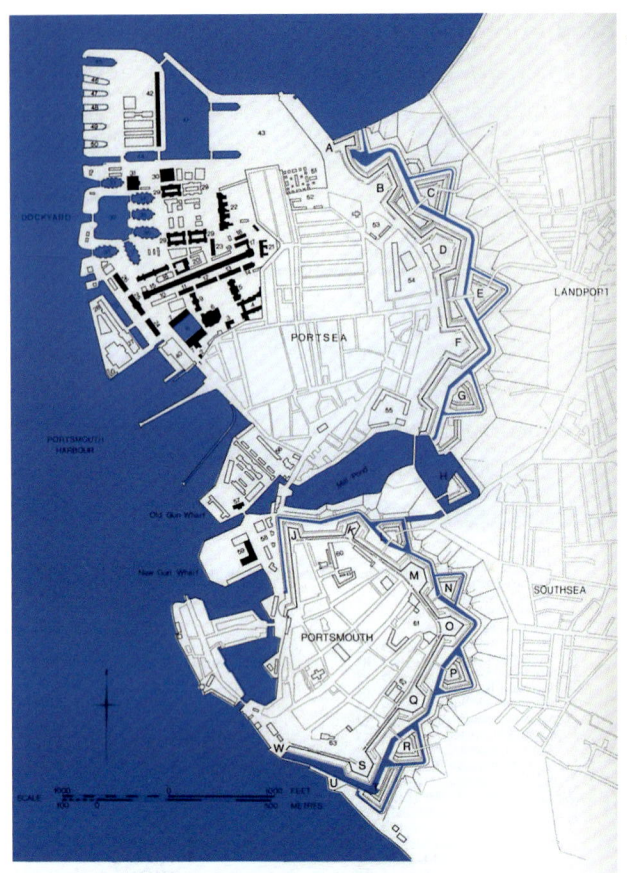

Figure 6. Portsmouth: Naval and Ordnance Establishments; Principal Buildings c. 1850
(Buildings in solid black still survive)

So by the middle of the nineteenth century, in total one third of the Portsmouth land area was dedicated to military functions: barracks, military hospital, chapel, Grand Parade, Governor's Green, the Lieutenant Governor's house and officers' quarters in the High Street. Colewort Barracks had opened in 1680 and was enlarged in 1828, Cambridge Barracks was rebuilt from 1856-8 and Clarence Barracks was enlarged in 1770 and 1824. Immediately outside the town walls Point Barracks was built in 1847 to house the gunners serving in the Round Tower and Long Battery. In Portsea was the Milldam Barracks of about 1820. In Gosport Anglesea Barracks and the Garrison Prison were first occupied in 1848. The Garrison Hospital was built in 1834, later joined by a convict prison. Between the two walled towns the Gunwharf was developed and extended on reclaimed land in 1662 and New Gunwharf added between 1797 and 1814 again on made ground to the south. So in total, by 1860 the Crown controlled two thirds of the total area of Portsmouth and three quarters of Portsea, and, in addition, Hilsea Barracks, originating in 1756 was enlarged in 1794 and 1854. At Tipner magazine of 1813 a small barracks was added. Point Barracks in Broad Street was built in 1847. St. George Barracks in Gosport was constructed from 1856 to 1859. Eastney Marine Barracks of 1863-1870 was protected in 1861-1863 by two small forts. Fort Cumberland defended the Langstone Harbour entrance. Apart from one small section the War Department then controlled the entire south coast of Portsea Island, as the Board of Ordnance purchased Southsea Common in 1785 to prevent enemy attacks on the fortifications from the land and to provide artillery with a clear sight of the waters at Spithead.[9]

The Ring Fortress: the Palmerston Forts

Portsdown's silhouette was dramatically transformed from the 1860s by construction of six huge red brick forts: Forts Fareham, Wallington, Nelson, Southwick, Widley and Purbrook a thousand yards apart along the crest of the hill. The older Hilsea Lines protecting the northern shore of Portsea Island at the bottom of the ridge were obsolete. Before the forts were built Portsmouth's defence had relied upon smoothbore guns positioned close to the shore.

A Royal Commission on the Defences of the United Kingdom was set up in 1859-60 to examine the organisation and strengthening of coastal defences, in response to a perceived threat from Louis Napoleon III, nephew of Napoleon Buonaparte. The doctrines of the "Jeune École" of French naval thinking were exemplified by the building of the first French ocean-going steam powered ironclad battleship, *La Gloire* in 1858. Britain produced the *Warrior*, the largest ironclad of its time in 1860 in response. Contributing to the height of the alarm was the much greater effectiveness of new artillery.

New rifled guns had a range of up to 8000-9000 yards, twice the range of the best guns. In 1858 a large basin was built to enhance the naval capacity of the port of Cherbourg – just across the Channel from Portsmouth, which was now extremely vulnerable to hostile ships firing from Spithead or landings to the east and west. Lord Palmerston said to Queen Victoria "It is better to lose Mr. Gladstone than to lose Portsmouth…" The Commission recommended that Portsmouth, Plymouth and Chatham dockyards must be protected by greatly strengthened defences; chains of forts, some of them in the seaward approaches were to be built at huge cost. Improvements at seven other ports were recommended. The estimated cost was to be nearly £12m at that time. Although supportive, the government reduced this by nearly £4m to ease the bill's path through parliament where there was much opposition to the foreseen threat from France as well as the cost.

The 'ring fortress' around Portsmouth dockyard was to consist of twelve land forts on Portsdown Hill, Portsea Island and in Gosport, as well as five sea forts on the shoals in the Solent to guard the anchorage and the inner approaches to Portsmouth Harbour from bombardment from the sea. The government acquired the northern slopes of Portsdown Hill. The farmland was cleared of trees to provide a clear field of fire - because the forts' guns were designed to fire inland to meet a landward invasion. It is a surprise to discover how completely they are hidden from the north – a tribute to the Royal Engineers who designed them. The whole of the crest of Portsdown Hill, including the forts and their connecting hilltop road built by navvies was War Department property. These forts' detailed history is explored in later chapters.

Bricks for the land forts' construction were generally manufactured in temporary brickworks as close to the point of use as possible. The area has a wealth of good brickmaking clay. At Burseldon Brickworks where some of the bricks used in the forts may have been made, there is a museum dedicated to the history of brick manufacture. Westbury Manor museum celebrates the famous Fareham Reds and chimney pots made by local brickworks.[9]

Releases of defence land

As already said in Chapter I, the maps in Ray Riley's study: 'Military and Naval Land Use as a Determinant of Urban Development' in *The Geography of Defence* demonstrate the extent of the fortifications as well as the military establishments they enclosed. The government could lay claim to no less than three quarters of Portsmouth town's land area. This made Portsmouth and Portsea no less company towns than those dominated by railways or steelworks.[10] In the mid-C20 the Ministry of Defence also owned 30% of Gosport.

The dominance of defence related activity on the harbour economy resulted in the city and borough having fewer civilian entrepreneurs, and hence few benefactors to provide civic parks, libraries, galleries or public art - as was likely to happen in other places. Instead the 'wealth' generated by the harbour labour force was garnered by the government.

As this book illustrates, some historic Ministry of Defence sites eventually offered long term benefits such as open space for Portsmouth and Gosport. Unusually, unlike most waterfront cities which are built up to the water's edge, this was not the case in Portsmouth, because the Government retained Southsea Common as a military assembly point and training ground into the early twentieth century. It was purchased by the Corporation in 1922 and laid out for recreation, forming a green lung in a city otherwise very short of public open space. Hilsea Lines and Gosport Lines are also now restored and mostly accessible.

Contemporary complaints about the fortifications constricting town development

Surprisingly, there's hardly any contemporary information about the disappearance of the two walled towns of Portsmouth and Portsea or of the parallel fortifications of Gosport. But there is comment about the constrictions the fortifications imposed on growing urban areas.

W G Gates in his *Portsmouth in the Past*, first printed in the *Hampshire Telegraph* in 1925 includes a detailed description of a 'PORTION OF THE OLD RAMPARTS' written in 1850: "a right pleasant stroll" around "the warlike envelope of the two towns". He asks: "How, it may be asked, do the Portsmouth people gain access to their green fields or their suburbs? Do they clamber over the fortifications, or dive beneath them, or cross them by a level?" Quay Gate, Landport Gate and Spur Gate were the access point for Portsmouth, and Lion and Unicorn Gates for Portsea. The serious disadvantage for local people was that:

"These gates and roads are so completely overlooked by lines of fortification, that the out-goers and in-comers, whether men or horses, or vehicles, are wholly at the mercy of those who govern the ramparts for the time being. The ramparts or terraces pass continuously over these roads; and there are at intervals flights of steps, or sloping paths, to lead down from the ramparts to the streets within the town, but none to the exterior. The interior and the exterior are certainly widely different in that respect; for while the former presents a mass of streets, cooped up within limits incapable of expansion, the latter presents much liveliness and openness of view."

John Webb's Portsmouth paper about the High Street describes the tensions between the military and the townspeople. The Board of Ordnance at first opposed any breach of the Portsmouth and Gosport defences by the railway lines on both sides of the harbour. As G H Williams describes, the railway had reached Gosport via Fareham in 1841, but the military authorities would not allow it to penetrate the ramparts, so the terminus was built in Spring Garden Lane. They also imposed restrictions on the height of buildings to minimize interference with the field of fire from the ramparts. In 1840 after Queen Victoria had purchased Osborne House in the Isle of Wight, leading Gosport inhabitants signed a petition complaining that the Double Gates, with a pavement of 20 inches wide on either side of the carriageway were dangerous to pedestrians. Plans were submitted for a pedestrian gateway and drawbridge, but as neither the Board of Ordnance nor the railway company would pay they were dropped. As late as 1922-23, the only way on and off Portsea Island was through two narrow masonry arches in Hilsea Lines – a penalty for being an island city still dominated by the military.

Pressure to break these tight bounds inhibiting urban expansion was the motivation for the demolition, clearance and redevelopment of the elaborate town fortifications in the 1870s and 1880s – though their sites mostly remained in military hands until the mid-twentieth century. The seaward fortifications survived because they were not impeding development.[11] Portsmouth City Council bought these at the end of the 1950s and early 1960s, partly with the aim of preservation. Southsea Castle became a museum in 1967. Eighteen Gun Battery was landscaped in the 1970s.

Dr. Ray Riley searched the local records and newspapers for a year in vain without finding much contemporary documentation about the demolition of the two walled towns' defences – or how people felt at the time about the disappearance of the complex ramparts, moats and ditches which had defended them for centuries. As we mentioned in chapter 1, it was not until over a hundred years later that academics at Portsmouth Polytechnic focused the Annual Conference of the Institute of British Geographers in the *Geography of Defence* on 'a city devoted in the past to the needs of the nation's defence, with a considerable residue of that function still present.' (p. ix). Ray Riley, the book's co-editor, along with Michael Bateman set out its aim an attempt to synthesize the geographical effects of national government defence policies, 'many of which have received little attention in the past.'

In his chapter Ray Riley first examines the development and evolution of military and naval land use in Portsmouth – dockyard, barracks and defences – following Henry VII's decision in 1496 to declare the town both a garrison and a Royal dockyard – which reached its greatest extent in about 1900. He then documents the nineteenth and twentieth century defence land releases around the harbour, describing the new uses achieved on them. This book draws extensively on his work and brings it up to the present.

Clearance of the early fortifications

By 1865 the two elaborately walled towns on Portsea Island: Portsmouth and Portsea, and Gosport's defences too had been obsolete for many years. The local population was growing rapidly. The government's Defence Committee recommended that all the Portsmouth Lines "except for those facing seaward and those overlooking the Common should be demolished." As a consequence, by the late nineteenth century most of the bastions, ravelins, walls, glacis and moats had disappeared as the towns expanded beyond these tight bounds. Portsmouth's landward defences had once stood at least as high as the seaward stretch, but only the Harbour approach defences were retained – from King's Bastion to the Round Tower.

Two of Portsmouth's town gates: King William Gate across Pembroke Road in Old Portsmouth and Quay Gate in the Camber disappeared, while one: King James's Gate, which constricted Broad Street in Old Portsmouth was relocated without its crowning cupola to a temporary site where the university's Nuffield Center now stands, in an area which was once a broad moat between the two walled towns of Portsmouth and Portsea. It was then re-erected in Burnaby Road where it flanks the MOD cricket pitch. Portsea's gates also migrated: Unicorn Gate to a roundabout inside the main city centre gate to the dockyard, while Lion Gate was inserted into the base of the rebuilt Semaphore Tower, visible from the harbour. Other relocations of military fragments are described in later chapters.

William Armstrong's 40-pounder gun of 1859 which incorporated both rifling and breech loading rendered Portsmouth's inner fortifications redundant, so they were the first to be cleared. Demolition began almost immediately, although the last sections of the Mill Pond were not filled in until 1876.[12] Riley proposes three motivations for this: the need for brick rubble as backing for the new docks and basin being constructed in the dockyard between 1867 and 1876, the pressure to provide barrack space for the growing garrison, and the decision to lay out extensive recreation grounds to help reduce the rigours of service life. "That the land might be sold off for non- military purposes seems not to have been a central issue."[13]

The first changes following demolition involved realignment and widening of roads that had breached the fortifications and the creation of new links to ease traffic flows between constituent parts of town. Park Road was built in 1875 over the Mill Pond between Landport and southern Portsea, meeting St. George's and Alexandra Roads, providing a direct link to Southsea from 1877. St. Michaels and Anglesea Roads linked the northernmost Southsea terraces and Queen Street. But much of the land released by demolition of the Portsmouth fortifications remained in military hands. Clarence Barracks were greatly extended to form Victoria Barracks between 1880 and 1897, the gunwharf annexed land reclaimed from the Mill Pond, and the United Services Recreation Grounds were opened near Clarence Pier in 1891, followed by the Garrison Recreation Ground adjacent to the sea. On the space between Victoria Barracks and the United Services Ground, Government House, the official residence of the garrison commander was built in 1882, followed by Gun House and Ravelin House in 1891 for other senior officers. Only Pembroke Gardens, leased to the Corporation for use as tennis courts and a small area sold in 1879 to Portsmouth Grammar School were made over to civilian use.[14]

In Portsea land was taken into the extension of the dockyard, while the Admiralty in consequence of the decision to house sailors acquired the adjacent Anglesea Barracks in 1891 ashore. The officer's mess gained further space for gardens and tennis courts in 1908 when the Garrison Hospital was transferred to Portsdown Hill, renamed Queen Alexandra Military Hospital. The railway companies jointly applied to breach Portsmouth's tight defences in 1876. They had been denied a direct link to the sea, impeding the transfer of holidaymakers to ferries to the Isle of Wight. The government insisted on the construction of sidings to the gunwharf and a railway viaduct to the dockyard across the Hard as part of the agreement.[15]

In 1814 when Napoleon was defeated the guns on the Gosport ramparts were dismantled. Although the railway reached Gosport via Eastleigh and Fareham in 1841, as we said earlier, in the same way as they did in Portsmouth, the military authorities would not allow it to penetrate the ramparts, so the terminus was built in Spring Gardens and buildings were restricted in height to preserve the field of fire from the ramparts. When Queen Victoria and Prince Albert travelled by train to Gosport Station en route to Osborne House, Prince Albert dropped a hint that the queen would be pleased 'If the Railroad were continued thro' the fortifications of Gosport to the water's edge in the Royal Clarence Yard'. Within a week a survey for a railway extension had started. The *Hampshire Telegraph* took up Portsmouth's case: if the Board of Ordnance "allow a hole to be made in the face of the Gosport works, we in Portsmouth may fairly ask to do away with all the useless fortification that exists in the heart of our town for no possible use or defence". The Gosport railway was taken through the rampart to a new terminus in Royal Clarence Yard, which the Queen and Prince used on their way to Osborne thereafter. Portsmouth's railway line penetrated Hilsea Lines in 1847 and in 1876 the line was extended through the town's fortifications to end at the Harbour Station.

In a densely built up town so lacking in open space, Portsmouth Corporation also purchased the ground on which Victoria Park – at first called the People's Park – was laid out. As we explore in our open space chapter it was designed by Alexander McKenzie was laid out in 1878 by the city council on land which had previously formed the glacis and open land of the defences of Portsea.

Just to the north a new Catholic church which became St. John's Cathedral was built in 1882 on land purchased from the War Department in 1877 because Portsea's defensive ramparts, 100 metres to the west, had become redundant and were demolished. The cathedral was designed to accommodate of the ever-increasing number of Catholic soldiers in the British Army, about 30%; Portsmouth was a major garrison town. It replaced a chapel built in 1796 in Prince George Street, half a kilometre to the west.

As already said, there appears to be little contemporary documentation about the demolition of the two walled towns' defences – or how people felt at the time about the disappearance of the complex ramparts, moats and ditches which had defended them for centuries. The authors met a similar opacity about why and how the landward extent of Gosport's town ramparts were destroyed and when the land was reused as playing fields and open space – apart from the modern on-site interpretation near Haslar Creek, quoted below.

The seaward fortifications survived because they were not impeding development. Portsmouth City Council bought these at the end of the 1950s and early 1960s, partly with the aim of preserving them. Southsea Castle became a museum in 1967. Eighteen Gun Battery was landscaped in the 1970s.

Gosport's Lines lasted longer than Portsmouth's, but in 1869 Walpole Park was created on the area known as the Horsefield, where the army had exercised their horses. In 1890 Walpole Road was driven through the lines to provide a direct route from the High Street to Alverstoke, and the Main and Double Gates were demolished. The millpond attached to God's Port, later called Gosport Creek is shown of maps of 1678. It became a cockle pond in 1751. A map of about 1865 shows the pond and the dam separating it from the town's moat. The cockle pond was renovated under the Unemployment Work Programme after WW1. Its opening as a model boating lake on 1 August 1921 was attended by about 3000 people. Skippers came from all over the world to race their yachts on the two lakes, where the British Empire Model Yachting Competition Championships were held from 1923. The lakes are 1.5m deep; they hold about 3.5m gallons of salt water which enters via a control valve from Haslar Creek. In 1971 they were designated sites of Special Scientific Interest and a Ramsar site wetland of international importance. The Millennium celebrations and development of Gosport Waterfront made funding available via the Heritage Lottery Fund for a new pavilion, the Compass Point which houses the club room and boathouse on the ground floor and a café above. Gosport Borough Council and the Model Yacht and Boat Club manage the lakes. The only part of the ramparts to survive, restored to their original form and height is the section to the south of Holy Trinity Church and the former vicarage, which extends to the seaward section of Haslar Creek, together with the landward defences to the west of Royal Clarence Yard.

Twentieth century defences

The remaining seaward defences were effective in guarding the harbour entrance until the advent of air and submarine warfare in the early twentieth century. Nineteenth century dockyard facilities were the infrastructure basis for the twentieth century wars. Ships based and repaired in the dockyard fought two world wars with Germany. Submarine obstacles were planted right across Spithead to the Isle of Wight and a missile battery was constructed at Fraser near the Langstone Harbour entrance. In the Cold War the threat was from the Soviet Union and more recently from terrorists and Russia. A siren warning of nuclear attack or accident was installed in the dockyard to alert the Portsmouth population, and headquarters facilities were constructed in Fort Widley for those who would run the city once attacked, though how they were going to get there inside the two-minute warning was unclear. The WWII air raid siren system was adapted for the Cold War and continued in use until the 1990s.

Horsea Island

Horsea was originally two islands, Great and Little Horsea, the former large enough to support a dairy farm. In 1804 a Royal Powder Works was established on Little Horsea in connection with the gunpowder magazine at nearby Tipner; by 1849, however, it was no longer in operation, and no above-ground evidence of the site remains to be seen. The islands were joined to form a torpedo-testing lake in 1889, using chalk excavated from Portsdown Hill by convict labour, one km to the north. The army constructed a narrow-gauge railway to distribute the chalk. Although the lake's length was increased from 800 yards (730 m) to over 1,000 yards (910 m) in 1905, rapid advances in torpedo design and range had made it all but obsolete by World War I. In 1909 the island became the site of one of the Navy's three high-power shore wireless stations, which saw it populated with dozens of tall masts. Parts of the Mulberry Harbours were built on the shore of the island, as they were near Clarence Pier and in Stokes Bay. You can still see remains of the ramps as you drive north along the M275.

In the 1950s the lake was used in the testing of improved Martin-Baker Ejection Seats, following catapult launch mishaps on carriers in which Fleet Air Arm aircrew often sustained serious compression injuries to the spine after ejecting from submerged aircraft. It was used by Southsea British Sub Aqua Club in the 1950s, and later by many other clubs and private training agencies for underwater swimming and diving. However, civilian access is still restricted for security reasons.

After closure of the telegraphy station in the 1960s the northern part became home to HMS Phoenix, the naval school of firefighting and damage control. A number of steel structures called trainers were erected to simulate three decks within a warship. Fires were set in the trainers for instruction in various types of firefighting. The kerosene and water mix burned in the trainers, known as *sullage* caused significant water and air pollution and created a health hazard for the staff exposed to the fumes for protracted periods. In 1994 the school moved to a modern gas-fired trainer on Whale Island as part of a consolidation and cost effectiveness initiative. The new facility is known as the Phoenix School of Nuclear, Biological and Chemical Defence, damage control and fire fighting. Responsibility for training and site management was contracted out to Flagship Training UK, which was taken over by Vosper Thornycroft in September 2008. The original island site continues to be used by the MOD, with a number of facilities on the site predominantly focusing on diving and underwater engineering. Infrastructure includes training facilities as well as workshops, decompression chambers and equipment testing capabilities.

Gains in public open space

The loss of local defence sites' military significance has led to considerable gains in local communities' access to open space. An example is Point Barracks of 1847-50 in Broad Street, which was demolished in the mid-twentieth century, creating a new open area with steps to the top of the battery, Point Battery and Round Tower. This high level walkway in Old Portsmouth, Southsea Common, Victoria Park, Hilsea Lines, Lumps Fort, Gosport's No. 1 Bastion, Walpole Park and Stokes Bay lines are all vital to our enjoyment, because both Portsmouth and Gosport are otherwise short of parks and gardens. As we explore in our open space chapter all of the undeveloped area to the southwest of Tipner lake, with the exception of the helipad, forms one of the few terrestrial parts of the Portsmouth Harbour Site of Special Scientific Interest (SSSI) because of the flora and fauna which flourish on the imported chalk. As a contribution to remedying the open space deficit, from 2012 the Paulsgrove tip to the north of Horsea Island has been transformed into Horsea Island Country Park.

It might be said that the surviving defence building assets and space becoming available for civilian use are in some way a substitute for the harbour communities' absent civic and cultural heritage - although what new uses are achieved and whether they meet local needs depend on critical factors such as the MOD's disposal rules, the local planning authorities' response and the market conditions at the time.

The dominance of defence in the harbour communities' local economy and landscape has bequeathed a great variety and wealth of historic sites to the twentyfirst century. As they become surplus to requirements, local people and developers are contributing to their civilian regeneration in many different ways - as the next chapters explore.

Sources

Tim Backhouse http://historyinportsmouth.co.uk/people/de-gomme.htm accessed 13 December 2018

Les Capon AOC Archaeology Group 2018 'Uncovering 350-year-old fortifications in Ravelin Park' Lecture University of Portsmouth 12 December

Arthur Corney 1983 'The Portsmouth Fortress' Journal of the Royal Society of Arts CXXXI pp. 578-86

Arthur Corney 1965 *Fortifications in Old Portsmouth: A Guide. Portsmouth, UK:* Portsmouth City Museums. OCLC 24435157

Edwin Course 1969 *Portsmouth Railways* Portsmouth Paper No. 6 Portsmouth City Council English Heritage National Survey of Ordnance Yards and Magazine Depots, pp10-12 William G Gates WG 1900 *History of Portsmouth* Evening News Portsmouth

William G Gates WG 1925 *Portsmouth in the Past* Hampshire Telegraph; republished 1972 Wakefield S R Publishers

William G Gates 1928 *Records of the Corporation 1835-1927* Charpentier Portsmouth

Hooper, S., Smith, D. & Tomlinson, N. (1991). *The COPSE* Report* (*City of Portsmouth Survey of the Environment). Portsmouth Urban Wildlife Group, Portsmouth. (Private publication)

HT Lilley editor 1899 *Guide to Portsmouth and Southsea and Neighbourhood* British Medical Association Portsmouth Meeting 1899 p. 68

T. Norriss et al. (eds). *Hampshire & Isle of Wight Butterfly & Moth Report 2010.* Butterfly Conservation, Lulworth, UK. Brian Patterson *A Military Heritage: A history of Portsmouth and Portsea town fortifications* Fort Cumberland and Portsmouth Militaria Society

Ray Riley 1985 The *Evolution of the Docks and Industrial Buildings in Portsmouth Royal Dockyard 1698-1914* Portsmouth Paper No. 44 Portsmouth City Council

Ray Riley 1987 'Military and Naval Land Use as a Determinant of Urban Development – The Case of Portsmouth' in *The Geography of Defence* Croom Eds. Michael Bateman and RC Riley pp. 52-81

John Sadden 2001 *Portsmouth in Defence of the Realm* Phillimore

Andrew Saunders 1989. *Fortress Britain: Artillery Fortifications in the British Isles and Ireland. Liphook, UK: Beaufort. ISBN 1855120003*

Basil Ripley 1982 *Horsea Island and The Royal Navy.* The *Portsmouth Papers*, No. 36 Portsmouth Museums Portsmouth

John Webb 1977 'An Early Victorian Street. The High Street Old Portsmouth" Portsmouth Paper No. 26

Williams GH *The Western Defences of Portsmouth Harbour 1400-1800.* Portsmouth Paper No. 30 1979 Portsmouth City Council

Duncan Williams 2018 "The Development of the Fortified Lines of Portsmouth and Portsea in the Nineteenth Century' *The Redan* Palmerston Forts Society Number 79 2018 pp. 20-50.p. 44-45

https://mapservices.historicengland.org.uk/printwebservicehle/StatutoryPrint.svc/1909/HLE_A4L_NoGrade%7CHLE_A3L_NoGrade.pdf : Fort Blockhouse Letter from Major General Whitelocke to Col Clinton, 2 July 1803

BH Patterson, *Military Heritage*, p.15, plan dated c.1700 'Gun Battery'

References

1. D Hodson D 1978 *Maps of Portsmouth Before 1801* Portsmouth Record Series City of Portsmouth

2. John Goodall 2008 *Portchester Castle*, London: English Heritage, p. 3, ISBN 978-1-84802-007-8 4 http://www.ecastles.co.uk/portchester.html; accessed 20 May 2020

3. https://historicengland.org.uk/.../church-of-st-mary-portchester-castle-portchester-6177 accessed 20 May 2020

4. David W Lloyd DW *Buildings of Portsmouth and Its Environs* 1974 p.55

5. Charles O'Brien, Bill Bailey, Nikolaus Pevsner, David Lloyd 2018 *Buildings of England Hampshire South* Yale University Press p. 503

6. http://www.ecastles.co.uk/gosportline.html). http://www.castlesfortsbattles.co.uk/south_east/old_portsmouth_defences_round_tower.html https://historicengland.org.uk/listing/the-list/list-entry/1008754;

https://www.escp.org.uk/long-curtain-moat-emergency-works-2014; https://historicengland.org.uk/listing/the-list/list-entry/1386892; https://www.portsmouth.co.uk/news/new- bridge-for-nelson-s-historic-walk-1-1261651 Alan Balfour 1970 *Portsmouth:* London: 1970-: 18 Arthur Corney 1965*: Fortifications in Old Portsmouth*: Portsmouth City Museums pp. 25-29 David Lloyd 1974 *Buildings of Portsmouth and its Environs*: Portsmouth: 1974 p. 59; David Lloyd & Nikolaus Pevsner1967 *The Buildings of England*: Pevsner N & David Lloyd 1967 *Hampshire and the Isle of Wight* Buildings of England Harmondsworth: pp. 422-424 GH Williams 1979 *The Western Defences of Portsmouth Harbour 1400-1800.* Portsmouth Paper No. 30 Portsmouth City Council

7   Ray Riley 1987 'Military and Naval Land Use as a Determinant of Urban Development – the Case of Portsmouth' in *The Geography of* Defence eds. Michael Bateman and RC Riley Croom Helm p.56 Fig. 1

8   Riley op cit pp.55-60

9   Geoff Hallett 2016 'Fareham Reds' *The Redan* Palmerston Forts Society Number 77 2016 p.66

10 Riley op cit p.59

11 www.fortified-places.com/portsmouth/ accessed 15 June 2016

12 Gates op cit 1925 p.132

13 Riley 1987 p.64

14 Riley p.66

15 Edwin Course 1969 *Portsmouth Railways* Portsmouth Papers No. 6 pp.16-17)

Chapter 3
The Dockyard downgraded; defence sites close

The last steel casting Portsmouth Dockyard 1982

Demise of Portsmouth Royal Dockyard

On October 1 1982, the Minister of Defence John Nott announced a profound diminution of centuries of proud tradition: Portsmouth dockyard was to be downgraded to a 'Fleet Maintenance and Repair Organisation' and the workforce reduced to 2,800.[1,2] This devastating news provoked a death notice, published in *The Times* on the same day: "HM Royal Dockyard Portsmouth passed peacefully away at 12 o'clock last night after nearly 800 years of faithful service. It will be sadly missed by many."

Celia Clark had sent it to the paper on behalf of the Portsmouth Royal Dockyard Historical Society, of which she was secretary.

A signal was sent from the Ministry of Defence to the Admiral Superintendent in Portsmouth enquiring if there were labour troubles at the Dockyard and asking him to explain the notice in *The Times*. There was however a problem for the Admiral – he hadn't seen the actual notice, neither had any of his staff or indeed anyone else in Portsmouth. Many London printed newspapers failed to arrive that morning due to a rail strike, resulting in the newspaper train to the south coast suddenly being cancelled.

Once they saw it, however, it displeased the admirals greatly, but it expressed the deep feelings of the civilian workforce. Brian Patterson, president of the Society - summoned to explain the obituary to the Admiral - said that the Society was not in the business of inciting the workforce into any kind of action. That would have been the job of the trades unions, but the Society's brief was to remember the city's contribution to the Dockyard – the workforce – fathers and sons – grandfathers and great grandfathers and those before that who made the Dockyard what it was. It appeared that future generations would not have the same incentive or patriotic opportunity to follow their fathers' footsteps – because of government decisions – not naval ones!

The honorary president of the Society at the time, Lord Mayor Councillor John Marshall promptly resigned from the society; he did not wish to be associated with the notice. The then Navy Minister, Keith Speed MP for Chatham who had 'spilled the beans' was promptly sacked. This was also the end of the three armed forces having their own individual ministers.[3]

When the dockyard was formally renamed the Fleet Maintenance and Repair Base a longer version of the obituary appeared in the local *Evening News*, adding "Died peacefully at midnight September 30th 1984, of considerable age. Has been seriously ill since June 25th 1981. Will be sadly missed by thousands." In 1998 the Fleet Maintenance and Repair Organisation was contracted out to the private sector as Fleet Support Limited.

Defence Closures in Portsmouth and Gosport

Regional Consolidation Plan   Prince's Trust Enquiry by Design 2008

Driven not only by a reconfiguration of the nation's military capability but also by financial measures designed to balance the books and a political programme to reduce the role of the state, the speed, scale and extent of disposals escalated, largely in response to the pressure exerted by the Strategic Defence and Security Reviews (SDSR) of 2010 and 2015.[4] From 2012 the MOD's freehold and leasehold landholdings decreased by more than 1,100 hectares, as explored in more detail later in this chapter.

In 2007 research established that the economic impact of Portsmouth naval base extended far beyond the footprint of the site itself. At that time the base supported nearly 35,000 jobs in south Hampshire; 21,600 of these civilian. In total this accounted for 8% of the jobs in south Hampshire generating £680m annually for the area's economy.[5] The Strategic Defence and Security Review of 2010 resulted in a 7.7% cut in the defence budget over four years and a reduction in naval personnel by 5,000 to 30,000.[4] In 2014, as part of the City Deal bid for government funding, the dockyard was still said to be "at the heart of the sub-regional cluster, providing, directly and indirectly, 20,000 jobs and contributing over £1.6bn added gross value added of output." The naval base was supporting the Royal Navy surface fleet, and together with the associated naval establishments and defence firms it offered integrated ship support, complex software engineering, advanced manufacturing management, equipment training, and estates and logistics services.[6]

Locally the MOD is still a major employer and purchaser, but new closures continue to reduce southeast MOD landholdings. In 2009-2011 according to *The Economic Significance of Military Activity in Oxfordshire and the Hampshire Economic Area Technical Report*[7] a number of MOD-owned sites in Oxfordshire and the Hampshire Economic Area were due to be released for a variety of uses, including housing (including demonstration eco-houses), mixed-use schemes, offices and employment and playing fields. These ranged from small plots (under 0.25 ha) through to large sites of over 100 hectares.

As publicly owned property, their redevelopment is a complex planning challenge, differing from ordinary changes of use in several ways, including the unusually wide range of interested stakeholders and their expectations of return to the public good, whether defined environmentally or in economic terms as well as financially. In the UK bringing these brownfield sites back into use took on a new urgency among policymakers, developers, planners, local governments and campaigning organisations in order to address a growing housing crisis, but the new land uses do not necessarily address local communities' desire to use the sites for other purposes, including new civilian employment to replace the lost military jobs. This raises questions about the trade-offs between short-term gains and long-term benefits, especially benefits that cannot readily be measured in cash terms.

The scale and extent of active and surplus MOD property in southern Hampshire is clearly demonstrated in this 2011 map from a joint report on the Economic Significance of Military Activity by Hampshire and Oxfordshire county councils. Sites already closed are not indicated. Hampshire planners used to meet the MOD regularly to discuss sites on the disposal list, but they no longer do so.

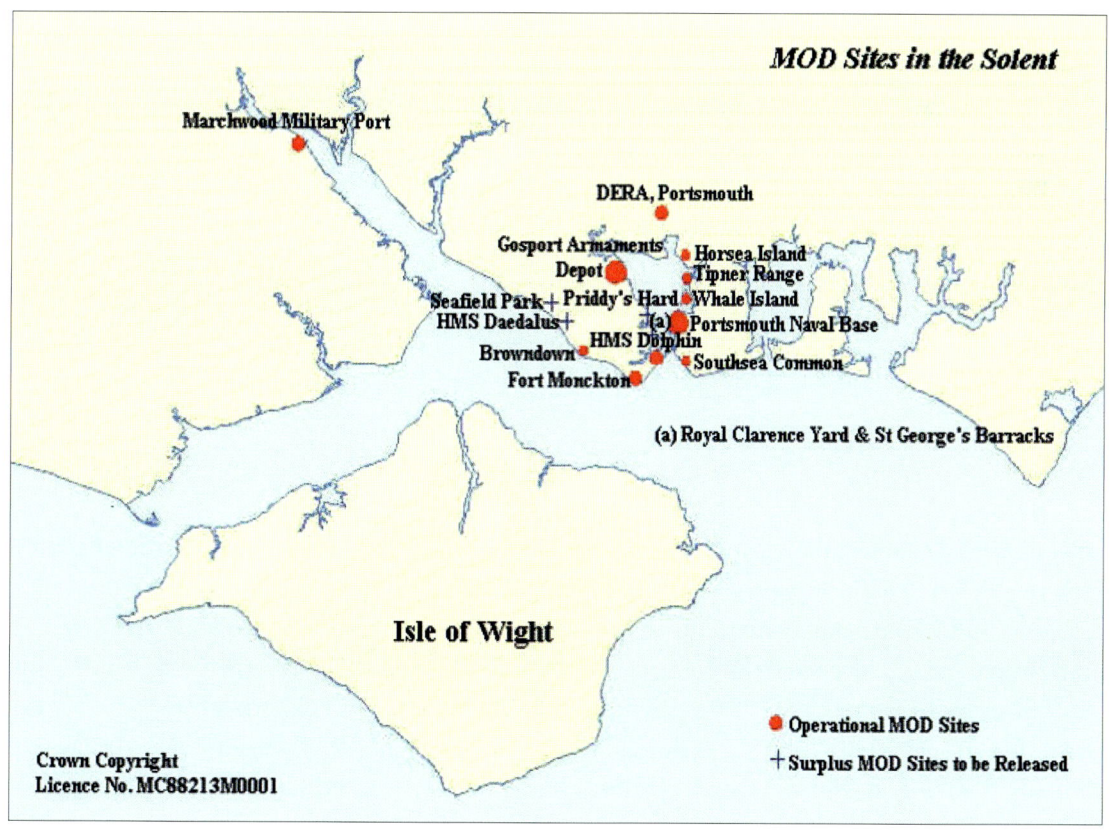

MOD Sites in the Solent.   Crown Copyright 2011

As already mentioned in Chapter 1, Gosport Borough Council and Portsmouth City Council produced the map below to show the extensive areas the MOD occupied or released in 2017 for the third regional seminar of professionals, developers and academics involved in the transformation of defences sites to civilian future, which we discuss later in the chapter. It was arranged by the Hampshire Buildings Preservation Trust and the Royal Town Planning Institute South East to hear from the many parties to these complex transformations and to point up best practice.

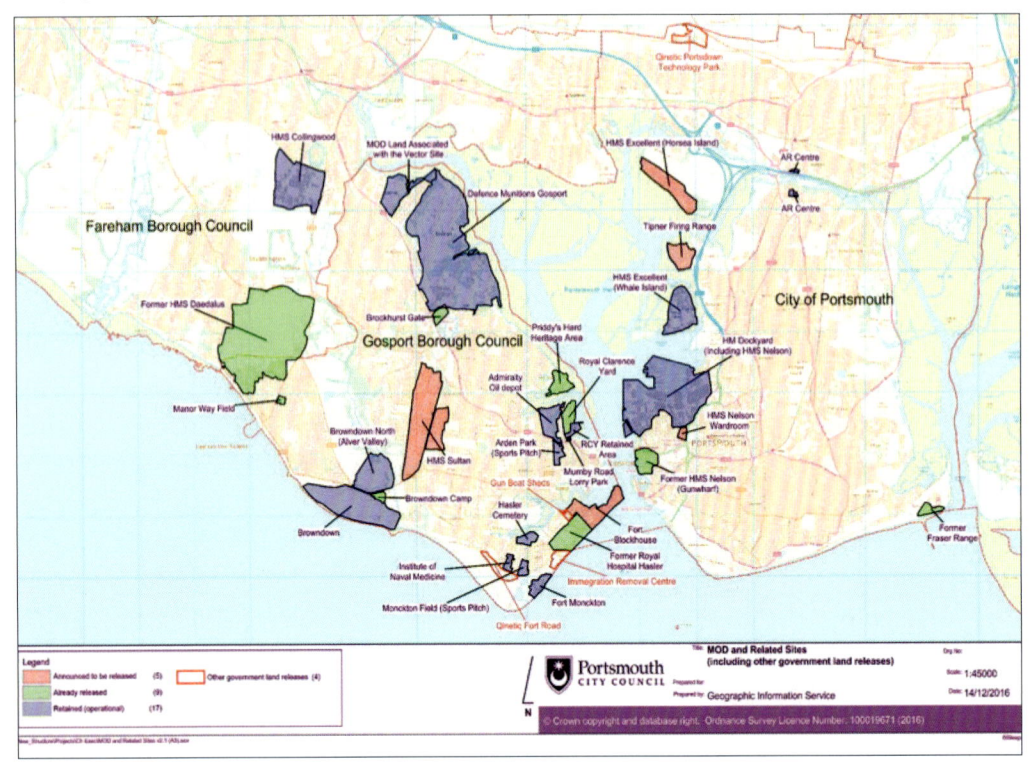

Portsmouth City Council Planning Department 2017

Red = Already released, Blue = still operational, White sites = other government land releases, and Green = sites to be disposed of.

Issues that arise from defence disposal include: urgency versus resource constraints, multi-designations adding to their complexity, flood risk, contamination, poor access, depressed or inflated land values, the unique form of many structures, the complexity of unravelling sites' significance, the many historic Buildings at Risk, the difficulty of accessing funds to unlock difficult sites and the failure to recognise the economic value of heritage. As a peninsula, Gosport has the added challenge of missing infrastructure, particularly transport.

Closures in Gosport

Gosport's defence legacy ranges in date from the medieval to the Cold War. A nineteenth century model of the defence structures in Gosport is stored in Fort Brockhurst. Multi-layered sites and landscapes include forts, naval airfield, armaments depot, hospital, barracks…. New discoveries about late nineteenth and early twentieth century local defence sites are still being made. Some of them contain remains that are both fragile and significant. What survives is protected in a range of ways: Listed, Conservation Areas, Scheduled Ancient Monuments, Non-designated Heritage Areas, Archaeological sites and Nature Conservation Designations, reflecting its age and importance to both building and nature conservation, also offering considerable challenges to sustainable reuse.

This next map shows the Gosport sites released by the MOD, announced for release and sites currently in military use.

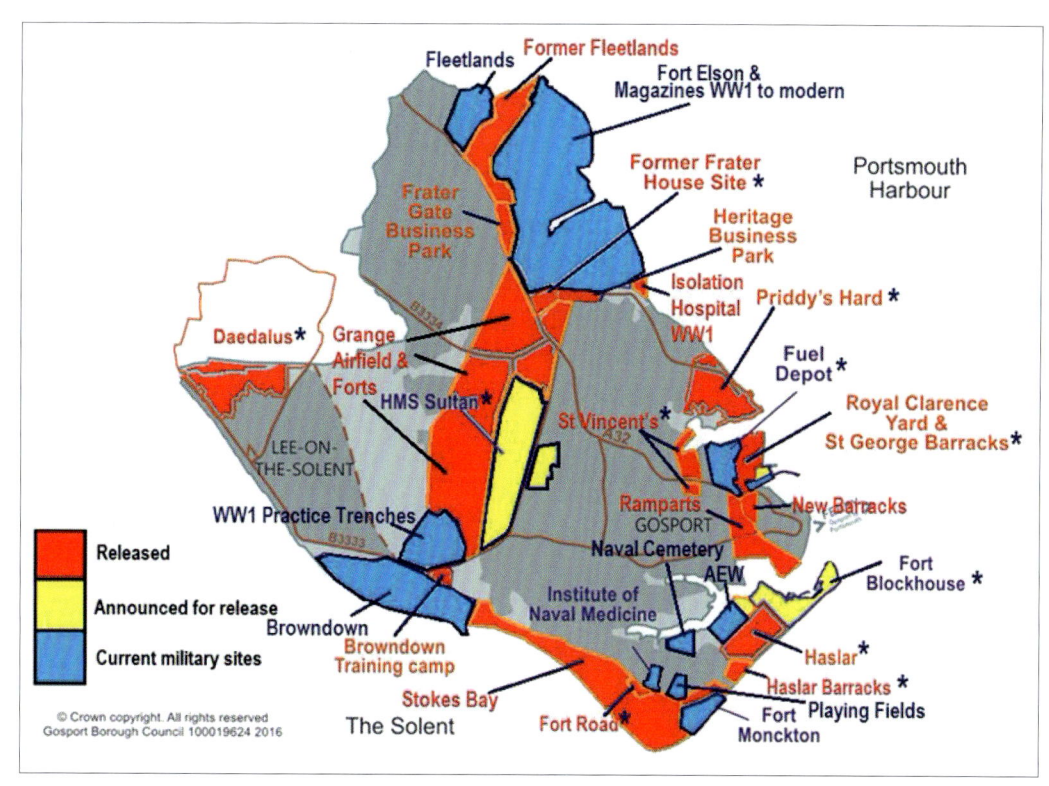

Gosport Borough Council 2017

This list of privatisations and closures in Gosport gives more detail of the range of sites which have changed hands and function in the last half-century. These closures are explored in detail in later chapters.

Closure dates – Gosport

Admiralty Engineering Laboratory (AEL) - part privatised in 2001, became part of a commercial company QinetiQ plc.

Bay House - In 1870 Admiralty leased Bay House land after War Office purchased it. Naval College set up there, later known as Ashburton House. In 1892 the house passed to Col. Francis Sloane-Stanley; now academy school. https://bayhouse.gfmat.org/school/school-history/

Browndown – seaward part still in use by MOD. Browndown Camp sold in 2011 to developer Jumbuck Ltd.

Cams Hall - In World War II, estate requisitioned by Admiralty until 1948 for the Engineering Department's drawing office when original office blitzed. Sold 1951; left to deteriorate, until reconstruction in the 1980s as business centre. https://www.camshall.co.uk

Daedalus - declared surplus to requirements in 2004. In 2006, the Maritime and Coastguard Agency (MCA) acquired the airfield.

Dolphin/Fort Blockhouse – still in MOD control but disposal plans being considered.

Fleetlands - In 2008 the DARA Rotary Wing maintenance, repair and overhaul facility acquired by Vector Aerospace.

Fort Elson - By 1901 all armament removed. Scheduled Monument but within active naval armament depot. In 1994 a detailed Strategic Proposal for Fort Elson drafted, outlining various future courses of action, from very expensive complete renovation through to "controlled ruination". Latter adopted; controlled deterioration with adequate records and safety.

Fort Brockhurst - released by the military in the 1960s. Handed over to Hampshire County Council, then to English Heritage.

Fort Fareham - The fort became surplus to requirements in 1965 when it was sold by the Ministry of Defence to Fareham Urban District Council. Now an industrial estate.

Fort Gilkicker - MOD declared it redundant; bought by Hampshire County Council in 1986; several reuse plans failed

Fort Gomer - released in 1964, sold at auction.

Fort Grange – absorbed into HMS Sultan in 1956

Fort Monkton – MOD(Army) as No.1 Military Training Establishment

Fort Rowner – retained within HMS Sultan, but mainly left to nature

Forton prison - surplus to requirements after about 1926; demolished in stages. In early 1960s remaining front buildings demolished for housing estate: hint at former use: 'Warder's Court'

Forts Widley and Purbrook - leased to Portsmouth Corporation in 1961. Sold outright to the council in 1972; current negotiations with Peter Ashley Activity Centre for longer lease

Frater House - site put up for sale in 2007

Haslar Gunboat Yard - under the control of QinetiQ since 2001; part leased to yacht club

Haslar Hospital - the provision of acute healthcare was transferred from the Defence Secondary Care Agency to the NHS Trust in 2001. Closed 2009 and sold later that year.

Haslar Barracks - closed in 1953 and lay derelict until 1962 when it was refurbished and reopened as a Civilian Detention Centre; put up for sale by Ministry of Justice 2019

HMS Sultan – MOD active use ongoing.

HMS St Vincent – Admiralty pulled out in 1969. Purchased jointly by Hampshire County Council and Gosport Borough Council.

Portsdown Oil Fuel depot - decommissioned in early 1990s

Priddy's Hard – gifted to Gosport Borough Council in April 1994

Royal Clarence Yard Clarence Yard – 16 hectares declared surplus in 1995. Berkeley Homes bid for the land in 1998. Remainder up for sale 2019

St George Barracks - Army left in 1991. Converted by Sunley Estates to housing in 2002

Stokes Bay Lines – No. 2 Battery - 1932 purchased by Gosport Borough Council
WWII: Requisitioned by the military. 1951 - released back to GBC and now the Diving Museum.

Reductions in Defence spending

Why did this huge process of redundancy happen? Significant international events including wars and acts of terrorism, political and technological change and redirected military priorities have inevitably had physical, social and economic impacts on defence facilities at local level. We survey these effects on the once defence-dominated communities around Portsmouth harbour over a long period, during which the UK economy was affected by the financial crash and periods of austerity in government spending – but also when property prices rose exponentially.

The evolution of Defence Estates policy outlined here needs to be viewed within this timeframe and in the local context. Specialist sites may change their function or become obsolete. The armed services provide many examples of serial reuse, both constructive and destructive. Changes in defence property management may be driven not only by financial cuts also by new objectives such as fuel economy or cybersecurity.

Two additional factors contributed largely to the scale of redundancy in the UK: the acceleration of the rundown of the defence industry, apparent since the 1950s, and the process of privatisation and the application of market values to disposal of publicly owned sites. Redundancy and pressure to reduce defence budgets to match Britain's reduced role in world affairs and weak economy in the 1990s linked to the requirement to maximise capital receipts on those for disposal have combined to concentrate defence activity onto fewer sites.

The new military necessity for rapid flexible response required quite different physical resources from the massive fixed positions of the Cold War, rendering huge tracts of land and buildings, some of them historic, redundant and looking for new uses. The closure and disposal of MOD bases accelerated. "The end of the Cold War led to a radical rethink about defence policy and the level of resources required by the armed forces in a less threatening world. This culminated in the government's 1990 discussion paper *UK defence policy: Options for Change* which reduced the size of the armed forces and closed more than 40 major military bases and other smaller installations. In addition, the US closed an additional 12 UK bases. Wiltshire was particularly affected; the county council worked with the Network Demilitarised partnership of 16 authorities in five countries "to address peace dividend issues"".[8]

The journalist Giles Worsley called this process "the greatest exchange of property since the dissolution of the monasteries". It was set in train in response to defence cuts and consolidation onto joint tri-service facilities. Defence Minister John Nott's 1981 Defence White Paper had already announced his decision to cut back on naval expenditure during the early '80s recession, which included the scrapping of the Antarctic patrol ship *HMS Endurance*, the reduction the surface fleet to 50 frigates and from three to two aircraft carriers. He transferred the resultant savings into nuclear submarines, naval weapons systems and air defence and the upgrading of the nuclear deterrent on board Trident.

According to Ron Tate, a former Portsmouth City planning officer, release and disposal of Ministry of Defence assets have been driven by two different kinds of reviews.[9] The John Nott review was based upon a desire to reduce the defence budget and to 'reap the peace dividend' (a phrase from the early 1990s at the end of the Cold War) across all the services. Other defence reviews were more focused on individual armed services, including changes to their requirements and technological advancements in ship construction, firefighting, weapon development, training and so on.

Ron Tate's understanding of overall MOD property strategy was that it was focused on training needs. For example, in the 2016 MOD estate report, surface ships were based mainly at HM Naval Base Portsmouth and HM Naval Base Devonport, with "some Minor War Vessels based at HM Naval Base Clyde, a submarine centre of specialization. Naval training establishments and accommodation centres were to be clustered around main operating base areas."[10] The intention was to have one third of the services in training for action, one third deployed after training and one third standing down ready for retraining; this group was often used for special humanitarian operations. It also meant that in the event of any emergency approximately half of the services were 'action ready' (i.e. close to ready or just off deployment to add to the deployed resources). The other priority was efficiency within a declining budget - which was sometimes achieved by tri-service consolidation onto one site. Inevitably, the effects of this defence property reduction process on local communities - and the responses by developers, conservation interests and planning authorities are considerable, as this book explores within the local context.

The National Audit Office report *Ministry of Defence - Identifying and Selling Surplus Property* (NAO 1998) and the White Paper *The Strategic Defence Review - Modern Forces for the Modern World* (1998) proposed to cut the defence budget by £2bn. The *Comprehensive Spending Review* by the Chancellor of the Exchequer in 1998 raised the issue of defence land. Between 1989/99 and 2008/9 the MOD raised £3.4bn from the sale of surplus property, which was invested in front line troops and equipment.

The sale process continued: in 2006/7, Defence Estates was charged with disposing of £151m worth of property. By 2012 the MOD's freehold and leasehold landholdings had decreased by more than 1,100 hectares.[11] The remainder contains more than 1,500 historic buildings and Ancient monuments,[12] a challenge addressed in the seminar arranged by the Hampshire Buildings Preservation Trust and the Royal Town Planning Institute South East held in Bursledon Brickworks in October 2017,[13] following two earlier regional ones in Oxford in 1996 and Winchester in 2002, all of which came up with recommendations for improving the disposal system. Sadly, until now, these have been mostly unheeded. We consider what needs to be done to make the whole process more locally beneficial in our concluding chapter. In 2003 the Association of Hampshire and Isle of Wight Local Authorities was concerned enough to approach the Deputy Prime Minister Expressing the continuing concern of local authorities at the Treasury's requirement to achieve "best consideration" in financial terms - which did not necessarily result in benefit to local communities.

How does the government dispose of land it no longer needs?

The Crichel Down case of 1954 gave rise to the rules on the disposal of Ministry of Defence land. It clearly showed the maladministration from which the previous landowners suffered, and that they were offered no legal redress. In 1937 725 acres of land at Crichel Down Dorset were compulsorily purchased. Local farmers were promised that when it was no longer needed for the country's defence they would be afforded an opportunity to bid for it. The promise was not honoured. After a public inquiry and the resignation of the Minister of Agriculture, the Franks Committee was set up in 1955 to investigate the fundamental problem of the conflict between private interests and public interests. Its reforms introduced a system of independent tribunals intended to ensure openness, fairness and impartiality, but they did not touch the huge area of decision-making in which no formal procedure was prescribed by Parliament.

Enterprising Portsmouth city council bid for the ornamental stonework from Crichel Down: the circular colonnade was set up in the 1970s in the garden behind the City Museum – although the central fountain was not reinstalled – and the elegant classical temple was re-erected in the grounds of Gatcombe House in Hilsea. Both were former defence properties.

For sites requisitioned by government in wartime or other circumstances the Crichel Down Rules updated in 2015 and 2017 require "government departments, under certain circumstances, to offer back surplus land to the former owner or the former owner's successors at the current market value. The MOD's Defence Estates department, now the Defence Infrastructure Organisation (DIO) is obliged to consider the potential application of the Crichel Down Rules to all of its surplus sites. If the rules apply and no exceptions from the rules are applicable to a site then DIO will offer that site (or the relevant parts of it) back to former owners, or their successors, ahead of any offer to the open market, in accordance with the rules."[14]

There are said to be exceptions to the rules if 'the site has 'materially changed' in character since it was acquired, such that there is no obligation to offer back. The 2017 Cabinet Office rules state that in such cases disposal to former owners … will be at a price reflecting current market value (including any development value), as determined by the Department's property advisers.'[15]

In 2020 this was a live issue at Southwick House built in 1800 to the north of Portsdown Hill whose grounds include the medieval ruins of Southwick Priory, the monks' fishponds, a formal historic landscape of stepped terraces and a lake. When the bombing of Portsmouth began in June 1940, the staff at the Naval Navigation School in Portsmouth who were housed in old buildings were up all night because the contents were inflammable and they were getting no sleep. So Col. Evelyn Thistlethwayte owner of Southwick House offered the head of Portsmouth Dockyard, Admiral William James the house for officers to sleep in. In 1941 the property was requisitioned under the War Act from the Southwick Estate. While much of the Allied planning for D-Day was conducted in London, the advanced plans for Operation Overlord (D-Day) were finalised at Southwick House, already the Allied Naval Headquarters. It became the Supreme Headquarters Allied Expeditionary Force as it was close to the ports from which the invasion was to leave. The whole of Southwick village was taken over by the Allied command, with Nissen huts to house the men and the officers' mess in the Golden Lion pub. Inside Southwick House a large plywood map was constructed to plot the progress of Operation Neptune, updated every hour to mark the position of the naval forces.[16]

After the war, in 1951 Southwick House and the surrounding land were compulsorily purchased from the Southwick Estate for £30,000 by the government. Barracks, service housing, sports grounds and other facilities were added over the years of MOD occupation. The navy moved out of what had become known as HMS Dryad in 2005. Since then, now known as Southwick Park, it has been home to the tri-service Defence College of Policing and Guarding which had moved from Rousillon barracks in Chichester which was developed for housing. Hampshire Constabulary have their forensic laboratory within the grounds of Southwick Park. In November 2016 the Ministry of Defence announced that the site was surplus to their requirements and would close in 2025.[17] The School of Policing was to move to Worthy Down near Winchester, but it was understood some limited new accommodation would first have to be built there. In January 2017 it was announced that the new £300m Defence College of Logistics, Policing and Administration was to be completed by the end of 2020, when navy, air force and army personnel were expected to be in occupation at Worthy Down.[18] They wanted to take the D-Day map with them - there is a Military Police Museum there, which includes accounts of D-Day, but after extensive lobbying from locals the map was listed as part of Southwick House (also listed) and therefore remained in situ.

Mark Thistlethwayte, owner of Southwick Estate whose property includes the surrounding village and extensive agricultural land continued to ask for the rest of Southwick Park to be returned to him - as the Crichel Down rules stipulate. Speaking to BBC Radio Solent he said: 'We were assured that as soon as it was not needed in the interest of the nation's defence, it would be offered back to us. Up until ten years ago they were always reassuring us that it would be the case. When we heard that they were selling the site we were delighted but then on the other hand the announcement appears to have got tangled up in the freeing up of government land for housing to try to raise as much money as possible. We hope it is just a case of a misunderstanding and we are completely opposed to plans of housing to go there." Mr Thistlethwayte added that if the family were able to buy back the land, they would look to build a sports academy on the site or find a 'community-based' purpose for it.

As we say in our open space chapter, in 2019 after much negotiation he was able to purchase 172 acres of parkland, the lake, property and golf course surrounding the main lake. But the MOD had declined to enter into further negotiations on any of the remaining 130 acres that had been compulsorily purchased. This area includes the derelict ex-naval equestrian centre, farmland, woodland and pony paddocks as well as the military base. With defence cutbacks the MOD then appeared uncertain both other when/if they would move out and whether to argue that Crichel Down should not apply anyway as they suggested the site had been materially altered, subsequently to the c.p.o. However, it was clear that development of the site had clearly already begun well before 1951. In 2020 much of the more recent building on the remainder of the site around Southwick House was in poor condition and a material amount would have to be demolished; however existing accommodation blocks were being used for people coming on courses.

Further to the MOD's announcement in 2016 Mark Thistlethwayte explored other potential uses for Southwick Park. These included possibly developing the house and the surrounding area as a campus for the University of Portsmouth department of criminology linked to the existing forensic laboratory, a sports academy and training facilities for Portsmouth Football Club, an equestrian centre linked to performance art and facilities for small businesses, with the business school installed in the main house. In 2020 the date for the move out was put back to 2026 or even 2029, so little progress has been made towards what appears to be an appropriate and sustainable new use for this important property.[19]

Government regeneration agencies take on redundant defence sites

An indication of the complexity of moving these sites on has been the creation and evolution of government agencies whose aim was to meet government objectives, especially housing, employment and sustainable development. Between 1998 and 2010 regional government agencies aimed at promoting sustainable economic development and regeneration and employment such as the South East England Development Agency (SEEDA) took on difficult sites such as HMS Daedalus. This is partly in Gosport and partly in Fareham. What was achieved there is examined in our aviation chapter. In 2010 these bodies were replaced by smaller partnerships between local authorities and businesses: local enterprise partnerships (LEPS). These were abolished in 2010.[20] In parallel, English Partnerships was the national regeneration body for England. It provided £7.5m to support the regeneration and renewal of the Rowner estate in Gosport, built in the 1960s to house service personnel. In his research for the Portsmouth based Bill Sargent Trust Julian Dobson found it "a case study of a disposal that, while it raised funds for the MOD, created continuing social and economic costs for the local community. It was sold in late 1980s and early 1990s to a series of private developers and landlords in a succession of piecemeal deals. In 2004 a BBC documentary described Rowner as "the worst estate in the country".[21] His research found that the Rowner situation was an extreme example of poorly executed asset disposals.[22] Between 2009 and 2015 English Partnerships working with Gosport Borough Council, Hampshire County Council, Portsmouth Housing Association and Taylor Wimpey Developments replaced over 500 homes at Rowner including 350 in private ownership. In 2008 English Partnerships also took on the former Coldeast hospital site in Fareham.[23]

In 2008 English Partnerships became part of the Homes and Communities Agency sponsored by the Department for Communities and Local Government. Its aim was to find government property on which to develop more housing. It funded the partial clean-up of the west side of the Tipner peninsula documented in our residential conversion chapter. In turn, this agency was replaced in January 2018 by Homes England and the Regulator of Social Housing to fund affordable housing.[24]

From 2013 to 2016 the government also had a Regeneration Investment Organisation to guide investment to meet objectives for housing and other aspects of economic regeneration.[25] However, under the government's Public Land for Housing programme launched in 2011 and relaunched in 2015, while enough public land for developers to build 131,000 homes has been sold, provision of affordable housing has not been achieved: only 2.6% of these were for social rent.[26]

In the 2015 UK general election the government returned to power had ambitious plans for equipping the armed forces and procuring a replacement nuclear weapons system, but with no equivalent commitment to maintaining or increasing military spending. Given this imbalance of ambition and austerity, something had to give.[27]

The 2015 Strategic Defence and Security Review (SDSR) provided an opportunity to rethink UK military spending in line with a greater long-term commitment to preventing, rather than suppressing armed conflicts abroad.[28] In 2015 the UK spent over £35 billion per year on its military: 2.0% of gross domestic product (GDP) or about 5% of total government expenditure. But military spending declined rapidly as a share of GDP under the Coalition Government from 2.5% of GDP in 2010, reflecting fiscal austerity as well as the withdrawal from Afghanistan. Despite a major increase in development spending, defence funds still represented 72% of the UK's internationally oriented budget.

To focus on naval disposals the MOD's *A Better Defence Estate* published in November 2016 said that "Continuing the consolidation of the Naval Estate that has taken place over the last 25 years, the strategy will see the Royal Navy focus ever more on Centres of Specialisation, with operating bases and training establishments locate predominantly around the port areas and Naval air stations." HMS Sultan, the navy's centre for mechanical engineering training was listed for closure in 2024 with training transferred to HMS Collingwood in Fareham and submarine engineer training to HMS Naval Base Clyde, Helensburgh. The Admiralty Interview Board was to be transferred to HM Naval Base Portsmouth in 2026. As we mentioned in the first chapter HMS Nelson Wardroom was also to be closed in 2021 and reprovided in HM Naval Base,[29] releasing a potentially valuable site in Queen Street Portsea with a large garden to the rear but this did not happen.

Furthermore, in March 2019 the defence minister announced that closure of HMS Sultan would be delayed until no earlier than 2029 – to the delight of armed forces veterans and local politicians who had fought to keep it open.[30]

Further defence cuts and changes to the defence estate

We now examine public expectations when government owned land is disposed of in the context of further defence cuts, and how the public interest is defined in these transactions. Should the financial receipts be spent partly in local economic reconstruction via the Department of Defense's Office of Economic Adjustment, as they are in the United States, or should they accrue to central government and the national defence budget?

Running the defence estate was estimated to have an annual cost in excess of £2.9 billion (NAO, 2010), prompting an assessment of whether these properties were needed or judged fit for purpose with some 6,500 of the properties managed by the MOD's estate management said to be vacant.[31]

Under the banner of greater efficiency the Review's aim was for a 30% reduction of the built defence estate by 2040, the sale of which was expected to generate £1 billion towards an increased equipment budget, and to free up development land for up to 55,000 new homes (HM Government 2015).[32] In response, under the direction of the DIO an increasing number of military sites were being declared as surplus to requirement and being disposed of.[33]

The *Strategic Defence and Security Review* published in October 2010 resulted in major changes to the defence estate and its management, although it contained no specific announcement about the closure of individual bases. Some vacant sites would be occupied by units returning from Germany, while others were retained for other purposes. Decisions about the future of the defence estate were to be taken on the basis of detailed investment appraisals and wider impact assessments.

In 2011 the Defence Estates organisation was amalgamated with other property and infrastructure functions within the MOD to form the Defence Infrastructure Organisation, which is responsible for around 80,000 hectares with more than 45,000 buildings, including more than half the Government's historic environment assets - a considerable challenge to reuse.[34] The DIO was effectively outsourced to the private sector through the selection of Capita, one of the largest government contractors, as 'strategic business partner' in a deal worth £400m over ten years.

The DIO now manages the defence estate: 1.8% of the UK's landmass[35] including 115,000 non-residential buildings and 50,000 houses. Its purpose was also said to be to be to reduce running costs and create commercial opportunities.[36]

An increasing number of MOD sites were being declared surplus to requirements and more were likely to be put up for disposal. These included airfields, barracks, training camps, research establishments and many miscellaneous parcels of land. As their historical significance is not well understood and many are inadequately recorded in national and local heritage databases because of "their closed and secretive nature", Historic England published a report and pilot project about sites in Wiltshire as part of the National Heritage Protection Plan. How this guidance worked out in practice around Portsmouth Harbour is explored in our later chapters.[37]

In December 2016 the MOD's *'A Better Defence Estate'* stated that: "The defence estate is where our people live, work, train, and ultimately operate and deploy from, where our equipment and weapons are stored, and where we carry out research to keep ahead of our adversaries. The ability of our Armed Forces to keep Britain safe and prosperous depends on a defence estate that enhances military capability."

"While in many areas we use our estate efficiently, overall it is still too big, too expensive, with too many sites in the wrong locations: it covers 424,000 hectares, about 1.8 per cent of the UK land mass; it costs £2.5bn a year to maintain; and 40 per cent of our assets are more than 50 years old. The Armed Forces are 30 per cent smaller than at the end of the last century but the estate has only reduced by 9 per cent."

"That is why the 2015 *Strategic Defence and Security Review* (SDSR) committed to invest in a better built estate that will reduce in size by 30 per cent by 2040.... Our plan builds on announcements made earlier this year by announcing a further 56 sites for disposal by 2040. Overall it will release 91 of the most expensive sites. The strategy also describes our re-provision plans. By getting rid of sites we no longer need, we will make an important contribution to our target to release land for 55,000 homes as part of the wider government housing policy."[38]

In January 2017 the Ministry of Defence announced that it intended to reduce its estate, still one of the largest in the country by 30% by 2040 in a 25-year plan which the National Audit Office said would be "extremely challenging" to implement, and also that there is a "significant risk' that the poor condition of the estate will affect defence capability. The plan was published in *A Better Defence Estate* (2016) already referred to. The Government said that an additional £4bn needed to be spent on the remainder over the next decade. So far £1bn had been committed. The estate as a whole covers 424,000 hectares (the MOD owns 220,000 hectares of land and foreshore and has access to a further 204,000 hectares) and the 'built estate' covers 30 million square metres.

The MOD spent £4.8bn or 12% of its budget on the estate in 2015-16 of which £2.7bn was spent by the Defence Infrastructure Organisation, which manages the majority of the built estate.

The varying details of just how much the DIO owns and manages is confusing, to say the least. In any case, according to Wayne Cocroft, Historic England's military specialist, the region most likely to first 'run dry' will be the southeast. This is where the majority of disposals have historically occurred and, perhaps not coincidentally, it is the area of the country known to generate the highest disposal receipts and where house building and development needs are most acute.[39]

What are the arrangements when surplus government land is sold? These were set in the 1990s.

The Treasury rules for disposal of government property

The Treasury guidelines of March 1998 gave general instructions for the disposal of surplus government property. Departments are required to dispose of surplus land and buildings at the best market price reasonably achieved by obtaining optimum planning value within a three-year timescale.[40] Any separate arrangements "where special policy considerations may apply" must be approved by the Treasury, for example on accepting a lower price than market value. This superseded the DOE/Welsh Office Circular 18/84 (1984) with the Crichel Down rules by which redundant land was to be offered back to former owners. This circular provided advice on the provision in the Town and Country Planning Act 1984 enabling the Crown, for the first time, to apply for planning permission prior to the disposal of Crown land - although the Crown Immunity exempting it from planning legislation was still in force, despite widespread opinions that its abolition was "well overdue." [40] It was abolished in 2006.

Critically for the historic defence sites around Portsmouth Harbour, the disposal system had slight modifications of timescale and price for historic sites. The 1995 DNH *Guidance Note for Departments on the Disposal of Surplus Historic Buildings* stated that all surplus historic buildings - particularly those which were vacant or only partially used - should be disposed of as quickly as possible; [however] maximising of receipts should not be the overriding aim in cases involving the disposal of historic buildings: "the aim should be to obtain the best return for the taxpayer that is consistent with Government policies for the protection of historic buildings and areas; these policies are likely to limit opportunities for the realisation of development value." [40]

Methods of disposal other than open market sale by auction or competitive tender might be considered where these would increase the chances of securing appropriate ownership and use of historic buildings; clawback covenants or other means might be incorporated to safeguard the taxpayer's interest. Surplus properties were first offered to other government departments. Disposal within three years might not always be achievable within three years; "in complex cases involving historic buildings"; because of risks of deterioration, vandalism and theft, it was better to keep them in full or at least partial use up to the point of disposal. The aim should be to obtain the best return for the taxpayer, having regard to:

> i. the provisions of the statutory development plan for the area;
>
> ii. Government policy for historic buildings and areas, as set out in Planning Policy Guidelines 15...which emphasises the importance to the nation of historic buildings and areas;
>
> iii. in particular, the clear recognition ...that the most appropriate long-term use for a historic building when account is taken of the need to protect its fabric, interior and setting) may not be the use which generates the optimum financial return;
>
> iv. the building's current state of repair, and the likely costs of future maintenance and repair.

This guidance was modified in 1999 by *The Disposal of Historic Buildings Guidance note for government departments and non-departmental public bodies* issued by the Department of Culture, Media and Sport. The 1992 Treasury guidelines acknowledged "the price realised on disposal of property by a public body can readily attract criticism". In February 1996 the Department of the Environment Transport and the Regions (DETR) commissioned research into defence disposals published in July 1998 because the DETR acknowledged that the planning guidance dealing with redundant MOD sites was outdated and did not recognise the plan-led system that had evolved in the UK since 1991, and also that there was a conflict between the market approach to disposals and the government's wider objectives.

Defining the public interest in public site disposals

In 1882 the first statute to introduce the principle that public interests should be given priority of private property rights was passed. It imposed the duty on property owners to keep historic buildings in repair, on the basis that they are national assets, which should be safeguarded for public enjoyment. However, the Crown exemption excluded publicly owned sites from the controls normally exercised over historic buildings in the public interest by government departments and local authorities, because in this instance the Ministry of Defence regarded defence of the country as paramount and a prudent use of diminishing financial resources - including little or no spending on unused buildings - as in the public interest.

In Celia Clark's PhD thesis she said that when sites were operational, it was possible to argue that national defence was as a clear public good; a country's forces provided a service for all. But does it follow that the sale of this publicly owned land is a public good, and is it a public good of the same value as national defence, which might justify the MOD's continuing control over the process 'in the public interest'? It might be said that such a contrived definition of the public interest is no more than a vague term used to justify actions of powerful bodies, excluding other legitimate ones. In the view of many of the participants, redundancy and sale of military land are expected to stimulate development of new civilian land values and disposal processes. But sales remain in the control of the Defence Infrastructure Organisation, which might be said to have little understanding of civilian expectations of regeneration on these sites. Disposals are controlled by the DIO in the interests of national defence and security. Internal collaboration - for example consultation of its historic buildings advisers when a historic site are to be sold - let alone external collaboration does not necessarily happen. Conflicted value systems between the MOD and other land use agencies is apparent. The Royal Town Planning Institute in its consultation on the DETR research said that although there is a "Defence national interest in the MoD acquiring and managing such (defence) land whilst in operational use ... there is no national interest in the MoD being responsible for the way in which such sites are disposed [of]. There are clear national planning and related policy objectives which are of no less importance than an obligation on the MoD to secure the best possible price for the land to meet its Treasury targets."[41]

A 2009 report by the Ministry of Defence identified a lack of community benefit from disposal of defence land. The rationale had been increased efficiency of resource exploitation, which has not even necessarily been the result, if the scandalous selloff of military housing to Annington Homes in 1996 is anything to go by.[42] The MOD sold its service family accommodation to this company for £1.662 bn. But of the 57,434 properties sold, 55,060 were immediately leased back on 200-year underleases. The service housing in Rowner Gosport and in Portsmouth was part of the deal. Extraordinarily, the MOD also retained all responsibility for maintaining and upgrading this accommodation, even though at the time of the deal it had expected its future commitments to Annington during the first twenty-five years of the deal to cost significantly more – at least £2bn more than Annington had paid for the properties. "In other words" this deal "was a terrible one for the state" and for the taxpayer. Annington continues to profit from it today – and by 2017 had already gained over £4bn. Brett Christophers' 2018 book: *The New Enclosure The Appropriation of Public Land in Neoliberal Britain* published by Verso sets out this and other massively poor returns on government land disposals in detail.[42] It also resulted in acute social problems in Gosport.

Christophers says that there have been no in-depth examinations of what public land disposal in Britain has meant specifically for those living in the vicinity of disposal sites, except for Julian Dobson's study *'In the Public Interest? Community Benefits from Ministry of Defence Land Disposals'* commissioned by a Portsmouth charity, the Bill Sargent Trust in 2009. Dobson's research into how to reconcile the two seemingly irreconcilable approaches – maximum cash return to the Treasury versus local gain in jobs, new facilities, open space, housing was conducted in three phases.[43] He encountered 'a lack of overarching academic research and little to suggest the issue has been high on the national policy agenda' and 'minimal interest in the issue from central government'. He identified a perennial tension between short-term budgetary exigencies of the public bodies selling land – and the long-term needs of the local community. Community benefits tend not to correlate with sale price or 'value for money'. Choices were made between the desire to maximise capital receipts from public land disposal and using the land for social benefit, and it was not surprising that benefit to local communities from a more considered but was ignored by the Ministry of Defence in favour of maximum financial return to the defence budget.[44]

Dobson identified a competing narrative to the logic of cost-saving and maximising the immediate financial return to the taxpayer, which is that the public good is best served by using public assets and their potential development to benefit the communities most impacted by them. The DIO's 2011 interim strategy for land disposal implicitly acknowledged the potential tensions involved, setting an objective of 'carrying out disposals in ways that achieve economic value for money in terms of receipts' but which 'also promote 'development, economic activity and growth... The DIO will work closely with interested parties to seek the best possible future for the site'.[44]

On the face of it this opens the door for creative partnerships between communities and the MOD on the future of vacated sites, allowing a broad interpretation of the idea of public benefits.

This approach was explored in the partnership between Rushmoor Council, the DIO and developers Grainger plc to redevelop land in Aldershot.[44] Another initiative was One Public Estate: where the Local Government Association and the Cabinet Office Government Property Unit was working with over 250 local authorities offering practical and technical support to deliver property- focused projects. By 2019-20 the aim was to create economic growth, 44,000 new jobs, 25,000 new homes via more integrated services and efficiencies financed by capital receipts of £615m and reduced running costs. One case study is RAF Mildenhall, the base for four different USAF commands since 1950. Its closure in 2023 was announced by the US Department of Defense in January 2015. Sale of the site was announced by the MOD in January 2016, but the ultimate decision was taken by the US government.[45]

There is other evidence in UK policy development that the notion of 'value' is beginning to be seen in terms of broad public benefits and not only in cash terms. Treasury guidance on asset ownership issued in 2008 defines value for money as 'optimising net social costs and benefits… based on the interests of society as a whole.'[46] While it assumes that assets are employed most efficiently in private ownership, it warns that 'externalities' affecting social welfare should be taken into account. Annex 2 of the Treasury's Green Book[47] makes it clear that policymakers should value 'non-market impacts' of policies or projects. 'The full value of goods such as health, educational success, family and community stability, and environmental assets cannot simply be inferred from market prices, but we should not neglect such important social impacts in policy making'. But of course, it is one thing to articulate a definition of value designed to encourage creative and long-term thinking about the future of land and property assets, and another to put such thinking into practice 'in the hurly-burly of negotiations with developers, target-setting by central government and continuing cuts in public finances.'[48]

Who influences the process and the eventual land use outcomes? Who gains and who loses?

The disposal/sale and redevelopment of sites, once important to once defence-dominated local economies and communities is often also vital to their future. Because state land belongs to the nation, the public expect the new civilian land uses achieved on them to contribute to meeting local needs. In addition, many different people and groups' interests converge in their disposal and regeneration, particularly in the case of historic defence sites. Unlike usual planning determinations, many more people have a stake in the process. But because the MOD as a government department is charged with national defence: the ultimate 'top-down' hierarchy, it has not been accustomed to collaborative working to meet local concerns.

The UK state land disposal process is often controversial, because these are no ordinary planning determinations. Before Google Maps, defence sites were just white holes on maps. Until 2006 when the Crown Exemption was abolished the MOD was not subject to civilian regulations including planning. Very often defence sites are closed to public access except for the people who work on them. So, locals, including local authority planners, know little about them – unless there are special public access events such as Heritage Open Days, which began in 1994. This lack of knowledge may limit or distort public perceptions about their potential and future contribution to the local economy, as also may the lack of, or incomplete knowledge of, local historic environment records about them.

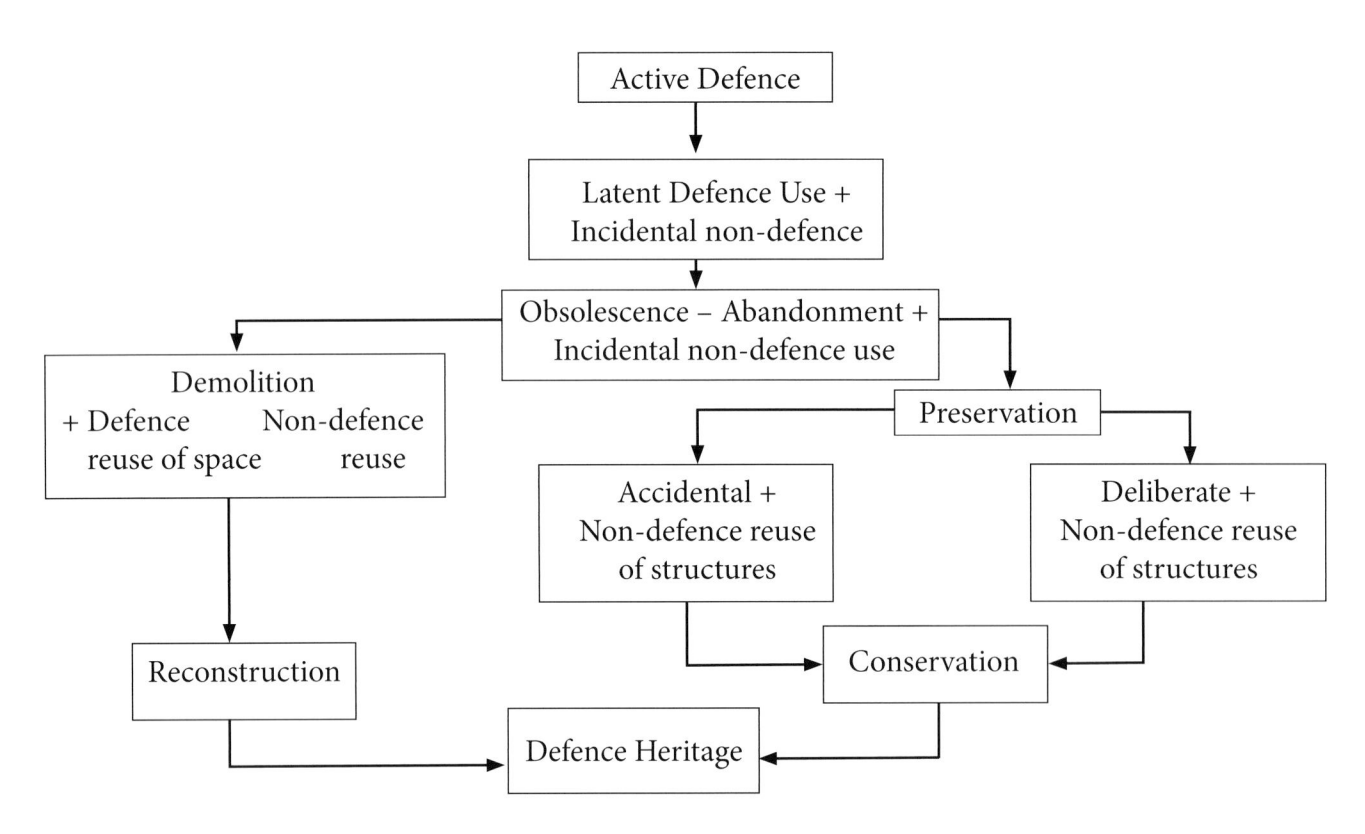

```
                    ┌─────────────────┐
                    │ Active Defence  │
                    └────────┬────────┘
                             ↓
                    ┌─────────────────────┐
                    │ Latent Defence Use +│
                    │ Incidental non-defence│
                    └────────┬────────────┘
                             ↓
                    ┌──────────────────────────┐
              ┌─────┤ Obsolescence – Abandonment +│
              │     │ Incidental non-defence use │
              │     └────────┬─────────────────┘
              ↓              │              ↓
┌──────────────────────┐     │         ┌──────────────┐
│ Demolition           │     │         │ Preservation │
│ + Defence  Non-defence│    │         └──┬────────┬──┘
│ reuse of space  reuse│     ↓            ↓        ↓
└──────────┬───────────┘  ┌──────────────┐  ┌──────────────┐
           │              │ Accidental + │  │ Deliberate + │
           │              │ Non-defence  │  │ Non-defence  │
           │              │ reuse of     │  │ reuse of     │
           │              │ structures   │  │ structures   │
           ↓              └──────┬───────┘  └──────┬───────┘
┌──────────────┐                 │    ┌────────────┐ │
│Reconstruction│                 └───→│Conservation│←┘
└──────┬───────┘                      └──────┬─────┘
       │         ┌────────────────┐          │
       └────────→│ Defence Heritage│←────────┘
                 └────────────────┘
```

Re-use of redundant defence works: a schema
War and the City Gregory J Ashworth Routledge 1991 p. 156

This diagram in Greg Ashworth's *War and the City* published in 1991 neatly sets out the possible routes for historic defence sites from redundancy to reuse, from active use to definition as 'defence heritage'. We explore local examples of sites that have experienced both paths.

Another critical factor is who has the most – and least – influence during the process and on the outcomes. In 2000 Celia Clark's research considered the many stakeholders involved in defence heritage disposals and their comparative power over the process. The Treasury and the Ministry of Defence dominated the disposal process through its Defence Estates Organisation and from 2011 its successor the Defence Infrastructure Organisation. These two government departments clearly have the most top-down influence on local planning authorities and on the eventual land use outcomes, along with the developers, while statutory bodies such as the environment and national heritage ministries might have an advisory role, with English Heritage/Historic England as a government agency. At that time voluntary advisory bodies: the Royal Fine Art Commission and SAVE Britain's Heritage were also involved in the case studies she examined: HMS Vernon/Gunwharf, Royal William Yard and South Yard in Plymouth and the Royal Gunpowder Works in Waltham Abbey. From the bottom up, local communities, military heritage groups, environment and conservation groups and individuals had much less influence on the eventual land uses.[48]

When plans are proposed for these complex sites how are local communities consulted? There's a small degree of experiment here – and relevant research[48, 49] ... but so often it's Publicise, Consult – and Ignore... just ticking 'public consultation' boxes, but with no real responsiveness to local needs or aspirations... Public consultation techniques used locally about the future of defence heritage have included both local initiatives and developer-led processes:

- Heritage Open Days – organised by the Gosport Society and elsewhere by planning authorities or the MOD
- Gunwharf site visit questionnaire in 1996 run by the Portsmouth Society
- Community Planning Events: Royal Clarence Victualling Yard Gosport - run by John

- Thompson Architects for developer Berkeley Homes
- Enquiry by Design: Haslar Hospital - run by the Prince's Regeneration Trust. The eventual developer was Harcourt Developments, but the recommendations in the trust's report and the entirely appropriate Veterans' Village plans were not implemented.
- Change.org petitions signed by local people and responses to public exhibitions e.g. Southsea Sea Defences affecting Southsea Common, Southsea Castle and Long Curtain Moat.
- Community Stakeholder Group – Southsea Sea Defences – 'cascade model' run by the engineers, East Solent Coastal Partnership for Portsmouth City Council.

Untried for the future of MOD sites in the local context are two other techniques: a Community Development Trust – a condition of planning permission at Graylingwell Hospital on the outskirts of Chichester, enabling the development of several buildings for community use – and Gosport's Community Engagement Plan as part of its Heritage Action Zone for the High Street. The Heritage Lottery's recent requirement that there should be measurable social benefit to communities as a condition of grants for building conservation schemes is a driver for change in this direction. Neighbourhood Planning, introduced by the Ministry of Housing, Communities and Local Government in 2014 "gives communities direct power to develop a shared vision for their neighbourhood and shape the development and growth of their local area… Local communities can choose to set planning policies through a neighbourhood plan that forms part of the development plan used in determining planning applications, and grant planning permission through Neighbourhood Development Orders and Community Right to build orders for specific development which complies with the order." There are many such initiatives in the southeast, but few in south Hampshire and, as far as is known, none relating to a former MOD property perhaps because of the public exclusion until they are released.

Often unacknowledged is the month after month scrutiny of key planning applications by local civic societies, commenting, where appropriate, on planning, design and sustainability. How this long-standing opportunity to influence local change would be affected by sweeping reshaping of the planning system proposed in 2020 is at yet unknown. In their discussion with local amenity groups, their central body, Civic Voice (the successor to the Civic Trust) called for a planning system with public participation at its heart with community-led participation driving the recovery from the pandemic that people should have the power and influence to decide what should happen. They asked for a national conversation, asking why, 50 years after the Skeffington Review, we were still having these discussions. As mentioned again in our chapter on ordnance yards, the 1969 Skeffington report on public participation in planning had recommended that planning should be a genuinely democratic process involving mediation between a wide range of competing interests, but this is by no means the norm.

The weight developers and planning authorities give to local responses varies... Top-down and Consult/Ignore is usual; Bottom-up: for example, Community Action Planning and Planning for Real are rare. More participatory processes involving contributions by those directly affected are proven to result in more publicly acceptable, long lasting and sustainable outcomes, but as this work shows, in most of the cases examined in this book, this potential was not acknowledged by the authorities with the most power over the sites.

Variations in disposal and Clawback

Although sale to the highest bidder is the norm, Portsmouth Harbour provides two cases where a redundant defence site was transferred to a local authority: Tipner and Priddy's Hard. The Hard was gifted to Gosport Borough Council free, but without a dowry to help tackle its severe contamination, as we explore in our residential chapter. In contrast, a £5.5m government dowry was given to the successor trust to tackle the even more serious contamination at the Royal Gunpowder Works at Waltham Abbey Essex.[48] The transfer of Tipner Range to Portsmouth City Council in 2018 came with a cost attached – relocation of the firing range. It was part of the City Deal Portsmouth and Southampton secured with government in 2013. Government gave the cities £48.75m; part of the deal was that Portsmouth City Council would then buy the range for a nominal fee and pay £3.75m to the MOD to relocate its facilities to the MOD's firing range in East Hampshire. The city's proposed development of 4,000 homes at Tipner might well pose expensive decontamination chalenges.

For a number of years there has been criticism of the Treasury's short-term outlook– the maximization of capital receipts, which may be incompatible with local authorities' strategic planning – and with housing developers' short term interest in rapid redevelopment and financial return.[50] Yet complex historic defence sites in particular, such as Haslar Hospital usually require long timescales and detailed negotiation if they are to be sustainably regenerated. Historic fabric does not always survive the transition to new owners, especially if the price they paid meant that they expect the maximum financial return on their investment - and the historic defence legacy is not protected by listing, scheduling or conservation areas. Sale to the highest bidder – or slightly less for historic sites – is likely to result in high return land uses – particularly high-end housing, shopping and leisure facilities e.g. cinemas, bowling alleys. As we explore in our ordnance yard chapter these were what replaced older structures on Gunwharf despite its designation as a conservation area with a few listed and scheduled buildings. Most of the site was cleared except for the few protected structures.

Nevertheless, there has been an increasing national focus on protecting and reusing historic defence buildings and structures as explored in more detail in the next chapter.

There is a provision for the public exchequer to gain if the profit from redeveloping former government property is well above the original price paid. The MOD can "reclaim or share any increase in value arising from either a granting of planning permission for a more beneficial use, or a change in the nature of the permission after disposal terms have been agreed. Clawback receipts negotiated for the last eight years total over £56m and account for a significant part of the overall disposal target." 2002-3 and the following year yielded particularly high returns.[51] The developer pays the MOD a proportion of the profit subsequently made over the first purchase price. Clawback happened at least twice during the redevelopment of Gunwharf - as we explore in our chapter on ordnance yards. Exponential rises in property prices and the intensity of development have resulted in the increased application of clawback.[52]

Sources

House of Commons Defence Committee 1994a The Defence Estate: Volume I - Main Report HMSO London

House of Commons Defence Committee 1994b The Defence Estate Volume 2 - Minutes of Evidence and Memoranda HMSO London

House of Commons Defence Committee 1995 first report The Defence Estate Vol I Report, together with the Proceedings of the Committee relating to the Report. House of Commons Session No 67-1 1994-5 HMSO London

House of Commons Defence Lands Committee 1973 Report of the Defence Lands Committee (The Nugent Report) HMSO London

House of Commons 1990 Options for Change Statement in House of Commons 25/7/90 Columns 468-486 HMSO London

House of Commons Public Accounts Committee 1987 Control and Management of the Defence Estate: Ministry of Defence: Property Services Agency HMSO London

House of Commons Parliamentary Question 1997 by Mr. Peter Viggers MP Answer by Mr. Raynsford MP Thursday 10 July 1997 (No.239)

House of Commons Defence Committee 1995 first report The Defence Estate Vol I Report, together with the Proceedings of the Committee relating to the Report House of Commons Session No 67-1 1994-5 HMSO London

House of Commons 1990 Options for Change HMSO London

Celia Clark 2002 "White Holes" Decision-making in Disposal of Ministry of Defence Heritage Sites PhD thesis University of Portsmouth

References

1. https://portsmouthdockyard.org.uk/timeline/details/1984-portsmouth-dockyard-becomes-fleet-maintenance-repair-organisation

2. Paul Brown 2016 *Maritime Portsmouth*. Stroud, Gloucestershire The History Press

3. Archie Malley 2013 Portsmouth Royal Dockyard Historical Trust Interview with Celia Clark 23 July

4. "Securing Britain in an Age of Uncertainty: The Strategic Defence and Security Review" (PDF). HM Government. 19 October 2010. Retrieved 18 November 2018

5. Prime Minister's Office, Cabinet Office, Department for International Development, Foreign and Commonwealth Office, Home Office and Ministry of Defence 2015 National Security Strategy and Strategic Defence and Security Review 2015 https://www.gov.uk/government/publications/national-security-strategy-and-strategic-defence-and-security-review-2015

6. J Grainger et al 2007 'Socio-economic impact assessment of Portsmouth naval base. Centre for Regional Economic Analysis, University of Portsmouth; Julian Dobson 2016 'From Crown to commons?' A UK perspective' in *Sustainable Regeneration of Former Military Sites* Samer Bagaeen and Celia Clark eds. Routledge p. 23

7. Southampton and Portsmouth City Deal n.d. https://assets.publishing.service.gov.uk/government/ uploads/ system/uploads/attachment_data/file/25 6460/2013111_Southampton_and_Portsmouth_City_Deal_ Document_and_Implementation_Plans.p df accessed 15 February 2020

8. Ministry of Defence 2011 The economic significance of military activity in Oxfordshire and the Hampshire economic area. Reports detailing the armed forces impact on the economy in Hampshire and Oxfordshire. https:// www.gov.uk/government/publications/the-economic-signifcance-of-military-activity-in-oxfordshire-and-the- hampshire-economic-area accessed 24 February 2020; Hampshire County Council, Oxfordshire County Council, SQW 2011

9. *"Defence (Options for Change)". Parliamentary Debates (Hansard).* House of Commons. 25 July 1990. col. 468–486; Roger Rankin 1995 'In Defence of the MOD' *Public Sector Building* August pp.8-9 9

10. Ron Tate 2000 September 24 Interview with Celia Clark

11. Ministry of Defence 2016 'A Better Defence Estate' November

12. Defence Infrastructure Organisation 2014 'Defence Infrastructure Organisation estate information' [Online] www.gov.uk/government/publications/mod-estate-information accessed 21 July 2015

13 Julian Dobson 2016 'From Crown to commons? A UK perspective' in Samer Bagaeen and Celia Clark 2016 *Sustainable Regeneration of Former Military Sites* Routledge p.29

14 Celia Clark 2017 Sustainable Regeneration of Former Defence Sites Report of Findings Royal Town Planning Institute/Hampshire Building Preservation Trust Seminar 13 October 2017 Burseldon Brickworks

15 Ministry of Housing, Communities & Local Government Compulsory purchase process and the Crichel Down Rules 29 October 2015 p143; Crichel Down Rules on land ownership. Rules that apply to the sale of sites previously acquired by government departments and in defence 12 December 2012; Ministry of Housing, Communities & Local Government Compulsory purchase process and the Crichel Down Rules 29 October 2015; https://assets.publishing.service.gov.uk/government/uploads/system/ uploads/attachment_data/file/68 4529/Compulsory_purchase_process_and_the_Crichel_Down_Rules_- guidance_updated_180228.pdf; Cabinet office 2017 Guidance for the disposal of surplus government land https://www.gov.uk/government/ publications/guide-for-the-disposal-of-surplus- government-land; Ministry of Housing, Communities and Local Government Guidance on Compulsory purchase process and the Crichel Down rules July 2019

16 Cabinet Office Guide for the Disposal of Surplus Land March 2017 3.2.5 Disposal at less than Market Value; 4.1.1 Disposal Objectives Updated 2 February 2017 from Ministry of Defence and Defence Infrastructure Organisation; https://www.gov.uk/guidance/crichel- down-rules-on-land- ownership accessed 10 December 2019

17 h t t p s : / /liberation route . com / great - britain / pois/ s / southwick - house?gclid=EAIaIQobChMIqsG4neXq5wIVB7DtCh3W2gYxEAAYASAAEgK9Y_D_BwE accessed19 February 2020

18 https://www.facebook.com/pages/Southwick-Park-Defence-College-of-Policing- Guarding/311249959065518; https://www.independent.co.uk/news/uk/home-news/jason-casts-a- cloud-over- naval-college-sale-1578790.html

19 Peter Riches 2017 'Worthy Down's Defence College of Logistics, Policing and Administration takes shape' 17 January 2017 Blog Inside DIO tps://insidedio.blog.gov.uk/2017/01/19/worthy-downs- defence-college-of- logistics- policing-and-administration-takes-shape/ accessed 25 February 2020

20 Mark Thistlethwayte 2020 Interview with Celia Clark, Martin Marks and Deane Clark 18 February

21 *"Local Enterprise Partnership Capacity Fund". Department for Business, Innovation & Skills. February 2011. Archived from the original on 9 November 2011;* UK Legislation website - The Regional Development Agencies Act 1998 (Commencement No. 1) Order 1998; UK Legislation website - The Regional Development Agencies Act 1998 (Commencement No. 2) Order 2000

22 Julian Dobson 2016 'From Crown to commons? A UK perspective' in Samer Bagaeen and Celia Clark 2016 *Sustainable Regeneration of Former Military Sites* Routledge p.33

23 Julian Dobson 2010 'In the public interest: Community benefits from Ministry of Defence land disposals'. Portsmouth Bill Sargent Trust

24 https://webarchive.nationalarchives.gov.uk/20090413233913/http://www.englishpartnerships.co.uk/corporateinformation.htm#SE

25 https://www.gov.uk/government/organisations/homes-and-communities-agency

26 https://assets.publishing.service.gov.uk/government/uploads/system/uploads/attachment_data/file/ 600615/RIO_Pitchbook_May_2016_withdrawn.pdf

27 NewStart https://newstartmag.co.uk/articles/scandalously-low-level-of-affordable-housing-on-sold- off-land/ February 2020

28 Richard Reeve 13 May 2015 Oxford Research Group 'Cutting the Cloth: Ambition, Austerity and the Case for Rethinking UK Military Spending' https://www.oxfordresearchgroup.org.uk/cutting-the- cloth-ambition-austerity-and-the-case-for-rethinking-uk-military-spending

29 Prime Minister's Office, Cabinet Office, Department for International Development, Foreign and Commonwealth Office, Home Office and Ministry of Defence 2015 *National Security Strategy and Strategic Defence and Security Review* 2015 https://www.gov.uk/government/publications/national- security-strategy-and-strategic- defence-and-security-review-2015

30 MOD *A Better Defence Estate* 2016 https://assets.publishing.service.gov.uk/government/ uploads/system/uploads/attachment_data/file/57 6401/Better_Defence_Estate_Dec16_Amends_Web.pdf

31 Byron Melton 2019 'Joy as HMS Sultan closure is put on hold' *The News* March 1 2019: https://www.portsmouth.co.uk/our-region/gosport/joy-as-hms-sultan-closure-is-put- on-hold-1- 8829888

32 NAO Report 2010 (HC 70 2010-2011: Ministry of Defence A defence estate of the right size to meet operational needs) The Stationery Office London

33 HM Government, 2015 https://www.nao.org.uk/wp-content/uploads/2016/11/Delivering-the- defence-estate-Summary.pdf https://www.nao.org.uk/wp-content/uploads/2016/11/Delivering- the- defence-estate-with- correction-slip.pdf, accessed 11 February 2020

34 Local Government Association 2014 cited in Nancy Tanner 2017 *FRIEND OR FOE? The MOD, Heritage and the Defence Review* MSc. Dissertation Oxford Brookes University School of the Built Environment with University of Oxford Department for Continuing Education May 2017 p.24; https://assets.publishing.service.gov.uk/government/uploads/system/uploads/attachment_data/file/33 228/20111005MODLandDisposalStrategy.pdf

35 Defence Infrastructure Organisation 2011 Defence Infrastructure Interim Land and Disposal Strategy October 2011; Defence Infrastructure Organisation 2014 'DIO welcomes new strategic business partner' www.gov. uk/government/news/dio-welcomes-new-stategic-business-partner, accessed 21 July 2014, in Dobson 2016 Julian Dobson 'From Crown to commons? A UK perspective' in Samer Bagaeen and Celia Clark eds. *Sustainable Regeneration of Former Defence Sites* Routledge 2016 pp. 19-31

36 Ministry of Defence 2016 Delivering the defence estate Report by the Comptroller and Auditor General https://www.nao.org.uk/wp-content/uploads/2016/11/Delivering-the-defence-estate- Summary.pdf accessed 24 February 2020

37 Jim Dunton 2018 "Defence Infrastructure Organisation chief Graham Dalton on its new role, the scale of his remit, and living through Carillion" https://www.civilserviceworld.com/articles/interview/defence-infrastructure-organisation-chief-graham- dalton- 21 May *Civil Service World* Retrieved 3 December 2018

38 https://historicengland.org.uk/images-books/publications/nhpp-plan-framework; English Heritage 2013 National Heritage Protection Plan 2011-15, October 2013

39 MOD 'A Better Defence Estate' November 2016 https://assets.publishing.service.gov.uk/government/uploads/system/uploads/attachment_data/file/576401/Better_Defence_Estate_Dec16_Amends_Web.pdf

38 Louisa Brooke-Holland 2017 House of Commons Library Briefing Paper CBP-07862 12 January 2017 *Defence Estate Strategy*

39 Cited in Nancy Tanner 2017 *FRIEND OR FOE? The MOD, Heritage and the Defence Review* MSc. Dissertation Oxford Brookes University School of the Built Environment with University of Oxford Department for Continuing Education May 2017

40 DAO(GEN)13/92 Annex A 1992), contained within Annex 32.1 of Government Accounting (Amendment No.8); Fuller Peiser/University of Reading 1998 p.19; DNH 1995 para. I iii p.1; *DETR Research Study Development of the Redundant Defence Estate Draft Final Report* Fuller Peiser/University of Reading 1998

41 D Rose RTPI 1998 Letter of 19 April to Peter Bovil, Fuller Peiser: Development of the Redundant Defence Estate DETR Research Project Seminar 28 April; D Rose RTPI 1999 email 10 March to C Clark on draft paper How the MOD's plan to privatise military housing ended in disaster". *The Guardian. 25 April 2017*. Retrieved 2 December 2018; "Empty MOD homes 'cost millions'". *BBC. 6 March 2008*. Retrieved 2 December 2018;

42 Brett Christophers 2018 *The New Enclosure The Appropriation of Public Land in Neoliberal Britain* Verso 2018 p. 276; p. 278

43 Julian Dobson 2010 *In the public interest: Community benefits from Ministry of Defence land disposals* Portsmouth Bill Sargent Trust; Julian Dobson 2012 *Forces for good: local benefits from surplus military land* Portsmouth Bill Sargent Trust; Julian Dobson 2016 'From Crown to commons? A UK perspective' in Samer Bagaeen and Celia Clark 2016 *Sustainable Regeneration of Former Military Sites* Routledge

44 DIO 2011: 8

44 Dobson 2012

45 *East Anglian Daily Times* 18 April 2017 Updates 4 August 2017

46 Joseph Lowe HM Treasury 2008 *Value for money and the valuation of public sector assets* July

47 47 HM Treasury 2011 http://www.cpppc.org/u/cms/ppp/201710/27090141j7ch.pdf p. 57

48 HM Treasury 2018 *The Green Book Central Government Guidance on Appraisal and Evaluation* https://assets.publishing.service.gov.uk/government/uploads/system/uploads/attachment_data/file/685903/The_Green_Book.pdf

49 Celia Clark 2002 "White Holes": Decision-making in Disposal of Ministry of Defence Heritage Sites PhD thesis University of Portsmouth p. 271

50 Taye Olukayode Famuditi 'Developing local community participation within shoreline management in England: the role of Coastal Action Groups' University of Portsmouth PhD thesis 2016

51 J Doak 1999 Fuller Peiser *Report Development of the Redundant Defence Estate* University of Reading; Fuller Peiser and University of Reading 1999 *Development of the Redundant Defence Estate: A Report to the Department of Environment, Transport and The Regions.* Thomas Telford London; Celia Clark 2002 "White Holes": Decision-making in Disposal of Ministry of Defence Heritage Sites PhD thesis University of Portsmouth

52 MOD 2004 *The Stewardship Report on the Defence Estate* p.47

53 Celia Clark 2017 The Search for Sustainable Futures for Historic Military Landscapes International Conference Military Landscapes La Maddalena 21-24 June 2017; Celia Clark 2017 Sustainable Regeneration of Former Defence Sites – Briefing 13 October 2017 Royal Town Planning Institute South East/Hampshire Buildings Preservation Trust P.7

Chapter 4

'Beware of Little Expenses: A Small Leak Will Sink a Big Ship'
Benjamin Franklin.

Decay, protection and conservation of the historic defence estate

Many of the local defence sites this book is about have long histories embodied in a significant legacy of buildings and industrial archaeology. Their physical condition has immense implications for their long-term future. Once no longer used and with no investment in regular maintenance, they decay. The announcement of disposal often throws light on their perilous state, victims of minimal maintenance and neglect - making the challenge of reuse – with its potential as a contribution to sustainability - more difficult. This chapter explores the slow development of conservation policy on historic defence buildings, the MOD's gradual acceptance of responsibility for them and how this acknowledgement has affected local examples.

Scheduling and Listing

It's important to remember that preserving old buildings and historic areas in Britain is comparatively recent. From the late nineteenth and twentieth centuries only a select few Ancient Monuments were listed by the Royal Commission on Historical Monuments. They were first defined as 'constructions connected with or illustrative of the contemporary culture, civilisation and condition of life of the people of England… from the earliest to 1700'. These were mostly castles such as Portchester and archaeological sites.

Perhaps in response to the destruction of war, the Town and Country Planning Act of 1944 introduced statutory listing of buildings of architectural or historic interest. The 1947 Act laid the foundation of the present planning and land use system, giving the Minister of Housing and Local Government the power to make preservation orders ('listed buildings'). Listing Grades 1, II and II* were originally military references.[1] But before about 1970 only really important structures were protected from demolition. It took 25 years to record the first list of 120,000 monuments, structures, sites and pre-1750s building in England. From 1968-70 in reaction to intensive post-war development the Ministry of Housing and Local Government initiated a re-survey of historic buildings, mainly in 39 towns and cities. The slow acceleration of national conservation legislation took hold from then on. The list has continued to expand; in 2020 there were over 374,000 listed buildings in England and approximately 460,000 in the whole of the UK. Defence buildings in the harbour area gained protection – although, as this chapter describes, legal protection did not always ensure their survival. As often happens, campaigning by pressure groups eventually resulted in responses by the government followed by legislation. SAVE Britain's Heritage's 1993 exhibition and book, *Deserted Bastions: Historic Naval and Military Architecture*[2] highlighted the losses of historic military buildings and the uncertain future many faced – as well as inspiring new campaigners to focus on defence heritage, including Celia Clark. The *Deserted Bastions* exhibition signalled the arrival of military heritage into the mainstream of national heritage policy discussion.

'The MOD estate, which housed the might that conquered and policed a global empire, is second only to that of the Forestry Commission. It contains 734 listed buildings and 111 ancient monuments, more than are owned by any other organisation in Britain" according to Martin Spring in the 9th December 1994 edition of Building Renewal.[3] The two MOD reports of 1990 and 1993 announced that 3600 ha of land and more than 100 listed buildings and ancient monuments were surplus to requirements and were to be sold off. As "an extraordinary array of peculiar sites and buildings, which had been developed for the purposes of waging war would not convert easily to civilian use." Their location was often deliberately cut off from civilisation, with buildings suffering from centuries of neglect, obscure legal restrictions and serious contamination from gunpowder and other materials of destruction. The Treasury rules that properties should be sold within three years of being declared redundant and for the highest possible price bound the MOD into an intractable dilemma.

The Royal Town Planning Institute in its submission to the House of Commons Defence Committee hearings in April 1994 said "The MOD portfolio is simply not going to be the 'pot of gold' which the MOD and the Treasury appear fondly to anticipate: in many cases they are precisely the opposite... There is considerable evidence of dissatisfaction in terms of inadequate liaison with both the local community in seeking alternative use...At the same time, there is considerable evidence that an incremental, ad hoc and short-term approach to land disposal is failing to secure the potentially greater returns available for the Exchequer if site assembly opportunities are maximised." In October 1994 a MOD Disposal Policy noted said categorically that "The MOD considers the local plan policy is the mechanism for resolving planning and development issues, and the MOD will abide by the outcome in respect of all surplus sites" – which was welcomed, significantly, by Ron Tate, assistant planning officer of Portsmouth City Council, who looked forward to 'increased dialogue on site disposals.'[3]

However, the Defence Land Agent in Portsmouth overseeing the sale of HMS Vernon in 1996 said in an interview with Celia Clark that the local authorities "expect us to work hand in hand with them and agree with everything they want to see done, planning the release dates totally in co-operation, and if necessary giving land away. We can't do that. It's totally contrary to the Treasury rules. We're under an obligation to get the optimum value for sites. We have to find a balance between these pressures."[3] This statement relates to MOD sites in general, but where defence heritage is involved, the crucial question is: as the largest and most diverse publicly owned estate, should the cost of its maintenance and disposal fall solely on defence funds - or if the price is too high - how else should it be funded? The Conservation Unit of the Department of National Heritage in its Annual Report on the Government's Historic Estate urged all government departments to treat historic buildings as valuable national assets - not just as things to be disposed of. English Heritage were responsible for liaison with the MOD and held quarterly meetings with the reshaped Defence Estates organisation as well as discussing particular sites with them. But the emphasis of the 1990 White Paper *Our Common Inheritance* on the importance of stewardship so that historic buildings can be passed on in good condition to those who can sustain them in the future, remains only rhetoric so long as it has little impact on this difficult reality.

The 'simmering row' between the heritage and defence lobbies was reported to the Defence Committee. Evidence was submitted to it by district and county councils and by the RTPI, Countryside Commission, CPRE and Dartmoor and Northumberland National Parks on the serious effect of military land redundancy on local communities and economies in the context of the importance of defence industry to the national economy. They laid particular emphasis on the damaging effect of delays in decision-making and lack of consultation with planning authorities. Michael Gwilliam, Bedfordshire County Planning Officer speaking on behalf of the County Planning Officers' Society, stressed the inevitably strong interrelationship between the procedural, land use, transportation, social and environmental issues stemming from the implications of rundown and closure of Ministry of Defence estate. In his view, the MOD needed to embrace the principles of sustainable development within a much needed but not yet apparent overall strategy of base closures, to reflect the government's commitment to the integration of environmental concerns into decision making at all levels as set down in *This Common Inheritance: Britain's Environmental Strategy* 1990. The Committee's recommendations included urgent discussions between the MOD and the Department of the Environment over general planning guidance on reuse of MOD sites and that the rule requiring the best possible return must be applied with due acknowledgement of potential non-cash gains to the community (HoC DC 1994a para. 42).

These complexes were too important in historical and economic terms not to be treated as special cases, rather than being marketed as ordinary sites. An MOD land agent said that there has been a shift in attitude from 'Here's the planning application. Too bad what you think of it; and we'll go to appeal if necessary,' towards trying to 'optimise' value, to be less provocative - to acknowledge 'the best commercial price the Department can achieve, having regard to the MOD's own policy on listed buildings and the aspirations of local authorities and local communities'. However, as the then Minister, Lord Henley, said: "The MOD is not an economic regulator" - a concept which some formerly defence-dependent local authorities find hard to accept when they compare the lack of support their areas receive compared with the comparatively generous state aid given to regenerate areas devastated by the loss of steelmaking, coal or shipbuilding industries. In this case it is the government that dealt the blow, but does not offer the compensation.

Heritage experts were said to be appalled at the condition of historic barracks, forts, docks and barracks owned by the Ministry of Defence. Following his attack on the ministry's stewardship to the Commons Heritage Committee, Jocelyn Stevens, chairman of English Heritage told them that action must be taken on buildings of architectural and historical importance. They must take strategic decisions "which are imaginative and radical" to save the buildings. "Redundant buildings have become dilapidated as the MOD comes under Treasury pressure to shed its stock of properties it no longer requires. According to the conservation group Save Britain's Heritage, the ministry spends little on maintenance once buildings are empty."

Charged with the disrepair of Woolwich Arsenal's Grand Storehouse in which the university of Greenwich wanted to house its laboratories, estimated to cost £30m to restore, an MOD spokesman said: 'We are not going to spend a lot of money on renovation'. Well-targeted criticism helped to bring about a change of heart in the MOD about its obligations to keep its historic defence structures in good repair. The Defence of Britain Project (1994–1995) and the Monuments of War Report (1998) followed.

Research

After the end of the Cold War the contraction and rationalisation of the Ministry of Defence estate also stimulated English Heritage into realising that wider public understanding of military sites was needed, and that, where necessary, the surviving structures on them should be conserved. Prompted by a growing awareness of public interest in military historic buildings English Heritage commissioned thematic reports on typical defence architecture and sites: dockyards, barracks, and military airfields, in order to identify what was historically significant across the country and therefore worth legal protection.

The Barracks Review launched with the associated book, *British Barracks 1600- 1914* by James Douet in 1998[4] raised barracks' profile at a critical moment, when they were being sold for redevelopment or reorganised for changing military requirements. English Heritage's message was that listing buildings could stimulate rather than constrain imaginative new developments that responded to their particular sense of place. As a result "many of the sites affected by their recommendations for protection were transformed from candidates for demolition into highly sought-after real estate."[5] Seventy-one barracks were added to the List. Barracks research continues: the historic significance of the army barracks adjacent to Haslar Hospital last used as a refugee detention centre has only recently been recognised.

A study of dockyards followed: a review of the protection through listing of the naval dockyards, initially to identify the best-preserved and historically important examples was produced by James Douet and Jeremy Lake in 1998. Wessex Archaeology made a map-based exploration of the layered archaeology of Portsmouth and Plymouth. More specialised studies were added. The Dockyards Thematic Review on the steam navy was published by English Heritage in 2004.[6] Jonathan Coad's *Support for the Fleet. Architecture and Engineering of the Royal Navy's Bases 1700-1914* followed in 2013. The Naval Dockyards Society's *Twentieth Century Naval Dockyards: Devonport and Portsmouth Characterisation Report* commissioned by English Heritage was published in 2015.

Ordnance yards, such as Priddy's Hard, were the next to be evaluated. They first supplied armaments to naval ships and then evolved from storage of great quantities of gunpowder into sites for the preparation and inspection of new types of propellants and projectiles during the arms race in the second half of the nineteenth century, first against France and then against Germany. In 2006 David Evans' *Arming the Fleet. The Development of the Royal Ordnance Yards 1770-1945* was published.[7] The development of military aviation from 1910 to 1945 on airfields such as Grange and Daedalus followed.[8] More recently Historic England produced a report about nineteenth century forts – of particular interest because of the twelve forts protecting Portsmouth dockyard. Together these impressive scholarly and well-illustrated volumes add immeasurably to our knowledge of dockyards and their associated facilities, not least about those around Portsmouth harbour. As this research made clear, "the viable long-term conservation of these sites is heavily reliant on co-operation and joint-working between a wide variety of interests and disciplines. These are, however, some of the most challenging historic sites to conserve for future generations, and a fine balance has had to be met between the needs for recording and in-situ conservation… they are often located in some of the most economically and socially disadvantaged areas of the country."[10]

Legal Protection

Historic military buildings' importance in the country's history is reflected in the highest levels of protection: in 1994-95 136 were scheduled monuments; and 20 Grade 1 listed buildings; 34 Grade II* and 493 were Grade II.

However, confusingly, a substantial number of historic defence buildings were Listed as Historic Buildings or Scheduled as Ancient Monuments – or sometimes both, which afforded them legislative protection, but in different ways. The procedures for applications to change or alter them are different too, and so is the range of parties able to comment or intervene. The preference for scheduling was not because it was the most appropriate designation: according to Cocroft, it was often the least, designed to protect archaeological sites and ruined buildings with little or no viable use.[10] It was based on security considerations: from identification to consent, procedures were carried out by government staff who had signed the Official Secrets Act. The granting of listed building consent for alterations or demolitions is the concern of local planning officers - who had not. Democratic input was thereby excluded. Scrase pointed out that the 1979 Planning Act suffered from hasty drafting, and that such acts had been drafted and administered from the narrow point of view that monuments are mainly untapped repositories of archaeological information. "This is probably true of buried archaeology but upstanding remains … are often cherished features of the local scene and occupy significant places in the local consciousness and folklore".[11] This needed to be examined in the public expectation of return to the community on disposal, but often little is known about them because active defence sites usually offer no public access.

According to the Department of Culture, Media and Sport in 2010: "The scheduling of Ancient Monuments began in the 1960s which was then considered more appropriate than listing in view of the restricted access required for active defence sites. It is now widely accepted that statutory protection should take into account the type of management that will best ensure the sites' long-term future. Scheduling is used where the future of sites as monuments is the preferred option. Listing is used when continuing or new use of structures is desirable and feasible. The MOD and English Heritage regard listing as a more appropriate means of managing buildings in the dockyards, and that an evaluation by English Heritage of the importance of dockyard buildings in their full architectural, technological and historical contexts provide the best means of assuring the significance of buildings within their boundaries."[12]

In 1999 this overlap between the dockyards' scheduled monuments and listed buildings was rationalised into one system, overseen by English Heritage. As the result of the Thematic Review, 170 buildings in Royal Naval Dockyards were recommended for transfer from scheduling as Ancient Monuments to listing. 18 others were upgraded. In the Royal Naval Dockyards Thematic Review Portsmouth's No. 1 Basin was recommended for listing at Grade I.

Upgraded buildings included No. 2 Ship Shop facing No. 2 Basin, now Grade II*, the dockyard wall in Queen Street also II*, Frederick's Battery, Unicorn Gate, Detention Centre (Royal Marines School of Music), the statues of William III and Captain Scott, the Fire Station (II*), Royal Railway Shelter, Iron Foundry, Block Mills (Grade I), Royal Naval Academy (II*), Short Row (II*), No. 6 Boathouse (II*) and the Vulcan Block in Gunwharf (II). Unfortunately, Boathouse 4, then under threat of demolition, was not included.

What this means is that at least in theory, the local authority and local people have a say in changes to military and naval buildings, because the listed building consent procedure is tied into local planning. Historic England is the chief consultee on work to Grade 1 and II* structures. Councils also have powers to issue repairs notices to owners of listed buildings in urgent need of repair – but they could not issue them to the MOD even after abolition of the Crown Exemption in 2006 – the implications of which we explore below.

Britannia waived the rules: the Crown Exemption

Land vested in the government had for centuries enjoyed a unique legal status, which reflected its overriding importance to the national interest. Because the Crown is deemed to act in the national interest, it was exempt from civil law - for all its functions, whether financial or spatial. The exemption of government property from normal planning procedures predated the modern town planning system. Though in practice it had existed for many centuries, this exemption first passed into law in the Ancient Monuments (Consolidation and Amendment) and Archaeological Areas Act 1913. Both church and state were declared exempt from new legislation to protect ancient monuments. The justifications put forward at the time were that in view of the enormous expense implied, since the majority of ancient monuments were churches, the Church of England could be trusted to look after its own buildings via its own internal system of development control which it still does. The MOD only consulted local planning authorities on change or demolition of scheduled monuments and listed buildings as a matter of courtesy, but it gave no commitment to attaching weight to any representations.

In the case of barracks, "in almost all cases, survival has depended on continuity of military occupation; once the military have left, most barrack sites are sold and redeveloped. There are few instances of long-term secondary occupation".[4] In fact Portsmouth can instance several - as can Plymouth and Chatham: Southsea Castle (1544, 1680s, 1814) a museum since 1960, Clarence Barracks (1897) which has served as Portsmouth City Museum since 1970 and Cambridge Barracks (c1850) which has housed Portsmouth Grammar School since 1926. Through pressure from Tony Whitehead the Sheffield city conservation officer who went on to be the MOD's conservation adviser Hillsborough Barracks retained most of its buildings in the conversion to a supermarket.

The Crown Exemption gave government property owners a freedom from controls over change or disrepair - which had no equivalent in some other European countries such as Italy. This freedom had important implications for historic structures and sites, particularly where they were redundant and subject to disposal - as well as for the sale price related to potentially profitable new land uses. Government-owned buildings were still exempt from normal planning procedures and from restrictions imposed by the Ancient Monuments and Archaeological Areas Act 1979.

Significantly, the stimulus to greater consultation was the changing status of defence lands the abolition of the Crown Exemption in 2006.[13] The Defence Estates/Defence Infrastructure Organisation now has to engage in statutory planning – whereas before abolition, they could agree substantial (and potentially profitable) proposals with their chosen developer. This may have been a factor at Gunwharf, where the MOD and their chosen developer agreed on the new land uses without reference to the local planning authority since they were not obliged to take local planning policies or public responses into account.

MOD Historic Buildings Advisory Group and Government Historic Estates Unit

From the 1990s the MOD had a Historic Buildings Advisory Group, which helped the Defence Estates/Defence Infrastructure Organisation to develop and implement its historic buildings policy and strategy and to liaise with the statutory bodies.[14] Sir William Whitfield who later became a trustee of the Portsmouth Naval Base Property Trust in 1983 was a member of the group, as were structural engineer Alan Baxter and property developer Sir Stuart Lipton. Its remit extended to considering alternative uses for redundant buildings as well as the importance of maintaining the setting of historic buildings.

Within the context of reducing budgets, government historic buildings policy and guidelines gradually became more specific. Defence Estates had specialist staff but the problem of unused and deteriorating historic defence buildings continues to be the reality. The government's 1990 White Paper *This Common Inheritance* about historic government property set standards for historic buildings in their care: "to aim constantly for the highest standards of conservation" and "to ensure that those responsible for historic buildings are aware of the importance of the heritage they hold in trust.[14] The aim of the report was to provide an overview of progress achieved in the care of the Government's historic estate, department by department. In 1993-4 the dedicated Conservation Unit was transferred from the Department of the Environment to the Royal Estate Division, later Heritage Division of the Department of National Heritage, a small specialist team. The Government Historic Estates Unit is now part of Historic England. Its main responsibility is the government estate in Greater London, although the head of the unit was also Historic England's principal point of contact with the MOD. His opposite number is the Senior Historic Buildings Advisor in the DIO's Environmental Support & Compliance Team, whom he meets twice a year. MOD sites are generally dealt with on a case-by-case basis through Historic England's regional offices.

"Selling the family silver"

The rationalisation of functions within the MOD in the 1990s led to the decision to combine all the armed services' staff training facilities at the Army's staff college at Camberley, Surrey. This raised questions about the future use of the Royal Naval College Greenwich, the most important historic naval townscape - apart from the dockyards - in the country. "The initiative to find an appropriate new use for this nationally important site is being promoted by the Ministry of Defence." In 1995 Michael Portillo, Minister of Defence, controversially, put up the College for sale - although removal of Jason, the nuclear reactor used to train engineer officers was a complicating factor.

A very public outcry about 'selling the family [nation's] silver', led eventually in 1998 to its transfer to the Greenwich Foundation, a charitable foundation with two educational occupants: the University of Greenwich in Park Row and King William Walk and Trinity College of Music, now the Trinity Laban Conservatoire of Music and Dance in King Charles Court.[15]

Defence land sales continue

In 1994 half of defence land and more than 100 listed buildings were in the process of being sold for conversion to civilian usage for an estimated £650m, according to the 1994 Defence Estates report *Defending our Heritage. Historic Military Buildings on the Defence Estate*. "On the larger sites, the MOD plans to enter into joint ventures with developers, housebuilders and contractors for phased development…. But developments for destruction do not lend themselves easily to peaceful purposes… Most of the MOD sites suffer from unprecedented problems that turn them into negative assets. They are often in isolated locations and blighted by massive military structures, buildings suffering from neglect, legal restrictions and severe contamination by gunpowder and even more noxious substances… What the sites are crying out for is vision and imagination from town planners, architects and engineers to convert them to new, socially appropriate, civilian uses. They also demand special expertise by builders to cope with the immense construction problems of these massive structures and contaminated sites. Unless the building industry can rise to the challenge… the much vaunted peace dividend could turn into a peace liability."[16]

Matching this aspiration with good practice was a laudable aim, but as we mentioned in the last chapter the supreme example of government mismanagement was the sale of service housing to Annington Homes in 1996.

Government guidance on the disposal of state owned historic buildings; responses by the MOD

The Parade or Long Row 1718 for chief civilian officers, converted to offices in mid-1980s.

Portsmouth Royal Dockyard Historical Trust

In 1994-5 the MOD asked the Advisory Unit for guidance on several Portsmouth dockyard buildings: Admiralty House; Block Mills: researching for original paint scheme; No. 2 Ship Shop – conversion to offices; The Parade – security measures and internal alterations; on Fort Monckton and on Fort Blockhouse, HMS Dolphin. Works on Residences 13 and 14 at Haslar Royal Naval Hospital were nearing completion.[17,18] Much more recent military structures were also added to the canon. The widespread disposal of newly obsolescent military sites at the end of the Cold War was the catalyst for their recognition as heritage in 1989.

Wayne Cocroft of English Heritage commented on the scale and extent of the disposals, fuelled by several sources: *Options for Change (1991)* and *Frontline First: the Defence Costs Study* (1994). These created massive challenges for conservation. Problems he identified included the lack of detailed information and understanding of important parts of the historic defence estate, which once researched were likely to lead the need for large scale listing and scheduling programmes, to be paralleled by the need to explain and justify this process to owners and the wider public. This became an increasingly important factor after the Secretary of State for National Heritage introduced consultation into listing procedures in February 1995. Both English Heritage and the Historical Monuments of England (RCHME) embarked on programmes of assessment of the defence estate.[21]

This new commitment was clearly expressed in the preface to *Defending Our Heritage. Historic Military Buildings on the Defence Estate* by Lord Cranborne, the Under-Secretary of State for Defence in the MOD's annual historic buildings report in 1994. "The Ministry of Defence is the single largest landowner of historic buildings of any Government Department. It owns over 700 of them, that are scheduled and listed. Many reflect the history and traditions of the Armed Forces, and we in the Ministry take great pride in them. The Ministry of Defence welcomes the policy the Government set out in its 1990 White Paper *"This Common Inheritance"*. Major groups of historic buildings, such as the Navy's at Portsmouth and Gosport" were involved. The MOD "must try and live up to the high standards [the White Paper] demands of us." The debate was underpinned by the concept of 'heritage at risk' when English Heritage began to produce annual reports of listed buildings in poor condition as we explore below. However, "The House of Commons Defence Committee report on the Defence Estate… noted that while many of the criticisms of the Ministry of Defence's record as owner and occupier of historic buildings were fair, it must be borne in mind that caring for historic buildings was not a natural function of a defence ministry and that funds devoted to it have to be found from within the defence budget."[19]

Several important sites around the harbour were described in detail in the report. Lord Cranborne refers to the Royal Naval Hospital at Haslar, Gosport having "avoided alterations through retaining its original use. Built solely for the treatment of seamen, the hospital in recent years has compensated for the dwindling numbers of Naval personnel needing treatment by taking in civilian patients. In this manner the Georgian hospital retains its use in the function for which it was built, and the hospital staff can keep practising their skills in readiness for operational deployment in an emergency."[21]

The report also refers to Gunwharf's Grand Storehouse of Vulcan building, built in the last years of the Napoleonic wars, "conceived with a finesse not usually seen in ordnance buildings. Lamentably, bombing during the Second World War destroyed the North wing and subsidence has seen to the remnants' demise. The Grand Storehouse, which has been disused for several years, is now on the market. It urgently needs a new use."

Thankfully, after about fifty years without a use, it found two – as the Aspex Art Gallery in the south wing ground floor and flats above. A replica north wing was added in the redevelopment of Gunwharf, as we describe in our chapter on ordnance yards.

The MOD's 7th Annual Historic Buildings report of 1998-99 said that certain historic sites, by virtue of their size, location or specialised design, could neither be easily sold nor remain in military use. The absence of dedicated funds to keep historic defence buildings in good condition within the overall defence budget continued to affect good intentions.

"Sites such as the Cambridge Military Hospital in Aldershot, Fort Southwick overlooking Portsmouth Harbour…will require a concerted partnership between the MOD, other agencies and the local community, if sustainable new uses are to be found."[19] The Cambridge Military Hospital was to be restored and converted by a housing developer Grainger within the large Wellesley housing development,[20] and Fort Southwick, a Grade I listed building because of its importance to D-Day was sold in 2003 to the New family of Portsmouth. The Historic Dockyard is given a whole page: "By day Portsmouth's historic dockyard is alive with people. The twelve and a half acre site, situated in the midst of the Navy's dockyard operations has become a major visitor attraction. Only ten years ago its important historic buildings were largely disused and in a sorry state of repair. Visited now by an average of three quarters of a million tourists every year, the historic dockyard has been vested with a new lease of life."

"The history of the dockyard on its present site dates back to the late fifteenth century but most of the dockyard's earliest structures, built of wood, were replaced during the dockyard improvements of the late eighteenth century. The vast majority of the Georgian brick structures have survived and now form the core of the historic dockyard."

"During their last years in naval use, several of the historic buildings were found to be ever more unsuited to the requirements of the modern Navy. For buildings in minimal use, their maintenance costs were unjustifiably high. As at Chatham, the termination of the Royal Navy's use of these buildings was announced in the defence review of 1980."

"The Navy's withdrawal from the heart of the historic dockyard came some months after the closure of Chatham dockyard. Valuable lessons learned in the handover at Chatham were applied at Portsmouth, making the transition considerably easier. A trust similar to Chatham was established to oversee the historic enclave's preservation and future management."

"In 1985 the responsibility for running the historic dockyard was handed over to the Portsmouth Naval Base Property Trust. Given an initial endowment of £6.5 million, the charitable organisation is expected to raise its own funds through tourism and reuse of the buildings. With a considerably smaller acreage and just thirteen scheduled monuments, the trust's task is significantly less daunting than that facing the Chatham Historic Dockyard Trust."[19] How this optimism worked out in practice in Portsmouth is analysed in our dockyard chapters.

In 1995 as the disposal of state property assets gained momentum, the Department of National Heritage issued guidance on the disposal of historic buildings. This was superseded in 1999 when the Department for Culture, Media and Sport published an update – with a picture of the Royal Naval College on the cover.[20]

This new advice said that all surplus historic buildings, particularly if they were vacant or only partially used, should be 'disposed of expeditiously' and 'regularly inspected and maintained in a secure, safe and stable condition pending disposal'. As far as historic buildings were concerned, government departments were instructed 'to obtain the best return for the taxpayer' consistent with Government policies for their protection but this was 'likely to limit opportunities for the realisation of development value'. However, finding new uses for specialised buildings, the design challenges their conversion pose and sourcing the finance cannot often be achieved within short time-frames, especially if there has been a long period when they were unused and not maintained.

Buildings at Risk

From 1998 English Heritage, now Historic England, continues to publish a national Heritage at Risk Register every year. In South Hampshire's case it's always Ministry of Defence buildings that predominate. Its purpose is to stimulate owners into repairing their neglected buildings, but the MOD tends to say that it is not funded to keep buildings no longer needed in good repair – their budget priority being defending the country.

Despite the government guidance, in the 1990s the MOD had 123 historic buildings for disposal, and 32 were 'At Risk', including Fort Rowner where "Casemates are now dry and habitable, and some have now been adapted for new uses." By 1999 44 MOD buildings appeared on the first national register of 'at risk' Grade I and II* buildings, representing 7% of the defence estate in England.[12] 14 listed buildings and two monuments were added, but there were 28 Buildings at Risk in the 2007 English Heritage Biennial Conservation Report. The MOD's slow responses to tackling this decay are documented throughout this chapter.

Unfortunately legal protections still do not extend far enough and are not the cold reality for unused dockyard buildings. As already said one huge obstacle remains: local authorities cannot insist that the MOD repairs its neglected buildings - if the base is in active defence use. They may not use their compulsory purchase powers to purchase neglected buildings, restore them and charge the owners for the costs, although, it has to be said, these powers are rarely exercised. Conservationists worried about the state of the empty Naval Academy and Block Mills in the dockyard find this a continuing source of frustration. The saga of Block Mills is described at the end of this chapter.

In 2001 there were 43 defence buildings at risk: 6.1% of the total; this was reduced to 32. However, 33% of the 9000 Scheduled Ancient Monuments on MOD land were in fair condition, while 21% were in poor or unknown condition. Examples include the Grade 1 Block Mills, the Navy Pay Office where Charles Dickens's father worked and No. 6 Dock.

Fort Southwick 2019

Neglect of historic property is not confined to government departments. The same problem occurs with subsequent private owners such as the New family who own Fort Southwick in Portsdown Hill - also on the Historic England's Buildings at Risk Register.

For historic properties still in MOD hands but not in use, despite the detailed and specific policies on their care, many local buildings continued to deteriorate. The guardrooms, workshops, boiler house and engine room at Haslar Gunboat Yard, as well as Fort Elson, the fortifications north of Mumby Road Gosport, the Iron and Brass Foundry, No. 6 Dock, Block Mills, No. 25 Store and 2-8 the Parade in the naval base were all listed as in poor condition according to English Heritage's *Biennial Conservation Report on the Government's Historic Estate of 2003-2005* – and except for some of the Haslar Gunboat Yard buildings their poor condition has not changed much since.

Changes in historic buildings policy and the MOD's response

The 4th Stewardship Report on the Defence Estate of 2005 acknowledged the need to take into account the wider interests of society – in particular sustainability and the environment - in defence historic buildings at risk. As Tanner says, in response to criticism of the condition of historic defence buildings in 2004 the MOD adopted the Department of Culture, Media and Sport '*Protocol for the Care of Government Historic Estates*' which required Departments to produce biennial conservation reports on the condition of their heritage assets. The MOD claimed that it continued "to make progress in further embedding these requirements within its management processes. [23] However, its report of 2005 revealed that, despite its earlier pledges, the condition of its heritage assets was still inadequate. This prompted the House of Commons Defence Committee to say in 2006 that it was 'completely unacceptable' that the MOD did not know the condition of 77% of the historic buildings it held, especially when it had a responsibility to maintain them. In response, as well as using specialist consultants and contractors, Defence Estates could draw on considerable in-house expertise. Their conservation officer was a top-level budget holder. They worked closely with English Heritage and conducted condition assessments on a four-yearly basis for listed buildings and a five yearly basis for scheduled monuments. "Heritage assets are identified within sites' Integrated Estate Management Plans and are accompanied by a maintenance programme." A MOD Buildings at Risk Officer was appointed in 2007, working to establish agreed costed plans to resolve the condition of Buildings at Risk. Oxford Archaeology produced reports on the Guard Rooms at Haslar, 2-8 the Parade and No. 25 Store in the Dockyard. DCMS guidelines on safeguards to historic buildings on the disposal list were applied, and efforts were made to ensure that the design quality of any new work enhances the historic environment.[24]

In order to address the issues identified by the House of Commons the MOD was instructed to produce quadrennial inspections of all its historic property and to implement a programme of repairs and maintenance. When its next Heritage Report was published in 2007, it confirmed all these measures had been put in place and that the condition of 84% of all listed buildings was known. By 2008 the target of establishing the condition of all had been achieved.

As the process of disposals continued, in 2010 English Heritage produced a guidance note: *The Disposal of Heritage Assets: for government departments and non-departmental public bodies.*[25] Key points were:

> a. Accepting the highest purchase offer is not always appropriate. Maximisation of receipts should not be the overriding aim in cases involving the disposal of heritage assets.
> b. Any options for reuse should be considered before deciding to sell. It may be possible to retain and adapt a historic building for a different use, instead of selling it.
> c. Unused heritage assets need to be actively protected. All vacant and non-operational historic buildings should be regularly inspected and maintained in a secure, safe and stable condition pending disposal (see section 5); important archaeological sites should be actively managed.
> d. English Heritage should be consulted at an early stage. English Heritage should be given the opportunity to comment regarding any site where there is potential for a significant heritage issue (see paragraphs 2.5, 7.4 and 10.3). Guidance on the handling of disposal cases is available from the Government Historic Estates Unit (GHEU).
> e. Departments should provide clear information for purchasers. Disposals should always be accompanied by information regarding the significance of any heritage assets and any constraints on change due to their significance (see paragraphs 10.5 and 10.6). Information about any repair, maintenance or management liabilities should also be made available, including an up-to-date condition survey (see paragraph 5.2).
> f. Heritage assets need sustainable ownership. Departments should take reasonable steps to ensure that purchasers of vulnerable heritage assets have the resources to maintain them. Alternative methods of disposal other than open market sale may need to be considered to ensure appropriate ownership (see section 8 and 9.4).

The MOD Leaflets 11 and 12 on the Historic Environment for those responsible for historic assets, setting out their policy and responsibilities in delivering the DCMS Protocol followed.[26] Paragraph 12-3 of No. 12 said "Our vision is to value and promote the sustainable use of our heritage assets, in recognition of the benefits they bring to the environment, the quality of life of defence communities, the Nation's cultural heritage and the role it plays in supporting defence capability on the development of the moral component." Perhaps this is a reference to expectations of public benefit? Further advances in how the historic estate was managed were signalled that year with the production of 'JSP 362', an estate managers' internal guidance document, containing a revised Section 12, the Historic Environment. This said that the MOD had a duty of responsible stewardship for heritage assets whether 'they receive statutory protection or not'.

Defence Heritage still At Risk – a continuing story

The *Historic Environment* (Ministry of Defence, 2010 paras. 12-52) declared that "the MOD is committed to resolving its HAR [Heritage at Risk] issues and ensuring assets do not become at risk. The MOD HAR Officer's role is to work with internal and external stakeholders to establish risks and develop a plan for the sustainable future of each MOD HAR asset. Performance on HAR is reported in the MOD Stewardship Report, Heritage Report and Sustainable Development Report. It is also a MOD Sustainable Development Key Performance Indicator."

The 2011 MOD Heritage Report noted that the importance of heritage assets was considered significant enough for them to be included within 'strategic estate decision making'.[41] This was endorsed by the MOD's Estate and Sustainable Development guidance of 2013, which noted that many non-listed historic buildings were of value to the heritage and ethos of the Services. Accordingly:

It is policy to sustainably manage and continually improve the estate, including the heritage assets. Heritage plays an important role in improving the quality of life for those who work and live on the estate and its role is recognised in enhancing the ethos of the Services.[27]

However, the MOD then said that 'austerity measures will continue to provide challenges for MOD heritage management. The effects were already being experienced with a decline in the condition of listed buildings and the scaling back of condition assessments as a result of budgetary constraints.'

With Crown immunity removed in 2006 the MOD must now comply with planning statutes. Doing nothing is no longer an option. The key historic buildings in Portsmouth naval base were discussed by the DIO and the local authority conservation officer, but these negotiations are not in the public realm.

In 2013 the Government Historic Estates Unit published English Heritage's *Biennial Conservation Report*.[28] The assessment of heritage issues of surplus sites prior to disposal identified in the Wiltshire pilot project was rolled out nationally as part of the National Heritage Protection Plan. In July 2013 Reserve Forces [territorial army] and cadet sites planned for closure were to be assessed.

In order to understand the government jargon one has to read this carefully…The Defence Infrastructure Organisation's Heritage Officer role is "split between the Estates Policy area within the Strategic Asset-Management and Programming Team (DIO SAPT) and the Senior Archaeologist and Senior Historic Buildings Adviser of the Historic Environment Team (HET) [which is part of the DIO's] Operations, Development and Coherence… Casework is taken forward by the HET. Policy and legislative matters are dealt with by the SAPT Estates Policy. There are frequent contacts between EH and DIO staff regarding casework and policy matters with biannual liaison meetings between EH and DIO held at officer level. DIO has not produced an illustrated report, as in previous years, but has supplied GHEU with updated information, including statistics for heritage assets and their condition, new designations and disposals." "DIO and EH have worked closely together to ensure that heritage assessments of sites affected by plans for disposal or major operational development are carried out."

Nevertheless, in successive years defence heritage buildings in poor condition came up again and again as still at risk. In 2014 the DIO reported that there were 846 listed buildings and 769 scheduled monuments on the Defence estate, although these figures were subject to change with disposals reducing numbers and new listings increasing them (DIO 2014). Many redundant MOD sites in the southeast region contain important historic buildings; but the local Historic England Buildings at Risk Register 2016 identified significant decayed defence heritage within our sub-region. These include Fort Fareham, defence and ancillary structures at RAF Bicester, Fort Cumberland and Hilsea Lines Portsmouth and Fort Southwick.

In 2015 English Heritage was renamed Historic England: "the government's expert advisory service" for the country's historic environment, giving constructive advice to local authorities, owners and the public. "We champion historic places helping people to understand, value and care for them, now and in the future" In recent decades interest in military sites has moved from specialist study onto the national heritage agenda, where it forms one of the core research points within Historic England's Action Plan 2015-18.[28]

In 2016 the South East Region Historic England Buildings at Risk Register identified significant decayed defence heritage in southern Hampshire. These included Fort Fareham, defence and ancillary structures at Fort Cumberland and Hilsea Lines Portsmouth and Fort Southwick "following on measures for arresting the damage caused by unchecked vegetation, action was taken to repair the roof and carry out repair on the casemates, the main fort and the moat."

The 2018 Register listed 18 buildings at risk in Gosport and 65 in Portsmouth. Further analysis reveals how many of these belong to the MOD, notably Block Mills, the Naval Academy, the Parade, the Iron and Brass Foundry, No. 6 Dock and No. 25 Store. Several of these have been on the Historic England At Risk Register since 2008. The Naval Academy was said to be at "Immediate risk of further rapid deterioration or loss of fabric; no solution agreed'.

The 2019 register reported further deterioration of key buildings in the dockyard. No. 6 Dock was suffering from rotation, mortar joints on the stonework treads had opened up and the stonework was spalling. The failure of the gates to No. 5 dock led to flooding, algal growth and saturation of the stonework, with the rick of accelerated decay. Inside the long vacant Royal Naval Academy the introduction of natural ventilation and background heating had reduced damp levels, water ingress levels had been reduced and dry rot appeared to be dying back, but asbestos had been found in the roof. Nos. 2-8 the Parade, the 1715-8 terrace of dockyard officers' houses, partially converted to office use in about 1995, had been empty for some time. The houses were prone to wet rot and some structural movement, but background heating had reduced damp levels. Problems persisted between the main building and the rear extensions and there was extensive decay in panelling; some plaster ceilings had collapsed. In late 2019 scaffolding was erected to examine the building's condition and to complete high level repairs to make it watertight. As we explore in Chapter 6 the deteriorating condition of dockyard buildings has long been a concern to the Naval Dockyards Society and to the Hampshire Buildings Preservation Trust.

In Gosport damaging vegetation had taken hold of Fort Elson but its removal was said to be only intermittent. A management plan has been drafted but not agreed or implemented. It could not be occupied because it lies within a munitions storage area. The condition of the Traverser System in Haslar Gunboat Yard was declining, while the gunboat sheds, which had ceased to be used in the mid-1970s have deteriorated, with only minimum maintenance undertaken. Historic England had recently grant-aided a feasibility study to help find an appropriate new use for the site, but the boundary walls, corner watchtowers, walkways and gates were at risk 'due to animal burrowing, defective roofs, eroding brickwork and pointing, missing windows and doors and vegetation growth.'

Outside MOD ownership, the unoccupied parts of Fort Fareham including the ramparts were suffering from decay and heritage crime. Fort Cumberland's casemates had water ingress and associated decay and the curtain defence and counterscarp had localised deterioration. Since the fort is in the care of Historic England, this was embarrassing; a programme of urgent works had been completed and options for sustainable reuse of vacant parts of the site were being considered, as we explore in more detail in our second forts chapter. As already mentioned, parts of the privately owned Fort Southwick were in a poor condition and the owners have carried out some repairs "but more are needed, together with a carefully planned maintenance regime.'

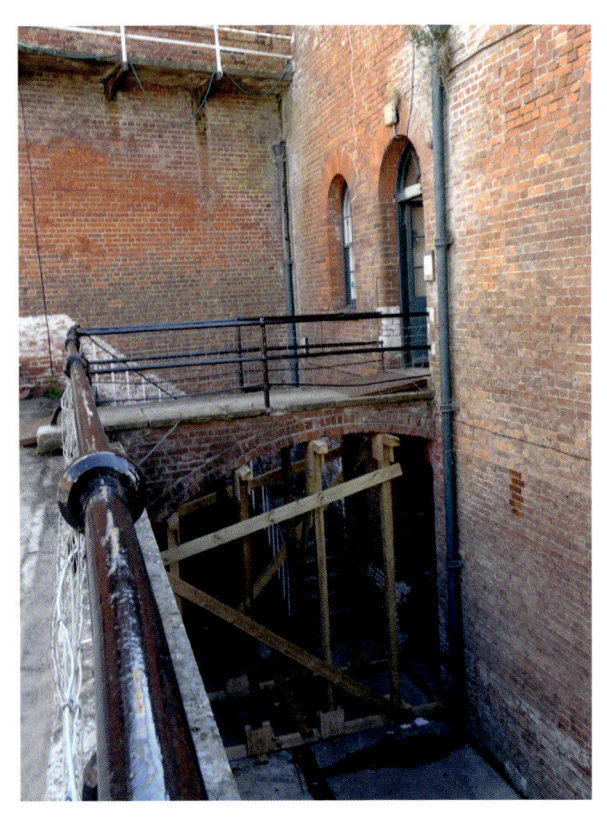

Poor condition of access bridge Fort Widley 2017

At Forts Widley and Purbook, which belong to Portsmouth City Council and are leased to the Peter Ashley Activities centre there are problems related to the drainage of the ramparts and the maintenance of the moat's retaining walls which put it at risk. Historic England grant-aided a conservation plan and urgent repairs were due to commence in 2020. Puzzlingly, there appeared to be no entry for Haslar Hospital, where as far as we knew the main hospital building had not had any work done on it.

A limitation of the register is that only Grade 1 and Grade II* and Grade II churches are included, not Grade II listed buildings outside London.[29]

National and local conservation groups: SAVE Britain's Heritage, the Georgian Group, the Naval Dockyards Society, the Hampshire Buildings Preservation Trust, the Portsmouth Society and others continue to be concerned at the longterm deterioration of forts, Haslar Gunboat Yard and key dockyard buildings no longer used by the MOD. But, as already said, local authority action such as the issuing of repairs notices cannot be undertaken in an active naval base. We now turn to the development of building conservation around the harbour.

References

1. Richard Harrison Gosport Society 2020 'Adding Value to the Democratic Voice: the Work of the Gosport Society Planning Sub-Committee' in *Gosport 2020 For the Record: Essays in Celebration of the Half-Centenary of the Gosport Society* pp. 24-26

2. Anthony Peers 1993 *Deserted Bastions: Historic Military and Naval Architecture* SAVE Britain's Heritage

3. Martin Spring 1994 'Post-MOD era' Building Reneal 9 December pp.4-5

4. James Douet 1998 *British Barracks 1600-1914 Their Architecture and Role in Society* English Heritage Stationery Office

5. Jeremy Lake 2005 'From Monument to Place: English Heritage and Military Industrial Complexes in England' https://www.birmingham.ac.uk/Documents/college-les/gees/Miscellaneous/CPPCultureWorkshop07Lake1.pdf accessed 17 February 2020

6. Jeremy Lake & James Douet 1998 English Heritage Draft Thematic Survey of English Naval Dockyards Summary Report Thematic Listing Programme; Jeremy Lake & James Douet 1998 *Thematic Survey of English Naval Dockyards Summary Report Thematic Listing Programme* English Heritage; *Building the Steam Navy. Dockyards, Technology and the Creation of the Victorian Battlefleet 1830 – 1906* by David Evans (Conway Maritime Press ISBN 0 85177 959 X; Jonathan Coad 2013 *Support for the Fleet. Architecture and engineering of the Royal Navy's Bases 1700-1914* English Heritage (ISBN 987- 1-84902-055-9); Naval Dockyards Society 2015 for English Heritage i *Twentieth Century Naval Dockyards: Devonport and Portsmouth Characterisation Report* ISBN 978-0-9929292-0-6

7. Jeremy Lake 2003 Thematic Survey of the Ordnance Yards and Magazine Depots. English Heritage; unpublished report; David Evans 2006 *Arming the Fleet. The Royal Ordnance Yards, 1790-1945*. English Heritage *2006* ISBN 10: 0-9553632-0-9). Explosion Museum Priddy's Hard

8. Jeremy Lake 2002. 'Historic airfields: evaluation and conservation'. In Schofield, J., Johnson, W. Gray, and Beck, C.M. (eds.) Materiel Culture. The archaeology of twentieth-century conflict. London: Routledge; Jeremy Lake 2003 Thematic Survey of Military Aviation Sites and Structures. English Heritage; unpublished report; Jeremy Lake, Dobinson, C. and Francis, P. 2005 'The evaluation of military aviation sites and structures in England', in Hawkins, Lechner and Smith (eds.), pp. 23-34.

9. https://research.hist orice ng land.or g.uk/Re port .aspx?i= 16215&ru= %2f Re sult s .as px%3fp%3d1%26n%3d10%26rn%3d87%26ry%3d2018%26ns%3d1

10. Wayne D Cocroft 2000 *Dangerous Energy: The archaeology of Gunpowder and military explosives manufacture*. Swindon: English Heritage. ISBN 1-85074- 718-0 p.265

11. Tony Scrase 1991 in Barry Cullingworth, Vincent Nadin *Town and Country Planning in the UK* Routledge

12. Department of National Heritage The Conservation Unit 1994-5 *Annual Conservation Report on the Government's Historic Estate* p.31; MOD 7th Annual Report Historic Buildings on the Defence Estate 1988-89 P.9; Defence Estate Organisation 1999 *7th Annual Report Historic Buildings on the Defence Estate* p.20; MOD Heritage Report 2007-2009 p. 3 https://assets.publishing.service.gov.uk/government/uploads/ system/ uplo ads/attachment_data/file/33360/heritage_rpt_200709.pdf

13. Removal of Crown Immunity from Planning Control July 2006 ISBN 0 7559 6183 8 https://www.webarchive.org.uk/wayback/archive/20180516214840/http://www.gov.scot/Publications/2006/07/21091434/0

14. HM Government 1990 This Common Inheritance: Britain's Environmental Strategy: A Summary Of The White Paper On The Environment 1 Jan. 1990

15. https://www.independent.co.uk/news/uk/home-news/jason-casts-a-cloud- over-naval-college- sale-1578790.html; https://en.wikipedia.org/wiki/Trinity_Laban_Conservatoire_of_Music_and_Dance

16. Defence Estate Organisation 1996 *Defending our Heritage Historic Military Buildings on the Defence Estate The Implementation* P. 6,7

17. The Conservation Unit 1994-5 *Annual Conservation Report on the Government's Historic Estate* p. 34

18. MOD Defence Estates 1994 *Defending our Heritage. Historic Military Buildings on the Defence Estate* p.1, p.7; Virginia Bottomley JP MP, Secretary of State for National Heritage The Conservation Unit *Annual Conservation* Department of National Heritage *Report on the Government's Historic Estate 1994-95*

19. MOD *7th Annual Report on Historic Buildings on the Defence Estate 1998- 1999* p.17

20. Department of Culture, Media and Sport 1999 *The Disposal of Historic Buildings Guidance note for government departments and non-departmental public bodies;* Department of Culture, Media and Sport 1999 *The Disposal of*

Historic Buildings Guidance note for government departments and non- departmental public bodies The Ten Key Points of the Guidance Note 20 David Casey 16 September 2016 'New Future for landmark Military Hospital' https://www. insidermedia. com/insider/southeast/new-future-for-landmark- military-hospital accessed 26 June 2019

21. MOD Defence Estates *Defending our Heritage. Historic Military Buildings on the Defence Estate* 1994 https://www.tandfonline.com/doi/abs/10.1080/13527250601119066 Ian Strange & Ed Whalley 2007 'Cold War Heritage and the Conservation of Military Remains in Yorkshire' International Journal of Heritage Studies Volume 13 2007 Issue 2

22. Regeneration Investment Organisation 2015 https://assets.publishing.service.gov.uk/government/uploads/system/uploads/att achment_data/file/600615/RIO_Pitchbook_May_2016_withdrawn.pdf

23. Nancy Tanner 2017 *FRIEND OR FOE? The MOD, Heritage and the Defence Review* MSc. Dissertation Oxford Brookes University School of the Built Environment with University of Oxford Department for Continuing Education May 2017

24. Government Historic Estates Unit 2006 *Biennial Conservation Report The Government's Historic Estate 2003- 2005* p.34

25. English Heritage 2010 *The Disposal of Heritage Assets: for government departments and non-departmental public bodies* https://historicengland.org.uk/images-books/publications/disposal-heritage-assets/

26. https://assets.publishing.service.gov.uk/government/uploads/system/uploads/attachment_data/file/33622/jsp362_leaflet11.pdf
https://www.gov.uk/guidance/defence-infrastructure-organisation-estate-and-sustainable-development accessed 26 August 2020

27. https://www.gov.uk/government/publications/mod-heritage-report-2011-to-2013 accessed 26 August 2020

28. Historic England 2015 Biennial Conservation Report 2013-15 https://content.historicengland.org.uk/images-books/publications/har-2015-registers/se-har-register2015.pdf/ accessed 14 September 2017

29. (Historic England Buildings at Risk Register South East 2019 https://historicengland.org.uk/advice/heritage-at- risk/search-register/.

Chapter 5
The conservation tide comes in: local responses

During WWII Gosport, Portsmouth and the dockyard inevitably suffered heavy bombing. Dr. White's *Story of Gosport* vividly describes how the centre of the town was left as "a decaying rubbish heap, jagged blocks of broken masonry … the new cinema remained a burnt out shell…The Trinity Church still stood, with fire bomb marks through its roof but the area around was a burned out slum, hideous to see, in the midst of which stood a few decaying almshouses. Clarence Square, once the home of Admirals, was a sordid wreck… In the claims for reconstruction, complete priority was given to housing – and municipal housing at that…" for the thousands of people left homeless or returning to the town. Many bomb damaged areas were cleared and redeveloped. *Smitten City The Story of Portsmouth Under Blitz* published by the Evening News in 1945 vividly illustrates the serious war damage to Portsmouth, epitomised by the burnt out Guildhall. Destruction of familiar streets continued all over the country until the late 1960s, when Portsmouth was drastically reshaped by new roads and clearance of historic areas in Old Portsmouth, Portsea, the city centre and Buckland. The Portsmouth Society dubbed this process 'the Second Blitz' in their book *Maritime City Portsmouth 1945-2005*.

Perhaps in reaction to this drastic redevelopment, public pressure to preserve buildings and areas gathered increasing momentum. The existing protections were felt to be too limited, so new laws were passed. The local application of Conservation areas, European Architectural Heritage Year, the Defence of the Realm policy, Heritage Open Days, Historic England's Portsmouth Harbour project and Heritage Action Zones are explored here, along with the parts played by key individuals and groups. Despite the increasing protection of historic buildings and areas there have also been some losses, while some historic structures migrated to new sites - and others remain vulnerable.

Listed buildings - and clearances

Soon after the Town and Country Planning Act came into force in 1947 the Ministry of Housing and Local Government listed a large number of Gosport buildings as being 'of architectural or historical interest'. The town was regarded as highly significant as a surviving Georgian seaport. However in the early 1960s the town's politics were dominated by a Labour council; their architect J E Tyrell was determined to replace the buildings of the old town with a series of tall 'system-built' blocks of flats, inspired by seeing similar buildings on a visit to Moscow.

As outlined in chapter 2 the landward part of the seventeenth century Gosport Lines was cleared in the 1960s and laid out as playing fields and open space. Gosport Borough Council gained national notoriety by razing 56 listed buildings illegally between 1947 and 1965, some of which were prominent in the harbour townscape or had naval or military connections.

Enraged by the demolition of the historic core of his home town, Portsmouth Grammar School pupil 16-year old Stephen Weeks took his protest directly to the Minister of Housing and Local Government, Sir Duncan Sandys, asking him to stop this obliteration of Gosport's history.[1,2] Weeks already had form in identifying the harbour's industrial and architectural heritage. In 1962 aged 14 he established the location of the King's Storehouse in Old Portsmouth – commissioned by King Henry VIII to store, amongst other things, the masts and tack of his new flagship, the *Mary Rose*. The Storehouse lay buried in later construction. Weeks tried unsuccessfully to fight its demolition for extension of Portsmouth power station for coal storage.

He also identified the first forge building of Henry Cort, the ironmaster whose twin inventions of the puddling furnace and the rolling mill fuelled the Industrial Revolution and supplied Portsmouth dockyard. His products were important in revolutionising the production of wrought iron for the Navy. His Gosport works were on the Green near today's High Street.[3] At Funtley Iron Mills near

Fareham in the summer of 1964 Weeks and a school friend conducted the first industrial archaeological excavation to uncover the site of Cort's first rolling mill. In October 1965 Weeks appeared in the first TV programme ever made on Industrial Archaeology, "What's It All About?" made by the BBC.[4] Weeks also tried in vain to save the Gilkicker, a unique 17th or 18th century triangular tower which was built as a navigational marker for ships at sea.

After three months' consideration, the minister declined to stop the Gosport demolitions. Sir John Betjeman, who up to that time had been one of the lone public voices of 'preservation', wrote to Weeks and they met. In 1972 Lord Kennet – the environmentalist Wayland Young – wrote up the case in his book on the early history of the preservation movement.[5]

This furore and others and marked a landmark change in attitudes towards building conservation in post-war Britain. Ordinary people were now taking a stand against the destruction of their familiar streets and built heritage in the name of modern development or redevelopment. In 1957 the Civic Trust chaired by Duncan Sandys was formed in reaction to 'comprehensive redevelopment': clearance of whole areas defined as slums; and to provide a central focus for local civic societies. Local responses to the Gosport demolitions led to the demand for a Gosport museum and to formation in 1970 of the Gosport Historical Records and Museum Society (later the Gosport Society) and of the Portsmouth Society in 1973. These groups – of which there are many in Hampshire – continue to promote high standards of town planning and architectural design, preservation of historic buildings and open spaces and public education about their home areas.

By 1965 the redevelopment of Gosport town centre had led to the demolition of groups of buildings at Clarence Square, Chapel Row and The Green. Both the former military governor's house and the Hall, a Regency house built for a shipyard proprietor in about 1830 in yellow brick with a gallery and cupola designed to catch views of the open sea beyond the harbour, purchased as offices for the Borough Engineer were demolished. In 1975 the council planned to build public housing to add to the existing high density Harbour Tower and Seaward Tower of 1963 on the remaining open space of Holy Trinity churchyard. The 'Battle of Trinity Green' erupted to prevent it being built on. The chairman of the Gosport Society was the Revd. John Capper, whose vicarage was near the church inside the last moated ravelin of the town's fortifications. The council's proposal met with a tide of public opposition.[6] Lesley Burton, activist member of the Gosport Society and also a town councillor for the ward which included Trinity Green had to be given police protection because her campaign for Trinity Green aroused such hostility. The scheme was abandoned.

The origins and achievements of the Gosport Society were celebrated in 2020 in two publications: *The Gosport Society 1970-2020 - The First Fifty Years* edited by Louis Murray and a series of essays: *Gosport Society 1970-2020 50 Years Promoting Local Heritage. Gosport 2020 For The Record: Essays in Celebration of the Half-Century of the Gosport Society.* The Society's publications also include *Naval Gazing in Gosport* which explores Gosport's waterfront with stories often bound up with the Royal Navy and Defence of the Realm. John Sadden published *Gosport in Old Photographs* in 1990 – a memorial to the old town from 1900 to WWII.[7]

In Portsmouth the 'comprehensive redevelopment' that involved demolition of Buckland, Somerstown and Mile End and the destruction of most of the buildings in Commercial Road, the main road to Portsmouth city centre stimulated the formation of the Portsmouth Society. It has campaigned for the preservation of the best of the city's historic architecture and for the highest design standards in new development ever since.[8] The Society successfully saved the armed forces recruitment office outside the dockyard gate and Boathouse 4 from demolition, as explored in our second dockyard chapter. In the 1980s Celia Clark represented the Portsmouth Society and Lesley Burton the Gosport Society on the national Local Amenity Societies Advisory Committee to the Civic Trust. Both societies urged the Civic Trust to put pressure on the government to acknowledge the importance of redundant defence sites – of which both had considerable acreage – to the future of their local communities.

Conservation Areas

The Civic Trust put pressure on the government to protect whole historic areas, and in 1967 the Civic Amenities Act introduced the principle of 'conservation areas' where there was a measure of control over demolition and rebuilding. The Department of the Environment Circular 8/87 set out local councils' powers and duties in the care of historic buildings and conservation areas, and the 1990 Planning (Listed Buildings and Conservation Areas) Act of 1990 consolidated the previous legislation. Local authorities were given power to designate conservation areas and work to enhance them. In 1983 the National Heritage Act transformed the Commission for Historic Monuments into English Heritage, Scottish Heritage and CADW, which became owners of historic properties and gave advice to government on listing and scheduling, as well as offering grants for town regeneration and rescue archaeology. From 1987 outstanding post-1939 buildings such as Boathouse 4 in the dockyard could potentially also be listed. Ironically, despite HMS Vernon/Gunwharf being designated a conservation area in 1992, most of the buildings on the site, except the few that were listed or scheduled, were cleared in the redevelopment. But conservation area status enabled English Heritage to intervene in the design – and to improve it.

A Portsmouth politician with a passion for military heritage

Freddie Emery-Wallis, a Portsmouth councillor and later Leader of Hampshire County Council had an enormous influence on conservation in the city and county, giving it an international reputation, also celebrated for commissioning outstanding public architecture. Large areas of Portsmouth needed rebuilding, and at the same time its role as the premier port of the Royal Navy was shrinking - in line with the loss of empire. The city council planned to close what was then considered the military Cumberland House, its only museum. So it was also to save this that Emery-Wallis waded into local politics - the start of his journey to preserve the distinctive defence structures of Portsmouth and Hampshire. Alarmed by the prospect of the severe damage to the city fabric through massive and insensitive redevelopment, Emery-Wallis began a political career that enabled him to radically modernise the city's direction.

Freddie Emery-Wallis

Freddie joined the city council in a by-election in June 1961, and he quickly became chair of the Development and Estates committee.

A key to his work was his passion for military heritage.

"Then there was a great campaign to save the fortifications which were heading for rapid destruction because the War Department had left them, the coastal artillery had finished... To do that, I combined with Bill Corney, who was the sole member of staff running the Museum Portsmouth Museum Service, which was Cumberland House... We formed the Portsmouth Museum Society. I wanted to form a Portsmouth society, he wanted to form a museum society, but we combined, we thought there is only room for one group - one pressure group - and we were very successful…. I'd got good friends in the Crown Commissioners, the state commissioners, who were very helpful - a pretty different organisation to how it is today."[9]

Freddie was always as interested in restoring and reusing historic buildings as well as being a much-praised patron of modern architecture. Under his leadership as a city councillor Portsmouth was an early adopter of the new conservation powers given to local authorities in the Civic Amenities Act 1967 and the Planning Policy Guidelines that followed. Nineteen conservation areas were soon designated, including the historic dockyard. In the 1960s once Point Barracks in Broad Street had been demolished, the high level walkway along the walls of Old Portsmouth was created. It is much enjoyed as a high level viewpoint for the harbour approach today.

Chris Hyson remembered that "During his time as Chairman of the Development and Estates Committee in Portsmouth [Freddie] ensured that good budgets were available for conservation work including purchase of historic buildings material and artefacts and ongoing maintenance. We first met when I was appointed in the early 1970's as Committee Clerk/Admin Officer to the Development and Estates Committee of Portsmouth City Council in the Town Clerk's Dept. Many happy days/years up until local government reorganisation in 1974 were spent working with [him] on this committee and its many offshoots (City Centre Redevelopment Working Party, Annual Liaison meetings with Royal Navy, all public utilities, health service, British Rail etc.). Planning in Portsmouth was pretty much sewn up and under control. I fondly remember working with him to ensure that Minutes of meetings said more than had actually been voted on - thus ensuring policy could be quoted in the future!" [10]

Freddie established "a network of good professional chief officers led by the remarkable Town Clerk John Haslegrave. Between them they forged development links within the city and with the services. They challenged the government of the day to fund decent new housing following slum clearance. So began the skilful regeneration of the city, bringing back pride to 'Pompey'. The dockyard had employed 20,000 workers but was in rapid decline so on reclaimed land, North Harbour became the base for IBM UK. With the new Portsmouth Polytechnic, the city was becoming a popular place." [11]

Freddie was also an early recycler of architectural salvage, which was stored in Fort Widley. The stone balustrade in the city museum and the classical temple in Gatcombe Park came from Crichel Down and the elaborate iron gates and railings installed behind the Guildhall from Knightsbridge barracks in London.

In 1970 two historic buildings architects: Brian Young and Deane Clark were appointed. Portsmouth City Museum designed by Deane Clark was created in the converted Clarence Barracks in 1973. Southsea Castle was restored and opened as a museum, the seafront walkway along the walls of Old Portsmouth was created, and Forts Purbrook and Widley were acquired by the city for leisure activities. Our first dockyard chapter explores Emery-Wallis's role in saving the dockyard storehouses and the establishment of the Royal Naval Museum.

Deane Clark's Portsmouth
Tricorn Books 2010

Gatcombe House and Temple from Crichel House Dorset (c.1780, extended 1877) 1983
Served as Officers' Mess for now demolished Hilsea Barracks. In the 1970s converted and extended by Warings contractors. The temple was part of Portsmouth City Council's acquisition of architectural details stored at Fort Purbrook. The temple was re-erected in Gatcombe Park in 1973.

European Architectural Heritage Year 1975

A particular stimulus to the growth in building conservation was European Architectural Heritage Year 1975, celebrated in Portsmouth with two awards and by an exhibition to which 'the local societies which interest themselves in our architectural environment, with details about their activities, [and] any specific projects in hand' would contribute.' [12] The conversion of the east facing bays of Storehouses 9 and 10 to the Colonnade on the Victory approach to the Royal Naval Museum won an European Architectural Heritage Year award. The work was designed by H Dobby of the Property Services Agency, with architects Stride, Treglown & Wyeth designing the shop and buffet and co-ordinated through the Naval Base Joint Planning Team. The contractors were Healy & Evans and stone and brick cleaning and repair was carried out by Peter Cox & Co. Ltd. The citation says: 'Portsmouth naval base as built between 1690 and 1880 remains intact, and most of its buildings are listed as Ancient Monuments or of historic interest.'

'Although three quarters of a million tourists visit *HMS Victory* each year, few refreshment facilities and little shelter had been provided, nor had much thought been given to security control.' '"The plan evolved to form a colonnade, with souvenir shop, buffet and display cases, out of the ground floor of two late 18th century storehouses opposite the dry dock, as part of a defined circulation route. Hard landscaping and provision of a coach park were also proposed.' 'The exteriors of the storehouses have been properly restored and new work has been carried out with careful regard to the existing buildings. The detail design is simple, modern and in an idiom reflecting naval influence, and the whole conception provided an appropriate and inviting tourist route to *HMS Victory*.'[13]

As part of Hampshire County Council's response to European Architectural Heritage Year, in 1976 Emery-Wallis also set up the Hampshire Buildings Preservation Trust, which has continued to pay an active role in conservation in the county up to the present. The Trust has had a longterm interest in military buildings including the Royal Victoria Military Hospital Netley acquired by the county council, the guardrooms at Beaumont Barracks Aldershot and Peninsula Barracks Winchester where it played a role in ensuring a conservation-led redevelopment. It also bought, restored and resold several key buildings in Gosport. Celia Clark was one of its founder members.

Policy: *Defence of the Realm* and *Hampshire Heritage* 1979

In 1979 Hampshire County Council commissioned a report by the Dartington Amenity Research Trust entitled: *An interpretive strategy for Portsmouth and the surrounding region* published as the City of Portsmouth/Southern Tourist Board DART publication No. 59. It identified tourism as a key activity for the surviving defence heritage resources in south Hampshire:

"For almost two thousand years, the Hampshire coast and the Isle of Wight have been a focus for invasion and defence of the Kingdom, and of counter-attack against the enemies of the realm…. The long saga of naval and military activity in the area has left a remarkable heritage of castles, fortifications, historic ships, guns, artefacts, documents and heroic stories… These relics have a power to stir the imagination of the residents of the region and of visitors from Britain and abroad. Some of the historic sites are open to the public; and the public response to them shows a continuing growth of interest in the historic heritage…. Portsmouth City Council has sought to diversify the economy of the city to reduce its historic dependence upon the navy. At the same time, the City Council has been actively engaged in conserving the historic defence-works … [It] now wishes to consider how these properties, and this investment, might be put to fuller public use, in order to bring recreational and educational benefit to its own residents and visitors from further afield and to assist the economy of the city through tourism."[14]

In the same year Hampshire County Council published the policy document "*Hampshire's Heritage - A Policy for its Future*" written by Mansell Jagger and Mike Hughes, which set out a policy framework for the whole county. All the districts and the two cities of Southampton and Portsmouth adopted it. It was based on an illustrated, in-depth survey of the historic resources of the entire geographic county area, including the military legacy. This document sat alongside the Hampshire Treasures Survey, carried out over a number of years at about that time, on behalf of HCC by the parish councils. It represented a "Domesday Survey" of all the local features, listed or not, which local people felt were worthy of protection. 'Together, these documents put Hampshire, its districts and cities, very much in the van of historic conservation strategy in UK and was part of the political momentum which saw Portsmouth later taking the lead in the preservation and conservation of the historic dockyards.'[15]

But as the government has made deeper and deeper cuts to local government over the past thirty years their ability to influence new land uses via planning has diminished. Recently developers have been in the driving seat. The creation of a South Coast Powerhouse in 2015-6 by Portsmouth,

Gosport, Havant, Fareham and Southampton, Eastleigh and the Isle of Wight was presumably intended to strengthen local authorities through a shared voice and expertise. Until 2020 when Portsmouth's conservation officer resigned, each local authority had a dedicated officer responsible for advice on listed buildings and conservation areas. Architects and developers were concerned that there was no in-house replacement in Portsmouth, which earns a substantial part of its living from heritage tourism.

Gosport

In Gosport across the harbour the Borough Council, in contrast to its earlier demolition of historic buildings, was also quite quick off the mark in adopting conservation measures. Its first conservation area was Alverstoke and Anglesey in 1975, followed by fifteen others, many of them military or naval: Priddy's Hard was designated in 1990, St. George Barracks North and South, Royal Clarence Hard and Haslar Peninsula in 1990, Daedalus in 1999 and Haslar Barracks in 2017.[16]

An example of the change of mind is the rescue and restoration of houses in Seahorse Street. In the late 1970s the street was awaiting demolition. Gosport Borough Council had acquired Nos. 19-22, a terrace of four listed houses together with No. 6 opposite, 'a substantial three storeyed house with a fine bow window for demolition and redevelopment as part of a new housing scheme.'[12] Deane Clark represented the Hampshire Buildings Preservation Trust at the public enquiry to oppose its demolition. The inspector visited the street and noted the quality of the 1920s terrace opposite which had purposely been vandalised. Key joists had been cut so they were uninhabitable. They had been built into the boundary wall of the local brewery which was retained although the brewery was pulled down. The inspector rejected the demolition of the brewmaster's house and proposed the listing of the Seahorse terrace.[17]

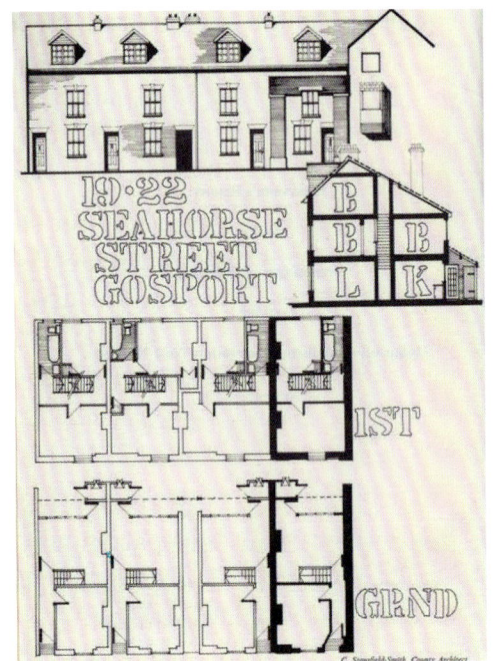

19-22 Seahorse Street Gosport. Elevation and plan
Hampshire Buildings Preservation Trust Annual Report 1978

"The Borough Council was only able to restore No. 6 within its housing budget and Nos. 19/22 were therefore offered to the Trust for a nominal payment of £5.00. The County Architect was asked by the Trust to prepare a detailed scheme and obtain tenders for their repair and modernisation. The external and internal walls and upper floors of the house were reasonably sound but the ground floors had been destroyed and the roof coverings had been badly damaged. The buildings lacked proper sanitary or kitchen fittings, or a safe electrical system, and there was no provision for heating other than open fires. The houses will require a great deal of work to be done but when this is complete they will offer a living room and dining room and kitchen on the ground floor, two bedrooms and a bathroom on the first floor and a third bedroom and a very useful store on the second floor. Each house will have a small private space at the rear. Listed Building Consent and Building Regulations approval have been obtained." With the maximum improvement grants the work was carried out by John Hunt Limited with completion expected in May 1979. By 1979 the houses were all sold and occupied.[18] The brewmaster's house was incorporated into the area's redevelopment. The Hampshire Trust also bought No. 1 High Street and commissioned conservation architect Louise Bainbridge to restore it in 1985/6. She researched the original façade. It was agreed to keep the existing fenestration pattern rather than reinstating Georgian windows. Access for the disabled requested by Gosport Borough Council was achieved by a secondary entrance at the side. The Chapman's Laundry shopfront was painstakingly replaced with special imperial sized blue glazed bricks made by Swanage Brickworks and a reproduction pedimented doorcase. The Trust sold the building to Gosport Borough Council and it has been in office use since 1996.[18]

Gosport has for a number of years the harbour's most active and scholarly conservation officer, Rob Harper, who anticipated the redundancy of Haslar Hospital, Fort Blockhouse, Priddy's Hard and the early naval barracks adjacent to the hospital by preparing policy guidelines, sometimes jointly with the Ministry of Defence property arm, before these sites were sold to developers.

The Palmerston Forts Society

Voluntary groups, local authority regeneration bodies and Historic England continued to work together to find new life for local defence sites. One of the most active is the Palmerston Forts Society.

In 1982 shipwright Brian Patterson and teacher David Moore ran a six-week course: "The defence of Portsmouth Dockyard' at Fareham Adult Education Centre, including visits to Priddy's Hard armaments museum, Fort Cumberland and the Old Portsmouth headquarters of re-enactors, the Fort Cumberland Guard. Hampshire County Council Clerk of Works Cyril Hathaway offered the group the opportunity to visit Fort Nelson. The group were amazed to find that most of the fort was overgrown and obscured by trees, but 'it had great potential and all agreed that it had to be preserved and restored."[19] In our chapter on museums and galleries we explore how the restoration of Fort Nelson was a catalyst to the huge subsequent and ongoing effort to restore and reuse the harbour's military heritage.

The Palmerston Forts Society plays a decisive role in revaluing the harbour's rich legacy of Victorian defences, - by research into their commissioning and construction, restoring Fort Nelson and Eastney Fort East through their own efforts, working with subsequent owners and users such as the Royal Armouries - and re-enactment of military drills.

A registered charity, it was formed in 1984 to bring together enthusiasts who have an active interest in nineteenth century military fortifications and associated artillery worldwide, but particularly within Hampshire. It is focussed on the ring of forts that protected Portsmouth. In the early 1980's Fort Nelson was quietly crumbling away but was rescued by the intervention of Portsmouth City Council and Hampshire County Council, together with a small number of dedicated volunteers, to preserve it for future generations. These local volunteers went on to become the core of the Society, which now has an international membership. Its authoritative publications, many of them by David Moore, are justly regarded as definitive histories. The Society has developed a remarkable knowledge and specialization on the interpretation of fortifications and how they would have looked when initially constructed or subsequently altered during periods of rearming. It has worked on many forts in the UK and elsewhere.[19]

The uniformed section of the Society is the Portsdown Artillery Volunteers (PAV) who dress as the 2nd Hampshire Volunteer Artillery from the 1880's to re-enact authentic Victorian gun drills at Fort Nelson. Geoffrey Salvetti led the excavation of Eastney Forts East and West and has lived in Eastney Barracks for many years.[20]

A national body, SAVE Britain's Heritage's report *Deserted Bastions. Historic Naval and Military Architecture* (1993) sparked Celia Clark's interest in historic military sites, and she worked closely with SAVE to prevent the demolition of Boathouse 4 – as we explore in our dockyard chapters.

Local initiatives to open up defence sites to the public

Once historic defence sites are released and either sold – for example Gunwharf, or transferred at notional value – Priddy's Hard to Gosport Borough Council – they were available for redevelopment into new, civilian land uses – which in many cases allowed public access for the first time. However there remains a problem for local authority planners faced with determination of new uses for these sites: nationally sanctioned security prevents locals from gaining any intimate knowledge of them until they are disposed of. During a tour of Royal Clarence Victualling Yard at a community planning event, the leader of Gosport council astonished Celia Clark by saying: "I've lived in Gosport all my life, and I've never been on this site before…"

Since their inception in 1994 Heritage Open Days have offered local people a first chance to see the MOD sites in their midst. The Civic Trust's statistics reflect this nationally in defence- dominated areas. The Gosport Society pioneered the opening of MOD sites in 1993, contributing considerably to stimulus of public debate about their future, with the help of the Civic Trust which subsidised an informative booklet on six of them. Other organisations followed suit: in the same year English Heritage opened Fort Brockhurst, Hampshire County Council opened Fort Gilkicker, the MOD the newly restored Fort Rowner, Gosport Borough Council Priddy's Hard Armaments Museum; and the Royal Navy Submarine Museum. The Open Days are now one of England's biggest voluntary cultural events, attracting some 800,000 people every year.

In Portsmouth opened places included Fort Cumberland in guardianship of English Heritage, HMS Excellent Drillshed and Quarterdeck and the HMS Nelson wardroom opened by the navy, as well as older military structures such as the Round Tower and the Square Tower owned by Portsmouth City Council. In 1996 of twelve sites, seven including the *Mary Rose* Trust Finds Bay, were associated with the city's naval and military past. 120 people visited Fort Cumberland, Eastney's 'sleeping beauty' and 50 went to HMS Nelson Wardroom.

Rediscovery of Portsmouth's historic defences

In 1996 the Berkeley Group commissioned an Environmental Analysis of the HMS Vernon/ Gunwharf site from consultants Mouchel. Their Excavation/Assessment report said "Over the past 10 years or so excavations have shown that in effect the whole plan of Portsmouth's fortifications, particularly in regard to any post-seventeenth century works, still exist in a researchable form below modern ground level. The limestone masonry of the massive bastions, ravelins, etc, remains virtually intact below the modern roads, buildings and parks, as does the extensive moat system. Redevelopment proposals have again presented the opportunity to examine a further section of the town's vast fortifications system."

The report concludes: 'Despite redevelopments over the centuries, twentieth-century war damage, and subsequent levelling, the investigations showed that evidence of the old town fortifications does still exist below ground. The fact that the remains lie 80 centimetres and more below the modern ground level, and are not in restorable condition without colossal excavation, indicates that the best approach would be to leave these remains buried. Perhaps the ground plan/outline of these works

London Illustrated News 6 February 1875

Portsmouth ramparts exposed 2018

could be incorporated at ground level in any future development, as a visible link with this part of Portsmouth's past."[21] Excavations in 1996 on the Gunwharf site uncovered significant structures, but eyewitness accounts in our ordnance yard chapter show that these remains were destroyed in situ.

People interested in local history were delighted when excavations in November 2018 briefly revealed the arrow-shaped low stone walls which were part of the defences of Portsmouth town built about 1667: Town Mount and the adjacent ravelin. The southern edge of Ravelin Park was being excavated before construction began on the university's sports centre. According to Les Capon of the archaeologists AOC Archaeology in his public lecture at the University of Portsmouth 'Uncovering 350-year-old fortifications in Ravelin Park' on 12 December 2018, earth excavated to make the moat and form the Town Mount was put back into it during the demolition of 1859. It is likely the dismantling of the Portsea and Portsmouth fortifications took several years. The *Illustrated London News* for 6 February 1875 has an engraving showing the demolition of the East Bastion and the curtain wall towards the Town Mount Bastion. The workforce is shown filling in the moat by pick and shovel, supervised by a man on horseback. The 1899 *Guide to Portsmouth and Southsea* states that the millpond that used to empty itself into the harbour through what is now Gunwharf was said to have been formed by excavating the earth in order to build ramparts for the town; it was filled up again in 1876 with materials from the demolition of these ramparts.

The 2018 excavations revealed the sloping layers of earth created in this process. What remained of the exposed stone fortifications had clearly been robbed for construction elsewhere. At the university lecture about the 2018 excavations we were assured that "This wealth of history will be incorporated into the building design". How remains to be seen; the huge excavation for the new swimming pool appeared to destroy what had survived below ground until then.

We now examine other losses of key historic structures, the continuing saga of Block Mills and others that have migrated to new locations.

Losses

Ropery in 1957   Deane Clark 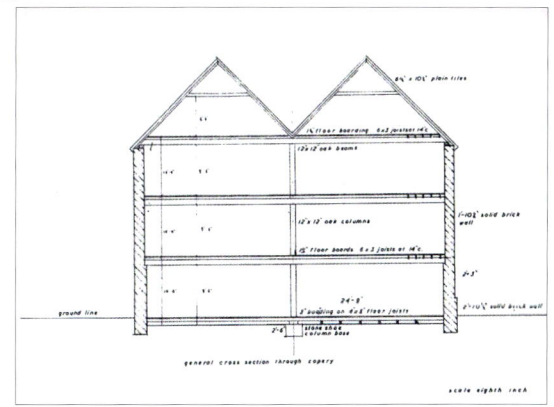   Ropery Measured drawing   Deane Clark 1957

Demonstrating the navy's past lack of interest in its historic legacy, the mutilation of the Ropery together with the loss of the 240-ton crane and of the majestic No.3 Ship Shop are still regretted by those who remember them. The gates in the modern end wall of the Great Ropehouse of 1770 at right angles to Storehouse 10 are often open, revealing the three-storey interior.

Ropery interior

In its time, at almost 1,100 feet it was one of the longest buildings in the country, purpose-built to produce the miles of cordage required by the navy's sailing ships. In the early twentieth century two arched passageways were cut across the building to improve traffic circulation blocked by its immense length. In 1960 – before ideas on conservation of historic buildings took hold - the Ropehouse was 'eviscerated' – to quote Inspector of Ancient Monuments Jonathan Coad. The attic windows and all internal floors were removed – to make one long tubular space for storage. Before this drastic surgery took place, architectural student Deane Clark made measured drawings of it. His dissertation is a historic document in its own right. By 2018 the ropery was within the remit of the PNBPT.

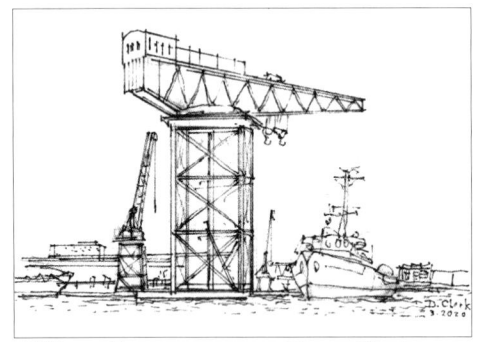

As the tallest structures in dockyards apart from ships, cranes sometimes acquire iconic status. But the British navy is unsentimental about destroying infrastructure once it has no further use for it. For many years Portsmouth dockyard's 240-ton, 200 foot hammerhead crane with a cantilever of 260 feet was an iconic landmark on Portsmouth's skyline - as well as the symbol used in the city's marketing. Celia Clark remembers standing underneath it with Ray Riley, historian of the dockyard when a high wind vibrated its steel structure like a giant harp. But when it had outlived its usefulness it was cut up. Many local people were sad to see it go. They felt the heart was being torn out of the dockyard.

240-ton Crane — Deane Clark

Another major loss in national terms was the destruction of the iconic Nos. 3 and 4 Ship Shop, shipbuilding slips that according to Ray Riley briefly had the widest span in the world. They were the largest structures on the harbour until the new ship halls and fleet headquarters were built. Their 1845 specialised cast iron roof by Messrs. Baker & Son of Lambeth pre-dated the famous railway station sheds.

No. 3 Ship Shop
National Monuments Record
Swindon

In 1974 David Lloyd described the building as "Even more dramatic [than the Fire Station] because of its sheer scale and the daring of its construction for the time when it was built, was the mighty No.3 Ship Shop which was built in 1845-6. This is a huge iron-framed shed of two main spans covering two slipways in which ships were built and repaired. The roofs of the two spans are supported by curved transverse iron braces, rising from complex vertical and horizontal iron framework on the sides of the building and between the spans. … it is still tremendously impressive as a piece of structural engineering, and an amazing building to have been erected in 1845-6. The parts were prefabricated at a foundry in London…The Dockyard buildings using iron in their structure: the former Pay Office (1798), No. 6 Boat House and the Fire Station (both 1843) – are all therefore notable in the history of metal construction, most notably the Ship Shop because of its unprecedented scale and daring, and the fact that the parts were wholly prefabricated, and assembled on the site… It is true that a contemporary and similar building in Chatham is to be preserved in situ, but the Portsmouth Ship Shop is equally important national, and of tremendous importance locally."[22]

The navy required the site, so in the 1980s No. 3's complex iron framework was dismantled and supposedly stored in the Chalk Pits Museum at Amberley. But the pieces are no longer identified as being there. English Heritage advised that there was a more comprehensive set of covered slips in Chatham, emblematic of their material development from timber, iron to steel. No. 3's disappearance was a great loss to the history of the harbour and to industrial archaeology. The Chatham series of covered slips – in wood, iron and steel - survive.

The Block Mills saga

A plain industrial Georgian building in outward appearance, it's what happened inside it that's reflected in Block Mills' Grade 1 status. In the early nineteenth century it was the world's first steam powered factory where a pioneering team of brilliant engineers developed specially designed metal machine tools to manufacture the many thousands of wooden pulley blocks needed for the navy's sailing ships. The MOD's 1998 report about it says: "Of the many important buildings in Portsmouth's historic dockyard, the early nineteenth century brick built block mills are the least prepossessing. However their sociological and historic significance is unrivalled. As Jonathan Coad says: "Compared with the great Victorian machine shops and sawmills, the almost domestic sized group of buildings seems tiny, yet they are of immense importance, not merely in the history of the Royal Navy, but also in the history of the industrial revolution."

At the turn of the eighteenth century a typical ship of the line needed about 1000 wooden pulley blocks of different sizes. The Royal Navy's fleet of over 500 ships required roughly 100,000 blocks per annum. The growth of the fleet and subsequent increase in demand for the wooden blocks put the workshops which hand-crafted the blocks under enormous pressure. In 1802 Marc Isambard Brunel, who had fled the revolution in his native France for America, came to England and married Sophia Kingdom, sister of the under secretary to the Navy Board. Having devised a method of making blocks whilst in America, Brunel persuaded the Navy Board to invest in his project. General Samuel Bentham designed Block Mills to house the machinery. The yard between the two wood mill buildings was walled and roofed to form a new workshop to house the blockmaking machines. By 1808 forty five machines designed by Henry Maudslay, Simon Goodrich and Brunel himself to cut and fashion the lignum vitae pulley wheels and solid elm shells could produce 130,000 blocks per annum. This was a technological first: a steampowered mass production line using metal machine tools requiring less than a tenth of the man power used when crafting the blocks by hand. The navy gained its first experience of steam power and realised the benefits of mass production, giving it a distinct advantage over its enemies. Block Mills attracted many visitors to see its innovative machinery.[23]

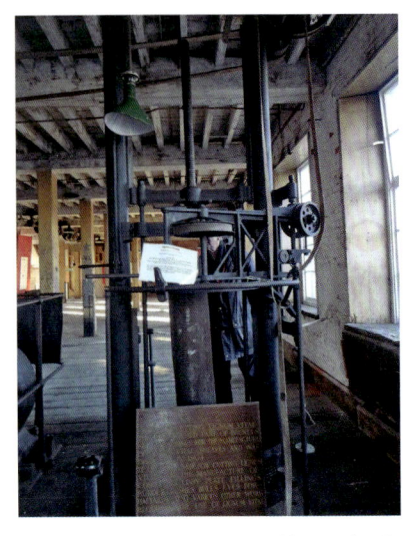

Block Mills Original machinery in situ    Block Mills 2016

A depressing example of how the MOD leaves its key historic buildings unmaintained if unused, Block Mills was in continuous use until the 1960s despite the north range being bombed in WWII; then it fell into disuse. Some of the machinery remains in situ, some is in Boathouse 5 and some is in the Science Museum in London. A scheduled monument as well as Grade I because of its importance in the history of technology, for many years the building remained in poor repair. It has been on English Heritage's Buildings At Risk Register, classed as a Building at Risk ever since the Register began in 1998. English Heritage and other groups put continuous pressure on the MOD to repair and reuse it. As long ago as the early 1990s the Portsmouth Naval Base Property Trust was negotiating to take it on. The English Heritage chairman Sir Neil Cossons visited the building in 2003 and 2005 and so did the Chief Executive in January 2006. English Heritage published Jonathan Coad's book on the Block Mills, which was launched in the building in 2005.

Horizontal boring machine Block Mills
Portsmouth Royal Dockyard Historical Trust

The Portsmouth Society said in the 2006 *Hampshire Buildings Preservation Trust Annual Report* that it was still in a very poor state. "Discussions, estimates for emergency repairs and plans for its future are on-going, but what is really needed is immediate repair work." Eventually only pressure from English Heritage, Prince Charles and national voluntary bodies persuaded the MOD to repair it. In 2006 a report on the reasons for the restoration of the roof of the north wing was produced. In 2008 the redesign and rebuilding of the north roof was finally finished, winning the Georgian Group Architectural Award in August that year: "The Block Mills Renovations Project team also won a certificate of commendation from General Sir Kevin O'Donoghue, the chief of Defence Materiel in recognition of their work. Now the restoration of the Block Mills is complete it will be removed as an entry in the English Heritage Biennial conservation Report and is in use as a store for the rigging of Nelson's flagship, HMS Victory - a fitting use for such a fine Georgian building."

In June 2009 the Portsmouth Society had two speakers from the MOD to talk about defence heritage buildings, when Block Mills was again on the agenda because it was still unused. Malcolm Ives outlined the national role of Defence Estates, as one of the UK's largest landowners, owning 1.8% of UK land, which contained 793 Listed Buildings and 720 Scheduled Ancient Monuments, holding the government's largest heritage portfolio. Their remit was to maintain them sustainably, find the best use and 'whenever possible open them to the general public.' Inevitably there are conflicts between military and heritage priorities. The speakers said: "Eighty per cent of MOD buildings and 89% of the Scheduled Monuments/Listed Buildings are in a 'good' or 'fair' condition. Twenty-eight of their buildings are on English Heritage's Buildings at Risk Register (BARR), but Defence Estates have appointed a BARR Officer. Four other structures within Portsmouth Naval Base remain on BARR: No.6 Dock, 2-8 The Parade, the Iron and Brass Foundry and No.25 Store. Defence Estates' annual outlay in the south of England is £40m, so Block Mills - estimated at £1-2m - was not a 'large' project. Up to £2m was allocated by Defence Estates in 2007, with further funds for internal works at the beginning of 2008."

"The layout of the Block Mills' north and south ranges runs west-east, three storeys high, with a single storey range infilling the middle space. The first floor corridor crosses the middle range north-south. But the main hazard throughout Block Mills' life has been damp rising from the Reservoir over which it was built. Ironically, open windows, roofs and doors, while letting in rain (sometimes in torrents), also produced an airflow which had ventilated the building fairly well since block-making ceased in 1983. Without it, internal timbers would have been ravaged by rot."[24]

Block Mills 2009

In 2020 so many years later, Block Mills is still unused and inaccessible, no doubt partly because it lies within the active naval base. It would, if released, contribute technological innovation to the tourist mix. We consider extension of the Heritage Area's boundary in our next chapter about the dockyard.

Migrating defence material
When army and navy sites close, the services often move symbolic objects such as regimental silver or paintings to still active bases. As towns grow and change, larger items are also repositioned. The biggest structures to migrate to new positions are three of the classical gateways which spanned the entrances to the two walled towns of Portsmouth and Portsea.

King James's Gate now in Burnaby Road

King James's Gate (1687), which controlled access to the Point outside the town jurisdiction was moved twice in the 1860s. Minus its crowning curved cupola it was rebuilt in a new position in Burnaby Road flanking the United Services Recreation Ground, once a broad area of water between the two walled towns. Portsea's Lion Gate (1770) was moved from Queen Street and rebuilt in 1929 as the base to the dockyard's Semaphore Tower. Its regal lion is in the garden of the wardroom HMS Nelson in Queen Street.

Lion from Lion Gate in grounds of HMS Nelson

The Quay Gate to the Camber and elaborate baroque King William Gate across Pembroke Road – added in 1833 for access to the new suburb of Southsea - were both destroyed later in the nineteenth century because rapid urban growth was constricted by the town's tight defences. However, the Royal monogram from the Camber Bastion with the date 1687 was installed, wrongly labelled, at the entrance to the 1970s development of Captains Row in Old Portsmouth.

Unicorn Gate

Unicorn Gate (c.1778) formerly at the end of York Place, Portsea was re-erected in 1865 as one of the principal entrances to the dockyard. Now it sits rather incongruously in the middle of a roundabout inside the modern dockyard boundary in Unicorn Road. Only Landport Gate (1760), designed in the office of Nicholas Hawksmoor which was the only access to the town until about 1873 is in its original position, but shorn of its context: facing a bridge across the moat between the two towns. It looks oddly orphaned within the modern buildings of the sports ground on reclaimed land where the moat once was, opposite the Mary Rose pub, now a Chinese restaurant. Other old survivors are also still in situ: the original gateposts still mark the entrance to the redeveloped Pembroke Barracks. How much Portsmouth's lost gates would now be prized, had they all survived!

As a memorial to the demolished Pitt Street Baths, rescued Art Nouveau ironwork was installed by the MOD in the blockwork wall of the new HMS Temeraire sports complex.

Pitt Street Baths. Royal Naval School of Physical Training 1910-11 Architect: probably GE Smith

Swimming baths converted to trapeze pits

Queen Victoria's Railway Shelter converted to a French restaurant for Navy Days

Queen Victoria's Railway Shelter relocated   Deane Clark

One of the largest structures which has been moved was Queen Victoria's Railway Shelter of 1888 at South Railway Jetty. It was at the end of a curved branch line across the Hard. In the twenty-first century it was raised on a platform and rollered through 90 degrees to a new position because the navy needed the dockside space.

Other smaller defence materials are also still extant. Once your eye's attuned, you can spot recognisable reused tooled and shaped Portland stone blocks around Portsmouth. Military engineers reused stone from the demolished fortifications to Portsmouth and Portsmouth Common/Portsea in the striped brick and stone perimeter wall of Colewort barracks in Old Portsmouth. Similar stone is reused in Southsea's seawall near Canoe Lake, in the wall to HMS Nelson in Anglesey Road, and in the southeastern wall of the dockyard – complete with watch turrets. Not only the military but Portsmouth Corporation and church authorities reused it - in the porter's lodge in Victoria Park and at Rivers Street Hall, a National School built in 1868 in Somerstown. It's also visible in the seawall of the promenade at Eastney and in the southern boundary wall of Cumberland House. The city council still relocates stone from the dockyard: in the central reservation of the main road into the city, in Warren Avenue and on the Langstone foreshore, where there's an iron mooring ring set on top of a curving granite block.

Mooring ring on Langstone Shore

Developers also reuse surviving structures. A cupola from the sickbay of HMS Vernon was relocated to the top of a new block of flats in the renamed Gunwharf and memorial stones from Colewort barracks are set into the entrance wall at Gunwharf Gate.

Claims for ships' timbers reused in older houses may not often be authenticated, except in the case of Chesapeake Mill Wickham which reused wood from the captured *USS Chesapeake*, and the dockyard storehouses where curved 'knees' and other pieces are visible in the floor. Civilian timber is also reused in defence contexts: the Hayling Island railway bridge was reused to bridge Long Curtain Moat in the 1970s. Pounds Yard at Tipner specialised in recycling defence materials including old naval vessels, buoys and other relics.

After the closure of HMS Vernon, the poop deck of *HMS Marlborough* was removed from the neoclassical wardroom Ariadne of 1922-5 to another naval site, HMS Sultan. As a result, Ariadne was no longer a listed building and it was demolished in the Gunwharf redevelopment.

As we explore in our chapter on open space, one war trophy, the historic Tanggu Bell was replaced by a replica made in the original Chinese foundry.[22]

We now consider recent initiatives to revitalise the harbour's historic defence sites.

Historic England's Portsmouth Harbour Project

From 2014 to 2017 Historic England had a joint project with the local authority group: Portsmouth Urban South Hampshire (PUSH) which aimed "to secure the sustainable re-use of historic buildings in and around Portsmouth Harbour, and in doing so, to safeguard their long-term maintenance; to ensure that the re-use of these buildings brings economic benefits to the area… and to maximise the social 'returns' of the project through high levels of community engagement…" Hilsea Lines, No. 2 Battery, Priddy's Hard, Haslar Gunboat Yard Guardhouse and Police Barracks, Royal Clarence Yard, Point Battery and Fort Cumberland were the defence sites on the list.[26] Work on Hilsea Lines is described in our chapter on public open space, Royal Clarence Yard in the residential chapter and Point Battery in the Make Art not War chapter.

In Gosport, as part of the Portsmouth Harbour project, a viability assessment looked at future potential for the Haslar Gunboat Yard with its group of Grade 1, II* and II listed buildings and Scheduled Monuments. This resulted in progress and in physical works on the site. The Grade II Wardroom was also assessed for viability to inform a wider assessment of the Daedalus Conservation Area. This work led to the Heritage Action Zone.

"Breathe New Life into an Old Place" Heritage Action Zone for Gosport in 2019-20

"Gosport is home to a number of naval buildings and places of historic interest. Whilst [it] is still home to HMS *Sultan* and a Naval Armament Supply Facility as well as a Helicopter Repair base, many of the historic buildings are derelict and in need of a new use and lease of life."[26]

A recent specialised regeneration tool, which began to be used in Gosport in 2019 is the Heritage Action Zones (HAZ). These were first launched by Historic England in 2017. They focus on places with special architectural and historic buildings which have lost their original use and urban landscapes which have suffered years of decay and economic decline. Twenty historic areas in England are being offered targeted help: research into historic significance, advice and grants for repair and capacity building, advice on finding new uses for buildings and on planning policy, condition surveys, historic area assessments, updating of listing entries, training to understand the significance of historic buildings, engagement with local communities and networks and contacts. HAZs are delivered over three to five years through partnerships with local authorities, private and voluntary organisations. Their aim is deliver opportunities for sustainable longterm growth.

In June 2019, as a culmination of the Portsmouth Harbour project Gosport was designated a Heritage Action Zone, focussed on regeneration of defence sites at risk. Over five years the ambition is to conserve and enhance Gosport's heritage through the built estate; to increase additional housing and employment spaces within the borough, to attract inward investment for business and the visitor economy in Gosport, rejuvenate the environment to improve the health, wellbeing and quality of life for local people and to pursue and support increased socio-economic opportunities for all in Gosport.

Led by Gosport Borough Council its aim is to create new civic pride alongside social and economic regeneration. Significantly in the local context, the Gosport Society is a partner in the project. Its scholarly publications including many on defence themes have contributed much to local knowledge. The other partners are the DIO, One Public Estate, the Portsmouth Naval Base Property Trust, Fareham Technical College, Hampshire County Council, Hampshire Cultural Trust, and East Solent Coastal Partnership. Meetings take place on a three-monthly cycle, reporting back on progress on each HAZ project. Detailed plans and progress on each site is recorded in the relevant chapters of this book.

Among the historic areas highlighted are the Gosport Lines, a Fort Rowner feasibility study, a Fort Blockhouse Master Plan, a Daedalus Wardoom Study, an evaluation of Haslar Barracks and a concise social and architectural history of Gosport. A detailed plan for the Heritage Action Zone was worked up by spring 2020, including grant funding provided by Historic England.

At the launch Clare Charlesworth, planning group lead for Historic England in the southeast region, said: "Gosport has so much to be proud of, from Priddy's Hard to Blockhouse. Some of its historic buildings are showing signs of neglect but with a little investment and imagination they can become assets for the people of Gosport to enjoy and a benefit to the town's economy. We are looking forward to working with Gosport Borough Council to increase awareness of what the town has to offer and to make it easy for investors and developers to bring jobs, homes and visitors here."

Mark Hook, Leader of Gosport Borough Council, said: "We're delighted that Gosport has been chosen as one of the country's Heritage Action Zones by Historic England. This confirms the borough as one of the richest areas for military historic buildings in the country. A lot has been done in recent years to make the most of our heritage and realise the potential of historic buildings, but there's more to do. We're delighted to be working in partnership with Historic England and look forward to some imaginative thinking to bring buildings back into use through restoration. This status will help us get funding for regeneration projects and give us more resources to work on our plans. We can't wait to get started on the project and to ensure that the town's character and heritage will truly benefit the area, and current and future generations."[27]

Michelle Lees was appointed to oversee the delivery of the project. A review of the Heritage Assets at Fort Blockhouse informed the masterplanning for the wider site, in liaison with the Defence Infrastructure Organisation as a key HAZ Partner. One Public Estate provided £130,000 funding towards the masterplanning of Blockhouse and of Haslar Barracks. The Ministry of Justice is also a partner in the HAZ.

Gosport Lines Bastion No. 1 (before restoration). Gosport Borough Council

Tree and shrub clearance and reseeding of No. 1 Bastion at the southern end of the Gosport Lines was funded by the Coastal Revival Fund. A detailed feasibility study about Fort Rowner and the complex issues surrounding its conversion and future use by architect Giles Pritchard was also funded by One Public Estate. Options for converting the existing historic fabric into a mixed use development of residential and light industrial uses supported by proposals for enabling development within the fort and on adjacent land were explored.[28] Historic England produced a detailed review of the historic significance of Stokes Bay, to form the basis of a Stokes Bay Conservation Area to be considered in the 2020-21 financial year.

A separate four-year Heritage Action Zone focused on the High Street and Stoke Road Conservation Areas began in 2020 with development of a Community Engagement Plan and the appointment of a project manager.

Key individuals and groups continue to play their part in this local renewal process. Important events were European Architectural Heritage Year in 1975, the Defence of the Realm initiative in 1979, and much more recently, Historic England's Portsmouth Harbour Project and Gosport's designation of two Heritage Action Zones in 2019 and 2020.

Sources

LFW White 1964 *The Story of Gosport* SWP Barrell Southsea

The News 1945 *Smitten City – the Story of Portsmouth Under Blitz*

Portsmouth Society 2005 *Maritime City Portsmouth 1945-2005* edited by Ray Riley Sutton Publishing

The Gosport Society 2020 *1970-2020 - The First Fifty Years* edited by Louis Murray

The Gosport Society 2020 *1970-2020 50 Years Promoting Local Heritage. Gosport 2020 For The Record: Essays in Celebration of the Half-Century of the Gosport Society*

Hampshire Buildings Preservation Trust Ltd. *Annual Reports* 1977, 1978, 1979

https://historicengland.org.uk/services-skills/heritage-action-zones/breathe-new-life-into-old-places-through-heritage-action-zones/

David Lloyd 1974 *Buildings of Porsmouth and Its Environs* City of Portsmouth pp.72-73

Jonathan Coad THE PORTSMOUTH BLOCK MILLS – THE START OF A REVOLUTION.pdf;

Jonathan Coad 2005 *The Portsmouth Block Mills: Bentham, Brunel and the Start of the Royal Navy's Industrial Revolution*. English Heritage

Deane Clark 1957 *Portsmouth Dockyard and Portsea up to 1850* Dissertation Portsmouth College of Art

Ray Riley 2005 Heritage and change since 1945' in *Martime City Portsmouth 1945-2005* Sutton Publishing p.121

References

1. http://www.gosportheritage.co.uk/history-boy-stephen-weeks/accessed 12June 2020
2. Roger Mills 1963 *Hampshire Magazine July*; *Portsmouth Evening News* – 5, 6, 9, 11, 12, 15 and 20 January 1965; *The Daily Telegraph* 9 January 1965
3. Prof, RA Mott, ed. P Singer 1983 *The Great Finer. Creator of Puddled Iron*. The Metals Society in association with Historical Metallurgy Society Library Catalogue
4. BBC TV programme: Industrial Archaeology, 1 – What's it all about? 1965
5. Wayland Kennet 1972 *Preservation* Maurice Temple Smith, London pp.153-162
6. Louis Murray 2020 *The Gosport Society 1970-2020 The First Fifty Years* The Gosport Society p.22
7. *Naval Gazing in Gosport* 2017 www.gosportsociety.co.uk; *The Gosport Society; the First Fifty Years* 2020; John Sadden 1990 *Gosport in Old Photographs* Sutton Publishing Ltd.
8. www.theportsmouthsociety.org
9. PORMG: 2382A Freddie Emery-Wallis Interview with Dr. John Stedman 11.4.1995 Portsmouth History Centre
10. Christopher Hyson 2017 Email to Jennette Emery-Wallis April in FREDDIE EMERY-WALLIS Tribute 2017 by Stephen Weeks, Celia Clark and many others; edited by Grahame Soffe
11. Letter from Rosemary Billett to Celia Clark April 2017
12. Letter from Lindsay Hoole Curator of Art Portsmouth City Council to Celia Clark 20th August 1975
13. *Heritage Year Awards* 1975 Civic Trust United Kingdom Council for European Architectural Heritage Year 1975 p 167
14. Dartington Amenity Research Trust 1979 *Defence of the Realm an interpretive strategy for Portsmouth and the surrounding region* City of Portsmouth Southern Tourist Board DART publication No. 59.
15. Email from Chris Williams to Celia Clark 3 March 2016
16. https://www.gosport.gov.uk/sections/your-council/council-services/planning- section/conservation/conservation-areas/area-designation/ accessed 29 May 2019
17. Deane Clark 2020 Recollection of action taken on behalf of Hampshire County Council and Hampshire Buildings Preservation Trust
18. Hampshire Buildings Preservation Trust Ltd. *Annual Reports* 1977, 1978, 1979; Sarah Strange Seymour & Bainbridge Architects 2020 Email to Celia Clark 2 September

19 Geoffrey Salvetti 2009 *Palmerston Forts Society Forts, Guns and Holes A Concise History of the Palmerston Forts Society 1984-2009* The Palmerston Forts Society; http://www.palmerstonfortssociety.org.uk/Home/1/

20 Geoffrey Salvetti Interview with Celia Clark 18 March 2019

21 The Berkeley Group Gunwharf Quays Environmental Analysis Addendum 1 December 1996 p. 2, 5

22 https://portsmouthdockyard.org.uk/timeline/details/1845-no.-3-shipbuilding-shop

23 Jonathan Coad 2005 *The Portsmouth Block Mills : Bentham, Brunel and the start of the Royal Navy's Industrial Revolution*, ISBN 1-873592-87-6; *"Portsmouth Royal Dockyard: History 1690–1840". Portsmouth Royal Dockyard Historical Trust. www.portsmouthdockyard.org.uk. Retrieved 7 October 2009; McNeil, Ian (1990). An Encyclopedia of the History of Technology. London: Routledge. ISBN 0-415-14792-1*

24 Portsmouth Society Newsleter 2009

25 https://www.local.gov.uk/topics/housing-and-planning/one-public-estate/about-one- public- estate

26 Historic England Portsmouth Harbour Project Methodology Alison McQuaid 2017 24 https:// historicengland. org.uk/services-skills/heritage-action-zones/gosport/

Chapter 6
Heritage tourism to the rescue? Portsmouth Historic Dockyard: landlord and tenants

"What can one do with a historic dockyard?" Sir Neil Cossons, Director of the Science Museum and chair of the Heritage Education Group at the Civic Trust set Celia Clark, their education officer, this challenge in the 1980s. This book, as well as her report on futures for European dockyards[1] and her wide-ranging research since then are attempts at answers, based on extensive regeneration research, both in Portsmouth and around the world. Defence heritage tourism was first identified for its local potential in 1979 - as we explored in the last chapter. The development of the Historic Dockyard within Portsmouth's active naval base over the last half century is a major example of heritage tourism in world terms.

HM Naval Base 2007

After historic defence sites lose their prime warlike purpose, tourism - in complete contrast to their earlier functions - or specifically 'defence heritage tourism' is often seen as the most appropriate new use for them. To preserve and celebrate these complexes as symbols of a national military or naval past including specialist museums means that they can continue to embody national identity, albeit in a new economic role that celebrates their history. Docks, boathouses, storehouses and barracks in historic dockyards around the world are being restored and converted in order to present them to tourists who are interested in them as places and in the role of the armed forces in defending their countries. Internationally, tourism is now the world's largest export as well as one of the world's most important sources of employment - 2% of total global GDP.[2]

Income from overseas visits to the UK in January 2017 was valued at £2,010m, and that figure does not include Britons exploring their own country, valued at £104.9 billion in 2013. Visit England's count of overseas day visitors in 2018 figures ranked Portsmouth/Southsea in position 20 for the whole country, with 95,000 trips and 46,000 holiday visitors in 2016. Portsmouth dockyard Heritage Area is the largest tourist attraction on the south coast of England. Portsmouth Historic Dockyard provides 2,750 full time jobs. Over 860,000 visitors to the National Museum of the Royal Navy were recorded in 2014 and its 'national reach' was said to be over 1.2m visitors in 2020, although these figures included the other museums under its umbrella. In 2018 this rich combination attracted

950,500 visitors a year, contributing £110.4m a year to the local economy.[3] Of course, tourism is not the only subsequent use of ex-defence sites around the harbour – as explored in our other chapters.

Inevitably the 2020 Covid-19 pandemic had a heavy impact on the tourism industry. Many countries introduced travel restrictions to contain the spread of coronavirus, airlines cut staff and fleets and museums and heritage attractions closed. The UN World Tourism Organization estimated that global international tourist arrivals might decrease by 20-30% in 2020. Extended disruption could lead to a global loss for the industry of between £25bn and £45bn on spending by international visitors – and might be even more pronounced.[4] In June 2020 the All-Party Parliamentary Group for Hospitality and Tourism's report: Pathways to Recovery revealed that just 11% of hospitality businesses in the UK had been able to operate normally during lockdown, highlighting the scale of the damage done.

The closure in March 2020 of the Historic Dockyard and all its attractions including the National Museum of the Royal Navy in line with government advice on the COVID-19 epidemic was unprecedented – and the effect on the national museum, historic ships and boats – was incalculable as it faced its deepest financial crisis in a generation. All but three of the Historic Dockyard staff were furloughed as were those of the museum, except for those performing essential tasks such as security checks and those working from home. Those still on site continued to maintain the museum's buildings, ships and collections and the museum's digital channels were refreshed The director general said that the challenge of making up the shortfall in admissions income had to be faced, while doing the right thing for staff, for the collection and for the museum as a whole.[5] On 23 July 2020 he announced that the closure had crippled the NMRN's revenue, effectively wiping out the entire summer season when most of its yearly income was generated. The museum was days away from declaring itself insolvent after losing £6.35m - when the Treasury approved an emergency rescue fund of £5.3m to save it from bankruptcy.

It would no doubt take the Portsmouth Dockyard Property Trust a long time to recover from the loss of income over a prolonged period, but by prudent planning they were better placed than many to weather the storm.[6] In October 2020 the trust was granted £698,000 by the government from the Culture Recovery Fund for Heritage and the Heritage Stimulus Fund to upskill its volunteers and carry out essential maintainance work. The Mary Rose Trust was given £655,304. These grants highlight the financial vulnerability of the country's heritage assets.

Historic Dockyard reopens to visitors – with supercarriers in the background September 5 2020

Once funding and distancing decisions were taken, the Heritage Area reopened with online booking only on 24 August 2020. Billboard adverts said "We've swabbed the decks!" and the website invited us to "Jump on board and explore our ships, museums and attractions with all the family – whilst knowing that as a charity your admission income helps us survive and protect our future." A £23 ticket for pensioners to visit the Submarine Museum might be too much for many. Free entry to Boathouse 4 was not immediately available but was likely to be reintroduced towards the end of October. Volunteers at the Portsmouth Royal Dockyard Historical Trust began to get back to their normal routine, but although they were classed as a 'workspace' they had to operate on a reduced numbers basis. With constant shifting of the guidance goalposts they had to cancel their 2020 AGM. The dockyard closed again for the second lockdown in November.

We now explore how this specialised branch of tourism has developed in the dockyard: the formation and development of the Naval Base Property Trust, the evolution of representation of the historic vessels on the management of the site, the development of the naval museum and the proposal to take in the Royal Marines Museum, the physical impact of the profound change of use from naval base to tourism inside the heritage area and in the wider context, the Millennium project and the World Heritage bid. Restoration of the storehouses, boathouses and the drydocks and the arrival of historic ships and boats and their restoration make up the next chapter. In 'Make Art Not War' we describe the development of civilian museums in other military buildings around the harbour. It's a complex story, with many twists and turns, but a constant thread is the condition of historic dockyard buildings both inside and beyond the area devoted to tourists.

Navy Days 1970s

Setting up the Portsmouth Naval Base Property Trust

Although visitors had been coming to see the dockyard for many years, in 1985 the navy no longer needed the Georgian core or certain Victorian and twentieth century structures and dry docks which were mostly disused. Lord Trefgarne, Minister for the Armed Forces chaired an MOD-led committee to plan to hive off a heritage area in the oldest part of the naval base. Rear Admiral Tony Tippett who had been appointed Flag Officer in 1981 took action. Richard Trist, the city's chief executive said: "The City of Portsmouth needed a new form of economy and a focus on tourism gradually developed. This tourism was largely focused on the heritage of the fighting Navy. No doubt with the backing of the Ministry of Defence [he] suggested to Councillor Marshall [Leader of Portsmouth City Council] that a meeting of all interested parties should be called to look at the possibility of releasing the historical part of the naval case so it became a tourism project…I estimate that it was late 1981 or early 1982. The work involved fell on my second Staff Officer Peter Goodship and thus started his long interest in the project."

"For this Naval Heritage Project the starting point was an historic area of the Dockyard containing some handsome buildings of varying ages from the late 18th century up to modern times. Inserted in this area were already the Portsmouth Naval Museum and *HMS Victory*. *Mary Rose* was just arriving from the seabed. *HMS Warrior 1860* was due to arrive in 1987. This constituted a significant start."[7]

The city and county council, navy, MOD, English Heritage, neighbouring local authorities, representatives of the three ships, the museum and other public bodies met. "There was a general enthusiasm for the project and everyone agreed we should go ahead. In the early stages there were no formal organisations or trusts formed but the project was seen to be very real in the minds of those involved."

Another stimulus was the raising of the *Mary Rose* in 1982 from the seabed amidst international publicity and her housing under a temporary cover in No.3 Dock next to the *Victory*. There had been plans to build a new museum for her at the eastern end of Southsea seafront, in sight of where she sank. Portsmouth City Council had also agreed with the Maritime Trust to construct a new pier just inside Victory Gate to berth *HMS Warrior*. This combination of factors led the city council to consider creating a tourist destination at the southwest corner of the dockyard. The English Tourist Board and Hampshire County Council were persuaded to fund a study of how this might be achieved. The Tourist Board also funded several study tours, including to the Rouse Corporation's innovative developments on the northeastern American coast, South Street Seaport in New York, Faneuil Hall Maketplace Boston, Mystic Seaport, Inner Harbour Baltimore with the *USS Constellation*, Colonial Williamsburg Virginia and Tobacco Dock Richmond.

The Ventures Consultancy of Beaulieu proposed a radical new approach. Their Broad Study considered how the many former military sites in southern Hampshire, most of which were constructed to protect and support the dockyard could be better marketed and ticketed on a collective basis. As already discussed, Cllr. Freddie Emery-Wallis, leader of Hampshire County Council, drove this forward; it resulted in the Defence of the Realm initiative.

Their Narrow Study focused on development of the southwest corner of the dockyard, recommending that the existing and proposed ship and museum attractions should jointly establish a charitable trust to co-ordinate the marketing and ticketing of their attractions under the leadership of an independent chair. They also recommended that an independent charitable trust should be established to hold the property and to grant concessionary leases to the ship and museum attractions and also to develop the other buildings commercially. The historic buildings would be vested in a separate organisation, while the ships and attractions would in practice be tenants but "should take charge of the project as a whole through some form of co-ordinating body" with "little or no role for the public bodies thereafter". The city council provided administrative support and money. "An informal body of representatives was soon founded to run the project but they did have difficulties in reaching agreement" – a problem that continued.

A precedent was the Chatham Historic Dockyard Trust, which had been set up in 1984 to take on the historic core of that dockyard. The Ministry of Defence had drawn up a Memorandum and Articles of Association for it in consultation with Kent County Council and the Department of the Environment. The MOD, through the Head of Secretariat Naval Staff produced similar draft documentation for the trust to be established at Portsmouth.

In 1985 the Ministry of Defence released the southwestern quarter of Portsmouth dockyard for development as a Heritage Area. Together with Portsmouth City Council, the Portsmouth Naval Base Property Trust (PNBPT) was set up - to take on a 99-year lease of the 11.25- acre (4.56-hectare) site on 14 November. Peter Goodship was appointed as its first chief executive, and Richard Trist attended many of the co-ordination meetings as a representative of the city council. Initially the MOD was to appoint the members of the trust, but because they wished to avoid being under any financial obligation in the longer term, "They got cold feet. If it was responsible for something that went wrong or there was a financial disaster, they would have a moral obligation to bail it out. So they sought to distance themselves."[8] The Department of the Environment (DoE) and the agency which was being transformed from the former Historic Buildings and Monuments Commission of the DoE into a new quango: English Heritage, were each asked to nominate three out of five trustees; Hampshire County Council, the English Tourist Board and Portsmouth City Council nominated the others. The city did so on the strict understanding that it would only appoint those recommended by the board, a principle that held until ten years later when political appointments were made.

Lord Montagu, then Chairman of English Heritage identified Robin Bishop, a senior partner in

Drivas Jonas who had served in the navy in WWII and was known by to be keen on historic buildings as a board member. He became the first chairman of the trust. English Heritage also appointed Professor Sir William Whitfield CBE, Historic Buildings Architect on the Ancient Monuments Board, Surveyor to St.Paul's Cathedral and from 1994 vice-chair of the MOD's Historic Buildings Advisory Group and John Roome, the senior partner of London solicitors Withers. The city council appointed David Thomson, a merchant banker with Lazards (later Director-General of the British Invisible Exports Council), and Admiral Philip Higham who had just retired as Keeper of *HMS Belfast*.

The Memorandum and Articles of Association took some time to be finalised and approved by all parties, but the MOD, English Heritage and Hampshire County Council shared a strong desire that the Memorandum should be widely drawn geographically to allow the trust to take on additional surplus MOD estate as it became available beyond the boundaries of HM Naval Base. The Minister for the Armed Forces at the time, Roger Freeman was particularly keen to ensure that the MOD acted responsibly in the disposal of its historic estate – and the trust was identified as the appropriate vehicle for this. The Trust's Articles of Association set out its objectives: "to preserve buildings, structures, vessels, docks and berths associated with the Naval Base at Portsmouth, and to protect them for the public benefit; to provide for access to the buildings and enjoyment of them for the purposes of education in naval history and other matters related to the Defence of the Realm."

This wide remit later enabled the trust to buy Horse Sand Fort in 1994, (subsequently sold to Mike Clare who also acquired and restored Spitbank and NoMans Land Forts) and the historic core of Priddy's Hard in Gosport, and to begin negotiations to take over Block Mills.

In accordance with Ventures Consultancy's Narrow Study a dual structure was recommended to reflect the double nature of the attraction. The Portsmouth Naval Heritage Trust (PNHT) chaired by Admiral Sir John Lea was established to represent the Mary Rose Trust, the Royal Naval Museum and *Warrior* once she arrived in 1986. The Royal Navy operated *HMS Victory*, which at the time was open to the public free of charge, which did not form part of the trust. With no funds of its own until 1988 when the PNHT persuaded the MOD to introduce an admission charge for *Victory*; all but £100k of the income per annum was retained to fund marketing of the attractions which were ticketed separately.

At first the property trust was London-based, banking at Coutts in the Strand and meeting in the MOD Main Building negotiating the terms of the endowment and the detailed terms of the Head Lease. Until the end of 1991 board meetings were held at the offices of Drivers Jonas in Suffolk Street.

Peter Goodship played a key part in developing the strategy of the property trust. He identified the development of the heritage area as a strategic project in terms of the local economy and the city's emphasis on tourism. The first task was to agree the level of endowment and the terms of the Head Lease. The MOD was reluctant to dispose of the freehold, given that part of the site was being used operationally. The principal difference between the government's treatment of Chatham and Portsmouth was that the Chatham trust was granted the freehold of its site, whereas the Portsmouth trust is an MOD leaseholder.

The Portsmouth trust was initially offered an endowment on exactly the same terms as the Chatham Historic Dockyard Trust. This had been set up by the government in 1984 to take on a much larger estate: 80 acres, the eighteenth century core of Chatham dockyard with 47 scheduled and listed buildings. Its aspiration is 'preservation through use', which like Portsmouth has been a particular challenge, given the state of disrepair. Kent County Council had accepted the Chatham terms: a dowry of £11.35m from the government Property Services Agency (PSA), much of it for the proper maintenance of the many scheduled buildings. It was not strictly an endowment, because the government also passed on to the Trust the responsibility of funding a £3.5m contract they had already let to repair the Double Ropehouse and renew its roof – a significant chunk of the whole amount. According to English Heritage the amount of the dowry was not related to the condition of the buildings and the cost of putting them in order, nor to the cost of converting them for the two options open to the trust: museum use open to the public, or conversion for letting to tenants who would generate income.[9] The Portsmouth endowment had been calculated on the basis of keeping the buildings wind and weathertight – with no allowance for the fact that they were historic buildings. Although the Chatham trust was already finding life difficult financially, as their dowry was proving woefully inadequate, the Portsmouth Trust with a much smaller area and fewer historic buildings was only offered £1.5m. This sum was calculated on the basis of the Property Services Agency's surveys and costings for temporary repair of the navy's unused buildings – merely making them wind- and weather-tight. These costs, according to Peter Goodship, were based on the capitalised annual market return on the basis of the use of the buildings – say, as a store – not as protected buildings.

Robin Bishop refused the deal offered by the MOD, threatening to dissolve the trust. The offer would have meant no chance of doing the extensive conservation work that was really needed in the face of a backlog of unfulfilled repairs and maintenance had the buildings not been declared surplus to operational requirements. The trust persuaded the Secretary of the Naval Staff that they should be put in funds so they could instruct a firm to do a proper building survey as the basis for renegotiating with the PSA. The Secretariat of Naval Staff offered £60,000. The contract was awarded to Watts & Partners of London. They produced figures of between £15m and £18m: £12.5m to put the buildings in a proper state of repair. Their recommendation was that for the purpose of restoration the trust needed £18.5m. After negotiation with the PSA there was a joint agreement in 1986 that the right figure was £12.5m – a pro rata rate four times greater than Chatham's, as the Portsmouth trustees were aware that Chatham was struggling to achieve its aims with a much smaller dowry to meet an even greater challenge of restoration. Since *Victory* was built there, nearly half the trust's endowment was spent on an exhibition of ship building in the Nelson era, but without iconic attractions such as *Victory*, *Mary Rose* and *Warrior* the attraction of an industrial dockyard was limited and the trust ran out of money.

This was at a time when the MOD was not allowed by the Treasury to roll over funds from one year to the next. There was £6m in the kitty. After meetings with Lord Arran, the then minister, PNBPT accepted £6.4m, plus £0.5m to build a fence and associated police posts. As Peter Goodship said: "They negotiated a lease reflecting half of what both partners accepted that we needed. This scenario absolved them from the obligation to put in representatives of the MOD, which would then have had to bail out the trust if it ran out of money. Ideally we should have received £12.5m at 1986, but we never will."

A 99-year lease was granted to the Property Trust under the Landlord and Tenant Act. Effectively the lease was for life – in perpetuity. As landlord, the trust leases space to its tenants, including the historic ships and the naval museum: a significant change of management style from other central government funded trusts. The pressure to operate commercially caused by the MOD's inadequate 'dowry' and therefore the necessity to make ends meet led to large rent increases over the years that were fiercely resented. The naval museum is the only national museum to pay rent for what is still a government building.

Improving visitor facilities and fencing

The two Portsmouth trusts recognised that the site lacked the most basic facilities. The only toilets were in the Victory Arena and there was a small café on the ground floor of Storehouse 9 run by the Royal Naval Museum's trading company. There was no car parking, no ticket office or reception. With the agreement of the PNHT part of the former HMS Nelson was leased for parking, existing workers' toilets in College Road were renovated and two tented structures were erected in the Port Arena to cover people queuing to visit *Victory* and *Mary Rose*. The security alerts in response to the

IRA bombing campaign detailed in our first chapter led to the decision to separate the working yard from the heritage area. In 1987 a security fence was designed and built with close cooperation from English Heritage on both sides of the narrow heritage area with two Security Offices (1/101 and 1/102) in Main Road. The trust's offices are housed in No. 19 College Road, a substantial house built in 1908 to house the superintendent of police, refurbished in 1988.

Fundraising

The MOD dowry was never going to be sufficient for major investment in tourism development: restaurants, shops and high tech leisure attractions. Raising the funds for installing enabling infrastructure and to supplement the shortfall in the initial dowries was a continuing task for both dockyard trusts. Maintaining, managing and finding new uses for significant historic buildings in the Heritage Area meant the endowment was far too small to bring the thousands of square feet of unused space back into productive new use. The trust therefore had to raise its own funding, relying on developers and financial institutions to implement its decisions and on professional conservation bodies for advice. A Conceptual Brief by Sir William Whitfield and Sir Neil Cossons differentiated between inalienable assets: the historic buildings and *Victory*, and alienable assets: *Mary Rose* and *Warrior*, which were considered as locatable anywhere – although today the *Mary Rose* in its permanent museum would be regarded as inalienable.

In 1985 the two trusts invited developers to submit proposals to develop the site for increased visitor attendance with the full concurrence of the city council. Several consortium's bid for the opportunity to invest in the heritage area, narrowed down to two leisure conglomerates, Allied Lyons - and Sea Containers - which was the successful bidder. It took several years of negotiations during which relationships between PNHT and PNPBT deteriorated before all the parties reached a consensus.

A series of agreements were made to invest in refurbishing Boathouses 4, 6 and 7, developing the storehouses and providing a new museum for the Mary Rose. In March 1986 Sealink British Ferries, which had been sold by the government to Sea Containers two years before, put forward proposals. In July 1989 they submitted an Outline Development Plan prepared with the architect John Winter to develop and manage catering and visitor activities, reception and retail outlets, which were to be "substantially themed and to reinforce the integrity of the project." They estimated that £26m was needed, of which £6m would be provided by Sea Containers, plus a similar amount from grants and sponsorship and the rest from the money market. The Royal Naval Museum was to move from the storehouses, a large restaurant and shops complex would replace the 1939 Boathouse No. 4 and the *Mary Rose* would be housed in a new museum. Sea Containers were to be offered a 25-year lease by the Property Trust, which would receive a management fee plus a share of the profits.

The debates that followed exemplify the tension between presentation of heritage on its own terms and the continuing need to break even via commercial exploitation. Opposition to Sea Containers' proposal came from the county council, the Heritage Trust representing the three ships, the trustees of the Royal Naval Museum, the Warrior Trust and the Portsmouth Society. The director of the naval museum did not believe the estimate of 3 million paying visitors was realistic. If it was not a financial success there was a risk of 'even more commercialisation'. In his view far more attention should have been paid to ways of preserving the traditional character of the site, for example exploring the possibility of boat repair and chandlering facilities.[8] Nordic historic dockyards: Suomenlinna in Finland and Skeppsholmen in Stockholm have adopted this complementary mix of uses.[1]

There were fears that a 'hands-on' facility with multimedia shows on the Royal Navy on Boathouse 6 would be a trivialisation of the site. Maldwin Drummond, chair of both the Hampshire Buildings Preservation Trust and of the Warrior Trust had visited other maritime heritage sites, and concluded that at South Street Seaport New York for example "You can't see the ships for the shops!"

Peter Goodship said that comparisons with South Street Seaport or Baltimore were unfair. Admiral Moreton, one of the museum trustees, did not like the interpretation of the great British Navy being handed to an American, James Sherwood, the founding owner of Sea Containers. Perhaps this was xenophobia: Sherwood had serious credentials: he owned the *Illustrated London News* and owned and ran the Orient Express and associated hotels. He was passionate about heritage.

Conservation interests including Sir Neil Cossons, the director of the Science Museum, were convinced they did not want commercial spoliation to happen in Portsmouth – which would ruin the city's greatest visitor asset. However, the editor of the local newspaper saw tourism as the only way to restore unused buildings in the heritage area. Both the Royal Naval Museum and the Warrior Trustees were reluctant to follow this line and in the end dropped out of the consortium mainly for the reason that they did not believe that the business plans were robust or sustainable. They also thought that the cost of servicing the vast borrowings required – some £25 million – would drive the partners into bankruptcy. But according to the director of the Naval Museum, there were other reasons for their reluctance to getting involved with this commercial partner. One trustee believed that to make the thing work would have required the whole place to be turned into an entertainment, a sort of fairground with ships. It was felt that these kinds of impacts on the character and atmosphere of the place were likely to damage the dignity and grandeur of the surroundings, the buildings and the ships, and that this commercial development would turn the Dockyard into just "another branch of the leisure industry!" The chairman replied: "Oh, for God's sake! Dockyards have always been in the leisure industry!" In recent years Dennis Miles, chair of the Portsmouth Royal Dockyard Historical Trust has been employed to take important visitors to see the yard.

Old and new M33 in drydock

Unlike Plymouth and Chatham, Portsmouth's heritage area has the great advantage that it is surrounded by an operational naval base. It also benefits from a well developed tourist industry - of which it is the focus. An admiral was heard to curse that he had to 'drive his ships through a bloody museum'. But as Peter Goodship, chief executive of the Naval Base Property Trust said, what visitors really come to see was not the museums or historic dockyard buildings but the modern navy steaming in and out of the harbour. At the same time, Portsmouth dockyard has become a major defence heritage destination in its own right. Its marketing makes the same point: *"Situated within a working Naval Base, Portsmouth Historic Dockyard is home to a collection of fantastic attractions, historic ships and museums... [It] is the only place in the world to see the Royal Navy past, present and future and is a must see for anyone visiting the south of England."*

But in the end, while criticism delayed a decision on the Sea Containers' scheme, the 1992 recession rather than opposition scuppered the deal: raising £26m on the market looked less and less feasible. Sea Containers withdrew, and a more gradual approach was forced on the PNBPT. In 1994 the trust funded infrastructure improvements including the construction of John Winter's Visitor Centre next to the Victory Gate, the refurbishment and conversion of Boathouse 7 to provide a new restaurant, a new shop, the Dockyard Apprentice Exhibition to tell the story of the civilian craftspeople who worked in the yard, a playship for young children, and new toilets. The trust installed new directional signage and site interpretation throughout the site via panels and an Acoustiguide in eight languages in a handheld wand as well as new seating. In 1993 the trust won a commendation for its Visitor Guide from Interpret Britain.

A berth was dredged for a new pontoon to introduce a harbour tour so visitors could see the Royal Navy's ships from the water. These new facilities increased visitor numbers significantly. They were funded through the trust's own resources, with several government tourism and regeneration grants as well as £500,000 from the EU and nearly £1m from the private sector for the new facilities in Boathouse 7.

It was felt that conservation of the site could not survive on tourism alone, so the trust pursued ideas of small business units and the introduction of an evening economy, as well as making the site more accessible to local people who perceive the dockyard as a special place, separate from the city, "a town within a town." The storehouses, boathouses, workshops have been gradually refurbished, mainly for conversion into museums and tourist facilities, although significant amounts of unused and unrestored space still remain.

Flagship Portsmouth

Members of PNHT and the Royal Navy suggested that the Property Trust should become a member of the 'site services' consortium. In 1994 PNBPT took the initiative to establish the Flagship Portsmouth Trust which consisted of the former members of the PNHT and PNBPT with the navy representing *HMS Victory* as observers. Flagship appointed a general manager and staff funded through a levy on ticket income to organise marketing and ticketing. This gave PNBPT a role in marketing of the site, which had previously only been the preserve of the attractions.

There was some concern when the governance and executive of Flagship changed in 2010-2 with a chair and chief executive working as an independent company marketing the heritage area. This coincided with the rapid growth of the National Museum of the Royal Navy – which was acquiring new responsibilities in remote locations at that time. Tension between the trust and Flagship's shareholders led to its demise and its succession in 2005 by the Portsmouth Historic Dockyard Trust, an independent company collectively owned by the attraction owners. The Property Trust joined it as a shareholder. The common services function was transferred to the National Museum of the Royal Navy.

Joint/separate ticketing

In 2018 the Mary Rose Trust decided not to continue with the joint ticketing arrangements, which until then had been offered to visitors. They wished to offer single tickets just for the *Mary Rose* instead, since their visitor numbers had dramatically increased on the opening of the new museum. The question of how to divide the ticket income between the different attractions then arose. So the ground floor of the Porter's Lodge was adapted to provide a separate ticket office. However, in summer 2020 the NMRN and the Mary Rose Trust agreed to work together again with joint ticketing.

Royal Naval Museum/National Museum of the Royal Navy

Mark Prescott-Frost, Secretary to the Admiral Superintendent had opened a dockyard museum in 1911 in the Ropehouse. It received around 17,000 visitors a year but was closed during WWI and ran down afterwards. The Society for Nautical Research decided during discussions about the new Victory Gallery that this should supercede the Dockyard Museum, while a substantial number of the museum's artefacts were incorporated into it.

In 1985 under the terms of the National Heritage Act 1983 the Royal Naval Museum was devolved from the Ministry of Defence to become an executive non-departmental body, supported by a grant-in-aid. Until then the museum had enjoyed a protected tenancy under a full repairing lease from MOD(N) which permitted it to carry on a modest but lucrative catering enterprise. In 1995/6 it was agreed with the MOD(N) that an increase in the Grant-in-Aid would meet the increased rent, with contributions towards rates and service charges. In 1996 after lengthy negotiation, the Royal Naval Museum secured a new 90-year lease from PNBPT permitting it to expand progressively into the upper floors of Nos. 11 and 10 as resources became available, but at the cost of its ability to provide catering services for the site. The physical changes to its buildings are described in the next chapter.

From 1999 the museum had a long term development plan to bring into use most of the thousands of square feet of empty space in the storehouses still above and below the public walkway, for which it had to raise the funds. Stage 1 was the development of Storehouse 11 and the Victory Gallery. Hampshire County Council provided the key investment of £500,000 in 1994/5 by the County Defence Heritage Panel, which unlocked the £2.8m grant from the National Heritage Memorial Fund. This helped boost the fund-raising effort launched by Princess Anne at the last official royal engagement on board the royal yacht *HMS Britannia* before she went to Leith. A further £2.1m was raised from Hampshire County Council, Portsmouth City Council, Wightlink, P & O European Ferries, the Princess Royal's Charities Trust, an anonymous donation of £300,000 and other sources.

The museum greatly expanded its influence by taking into its orbit *HMS Warrior 1860*, and the WWI survivor *HMS M.33*. It is also responsible for the replica dockyard surrounding *HMS Trincomalee* in Hartlepool and for *HMS Caroline* in Belfast. It advised the Coastal Forces Heritage Trust for which a new museum is planned at Priddy's Hard alongside Explosion. In 2014 management of the Portsmouth Historic Dockyard site was transferred to the museum. In 2018 it also developed the Devonport Naval Heritage Centre with the Naval Base Commander there and Plymouth City Council – to secure the Devonport heritage collection. It also undertook a feasibility study into the future of the Fleet Air Arm Museum at Yeovilton in Somerset. In addition it had a three-year memorandum of understanding with Malta Maritime Museum to identify opportunities for collaborative research and understanding of naval heritage at regional, national and international levels.

The four original museums now under the aegis of the National Museum of the Royal Navy (NMRN) were the Fleet Air Arm Museum, the Royal Marines Museum, the Royal Naval Museum and the Royal Navy Submarine Museum with the Cold War submarine *HMS Alliance*. Each had a separate board of trustees. Explosion: The Museum of Naval Firepower was added in 2013. Financial control and strategic direction is now provided by the board of trustees of the NMRN, from 2008 the new name of the Royal Naval Museum. Visitor figures to these museums increased from 357,048 in 2009 to 765,810 in 2014, as well as increases in learning visits from 31,792 in 2009 to 52,600 in 2014.

Mary Rose Trust

The Mary Rose Trust office and store are housed in two buildings. The older is the original Old Pay Office of about 1754. It was here that Charles Dickens's father John Dickens worked as a clerk when his son was born in Mile End Portsmouth in 1812. The Pay Office's fireproof construction consists of unusual swelling shaped cast iron columns. Contractor PMC LTD refurbished it in 1986 at a cost of £50,000. As a Building at Risk it is again in need of repair. The main Mary Rose offices are in Building 1/10 in College Road, a guardhouse constructed against the dockyard wall, later incorporated into a 1909 building in College Road and. From 1951 Building 1/10 served as the Police quarters until the Mary Rose took occupation in 1986.

Project Prime

In 2005 Naval Base Commander Paul Boissier and his successor Commodore Amjad Hussain, a graduate of London Business School who combined navy and business expertise approached the PNBPT to deliver Project Prime, a scheme to dispose of the remainder of the Georgian dockyard and bring its many unused historic buildings back into use. In his announcement of the development plan for Portsmouth Naval Base – some 30 acres of the western part of the base to be released for private development - Project Prime, Commodore Hussain said

"British Naval history is in the bricks and mortar. It represented a new model of naval base renewal whose aim was to generate income [from the private sector] to invest in a stream-lined modern Naval base, directing resources in support of the [national defence] front line, by the creation of residential, leisure and commercial opportunities, providing employment opportunities within the local area, and preserving and enhancing buildings of National Historic Importance"

Led by the MOD's property agency, Defence Estates, a private sector partner was sought by project teams from the Naval Base Commander, with the support of Portsmouth City Council, English Heritage, Portsmouth Naval Base Property Trust and Partnership UK, a Treasury led organisation whose aim was to improve practice among government departments via Private Finance Initiatives to advise and help deliver and achieve enhanced value by going to the market. The Property Trust and the MOD's assets would be taken on by a PPP (Public Private Partnership) to provide new facilities for the naval base and open up 60 listed buildings into the public area, offering them to a developer. Very large financial resources are needed for restoration and repair, and for conversion to new uses. Instead of the tight security of a naval base, there would be open access to the public – to buildings locked away for 300 years. The intention was to appoint one developer to take on all the buildings – only some of which would be profitable – for example residential or hotel conversion of the Naval Academy - which might provide cross-subsidy for others.

Advice on Project Prime was offered by Defence Estates' Sustainability and Conservation group based in Warminster, responsible to MOD for conservation of buildings, wildlife, sustainability, energy efficiency, and green eco-building regulations. Project Prime and Defence Estates MODE London, which was using similar scheme to reduce MOD's footprint in London from 13 sites to three or four, were to be showcase projects for this type of procedure for the MOD.

Until recently the MOD's tactic had been to sell redundant property, but according to Defence Estates, this did not necessarily bring in the best return. Although the MOD had not done so before, Public/Private Partnerships have been used by other government departments: for example British Waterways' Isis project, where the whole of the waterside estate including lock keepers' cottages and land was leased away to a developer. Another example is One North East where Newcastle City Council and Gateshead passed over their properties.

The Project Prime developer was to be engaged on a contractual basis. The MOD would retain the freehold; the developer would have a 125-year lease. Owners of residential development would have long leases, which might have to be reviewed in the light of new legislation on tenants' right to buy. The Naval Base Property Trust which managed the Heritage Area within the larger area of Project Prime, as an existing tenant might take up the management duties of the leasehold. There would be only limited facilities for rebuilding and new construction. 67 companies expressed interest in response to the announcement in the EU official journal in September 2004: contracts had to be open to all member states. In the event all the companies were British. Berkeley Homes, which took on Gunwharf Portsmouth, Royal Clarence Yard Gosport, Woolwich Arsenal and other military sites was one of the applicants. The developers were shortlisted via a questionnaire and evaluation in a formal procurement process – on their experience, track record, historic conversion, MOD contracts, health and safety, sustainability and long term issues.

Maintenance of dock walls, culverts, basins, caissons, cranes and other infrastructure would have to be separately funded via a sinking fund which service charges to the new occupiers would not cover. A precedent for this was the Chatham Maritime Trust which was established to look after and collect charges for maintenance – on the advice of Price Waterhouse who analysed commuted maintenance charges which service charges would not cover. This advice was commissioned by the South East England Development Agency and English Partnerships which took on part of Chatham dockyard.

Gavin Marshall, Commercial Officer for Project Prime based explained it in more detail in an interview with Celia Clark in the Semaphore Tower in January 2005. The navy was restructuring – bigger vessels and smaller numbers. The project was three years in the gestation. A blue line defined the project's area, reflecting construction and practicalities. The PNBPT entered into a partnership agreement with the MOD and the South East England Development Agency (SEEDA) whereby the trust would act as the head leaseholder of a mixed-use development by a single developer.

With participation by English Heritage considerable time and effort was expended, negotiating new accesses into the base via Marlborough Gate as well as contributing towards a development brief which was not prescriptive about uses or potential value, although it was likely to be residential led. This approved by the local planning authority. The MOD expected the project to be self-funding and to sit with other developments in Portsmouth: phase 4 of Gunwharf, Portsmouth Harbour Regeneration and Portsmouth City Centre North. Interested parties were to be invited to submit draft proposals. In contrast to Gunwharf this was to be a refurbishment and regeneration project – not another Gunwharf. The six contending developers were graded and were to be invited to submit further details.

At the last moment the MOD pulled the project. The navy concluded that it was too risky to go ahead with big development when Portsmouth was likely to be where the new breed of supercarriers would be berthed. Private industry already operating inside the dockyard was not considered a threat. In 2003 VT Group recommenced shipbuilding on the site of No.13 Dry Dock, and BAE Systems which had subsumed Fleet Support Ltd. continues to manage ship repair and maintenance facilities around No. 3 Basin. There were said to be fears that residential development within the dockyard could lead to closure of the naval base if residents claimed that their peaceful enjoyment was being fettered by naval operations. Threats to security were also mentioned, but since missiles could already be fired from nearby tower blocks, this seemed unlikely. In reality the project failed through naval opposition.

The Millennium Project: the Renaissance of Portsmouth Harbour

A major stimulus to change in the historic dockyard and around the harbour was the Millennium project – also discussed for its effect on Gunwharf in the chapter on ordnance yards. It was decided that the best way to revitalise the redundant MOD sites around Portsmouth Harbour was to link them under a common theme of regeneration. The city council claimed that with matching funding and local contributions this would add up to £112m, which would be injected into the local economy damaged by the rundown and closure of many Ministry of Defence establishments. By "dramatically increasing public access… and acting as a catalyst to unlock their potential", 3,500 new jobs might be created on these closed sites. They were to be linked via 5km of new public walkways: 2000 metres along the Portsmouth waterfront and 3000m in Gosport. Blue street lighting would highlight their common identity at night. This was a visionary idea that continues to be much enjoyed.

The Commission preferred to deal with a single entity so, in response to the government's pressure on local authorities which were directed to form partnerships with the local business community, a limited company: Portsmouth Harbour Renaissance Limited was established. Its members were the city and borough councils, PNBPT and the Portsmouth and South East Hampshire Partnership, a forerunner to Shaping Portsmouth. They put together a bid for funding an ambitious £80m scheme to establish Portsmouth harbour as a destination in its own right, uniting the communities on both sides of the harbour. The Renaissance Company was chaired by Ben (later Lord) Stoneham, the chair of Portsmouth and Sunderland Newspaper Group and Chief Executive of PPP Publishing, owners of the local newspaper *The News* and the free *Journal*. Other members were Roger Ching, city council Head of Finance and Resources, Peter Goodship of the Portsmouth Naval Base Property Trust, and Ron Wilson of Gosport Borough Council. Two other non-executive directors appointed were Lord Judd of Portsea, former Minister of Defence and MP for Portsmouth West and Hugh Siegle, Managing Director of Whitbread Property, who later chaired PNBPT.

Inside the dockyard it was proposed to convert Storehouse 9 to house a new Museum of the Dockyard in partnership with the Portsmouth Royal Dockyard Historical Trust of former dockyard employees. New galleries were planned for the Royal Naval Museum. A new Mary Rose museum was to be built to unite the recovered artefacts in Boathouse 5 with the hull of the ship. New landscaping would be created, and Action Stations: new interactive attractions promoting the modern Royal Navy's current role to contrast with stories of its past would be installed in a refurbished and converted Boathouse 6.

Jennie Page, formerly head of English Heritage, who then headed the Heritage Lottery Fund came down to Portsmouth to tour the historic dockyard and to discuss this basket of projects. She liked the excitement of Action Stations and the concept of a new landscape and processional walkway to *HMS Victory*, but considered that the Heritage Lottery Fund should fund the 'museum stuff' in the dockyard. However, she agreed to fund the Museum of Naval Armament, later the Explosion Museum at Priddy's Hard. The Millennium Project proposed, instead, that the museum projects in the dockyard should form part of the scheme and that it would oversee their delivery with the HLF contributing the funding. In the event the HLF considered each museum application on its merits.

In 1995 the Millennium Commission approved the Renaissance of Portsmouth Harbour project in principle and agreed to give up to £40m of lottery cash towards its cost of £90m. This was one of five or six big schemes that were amongst the first to gain approval.

Not all the PNBPT's proposals succeeded. The dominance of naval history distorts the reality that while the navy comes and goes, over the centuries the dockyard's civilian workforce built, maintained and repaired the navy's ships, reaching peak numbers in WWII when 25,000 civilians worked in the 'yard. To the disappointment of many, funding was not offered to create a new Museum of the Dockyard. As we mention in chapter 3, in 1982 skilled craftspeople had formed the Portsmouth Royal Dockyard Historical Society at the time of John Nott's review, preserving specialised tools, equipment and documents, but their collection had so far found no permanent home. Their publications record the dockyard's physical and social history. PNBPT had recognised and celebrated this key strand in the history of the harbour communities, creating the Dockyard Apprentice Exhibition in Boathouse 7 in 1994 and putting on several dockyard related temporary exhibitions as well as acquiring vessels that contribute to the civilian story: *Vic 56* and *HDL 49*. Brian Patterson, shipwright liner, founder member of the Portsmouth Royal Dockyard Historical Society and of the Fort Cumberland Guard, historian of the dockyard and author of publications about its history served as PNBPT's Curator and Keeper of Historic Boats for nineteen years. He was largely responsible for building up the trust's collection of historic boats. He is commemorated in Boathouse 4. The trust had commissioned studies on how best to establish a more comprehensive exhibition, building on the success of the Apprentice Exhibition from Lord Cultural Enterprises.

This was a major opportunity lost, but PNBPT did reasonably well to secure a new landscape including the new processional walkway from Victory Gate to *HMS Victory*, the creation of the Porter's Garden, new site orientation and interpretation, and, most important of all, the refurbishment of Boathouse 6 and the creation of Action Stations. This project also included the installation of two pairs of lock gates within the Scheduled canal connecting the Mast Pond to the harbour so that Boathouse 6 did not flood at high tide and to keep a minimum amount of water in the Pond at all states of tide.

The Mast Pond, the first body of inland water tourists encounter on their way into the 'yard, was constructed in 1665. It was used to store timbers to prevent them from seasoning unevenly or cracking before they were made into masts. Dutch prisoners of war who were paid one-and-a-half pence a day began work on it, but following complaints by unemployed townspeople, the Dutch were dispatched to Portchester Castle and local labour hired in their place. The Mast Pond is a Scheduled Monument and Listed Grade 1. It is linked to Portsmouth Harbour by a tunnel, which runs from its western end underneath Main Road and Boathouse 4. Tidal flow was originally partly controlled by a drop-gate at the western end of the tunnel. The new hydraulic lock gates installed by the Trust in 2000 and designed by Posford Duvivier now maintain a minimum level of water in the Mast Pond and prevent Boathouse 6 flooding at spring tides. It is used for miniboat rides and historical re-enactments and acts as a mirror to the surrounding boathouses.

Peter Goodship had always been conscious that while visitors could see and hear the modern navy as the heritage area is embedded within the active base, there was very little explanation of what it did day to day – in contrast to the story of its role in days gone by being told in the museum. So the idea was born of making a short James Bond style movie about the Royal Navy in action and also providing relevant activities for the public to enjoy. The idea was embraced by the Director of Public Relations and Recruitment (Navy), a Royal Marine. The film by BBC Resources for a fixed sum of £2.1m directed by Peter Hutchinson of Star Wars fame featured a Type 23 frigate *HMS Monarch* (made available by the navy for three weeks at no cost) in combat with gold bullion pirates in the Bahamas is shown to Action Stations visitors. There had been very little for children to enjoy apart from play equipment put in by the trust in Boathouse 7, so the ship's bridge simulator, Weapons Island, Merlin helicopter, Operations Room, Royal Marines island with the popular climbing wall enjoyed by adults too were installed in Boathouse 6.

Despite the financial difficulties caused by the trust's finances by the overspend on Boathouse 6 described in the next chapter, the realisation of the Millennium Scheme was another massive milestone in the development of the Historic Dockyard as we know it today. In the same way as the investment the trust made in 1994 it made a huge difference to the quality of the visitor experience and for the first time made the site a big attraction for families whose children loved Action Stations, the venue of choice for children's parties for many years.

New buildings and exemplary conversions

Architect tenants have long occupied the 1704 Porter's Lodge to the north of Victory Gate. Over the years the Property Trust's board has had expertise in finance and property – and design – particularly in the notable contributions of Sir William Whitfield and Sir Colin Stansfield Smith, RIBA Gold Medal winning chief architect of Hampshire County Council. The trust selects the best architects in the UK for its own projects, and the two knights were instrumental in the choice of Wilkinson Eyre to design the new Mary Rose Museum. In 2018 the Trustees included architect Michál Cohen, director of Walters and Cohen, the practice which transformed Boathouse 4, as well as David Butters and Robert Palmer, chartered surveyors, Roger Ching, retired city council chief financial officer who played a major role in development of the Spinnaker Tower and the Mary Rose Special School, Rear Admiral Neil Rankin CB CBE, Philip Marriott, Managing Director of Portsmouth Historic Dockyard, Mike Ridley who was consultant advisor to the trust on the condition of the Historic Dockyard estate when it was transferred from the MOD in 1966, Councillor Steve Pitt and Terry Hall who was a city councillor and tax manager.[16]

The Trust has an excellent record in commissioned new buildings, such as the Ticketing Office designed by John Winter & Associates and built by Heaton Waring Ltd. in 1993-94. Inside the Victory Gate, it is symbolic of the heritage area's change of use from functioning naval base to the major tourist attraction. Its galvanized steel frame with a glazed outer skin and corrugated metal roof was designed to echo the simple lines and functional design of other dockyard buildings. The roof's supporting beams in steel echo the underpinnings of the wooden roofs nearby. It has space for the large queues of visitors to buy tickets and gather information, as well as a coffee shop on the seaward side with a nose-to-nose view of *HMS Warrior* and the harbour.

The trust also asserted its design credentials over the dockyard stretch of the Millennium walk. Sir William Whitfield and Sir Colin Stansfield Smith were insistent that the final part of the Millennium walkway from Victory Gate to HMS Victory should not be expressed by the engraved concrete chain motif of the rest of the proposed walkway on both sides of the harbour, because over time it would deteriorate. The trust had to fight hard to persuade the city and borough councils that

Recruiting Office and Cell Bock

Cell Bock interior 2016

rather than the poor quality chain-engraved concrete paviours marking the walkway on the Portsmouth side of the harbour, they should install their own design. Gosport's paviours must have been to a higher specification: the chain motif and the names of their sponsors are still clearly readable, but the Portsmouth ones are now badly worn. They wittily end on the Hard, where a chain arises from the pavement, anchored around a bollard. Instead the historic dockyard the Millennium Walk is marked by neat metal studs, still intact!

Widening of the Victory Gate for lorry traffic in WWII removed the staircase and access to the upper floor of the two-storey brick Cell Block of 1882-1883, which forms the harbour-side end of the dockyard wall. It has twelve cells with steel doors to imprison people stealing materials from the dockyard as well as drunken sailors. At the harbour end it has a wonderfully ornate cast-iron lavatory. It had an adjoining police post and search area – important to dockyard security and the prevention of theft. The pile driving for the Warrior's berth cracked its structure across – as reported to the Hampshire Buildings Preservation Trust in July 1987. Repair and refurbishment of the exterior in 1993-4 at a cost of £100,000 by contractors T Coleborn & Sons included replacement of the parapet and chimneystacks.

In 2017-8 a new glass link containing a staircase joined the Cell Block to the adjoining Ticketing Office. The university of Portsmouth funded the work to provide hireable space for starter businesses. These were proving difficult to let, perhaps because working in a windowless cell is a minority taste, though some occupants clearly enjoy the quirky spaces.

Cell Block glazed link connected to Ticketing Office 2016

New infrastructure; new landscaping symbolises the change of use

PNBPT now manages Boathouse 4 and Action Stations – as well as a long list of other buildings listed on its website: the Porter's Garden, Mast Pond, landing station and 4 and 5 Docks in the Heritage Area and at Priddy's Hard listed. Inside the Heritage Area major investment in essential infrastructure has been necessary. The main drainage of the site is still the responsibility of the MOD, but the connections between the pipe on the roadway and individual buildings are the trust's responsibility. The trust installed its own gas main as part of the Millennium project between Storehouse 9 and No. 7 Boathouse and to Boathouse 6.

The trust functions in a similar way to a commercial property development agency - as well as a historic buildings trust specialising in the preservation and re-use of historic buildings and structures within the heritage area, modifying its ambience to suit the new tourist activity. The radical change of use from heavy industry to tourism was exemplified in the new landscaping designed by Camlin Lonsdale. The northernmost section of the *Victory* approach had a steep change in levels – from the layers of tarmac built up over the years. The new lower stepped stone landscaping on the heritage area side of the security fence replaced the traditional tarmac with stone steps in order to create level surfaces. Original paving materials were uncovered in the process. English Heritage insisted that they should remain on site, so they were incorporated into the new landscaping, which also exposed the cobble footings of No. 11 Storehouse.

But replacing the traditional tarmac and railway lines with expensive stone and gravel "bears as much relationship to the historic dockyard as a golf course does to real countryside" according to an insider. Evocative industrial archaeology, particularly at ground level was lost in the process. 27 miles of railway track once threaded though the dockyard, essential for moving goods around. These were destroyed by the MOD, who reused the rails as fences around drydocks. The wagons were moved to No.1 Basin where they are in a poor state, out of public view. They could do with conservation. Only a short stretch of railway line remains in front of Storehouse 10 on which railway trucks once were parked. The grooves between the rails have been filled in with tarmac, in case tourists trip... But inside Victory Gate exposed original stone cart tracks, which facilitated goods movement before the railways were left in place and in their original position. In 2018 the property trust had a presentation about the recycling of historic industrial archaeology. The navy has always reused hardware - including cannons to protect the corners of buildings – and the trust has carried on that tradition. When No. 5 Boathouse was restored the timber taken out was recycled.

Dockyard railway wagon

Also to celebrate the millennium the trust commissioned Camlin Lonsdale to design a new garden on the site of the former Porter's Garden as a place for dockyard personnel and visitors to enjoy. Trustee Sir William Whitfield wished to avoid an archaeological reconstruction of the original but instead to commission a modern design, since the new garden had a different purpose from the domestic character of the original. He also insisted that the hedge should be allowed to grow to a height to create a sense of enclosure. The wrought iron gates were specially commissioned from Peter Clutterbuck, a local artist blacksmith. The granite and iroko seats against the dockyard wall of 1711 echo the cross sections of *Mary Rose*, *Victory* and *Warrior*. Salisbury-based sculptor Roger Stephens made them using granite slabs salvaged from the old dockyard cart tracks. The garden was funded by a grant from the Onyx Environmental Trust under the landfill Tax Credit scheme. Planted and maintained by volunteers with species relating to the dockyard's history, it is a beautiful enclave, enhanced by two statues. The gilded William III as a Roman emperor is by Van Ost. William's wars against Louis XIV led to the expansion of the dockyard in the 1690s and he was also an enthusiastic gardener. Sculptor Kathleen Scott's 1915 brooding figure of her husband Captain Scott looks out towards the harbour. The Portsmouth Society awarded the garden its Best Landscaping plaque in 2003.

William III statue 2009

Impacts on the wider area

Since 1985 the Heritage area has developed into a major tourist attraction, joined by waterbus to Priddy's Hard and *HMS Alliance* submarine museum across the harbour. The Dockyard Heritage area is one of the most popular tourist attractions in the south of England. To support such a large expansion of tourist activity has also had a considerable visual and archaeological impact on the adjacent residential area, Portsea.

Despite its proximity to Portsmouth's major public transport interchange, many visitors to the city's naval attractions arrive by car. As already mentioned, part of HMS Vernon had been leased from the MOD for carparking. The 4.5-acre Whitbread brewery site just outside the dockyard wall in Queen Street came on the market in 1991. With £2.6m from its own financial resources, PNBPT Property bought it to clear for a car and coach park. The brewery and two listed buildings and archaeology below ground including the remains of two terraced Portsea streets that preceded it were destroyed to lay out the site. The main office on the corner of Admiralty Road and Queen Street was at first retained to let out and to provide an entrance/exit to the carpark, where visitors could purchase an award-winning Historic Dockyard guidebook in six languages. Income from parking fees and the guidebook provided a return on capital approaching 20%. The trust had a reputation for commissioning excellent architecture. When it decided to redevelop the site, the board's two architect-knights: Sir William Whitfield and Professor Sir Colin Stansfield Smith suggested that the trust should approach the RIBA and the Royal Fine Arts Commission with a view to organising an architectural competition for its design. Portsmouth planning authority and English Heritage formed part of the judging panel. Deane Clark persuaded David Richmond Associates to enter the competition, which they won with a high density scheme. The path to planning permission was smoothed by the city council's participation in judging the design competition. Builder Crest Nicholson adopted the winning design.

Admiralty Quarter roof garden above two storeys of carpark 2009

Constructed in 2006-8, the 'Admiralty Quarter' rises to nine storeys. It contains 566 apartments, towers and a 22-storey oval point block on Queen Street. Its centre encloses two layers of carpark, topped by a central private garden. The project was valued at £63m. Four holiday apartments are offered by the PNBPT for rent at £155-£200 a night. The Admiralty Quarter won the CABE Life Gold Standard, the Portsmouth Society Best New Building and the RTPI Regional Planning Award in 2009.

Adjacent civilian development may also dramatically impinge on naval heritage areas. Portsmouth Football Club's proposed football stadium filling in most of the space between the dockyard and the bus station provoked a strong reaction, not least in the PNBPT's press statement of 25 April 2007:

'As landlord and guardian of the land and buildings in the Historic Dockyard established in 1985, the Trust's role is to find new economic uses for former Naval buildings and docks so that they can be preserved in perpetuity. Most of the buildings adjacent to the proposed development are either Grade 1 or Grade 2* listed'.

Response by architect Mick Morris showing the huge scale of Portsmouth Football Club's Stadium at the Hard in relation to Portsmouth Harbour Station, Boathouse 4 – with the *Warrior* shown as a yacht!

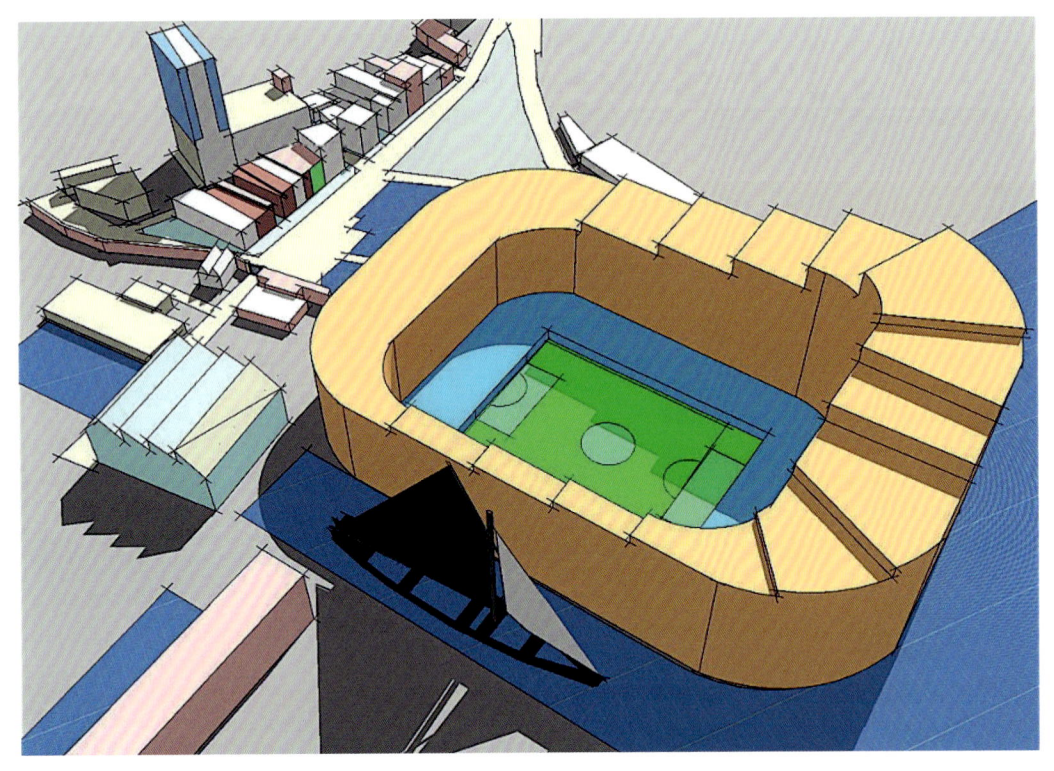

Proposed football stadium on the Hard 2007　　　　　　　　　　　　　　Mick Morris

It was difficult for the Trust to judge what impact such a dramatic and innovative scheme would have. The stadium's sheer scale and ambition would have been challenging for all involved in the Historic Dockyard. They needed time to digest the implications for the trust's own plans, particularly in relation to the re-use of Boathouse 4 and neighbouring buildings which had been about to be marketed, via a scheme prepared by London-based architect Eric Parry. The trust said that it would have to consult closely with its colleague ship and museum trusts, as well as their commercial tenants operating in the Historic Dockyard. The pioneering *HMS Warrior 1860* which currently dominated the Hard waterfront would be in a much less prominent position if the stadium were to be built. The Hard, an ancient public landing place used by fishermen and small boat owners for millennia would be destroyed. In the event, the navy's objections including those of the Queen's Harbourmaster stopped the proposal. In 2008 a new stadium site was proposed on made ground bordering Horsea Island, home of the navy's diving school but this was not built either. In the end, the club refurbished its existing ground in Fratton, partly financed by development of a two-storey Tesco store on its carpark.

Boundaries and Dockyard Heritage at Risk

The boundaries of the Heritage Area continue to be controversial, because they exclude large areas of important historic structures still within the operational naval base - which are not as well protected or maintained as those inside it, or as available for public access - except on the once annual Navy Days, which no longer happen. Their deteriorating condition is a continuing concern, because of course the MOD has other priorities – and, as it says, it is not funded to look after buildings it has no use for.
In 2018 preparations for dockside facilities for the second supercarrier the *HMS Prince of Wales* by the Kings Stairs were absorbing major defence funding, but the reduced size of the navy means that large areas are no longer needed for national defence. In 2018 it was announced that the eight Type 26 frigates specialised in anti-submarine warfare due to be delivered from the mid-2020s will be based in Devonport, along with many of the navy's submarine fleet, rather than Portsmouth. Five Type 31 Frigates would be based in Portsmouth, along with the supercarriers.[15]

No. 6 Dock being surveyed by Sarah Roberts and Martin Crichley

While there have been examples of reuse of dockyard buildings such as No. 2 Ship Shop, the former Steam Factory of 1847-9 alongside No. 2 Basin converted in the 1990s into offices, and the main building of the Iron Foundry (1/140) into offices for BAE Systems, heavy office floor loadings severely had damaged 2-8 the Parade of 1715-19. Where families of six once lived, sixty office workers occupied the houses. Although major works repaired the Parade in 2019-2020, there continues to be serious concern about the decay or 'vandalism by neglect' other key dockyard buildings outside the heritage area, including Block Mills, No. 6 Dock and the Naval Academy – all no longer of use to the navy. Historic England lists both 5 & 6 Docks as at risk because of their poor condition.

An ingenious idea to reverse the MOD's progressive abandonment of unneeded structures is designer Sarah Robert's 2018 proposal for reinventing No. 6 Dock as a multifunctional innovative hub for joint use with the PNBPT, in order to redress the absence of the future dimension currently missing (or outdated) from the dockyard's interpretation. Apart from the navy's ships coming and going, the only place where current service experience is visible is in Action Stations in Boathouse 6 where as we've said proxy experiences are offered to children and adults, partly as a recruitment tool. Despite the abolition of the Crown exemption in 2006 which meant that the MOD now has to apply to local authorities for planning permission, it can still construct new buildings 'in time of national emergency' without it. In addition, perhaps the MOD prefers newbuild and is not of a mind to reuse existing structures, especially if they are listed or scheduled, because they require complex and time-consuming planning applications and funding. Roberts suggests that if these regulations were somewhat relaxed for the MOD, there might be an incentive to consider reuse, although the VAT on repair as opposed to zero on newbuild would still be a deterrent. No. 6 Dock first built in 1698 and later reconstructed, is currently in a very poor state, its northern wall bulging inwards and the caisson leaking badly. Repairing it and reusing it in a creative way by inserting conference facilities into it as she proposes would serve the navy as a place for planning the future.[16] There are precedents for infilling a dock with new facilities, for example in Copenhagen.

Important protected buildings and structures outside the heritage area remained empty for many years; unoccupied, their condition deteriorated significantly, in some cases close to the point of no return. The Naval Dockyards Society (NDS) has taken a close longterm interest in their condition. Its primary aim is to stimulate research and exchanges of information about naval dockyards. In correspondence with the Second Sea Lord & Deputy Chief of Naval Staff, the Portsmouth Naval Base Commander and Portsmouth city council planning committee in 2017 it called for the MOD to carry out immediate remedial work on 2-8 The Parade, the former Royal Naval Academy, the east wing of the Iron and Brass Foundry, No. 6 Dock and No. 25 Store, to prevent further deterioration. Several of these have been on the Historic England At Risk Register since 2008. "Many of the most important buildings of the Portsmouth Naval Base are falling apart – from lack of or bad maintenance. They include the World's first steam powered factory of mass production, Marc Brunel's 1802 Block Mills, listed Grade 1, the highest category of historic building conservation. Also Grade 1 is No 6 dock, which was built in 1700. The Boulton & Watt engine house is collapsing. The Academy, which has been empty six years, needs a new use. The porches of Long Row/The Parade are falling down." [17] At the Annual General Meeting of the Hampshire Buildings Preservation Trust (HBPT) on 4 November 2016, members welcomed the good news that the Portsmouth Society had made a Restoration Award for the splendid reinvention of Boathouse 4 in Portsmouth Dockyard for its original purpose. Also welcomed was the conversion of the Cell Block for starter businesses.

"But of great concern was the bad news about key decaying buildings, notably the Grade I Block Mills and the nearby historic docks. The Block Mills buildings, historic dry docks and land were to be transferred in 2016 to the Portsmouth Naval Base Property Trust but a Ministerial answer in Parliament indicated that delay is due to "...security and safety aspects of operating two carriers in the vicinity of what could potentially become a tourist zone." The location was no closer to where the new aircraft carriers are to be berthed than *HMS Victory* and *Mary Rose* which already attracted three quarters of a million visitors a year and so the Minister's explanation was hard to understand.

The Hampshire Buildings Preservation Trust supported the NDS. It recognised that "the MOD's main purpose is to defend the nation rather than to preserve our nation's heritage, but Government should live up to its responsibilities as owners of important protected buildings'. [18] 'What is really needed is a positive discussion about how the MOD might enable new, creative new uses to be found for them. This would do much more to ensure their survival - than just keeping them 'wind and weatherproof' - but empty.' [19]

A more hopeful scenario for the Old Naval Academy of 1729-32, unused and empty since 2007 was on the cards in 2019. The current very large and potentially valuable wardroom in Queen Street was on the MOD's 2018 disposal list, and there was a possibility that the Academy might be reopened for naval officers – although upgrading the 'cabins' in the hollow square behind the main block with its distinctive cupola would be very expensive - and the MOD's assessment was that the Academy's restoration costs would be higher than the return on the sale of the wardroom might prevent this happening. A further blow in autumn 2018 was that asbestos was discovered in the roof and it was airborne through the whole building. It was likely to cost approaching a million pounds just to clear it before any restoration or conversion could begin.

The Second Sea Lord Vice Admiral Jonathan Woodcock OBE replied to the NDS on 27 February 2017 that 'we are looking at how best to optimise our estate and where possible to bring historic buildings back into use, the Old Naval

Naval Academy from Admiralty Quarter point block 2013

Academy is an example where I expect to see future investment.' By autumn 2018 the navy was beginning to acknowledge that it had dropped the ball in the way that it had let its historic assets decline, hopefully with support at senior level. Simple measures like taking up the carpet and airing the spaces would have helped to delay deterioration years earlier.

After two years the NDS wrote again to the Second Sea Lord and Deputy Chief of Naval Staff and Portsmouth Naval Base Commander on 3 March 2019, asking 'for new and appropriate uses to be found for these buildings, "so that operational budgets could also finance conservation." In their view this was the best way to secure their conservation and future. Refurbishing the former Royal Naval Academy and The Parade would be logical projects to meet naval accommodation needs and go some way to restoring the Royal Navy's ethical and official reputation, their refurbishment meeting naval ethos, conservation and operational requirements.

Concerned about the longterm deterioration of key dockyard buildings, in 2019 on behalf of the Naval Dockyards Society Dr. Paul Brown requested detailed information under the Freedom of Information Act about the expenditure on surveys, maintenance and conservation of 2-8 The Parade (Buildings 1/125-131) and on No. 5 Dock and No. Basin over a ten-year period. The reply from the Navy Command Secretariat showed that survey and inspections and maintenance on all three amounted to £120,120 each.[20]

The Society obtained further details of surveys and repairs by Freedom of Information requests. These showed that after years of neglect the MOD was now responding to the calls for action and is carrying out repairs and reactivation of four buildings – the Old Naval Academy (1729), The Parade (1717) which they intend to use as housing, the Old Iron and Brass Foundry (1854) and Storehouse No 25 (1782). Dr. Paul Brown, commented, "We applaud the work that the Ministry of Defence is now undertaking to safeguard the future of these structures. They are part of not only Portsmouth's heritage but also the nation's. Within their walls lie hundreds of years of history." The very positive reply in April 2019 From Navy Command HQ Policy Secretariat did not address all the points raised by the NDS, but they 'clearly present a visible marker of the RN's intent to care for the heritage infrastructure in its charge' – as their letter stated. 'The Royal Navy (RN) acknowledges that we are in a period of particular challenge in balancing available funding investment between the operational infrastructure, improving the outcomes for the ships and our people, and preservation of our heritage.' As funding allowed, investment was being used to stabilise structures and arrest decline. Active studies were going on into 'development options, including further targeted disposals, to bring the historic buildings into sustainable use'. The navy was 'actively investing, as funding allows, in stabilising structures and arresting decline'.[21]

In 2018 the upper floors of the Rodney Bock had been refurbished and reoccupied in the Personnel Centre. Stabilisation work on the structure of 25 Store and Long Row (The Parade) 'will continue into 2020.' The MOD would be installing a tidal cofferdam around No. 6 Dock gates to protect them and the dock from threatened tidal impact, thereby enabling repairs to the masonry and eventually to replacement of the gates. Having stabilised the decline in the Old Naval Academy, a programme of asbestos removal would make its 'subsequent development and regeneration both quicker and safer'. The navy also intended to 'bring historic buildings across Watering Island back into full use as the new regional Royal Naval Reserve Headquarters… The Naval Base at Portsmouth is at the beginning of a once-in-a-century period of regeneration. The arrival of *HMS Prince of Wales* at the end of this year, with her berth only 100 metres from *HMS Victory*, herself under long term restoration, will be a clear symbol both the RN's pride in its heritage, and a real optimism for the long-term future use of this historic site.' The chair of the NDS commented that these measures were beneficial not only in refurbishing these structures but ensuring an operational use, apart from No 6

Dock, which was admittedly challenging.

However, a much more co-ordinated approach was clearly needed. The NDS has called several times for a collaboration of stakeholders in Portsmouth Dockyard to set up a MOD Conservation Group or a Heritage Partnership Agreement to accomplish what had not been achieved by the MOD in the last ten years, starting with a site meeting with interested parties. This was recommended in an MOD leaflet of 2010, but to date this potentially positive proposal, which could work towards an effective Conservation Management Plan has elicited no response.

Extension of the Heritage Area

PNBPT has twice come close to taking on unused buildings outside the area to restore and reuse them – in Project Prime, and again in 2013 when the MOD was again in negotiation with the Property Trust to extend the Heritage Area. In 2015 a Portsmouth Historic Dockyard Strategic Overview was produced by the site's partners about the potential for the transfer of land and buildings to enlarge the area of the Historic Dockyard. From 1986 to 2014 the partners had together invested £136,500,000 while visitor numbers had risen dramatically. [10] It was clear, however, that even at the current level of visitors the £7.5m generated each year in ticket income for the Heritage Area partners was not enough to sustain the site. It was only financial resources from outside visitor income which allowed the Historic Dockyard to survive. The new development masterplan aimed to create a site-wide integrated visitor experience and improve and extend the infrastructure to support it.

In 2016 the PNBPT's long gestated proposal to secure the release and future preservation of protected scheduled and listed buildings within the remaining part of the former Georgian and Victorian yard was being negotiated. A letter from the DIO Secretariat to Pam Moore of Hampshire Buildings Preservation Trust who had expressed concern about five dockyard buildings on the Heritage at Risk List dated 18 April said that "The Ministry of Defence is conscious that the long term sustainability of these buildings can only be secured if they are put back into use", and that "No. 6 Dock was "part of a collection of docks, basins and sea-walls that include Marc Brunel's Block Mills and are both a Scheduled Ancient Monument and Grade I Listed. The dock in land-locked and therefore cannot be used operationally, even if its condition allowed. The dock, along with Block Mills, Docks no. 1 to 5 and No. 1 Basin are part of ongoing negotiations aimed at transferring them to a new owner."

The boundary of the Historic Dockyard was to be extended northwards. This would have enabled the PNBPT to restore Block Mills and the oldest dry docks in the 'Yard, as it would be eligible for grants not available to a government ministry. A new lease was fully drafted and the level of the endowment - £13.5m - was also agreed. But in 2018 this long awaited extension was ultimately cancelled by the MOD - ostensibly because of a notional threat to the new aircraft carrier. A different reason is used to justify leaving the Naval Academy empty and decaying - that it's within the 'arc of fire' from the harbour... but so are all those tourists in the Historic Dockyard! The MOD has gradually accepted their responsibility for maintaining its historic defence estate in good repair, but as these debacles show, where the structures are in the active base, it's a different story.

However, extensions of civilian activity into buildings beyond the security fence guarding the active base have recently been allowed. In March 2018 South Office Block was made available for use as offices by the NMRN. This necessitated a new pedestrian security point. The western end of the Ropery and Store 12 were transferred to the Naval Base Property Trust, and they also negotiated to take a lease on the Chain Test House, currently used by the Mary Rose for conservation work in tanks and air dryers; it would however remain within the operational naval base, not open to public access. Uncertainty remains about the timescale when the other docks in No. 1 Basin would come across to the PNBPT.

The NMRN's SeaMore Project

The SeaMore Project in 2014 aimed to combine the NMRN's archives and nearly 2,500,000 artefacts housed in 30 separate stores in 14 buildings on six sites inaccessible or hidden from view collections. They were to be brought together into a Centre for Discovery in Storehouse 12 and in Boathouse 6 where the Royal Marines Museum was to be relocated, together forming Britain's most comprehensive collection of Royal Naval heritage. SeaMore was regarded by the museum as a key stepping stone in the site's strategic masterplan, leading it toward it towards financial sustainability. In 2915 Portsmouth University business school produced an Economic Impact Assessment of SeaMore, while a Social Impact Assessment was produced by Deborah Hodson, Head of Engagement and Learning at the NMRN. According to the Project Outline: SeaMore: Sharing the Newest National Collection it was hoped to attract 7-8% more visitors to the Heritage Area: 200,000 to the Royal Marines Museum in Boathouse 6 and 120,000 to the Centre of Discovery and a million a year in total.[11] The Heritage Lottery Fund awarded £13.85m to support the project. The development phase was to cost £548,726 and the delivery phase £17,021,237. NMRN asked for a 79% grant. The matching funding requirement was £3.7m and the museum had already secured £2.1m. The disposal of the former Royal Marines Museum building was expected to raise a further £1-£1.5m, but this potential was not included in the financial assessment.

A mezzanine floor was to be inserted into the high spaces of the ground floor of Storehouse 12 with provision for hand-on investigation of artefacts, a display of the collections rising vertically through the building, a working area and research on the first floor. Staff and volunteers would work on the collections and interact with visitors. Outline plans and images of how the Royal Marines' 350-year history would be displayed in Boathouse 6 were shown, accessed via a new walkway right across the Mast Pond. The auditorium was to be reconfigured to enable the Royal Marines Band to play there.[11]

However, the senior museum's second application for £12.9m for the SeaMore project from the Heritage Lottery Fund was refused in September 2018, despite their Round 1 approval of the earlier offer of £433,500 in 2016 to kick-start the plan.[12] Fewer people are now gambling on the lottery, which reduced the funds available to pay out to heritage projects. The move was delayed for at least two years. The NMRN sought to raise the funds themselves, £10m being needed, of which some £3m had already been raised and £2m would come from the sale of the Eastney site. A revised £3.9m lottery bid was submitted to the Heritage Lottery Fund in November 2019, but, as was said at the time, if this funding did not materialise the museum's plans would be set back by four or five years – or scrapped completely. The Royal Marines' 355-year history would then no longer be celebrated in Portsmouth at all.[14] This became even more likely in April 2020 when the National Lottery Heritage Fund rejected the second bid. In Autumn 2020 the Royal Marines Museum site at Eastney was still for sale.

The HLF regional director said that the fund had already invested £55m into Portsmouth Historic Dockyard. It is true that by 2016 a number of local defence sites, historic ships and boats and buildings have received lottery funding, including:

- Priddy's Hard initial funding of a £1.9m bid 2015
- HMS Warrior Preservation Trust £2.6m 2015
- RML 497 Fairmile Coastal Forces Motor Gunboat £90,600 2015
- LCT 7074 Landing Craft £916,149 2014
- HMS M33 £1.79m 2014
- Boathouse 4 boatbuilding skills project £240,800 2012; £3.75m 2013
- Mary Rose £21m 2012; £3m 2016
- Storehouse 10 £110,00 2010; £1.4m

Professor Dominic Tweddle, director of the NMRN was bitterly disappointed by the decision not to fund the Royal Marines Museum's move into the dockyard, "particularly at a time when the excellent work of our armed forces in supporting civilian services is never more apparent." In February 2020 he said that a new Marines museum might not be based in Portsmouth, but instead at one of the museum's sites in the north of the UK. As Portsmouth was one of the Royal Marines' home bases – the others being Plymouth and Chatham, it did not seem appropriate for their museum to be moved to the north of England, according to Ann Coats, chair of the Naval Dockyard Society.

In the meantime, the Royal Marines archive and library and all the collections were transferred to the National Museum of the Royal Navy in the dockyard in summer 2019. According to Matthew Sheldon the plan to refurbish the NMRN's library in Storehouse 12 between September 2019 and January 2020 was to go ahead at a cost of £1.2m. Reopening in Easter 2020 would have allowed the Royal Marines Museum archive and others to be accommodated in a single building, offering better integrated access to a wider range of collections.[13] In March 2020 in response to the Covid19 outbreak the museum accelerated the decanting from the old Royal Marines Museum to get the whole collection safely into the NMRN before any further lockdowns. "As a result the last objects were moved on 20 March and stored safely in the Nelson Gallery. Appropriately enough one of the last objects to come off the truck was the old garrison bell from Eastney."[14] But the NMRN museum itself then closed in response to Covid-19, not reopening until late August.

The bid for World Heritage designation

Proposed boundaries of Portsmouth Harbour World Heritage Site

In 2004 David Michelmore, an expert on world heritage sites asked Celia Clark why Portsmouth Harbour was not a World Heritage site while they were both working on preparation of a bid to inscribe Xingcheng, a Ming walled city in Liaoning Province in China onto the WH List. Proposed bids supported by the local authority have first to be accepted by the cultural ministry onto countries' Tentative List, from which one a year can be put forward by national governments for acceptance to UNSECO's World Heritage Committee. Because each place has to be unique, Michelmore proposed that the focus should be on water, a "cultural seascape": Portsmouth Harbour and Spithead and the surrounding defence installations and the associated technical innovation over centuries, a focus which had not been proposed before. A steering group chaired by Peter Goodship of the PNBPT was set up, with Portsmouth City conservation officer John Pike, Chris Dobbs of the Mary Rose Trust and others, with Celia Clark working on the bid.

The team considered that Portsmouth Harbour and Spithead might fit categories (ii) for technology and architecture, (iv) or (v) as well as being a cultural landscape. These specified that a WHS should:

- (ii) exhibit an important interchange of human values, over a span of time or within a cultural area of the world, on developments in architecture or technology, monumental arts, town-planning or landscape design;

- (iv) be an outstanding example of a type of building, architectural or technological ensemble or landscape which illustrates (a) significant stage(s) in human history;

- (v) be an outstanding example of a traditional human settlement, land-use, or sea- use which is representative of a culture (or cultures), or human interaction with the environment especially when it has become vulnerable under the impact of irreversible change." There is also a category for moveable objects eg historic ships.

The bid by the historic dockyard trust in Chatham including the submarine *HMS Ocelot* and a destroyer had been on the UK's Tentative List since 1987 without making any progress towards being nominated by the UK to UNESCO since then. Because western Europe was heavily over-represented on the WH list, a combined bid with Chatham was suggested, but in a 2009 letter to Celia Clark, the Chair of their Steering Group declared that "Chatham and Portsmouth are making very separate cases for their undeniably unique qualities, and that distinct nominations are viable and desirable... It therefore seems sensible to work up our own individual bids at this stage." The Chatham bid has still not progressed beyond the Tentative List.

The Portsmouth bid's case was that Portsmouth Harbour and Spithead would be the first to include objects and underwater heritage. According to the Nara document: a WHS should also demonstrate authenticity of materials, design and setting - of worldwide significance. The bid was ambitious because support had to be secured from no less than five local authorities: Portsmouth, Fareham, Gosport, Hampshire County and the Isle of Wight, as well as from the Ministry of Defence, Queen's Harbour Master, Defence Estates, the PNBPT, the Royal Navy Museum and the English Tourist Board. The naval base commander agreed to chair the Management Committee – to ensure that national defence had priority. Many local and voluntary organisations including the Portsmouth Society and its equivalents also supported the bid.

The PNBPT hosted a conference of over 100 potential stakeholders in Action Stations in 2006 to launch the bid, which, among other things, would have established Portsmouth Harbour and Spithead as a destination in its own right. Admiral Band, Peter Goodship, David Michelmore and Celia Clark were the speakers. But despite assurances that UNESCO respected the concept of history being a continuum and thus would accept change to accommodate the changing needs of the modern Royal Navy, these assurances were not accepted by DIO or the City Council politically, and thus the bid failed. After six years' work the leader of Portsmouth City Council had cold feet. In October 2006 Cllr. Mike Hancock MP said in a letter to Celia Clark that the city's reputation was already high and he was "at a loss to seriously understand what significant benefits we would gain from it." Cllr. Vernon Jackson feared that designation would inhibit development. He cancelled the essential support for the bid, which has to have a lead local authority. In contrast to this response, in 2018 the minister for Wales supported the UK's nomination of Gwynedd's slate landscape for world heritage status, which she said "acts as a catalyst to investment and tourism."[22]

With the Grade 1 Block Mills at its centre, the Portsmouth Harbour proposal still has validity. It would have unlocked funds to restore unused buildings and docks and also shine an international focus on the dockyard's importance in world historical terms. In the context of World Heritage status, however, English Heritage published an assessment of all former Royal Dockyards throughout the world for the DCMS to determine which should be nominated for World Heritage status. In the UK Chatham Dockyard was nominated as the dockyard best equipped to represent the story of Royal dockyards given its excellent state of preservation and the fact that Portsmouth and Plymouth had been badly damaged during WWII as well as being subject to insensitive postwar development, for example, the Ropery in Portsmouth which had been compartmentalised vertically, destroying its original configuration.

However this assessment ignored the wider significance of Portsmouth Harbour as well as the need to tell the local dockyard story, hence the efforts the PNBPT made to present it for the benefit of the visiting public as well as the local population of the City many of whose families have had such strong connections with "the Yard".

Even if, in Peter Goodship's view it is most unlikely, given the enormous shift of importance throughout the world towards the environment and climate change that a bid of this nature would ever succeed, that should not prevent the Trust from continuing to pursue the narrower goal of establishing Portsmouth Harbour as a destination in its own right - a harbour which deserves to be recognised as every bit as iconic as Sydney Harbour and have international recognition; not just because of the important role it has played in the history of the Royal Navy, but for the centuries it has played before that in the defence of the nation since Roman times. Emperor Vespasian left the Harbour from Portchester to take part in the siege of Jerusalem. Portchester has the most intact walls of any Roman fort in Western Europe.

Portsmouth Harbour has the potential to change fundamentally the public perception of Portsmouth as a city and Gosport as a borough: something recognised by those in political control of both councils at the time of the Millennium. It is sad that subsequent administrations failed to capitalise on the £80m investment made at that time in an attempt to unite both sides of the Harbour.

Portsmouth Dockyard Heritage Area as a national and international exemplar

The achievement of the trust is as much financial as it is an exemplary restorer of historic defence buildings. In response to their requests for funding the Heritage Lottery Fund has invested £55m in the Historic Dockyard – notably the very large grant to build the new Mary Rose museum and £3.75m in 2013 to support traditional boatbuilding in Boathouse 4. Its work in Priddy's Hard is described in the chapter on ordnance yards.

In 2013 Portsmouth Naval Base covered 300 acres of land, with 62 acres of basins, 17 dry docks and locks, 900 buildings and 3 miles of waterfront, according to the magisterial study of the dockyard in the twentieth century by the Naval Dockyards Society which was commissioned by English Heritage and published in 2015. Portsmouth City Council's Conservation Area No. 22 covers the southeastern area enclosed by the original dockyard wall and Victoria Road to the West. In terms of operations within the Dockyard, the Royal Navy stands on an equal footing with the Defence Infrastructure Organisation: DIO (formerly Defence Estates), as they are both MOD. However as the DIO deal with the whole defence estate it has set up a specific sub division to deal with Portsmouth Dockyard - the Intelligent Consumer Group. The ICG and the RN both instruct BAE, which looks after the buildings. BAE converted the old Iron Foundry into their offices in two phases from 2010, designed by Robert Benn Associates.

As a microcosm of the complex challenges and problems presented by such a high concentration of defence heritage sites compressed into a small compass, the harbour's successful development for defence heritage related tourism is both a microcosm of achievement and also an international beacon of experience. It demonstrates how the search for new civilian futures on formerly defence-dominated areas can be successfully met.

An indication of how rich this local experience is and how much it is recognised internationally is that the chief executive of the Portsmouth Naval Base Property Trust is often asked to share the trust's considerable expertise elsewhere. Avoiding the trap of imagining that establishing a few museums will lead to meaningful visitor income without an iconic attraction such as *HMS Victory* or the *Mary Rose* is a key to success. Peter Goodship and Celia Clark were invited to advise the Plymouth Development Corporation (1993-8) on how to regenerate the historic buildings in Royal William Yard. In 2008 they were also delegates to a seminar on the future of historic dockyards in Helsinki organised by the governing body of Suomenlinna to share experience with representatives from the governments of Norway, Finland and Sweden and participants from St. Petersburg and Rochefort. Peter also advised those responsible for the care of the historic buildings within the former dockyard in Pembroke. Delegations from several European cities visited the Historic Dockyard and the PNBPT networked with and hosted visits from other Royal Naval Dockyards further afield including Gibraltar, Antigua and Bermuda. The EU-sponsored networking programmes for former naval bases to share experience and these or direct invitations financed his advisory visits to Den Helder Holland, Karlskrona in Sweden, Rochefort and Toulon in France, Rostok in East Germany, Sevastopol in the Ukraine (now annexed by Russia) and the former naval dockyard and hospital in Mahon, Mallorca.

Sources

Admiral GA Ballard l980 *The Black Battlefleet*. Annapolis, Maryland: Naval Institute Press. ISBN 0-87021-924-3

All Party Parliamenttary Group for Hospitality and Tourism 2020 *Pathways to Recovery A report from the All Party Parliamentary Group for Hospitality and Tourism* 12 June

Michael Asteris, David Clark, Shabbar Jaffry 2016 The economic effect of military facility contraction: A Naval case Study. p. 14 https://doi.org/10.1080/10242694.2015.1122281

Michael Bateman & Ray Riley eds. 1987 *The Geography of Defence* Croom Helm Ltd. reprinted 2016 by Routledge Library Editions: Human Geography Volume

Naval Dockyards Society 2015 *20th Century Naval Dockyards: Devonport and Portsmouth Charactisation Report* for The Historical Monuments Commission for England p.131

Mark Brown 2018 'Welsh slate landscapes up for Unesco world heritage status' *The Guardian* 24 October p.6

Paul Brown 2018 *The Portsmouth Dockyard Story* The History Press Stroud

Celia Clark 1994 The Future of Dockyard Heritage Conservation, Community and Economic Aspects of the Transition of Naval and Military Sites to Civilian Use in Four Dockyard Towns: Chatham, Portsmouth, Plymouth and Venezia MSc. Dissertation Oxford Brookes University

Celia Clark 2000 *Vintage Ports or Deserted Dockyards: differing futures for naval heritage across Europe*. University of the West of England Centre for Environment and Planning Working Paper 57 July 2000

Celia Clark 2002 "WHITE HOLES": DECISION-MAKING IN DISPOSAL OF MINISTRY OF DEFENCE HERITAGE SITES PhD thesis University of Portsmouth

Deane Clark 1956-7 *Portsmouth Dockyard and Portsea Up to 1850* Thesis for Diploma in Architecture Portsmouth College of Art

Jonathan Coad 1989 *The Royal Dockyards 1690-1850 Architecture and Engineering Works of the Sailing Navy* Royal Commission on the Historical Monuments of England Scholar Press Aldershot

Dartington Amenity Research Trust entitled *Defence of the Realm an interpretive strategy for Portsmouth and the surrounding region* City of Portsmouth Southern Tourist Board DART publication No. 59.

DLO and Defence Estates n.d. *Project Prime Portsmouth Regeneration and Investment in the Maritime Estate* Insertions in square brackets by P Goodship

http://www.hmswarrior.org/history; Winton 1987, p. 76 Goodship

Peter Goodship 2018 Interview with Celia Clark 6 November 2018

HANSARD 1803–2005 → 1970s → 1975 → June 1975 → 11 June1975 → Commons Sitting → ORDERS OF THE DAY EUROPEAN ARCHITECTURAL HERITAGE YEAR HC Deb 11 June 1975 vol. 893 cc621-32

Andrew Lambert 2010 *HMS Warrior 1860: Victoria's Ironclad Deterrent* (2nd revised and expanded ed.). Annapolis, Maryland: Naval Institute Press. ISBN 978-1-59114-382-6.

David Lloyd 1972 *Buildings of Portsmouth and Its Environs* City of Portsmouth

Archie Malley 'The Obituary Notice that Caused a Rumpus' *Dockyards* Naval Dockyards Society

https://www.ons.gov.uk/businessindustryandtrade/tourismindustry accessed 7 April 2017).

Campbell McMurray 1994 *Interview with Celia Clark for Oxford Brookes MSc. Historic Conservation Students*

Campbell McMurray 199? *Forward to Royal Naval Museum Corporate Plan*

McMurray Campbell 1999 *Letter to C Clark 16 March 1999*

McVicar L1999 'Princesstoopenrebornnavymuseum'*TheNews*PortsmouthMarch18https://navaldockyardssociety. files. wordpress.com/2016/02/part3-twentieth-century-naval-dockyard devonport-portsmouth-characterisation.pdf

Dennis Miles 2019 Oral History Recording by Celia Clark 15 March

Naval Dockyards Society 2017 Twentieth Century Naval Dockyards Devonport and Portsmouth: Characterisation Report: https://navaldockyardssociety.files.wordpress.com/2016/02/part3-twentieth-century-naval- dockyards- devonport-portsmouth-characterisation.pdf

Nikolaus Pevsner & David Lloyd 1967 *Buildings of England Hampshire and the Isle of Wight* Penguin Books Harmondsworth

Charles O'Brien, Baily B, Nicolaus Pevsner, David Lloyd 2018 *Hampshire; South* The Buildings of England Yale University Press The News April 8 2017 'New deal secures the future of HMS Warrior' p.5

Portsmouth Naval Base Property Trust 2006 *Portsmouth Naval Base Proprty Trust 20 Year Review 1986-2006*

Project Prime Portsmouth Regeneration and Investment in the Maritime Estate HM Naval Base Portsmouth, DLO and Defence Estates n.d. "Restoration — Homecoming". HMS Warrior Preservation Trust. Retrieved 28 March 2013. Winton 1987, p. 84

John Winton 1987 *Warrior: The First and The Last*. Liskeard, Cornwall: Maritime Books. ISBN 0-907771-34-3.

"HMS Warrior". National Historic Ships UK. Retrieved 9 April 2013.

"HMS Warrior 1860". Portsmouth Historic Dockyard. Retrieved 9 April 2013.*Statistics newsletter* (PDF) (61),

Portsmouth City Council, Winter 2000–2001, p. 16, retrieved 28 January 2010

"About Us". HMS Warrior Preservation Trust. Retrieved 22 May 2013.

"Wedding Brochure" (PDF). HMS Warrior Preservation Trust. Retrieved 28 March 2013. Lambert 2010, pp. 203– 04

Ray Riley 2005 *Maritime City Portsmouth 1945-2005* Sutton Publishing Limited Stroud

Sarah Roberts 2018 *What Happens When Buildings Die? Military Buildings Case Study: No.6 Dock, HMNB Portsmouth* MSc. Dissertation University of Portsmouth

Richard Trist n.d. *Project Visit Portsmouth Naval Heritage Britain Summary insights of overseas holiday visitors who take day trips to and from the UK's cities and towns* July 2018 bdrc continental

https://www.historicdockyard.co.uk/tickets-and-offers?gclid=EAIaIQobChMI_p6c38Sp6wIVQe7tCh0lywUnEAAYASAAEgJ9K_D_BwE accessed 19 August 2020

The Art Newspaper 2020 'First government grants for UK art spaces announced…' 12 October https://www.theartnewspaper.com/news/museums-and-galleries-finally-receive-lifeline-government-cultural-recovery-grants

The News 2020 '£698,000 for 'under threat' naval trust. Grant a lifeline to preserve 'unique' historical boats and buildings October 9 p.7

https://www.archdaily.com/440541/danish-national-maritime-museum-big

References

1. Celia Clark 2000 *Vintage Ports or Deserted Dockyards: differing futures for naval heritage across Europe* University of the West of England Working Paper 57

2. Molly Blackall 'How many tourists are there – and where do they go? *The Guardian* 1 July 2019 pp.10-11

3. M. Asteris, D. Clark, S. Jaffry The economic effect of military facility contraction: A Naval case study, 2016, p. 14; https://www.nmrn.org.uk/news-events/nmrn-blog/best-ever-year-national-museum-royal-navy accessed 9 April 2020; Tom Cotterill 2020 'City visitors 'key to very suvival' of naval museum' *The News* 24 July 2020 p.4

4. Impact of the Covid-19 pandemic on tourism. Business News – BBC News retrieved 25 May 2020; Tourism and Covid-19 www.unwto.org retrieved 21 April 2020

5. Dominic Tweddle COVID-19 Update 9 April 2020;

6. Peter Goodship Email to Celia Clark 20 April 2020

7. Richard Trist, former Portsmouth City Council Chief Executive 1978-1993 *Portsmouth Naval Heritage Project* n.d.

8. Peter Goodship June 1994 *Interview with Celia Clark for Oxford Brookes MSc. Historic Conservation Students;* Celia Clark 1994 "*The Future of Dockyard Heritage: Conservation, Community and Economic Aspects of the Transition of Naval and Military Sites to Civilian use in four former dockyard towns: Chatham, Portsmouth, Plymouth and Venezia*" Oxford University/Oxford Brookes University MSc. Dissertation Historic Conservation p. 99

9. Peter Goodship Interview with Celia Clark 6 November 2018

10. https://www.nmrn.org.uk/sites/default/files/3_project_outline_master_document_final_to_print.pdf accessed 29 July 2020

11. 'Big plans Lincoln Clarke at Portsmouth's Royal Dockyard' by Michael Powell 19 February 2015 http://www.portsmouth.co.uk/news/business/local-business/portsmouth-s-historic-dockyard-aims-to-get-a-million-visitors-a-year-1-4412179

12. *The News* Portsmouth 2020 'Bitter disappointment as lottery fund snubs marines museum bid' April 15. p. 4

13. Matthew Sheldon Email to Paul Brown Naval Dockyards Society 8 July 2019

14. NMRN Internal website 20 March 2020
20th-century Naval Dockyards characterisation report". *Historic England*. Historic England. Retrieved 5 February 2017

15. 'Defence secretary dashes hopes as home of frigates is revealed' *The News* October 9 2016 p. 11

16. Sarah Roberts 2018 WHAT HAPPENS WHEN BUILDINGS DIE? *Focus: Military Buildings Case Study: No.6 Dock, HMNB Portsmouth* MSc. Dissertation 2018 University of Portsmouth

17. Ann Coats, Naval Dockyards Society NDS briefing document January 2020;

18. Hampshire Buildings Preservation Trust Press Release 17 March 2017

19. Celia Clark *The News* Portsmouth 30 March 20i7 p. 24

20. Navy Command Secretariat – FOI Section Email to Paul Brown 26 June 2019 10 year 1FMP Expenditure

21. Navy Command HQ Policy Secretariat, HMS Excellent Portsmouth Letter of 24 April 2019 to Dr. Paul Brown

22. Welsh slate landscapes up for Unesco world heritage status' Mark Brown T*he Guardian* 24 October 2018 p. 6

Chapter 7
Portsmouth Historic Dockyard: buildings, ships, boats – and one that got away…

Over the last fifty years restoration and conversion of historic buildings in the Historic Dockyard and the addition of tourist infrastructure has been a continuous process, which has also deservedly won awards for good design and excellent restoration. This chapter describes in detail changes to the storehouses, boathouses and drydocks and the arrival and restoration of the historic ships. It draws on the extensive PNBPT website and other publications as well as personal recollections. Within the Historic Dockyard area, as well as smaller vessels, are the four historic ships: *HMS Victory*; the new £35m museum housing the *Mary Rose* which was raised from the seabed at Spithead in 1982; *HMS Warrior* (1860) which arrived from restoration in Hartlepool in 1987; and *HMS M.33 Monitor* (1915), restored by Hampshire County Council and opened to the public in 2015. Land-based activities in restored buildings include the National Museum of the Royal Navy, Action Stations and the arthouse cinema in Boathouse 6 and the School of Traditional Boatbuilding in Boathouse 4.

The Boathouses

Boathouse 4 West elevation 2014

The utilitarian design of **Boathouse 4** is a considerable and some would say incongruous contrast to its Georgian and Victorian neighbours. Its lightweight construction of 1939-40 is an excellent example of a 20th century boathouse, the only one constructed in a home dockyard during the rapid rearmament build-up to WWII. Four engineers, led by AE Scott, designed this 'cathedral to small boat building'. Scott was responsible for the elevations, while the structure was designed by KF Buchanan, JDW Ball and J Angell. It was constructed around four parallel electric Vaughan cranes, designed to move boats and machinery across the width of the building. The steel structure was erected by Dorman Long Ltd., which, seven years earlier had built Sydney Harbour Bridge. The boathouse's orientation and skylights was designed to light workers on double time in the summer. As this drawing shows, it was designed to extend as far as Victory Gate, but building the twelve bays stopped at the outbreak of WWII. In May 1940 a temporary corrugated steel cladding was added to the southern elevation to enable the new boathouse to be partially operational until construction of the second section was completed. This is still in place today. The whole structure is supported on concrete piles. With its saw-tooth roof and horizontal Modern Movement glazing it is a distinctive and rare twentieth century survivor facing Portsmouth Harbour in the naval base.

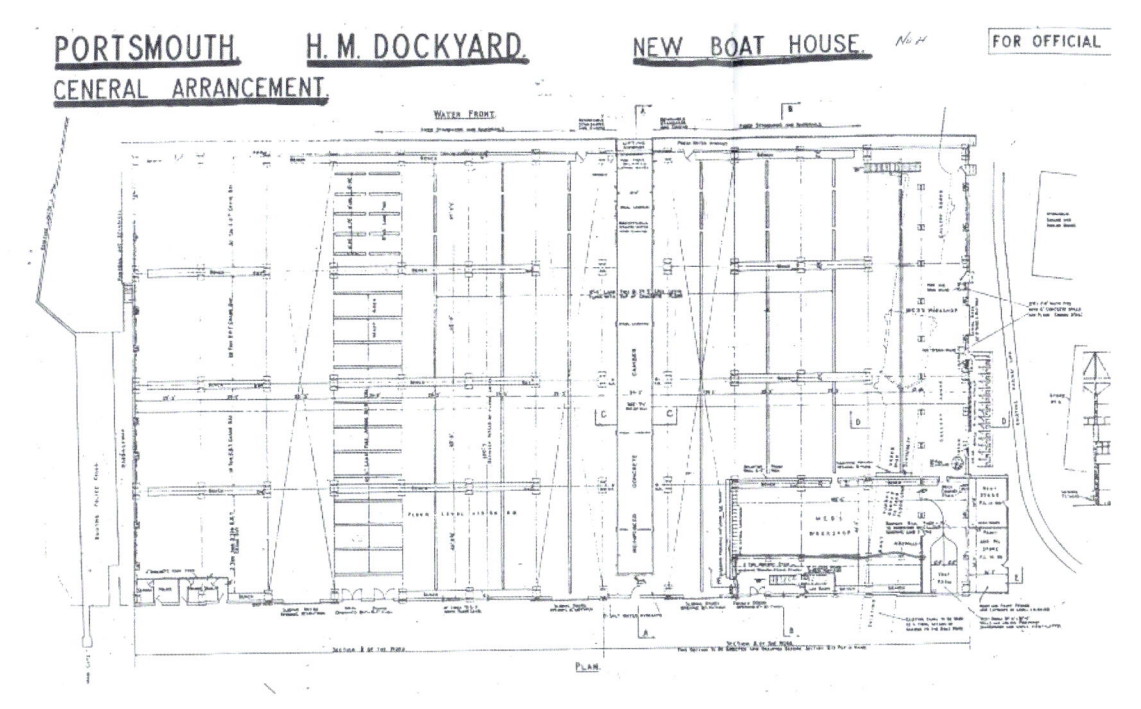

Dockyard wall/Victory Gate Not constructed section Existing structure

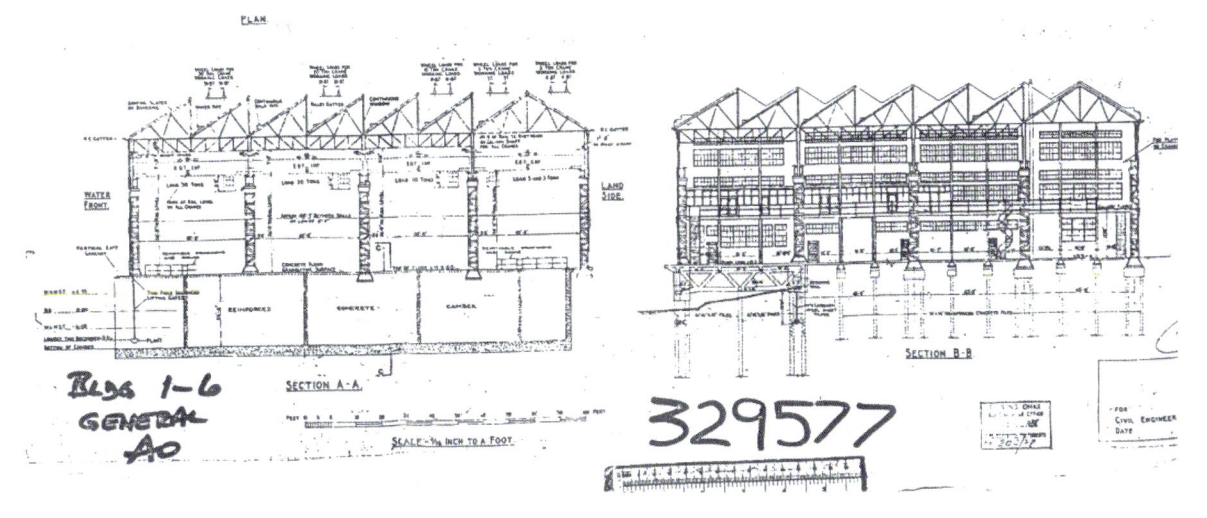

Boathouse 4 Construction plan

Boathouse 4 Section and elevation

Boathouse 4 South elevation showing temporary wall 2005

135

It continued in operational use by the MOD until 1996 when it was transferred into the care of the PNBPT. At that time the trust also saw it as an offence to its more gracious neighbours. The local planning authority published a formal Development Brief proposing its demolition and replacement by a building of equal mass and volume. English Heritage was also happy to see the building removed and turned down attempts to spotlist it. Several schemes were suggested, the earliest of which was developed speculatively by architect John Winter who had designed the Visitor Centre. He proposed a hotel/conference centre on the site designed to attract more income generating visitors.

But there was an opposing view that the boathouse was a very good example of 1930's austerity architecture and very much part of the history of the naval base. When it was under threat of demolition the secretary of the Portsmouth Royal Dockyard Historical Society and the Portsmouth Society, Celia Clark identified the designers responsible for its original construction from the drawings she found in the Unicorn Consultancy's office in Murray's Lane in the dockyard.[1] This was to support her campaign to get the building listed as of architectural and historic interest and thereby prevent its demolition. At her request, SAVE Britain's Heritage took up the cause. They commissioned architect Huw Thomas to design a scheme for reusing the boathouse for the exhibition of small boats. The internal dock with a gate opening directly into the harbour was to be used to moor historic boats. As part of SAVE's campaign, for four months in 1990 the historic *Victory* sail with its shot holes and burns was unrolled and put on display in the boathouse with the help of the Royal Naval Museum.[2] There were discussions about moving the *Mary Rose* hull complete with its container, inside which she was sprayed continuously with preservative, into a concrete cocoon fitted inside the boathouse – a model showing how this might be done commissioned by Hampshire county architect, one of the PNBPT trustees, is in the trust's archives. But the ship was too tight a fit and moving its 400 tonne weight inside its hefty cradle from the dock to the boathouse would have destroyed the underlying archaeology. Although the Secretary of State refused to list it, the Property Trust began to change its view. The building's large spaces were also useful to the trust for storage and for staging exhibitions and events including Christmas fairs. Architect trustees Sir Colin Stansfield Smith and Ed Jones advocated its retention, so the trust began discussions with the Osborne Property Group who were working on the conversion of Oxford prison to a hotel. The catalyst to keeping it was Brian Patterson, the charismatic leader of the group of dedicated skilled volunteer craftsmen. So it survived. He proposed that the care and maintenance of the trust's boat collection should be professionalised and enhanced by establishing a traditional boatbuilding college and the creation of an exhibition telling the story of small craft in the navy's operations. Discussions about recreating the generations of shipwrights' skills began in 2010 with a Portsmouth outpost of the International College of Traditional Boatbuilding (ICTB) in Lowestoft. After three attempts, in 2014 a Heritage Lottery Fund bid to finance a complex mix of new uses succeeded. As well as housing the trust's collection of small boats, teaching spaces were provided for Highbury Further Education College and a harbour facing restaurant inserted above them.

The easternmost 10-ton Morris crane, a later replacement, did the heavy lifting for the new construction. The total cost was £5.5m; the HLF gave £3.75m and the EU Regional Growth Fund £479,000. With these funds, architect Michàl Cohen of Walters & Cohen brilliantly reimagined No. 4 in 2015-16.[3] The lofty workshops for the ICTB and Solent Marine Academy below the restaurant are light, airy and functional. A masterstroke was cutting a window into the temporary wall to reveal the boatbuilders at work – lit in the evenings and in winter. The existing mezzanine was extended and reached by a new staircase and lift. It fits round three sides of the tall volume of space. Here the trust's small boat collection is displayed - including boats hung over the entrance and suspended on cradles over the busy workshop space below. The first floor also has a classroom, meeting room, toilets, and popular children's facilities including a climbing mast. A film projection room behind an angled wall is unobtrusively tucked into the northeast corner.

Seeing and hearing wood being shaped, smelling wooodshavings and tarry rope and seeing varnish and paint being applied to historic boats and the construction of new ones by busy workpeople is a wonderful reversion to the boathouse's original purpose. The kitchen, bar and popular restaurant complete the harbour-ward side, with offices and workshops underneath. A glazed acoustic screen shields the restaurant from the main workspace. The unique view down into the workshop and harbourwards to the ferries, commercial and naval shipping make dining here a very special experience. The trust's in-house conservation team made the restaurant's rope partitions and woodwork details; timber from Beaulieu in the New Forest, where timber framing still continues, was used for the cabinets. It is the only building in the property trust's estate with direct access to the harbour: via the dock at the southern end, which houses the restored naval steam pinnace - described in detail in relation to the Mast Pond. On the northern end two locks are actually inside the building. These control the flow of seawater into the Mast Pond and also provide boat access to it through the existing historic tunnel. As SAVE anticipated historic vessels are moored in the lock, with direct access to the sea. Notably, Boathouse 4 is also the only free attraction in the heritage area, with special activities for children.

Boathouse 4 Interior restored and converted 2016

Three of the sets of steel columns supporting the roof are painted a vibrant yellow. Detailing throughout is plain, robust and industrial – entirely appropriate for what is still mainly an industrial building. Materials are recycled as far as possible - as in the surrounding active naval base. Heating the 2,700m$_2$ space was not feasible, so only the enclosed areas are heated, while timber stores season in the main space under cover. The Portsmouth Society awarded No. 4 its Best Reuse award in 2016.
[4] It also received awards from RIBA South, the British Construction Industry, the Civic Trust and a commemorative visit from the Princess Royal.[5]

The colleges offer participation in education at many levels – family, school and college and adult. From 2013 the ICTB and Solent Marine Academy trained a new generation of students in the techniques of traditional boatbuilding, passing on skills vital to the ongoing preservation of Britain's historic wooden vessels. Reversion to the original use was not only a gratifying return to the boathouse's original function, but it also addressed the continuing loss of traditional dockyard trade skills. The workshops make an excellent and unmatched contribution to technical education locally and in the sub-region.

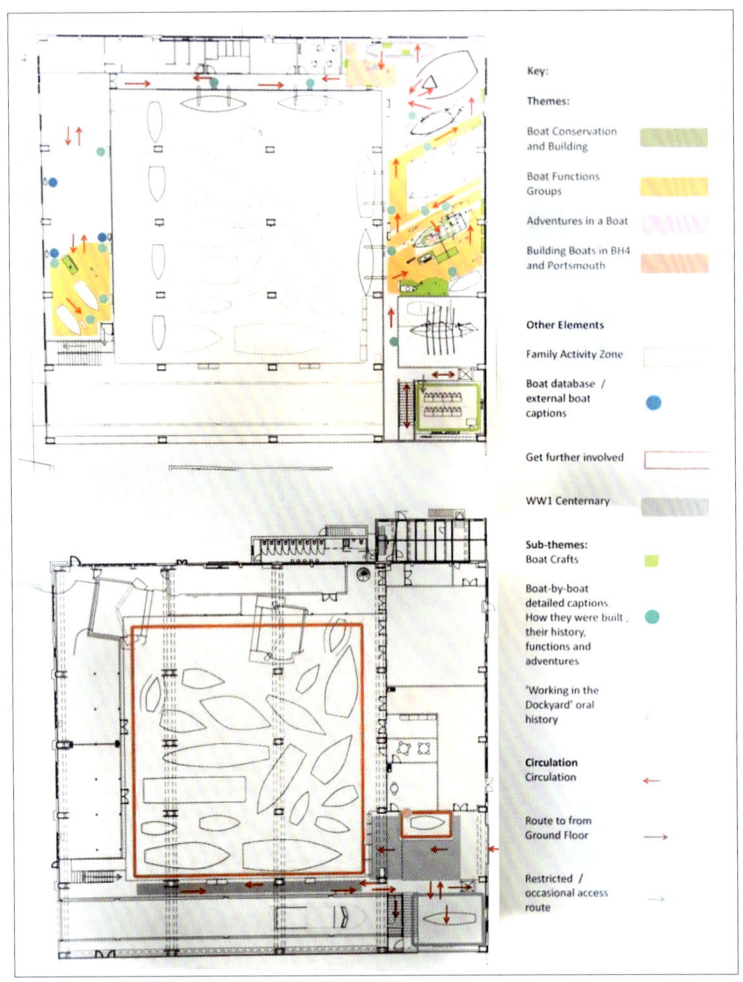

Boathouse 4 Conversion plans Portsmouth Naval Base Property Trust 2014

Boathouse 4 Axonometric RealStudios 2014 Portsmouth Naval Base Property Trust

Traditional boats built in workshop lit by new opening in Boathouse 4 October 2016

Boathouse 4 decorated for Catherine Clark's wedding to Noam Fast. September 2008

Sadly, in October 2018 the ICBT wound down; charity operations ceased in August 2017.[6] An Asset Purchase Agreement was signed, transferring the business to the Portsmouth Naval Base Property Trust, which continues to pursue the original aims: to train students to maintain traditional wooden vessels, collaborate with other charities and provide bursaries to students who lack means, as well as offering community participation programmes.

Boathouse 6

The three-storey buff brick classical façade of Boathouse 6 facing the Mast Pond gives no clue to the pioneering structure inside. Captain James Beatson of the Royal Engineers designed the massive cast iron columnnar structure in 1845-8. It is one of the first examples of a brick building constructed around a metal frame, with massive iron girders and underslung trusses to support boat storage on the upper floors. Its cast iron beams are inscribed with their load-bearing capacity.

In the 1980s Celia Clark interviewed the dockyard's last block maker Tom Birch there. He was working on a versatile boring machine first displayed in the Great Exhibition in 1851 - still in use in the late twentieth century. The first floor's huge space was empty apart from the two of them. The wooden floor was worn in patterns around other vanished machines that once were there by generations of dockyardmen's boots. She asked him what would happen if a fire broke out. He walked her to the southwestern corner and pointed to a notice. It read: "In the case of fire, shout Fire…" The rear section was badly damaged by bombing in WW11: the upper level remained open to the elements for many years. Celia Clark remembers Brian Patterson showing her, outside, how lignum vitae, the hardest known wood, was steeped there before it was machined into ships' fittings. Peter Goodship remembered a forest of trees growing in the ruins.

Southern elevation showing new glass access staircase 2011

As part of the Harbour Renaissance project, the refurbishment of Boathouse 6 was conditional on the trust finding matching funding. They raised £2m and the city council guaranteed a further loan. Sir Richard McCormack was appointed as architect and Ove Arup as structural and mechanical engineer. There were debates about how to access the top floor which was rented out to the university and later as offices. Rather than insert new access through the internal iron structure, an external circular brown glass walled staircase was added to the southern elevation, clearly marking the modern intervention into the robust Victorian structure.

Auditorium supports inserted through the original structure 2014

Boathouse 6 View through original structure to café

The war damage did enable the dramatic insertion of a new auditorium/cinema supported on slanting purple-lit columns. But when trial pits and bore holes were undertaken, significant granite obstructions in the ground were missed where its piled foundations were to go. This caused massive delay and a significant redesign. The 12-month building contract took 24 months and the contractor claimed an additional £5.5m. The landscaping works also hit many ground obstructions, which required direction from English Heritage as to the solution. This also resulted in additional expense. The total cost of restoration and conversion of Boathouse 6 was about £16m, and with the landscaping the trust spent about £20m. The Millennium Commission awarded the trust extra grant but it was insufficient to meet the total cost. The whole of the Historic Dockyard became a building site for over 12 months, but the transformation was significant and led to a considerable increase in visitors. The Trust also secured a 15% share in total site income for Action Stations. However the financial impact of the overrun was enormous and placed the trust in a difficult financial position. The city council loan guarantee was called in and the trust was obliged to sell Horse Sands Fort.

The auditorium has a triple function. The Ministry of Defence uses it for conferences with foreign navies and contractors. As mentioned in the last chapter a film showing the navy in action is shown to Action Stations visitors. And since 2006 No. 6 Cinema has screened both popular general release films and art house movies. It stimulated the beginnings of an evening economy – with early supper offered in Boathouse 4, but unfortunately there was not enough take-up to continue. The ground and first floors of the Boathouse are currently occupied by Action Stations about the modern navy based on a Type 23 frigate. It puts visitors inside a modern warship – in the operations room, on the bridge, in the gun turret, in the cockpit of a helicopter and on the training ground of the Royal Marines. Children and adults enjoy the climbing walls and assault courses. The place heaves with excited youngsters and their parents in the school holidays. As mentioned in the previous chapter the NMRN's plans to move the Royal Marines Museum into the space, delayed by failure of its lottery bid in 2018, would entail removal of Action Stations as well as installation of a lift to the auditorium - which would be altered to enable the band of the Royal Marines to play there, with implications for the future of No. 6 cinema, mentioned in the next chapter.

Boathouse 6 Auditorium 2014

The architects who converted No. 6 won many prizes for their dramatic interventions inside the building. This exhilarating combination of magnificent nineteenth century structure and elegant modern adaptation won an award from English Heritage, the Portsmouth Society's Best Restoration Award in 2002 and a Civic Trust Award in 2003.

Boathouses 5, 6 and 7 right to left and the Mast Pond 2018

Boathouses 5 and 7 were built on timber piles over the Mast Pond where timber was seasoned in seawater as mast houses but later used for the construction, repair and storage of small boats. Before No. 7 was converted Celia Clark interviewed craftsman Peter Tolfree, who was shaping the spars for restoration of the *Victory*. "Treat me like a dim apprentice" worked well! He explained how many thou of an inch had to be taken off the spar as he shaped it. This recording and others made by dockyard craftsmen are in the Portsmouth History Centre. Terry Wren, Conservation Officer of

Portsmouth City Council supervised the rapid and exemplary conversion of Boathouse 7 in 1993-4 to provide a new restaurant, tell the story of the civilian dockyard in a new exhibition, a playship for young children, toilets and baby changing facilities and a shop. When the boathouse plans were drawn on a computer, it was shown to lean significantly towards the water. The computer could not cope with a crooked building - but luckily the lean was exaggerated in the machine drawing and was not in reality damaging to the stability of the structure. Celia Clark was pleased to see that the floor still has the paint splatters from its industrial use. The work cost £2.6m and the builder was Heaton Waring Ltd. Several government tourism and regeneration grants, the EU's £500,000 and close on £1m from the private sector financed it.

Boathouse 7 before conversion

Boathouse 7 with Boathouse 6 in background 2016

Appropriately No. 7 houses the Dockyard Apprentice display. A visitor can clock in as an apprentice and learn about the many different trades involved in building and repair of the navy's ships. The exhibition was designed by the Portsmouth Royal Dockyard Historical Trust, which maintains it and adds new topics. A member was sometimes on hand to talk to visitors. Their specialist tools and equipment and the apprentice pieces made as part of their training vividly evoke the many skills of generations of dockyard craftsmen. As well as some of the original cast iron machinery from Block Mills, the first interactive museum display in the 'yard was designed to explain the function of pulley blocks once manufactured by the thousand in Block Mills. Lifting the same weight directly and then via pulleys explains how they lighten the load. Boathouse 7 now houses the dockyard's main restaurant and the Nauticalia maritime-themed gift shop. In 1995 the conversion won the Portsmouth Society's Restoration Award, Portsmouth City Council's Good Planning & Conservation Award - and a Good Loo of the Year Award!

No. 5 Boathouse PNBPT Small Boat Collection

Boathouse 5 is also a good example of the timber-framed buildings once common in naval dockyards. For many years it housed the extraordinary finds retrieved by divers led by Dr. Margaret Rule from the wreck of the *Mary Rose*, before the new museum was built which unified ship and contents. As a researcher on the future of historic dockyards, Celia Clark remembers introducing Dr. Margaret Rule to Lady Frances Clarke, Life President of the Venice in Peril Fund, who was seeking advice about preserving the historic Arsenale in Venice, a case study in Celia's masters degree. She left the two grand dames sitting together in the boathouse: they had a lot to talk about! Since the *Mary Rose* collection moved to its permanent home, No. 7 has been used for temporary exhibitions such as displays of the PNBPT's small boat collection and the commemoration of the centenary of the Battle of Jutland. The annexe to the south was a sail loft, and there were originally brick-lined sawpits to the west, where seasoned timber for masts was shaped.

Storehouse interior

Storehouses 9, 10 and 11
The Storehouses and the National Museum of the Royal Navy

It was only luck, timing, an American gift and naval and political intervention that prevented the total disappearance of Portsmouth's eighteenth century storehouses, now so cherished as the spine of the Naval Base Heritage Area: in red brick, three storeyed, thirteen bays wide in a simple, dignified classical style with pediments over the centre bays and round-arched openings in dressed stone.[8]

144

They are now listed Grade 1. The prototype was the Present Use Store, now No. 11 Store in 1763. The Middle Store or No.10 followed in 1776 and South Store or No. 9 in 1782.[9]

They were built to house a huge variety of raw materials and goods needed storage in dockyards: "Hemp from the Baltic, iron from Sweden and the Forest of Dean, timber from the royal forests and private estates, from Scandinavia and North America, glass from the Midlands and lead from the Mendips". Equipment processed in the yard: sails, masts, hemp, paint, rigging and timber, iron and many other items all needed to be stored near to the quays, for fitting out or for replenishing stocks, or as reserves of essential items in the event of war. No. 11 served as a template for the later two storehouses, although the internal fittings in No. 11 were of higher quality. Stores were taken in from the eastern side facing the road and taken out from the western side facing the Camber dock, where they were loaded onto small boats for onward transmission to ships in the harbour.[10]

In the late 1960s the General Manager of the Dockyard issued an order for the storehouses' demolition. The MOD had no further naval use for them, because they were not in keeping with modern storehouse design; and there was a plan to push a road to the harbour edge across the site. Councillor Freddie Emery-Wallis who was then chairman of Portsmouth Corporation Development and Estates Committee and Lord Mayor in 1968 intervened. Although Government-owned buildings were still exempt from normal planning procedures until 2006, the proposed demolition was picked up as a result of the courtesy consultation of local planning authorities imposed on government departments. Emery-Wallis contacted the Port Admiral and set up links between the navy and the city, which provided a stay of execution - until the appointment of Admiral Sir Terence Frewin, Commander in Chief, Naval Home Command who was instrumental in ensuring their preservation. He was succeeded by Admiral Sir Horace Law who was equally supportive of the development of the museum. American collector Lily Lambert McCarthy donated her collection of Nelson memorabilia to the Royal Naval Museum, which was moved into Storehouse 11.

But soon afterwards, the Ministry of Defence planned to move the museum's contents to Greenwich. Emery-Wallis found himself up against Dr. Basil Greenhill who was the Director of the National Maritime Museum, Greenwich from 1967 to 1983. Freddie said he argued for the collection to stay in Portsmouth, with the support of the City Council. Another battle was fought, and the collection stayed where it was, becoming a key plank in the whole revival of the Historic Dockyard as one of the most successful maritime museum complexes in the world, now including *HMS Victory*, *Mary Rose*, *HMS Warrior* and *M33*. Early on he arranged a small grant to the chair of fundraising of the Mary Rose Trust, as he saw the potential of its raising from the seabed for the city. The first plans were for the ship to be housed in a new museum on the beach at Eastney, but he persuaded the Joint Planning Team in the dockyard to fund her home in a drydock off Basin No, 1 inside the dockyard.

Some use had been made of the upper floors for storage, but according to the PSA report, most were evacuated in 1976 because of structural and fire precaution problems. Notices are still posted that the maximum floor loading is 56lb and in places only 40lb per square foot - a bit low for most of us! Climbing the wide timber staircases as the storemen once did is a special privilege for anyone interested in historic buildings. There were still some museum stores upstairs in No. 10 but most of the upper floors remained empty for many years.

Once saved, their subsequent reuse has only been achieved by a combination of factors: inventiveness of fundraising, innovative design, allied to increasingly rigorous standards of conservation. The state (the MOD) retained ownership but leased the site to the PNBPT, with physical control of change through the agency of English Heritage. The three storehouses are amongst the earliest dockyard buildings in Portsmouth to be turned to heritage uses, but modern stores designed for keeping objects need considerable physical intervention to provide for human occupation and public access: lifts, toilets, heating, lighting, ventilation, museum standard environmental controls, fire escapes…

because of their deep plan and lack of services.[8]

The storehouses have been much more sensitively handled than the ropewalk at right angles to them, which was 'eviscerated' in the 1950s.[10] Designers experienced in conservation require greater ingenuity than newbuild requires to insert new uses. The increasing rigour of English Heritage over what changes may be permitted was related to the rise in popular support for conservation from the 1960s, resulting in new legislation and the volume of buildings and areas preserved.[11] In 1972-3 the museum converted the ground floors of the three storehouses to museum and visitor use. It expanded along the ground floors of Nos. 10 and 9, converting them into galleries, a café and a shop. As already mentioned in 1972-3 the landward side of each storehouse was opened up, creating an arcade by removal of the windows to provide a covered approach to the *Victory*.

No. 9 Storehouse

This was the last of the Georgian storehouses to be built during the dockyard expansion from 1760. Contractors Templar and Parlby constructed it in 1782. A blue plaque to Thomas Telford, famous Civil Engineer (1757-1834) who worked in Portsmouth dockyard from 1784 to 1786 was unveiled in 2007 in the centre of the ground floor to commemorate the 250th anniversary of his birth. He designed Short Row in 1787 as houses for officers of the dockyard.

In the late 1980s planning permission was given for the conversion of the upper floors of No. 9 Storehouse as a hotel, though considerable concern was expressed by conservationists at the degree of new insertions and services required, particularly on the top floor, which has large diagonal roofing timbers dividing it into bays. These would have had to be encased and enclosed to make individual hotel rooms. The scheme did not go ahead. Planning restraints such as parking and goods supply have still not been entirely resolved for the Heritage Area. These are particularly difficult to arrange in what is a long narrow sliver of 'heritage' in a working fleet maintenance and repair base.

Between 2003 and 2009 the eastern a Southsea couple converted half of the ground floor of Storehouse 9 into the Artists Harbour art gallery with a printmaker/publisher. Their aim was to take over the entire building including the defunct pub in the western half, and turn that into an art café, but the Portsmouth Naval Base Property Trust refused this for fear it would compete with their own canteen-style restaurant opposite.

"The art gallery and printmakers' businesses were successful on a cultural and social level, winning the support of many of the best artists both in Portsmouth and across Hampshire and West Sussex, plus many local art lovers. It became a favourite place for Royal Navy personnel to buy leaving gifts for their comrades. Images published by Artists Harbour became favourites for sale in the shops of the Royal Naval Museum, the Mary Rose Museum, St. Thomas's Portsmouth Anglican Cathedral and Chichester Cathedral as prints, posters and greetings cards – some reproduced from antique maritime artworks but many newly created by local artists."

"Additionally, Artists Harbour provided picture framing and art conservation services to those preserving the Royal Navy's own huge collection of historic artworks and artefacts, including such gems as an illuminated calligraphic certificate commending Royal Navy heroism at the Battle of Sirte in World War II, when an overwhelming enemy force was defeated by a handful of smaller British warships… signed by no less than the then Prime Minister, Winston Churchill. The historic certificate, previously nailed to a piece of fruit box wood, was framed because it was due to be loaned to the British embassy in Washington."

"Other innovations included Gallery Gourmets – fine dining at night in the gallery space. Artists Harbour also employed a full-time artist in residence, noted Hampshire artist Julian Bond, who produced a number of fine images of the dockyard including an outstanding painting – *HMS Victory in Dry Dock* – and also ran screen printing workshops for numerous schools' talented pupils and even for Hampshire art teachers."

"However, as the overwhelming percentage of Historic Dockyard visitors were children on school trips, without the added turnover and attraction provided by the proposed art café the business proved not to be financially viable long term at that location… so when the Property Trust proposed to increase the rent in 2007 the gallery did not want to remain there – even at the old rent."[12] There have been plans to offer space to small businesses, but the costs of conversion were perhaps considered too high. In 2020 a textile workshop was installed on the ground floor.

No. 10 Storehouse

Built in 1776, also by contractors Templar & Parlby, the rusting of the underfloor iron plating and a continual spalling of the soft facing bricks of the solid walls into brick dust also evident in the other storehouses are some of the serious problems facing their conservation and reuse. The walls were once limewashed, which may no longer be permissible because of a health hazard. There were various suppositions as to the cause of the spalling: salt leaching through the walls from seaspray, internal salt damage from the wartime inundation to douse fires, or the removal of the timber lining..

No. 10's clocktower, cupola and southern part of the roof were burnt out and collapsed in the March 1941 blitz, when 35,000 incendiaries were dropped on Portsmouth. The first radar sets to be installed in ships were stored inside. The main water main into the city was cut but the dockyard had its own firepumps, which flooded tons of seawater into the storehouse. The iron sheathing formed a seal that prevented it from escaping, so that the walls were in danger of collapse. Huge holes were hastily hacked in the floor, which saved the southern half of the building and the radar sets. The roof was covered with temporary corrugated sheeting and the north half was rebuilt with a gabled mansard with steep tiled faces on a light steel frame. In 1956 Deane Clark, an architectural student, remarked in his thesis about the dockyard's historic buildings that the drawings of the cupola and clocktower on No. 10 Storehouse were in the Dockyard Engineer's Department in the North Office Block facing the Victory, "but it is unlikely that this dominating feature will ever be replaced on the building".

The greater value placed by heritage use on care and enhancement of surviving historic structures is such that in 1992 the Naval Base Property Trust took the bold decision to restore the bomb damage of No. 10, as the centrepiece of the three storehouses. Terry Wren, Portsmouth City Council historic buildings architect and Deane Clark's successor, designed this re-creation. A photograph and a surviving drawing differed considerably from each other, so the photograph was taken as a basis from which Wren used classical precedent to create a Doric cupola. Though charred, the surviving strength of the fir beams was sufficient for several to be retained, in some cases lined with steel plates. The new supporting structure for the cupola is honestly and directly expressed in a marriage of old materials and new technology. The new timbers are joined with galvanised steel bolts in compression, which is quicker and cheaper than the traditional wooden trennails, which are wedged up from inside, which might have allowed the structure to flex. There have been some leaks, which have been tackled with plastic fillers. There were protracted negotiations with English Heritage over the gauge and runs of lead on the new roof, which the property trust wished to reinstate as near the original as possible. Terry Wren installed a Gillett and Johnson clock of 1878 from Clifton Grammar School and the same Croydon firm that made it, which is still in business, refitted it. In accordance with naval tradition, only the "Ting Tang", not the full chime of the five bells ring out. The original weathervane rescued from the war rubble was mounted on top of the new clocktower. Local builder John Lay & Co. carried out the work at a cost of £750,000.

This important project has enhanced the look of the whole area, by land and sea. This achievement won the Property Trust a Europa Nostra prize and a Best Restoration Award from the Portsmouth Society in 1993.

Dockyard craftsmen Portsmouth Royal Dockyard
Portsmouth Royal Dockyard Historical Trust collection

The trust's base in Storehouse 10

Portsmouth Royal Dockyard Historical Society and Trust

Appropriately, the Portsmouth Royal Dockyard Historical Trust (PRDHT) now occupies a bay on the first floor of Storehouse 10, a welcome haven after many moves around the yard when important items were lost or stolen. It was established in 1982 following the announcement of the run-down of Portsmouth Dockyard under the terms of the 1981 Defence Review. For two years Celia Clark was the society's first secretary, Ray Riley the first chairman and the president was Brian Patterson. Most of its members are former Dockyard employees. In the context of the rundown, they made intensive efforts to collect representative examples of machinery, tools, drawings, photographs, scientific testing apparatus and other important examples of the material culture of the dockyard before they were lost to posterity and the skills and trades they represented had disappeared. As already mentioned, the Property Trust had long term plans to develop a new Museum of the Dockyard in No. 9 – or in Block Mills - telling the story of the of the men and women who spent much of their working lives in the dockyard through peace and war, to complement the naval story told by the Royal Naval Museum and the historic ships. This was to be based on the collection of tools, artefacts, equipment and documents collected by the PRDHT. Sadly, the Property Trust did not win the hoped-for Millennium grant, because the Commission said they did not wish to fund another museum. The PRDHT has equivalents in Plymouth and Chatham. The Chatham group's collection which was once housed long term in the Lead Rolling Mill but now it is incorporated into the Chatham trust's interpretation of the 'yard in the Anchor Storehouse. In 1994, to formalise the Portsmouth group's status particularly in relation to other trusts in the historic dockyard and to facilitate access to grants and other sources of development funding the Society was converted into a charitable trust (PRDHT), with representatives from the property trust, city council, Portsmouth University and Fleet Services Ltd. The Support Group ensures the safe storage and maintenance of artefacts and arranges displays for public interest and education. The trust's impressive collection of tools, photographs, documents and equipment began to be moved to Storehouse 10 on 1 February 2016. They also have considerable hardware stored elsewhere in the dockyard, including their larger items in the Fire Station. Their new space has a working area for their knowledgeable volunteers.

This group's intimate knowledge of the dockyard proved useful to the navy when water pooled in the filled in C-Lock north of Block Mills. As their secretary Celia Clark requested the dock's plans from the Royal Engineers Museum at Chatham; these showed the navy that they had not kept the drainage channels at the base of the lock free, which had caused the problem. The Society also published a detailed study of the cast iron caisson which once controlled access to No 2 Basin, opened in 1945 by Queen Victoria, when its top was cut down by the dockyard management to provide a carpark in the 1980s. Historical artefacts are sometimes now returned by the Society to their original owners. The Royal Marines School of Music installed their honours board in the former prison they now occupy.

As well as the recordings made by Celia Clark in the 1980s, an oral history programme was initiated. A wide cross-section of former Dockyard employees was recorded with assistance from Portsmouth University. The 400 interviews are available in the Portsmouth History Centre which transcribes and digitises the recordings and is compiling a website catalogue.[13]

On the top floor of No. 10 the Victory sail is partly unfurled and interpreted in a dramatic wordless video of the horrors experienced by sailors at the Battle of Trafalgar, and in a static display about the sail's preservation. On the second floor Celia Clark carried out a first sorting of the PNBPT archives, which were transferred here from the basement of the Porter's Lodge after a flood. Amongst the archive files she remembers seeing the fragile egg-shaped model of Sir Colin Stansfield Smith's proposal to house the *Mary Rose* inside a cocoon. The windows in Storehouse 10 were reinstated when the colonnade was removed for creation of the Babcock Gallery on the ground floor to coincide with the centenary of WW1 in 2014.

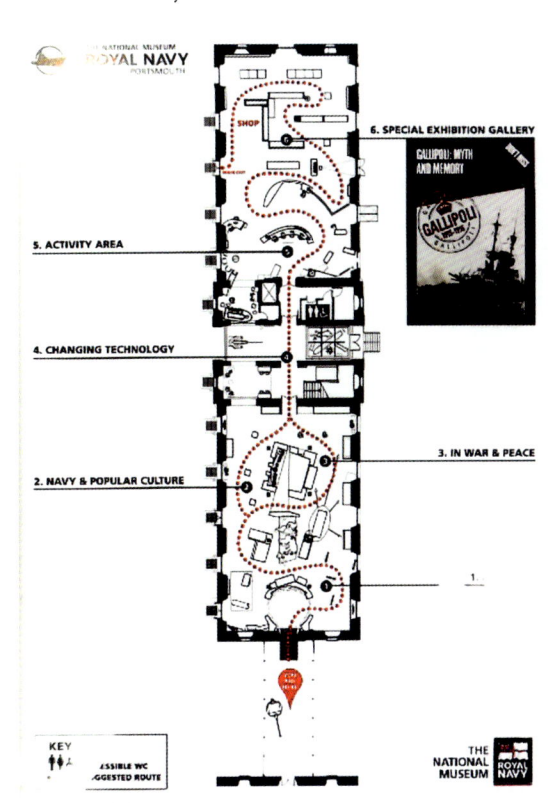

Babcock Gallery Layout 2016   Babcock Gallery interior 2016

The £4.5m refurbishment of the Babcock Galleries is linked by a new steel and glass walkway connecting the two new galleries which tell the story of the navy over the past hundred years.. The architects Purcell and contractors Warings brilliantly reused the space, installing modern museum displays at the same time as preserving the physical experience of the brick walls and timber floor and strong supporting columns of a naval storehouse. Before the main work began the eroded brickwork was repaired with about two thousand bricks and the timbers were stripped back. One of the space's most significant characteristics is the floor, which is made up of Spanish ships' timbers in places. A raised floor has been installed above the planks to protect them; glass panels were inserted over them so that they can still be seen. This fully reversible solution required by English Heritage usefully created a void for all services to run beneath the floor and ensures level access.[15] The galleries were opened in June 2014 with a permanent exhibition: HMS Hear My Story. The event was celebrated with a dinner on board *HMS Victory*. The Portsmouth Society was so impressed by the excellence of the design, it gave the museum its Best Restoration Award in 2015. Enjoyable temporary exhibitions in 2018 were Lachlan Goudie's vivid paintings and drawings of Rosyth Dockyard and dockyard workers and the history and iconography of tattooing.

No. 11 Storehouse

Built in 1763 by Templar and Parlby, Storehouse 11 was refurbished and converted by the PNBPT in 1988-92 as galleries, specialist stores, offices and a library for the Royal Naval Museum at a cost of £230,000 for the refurbishment and £111,000 for the conversion. The Lily McCarthy Gallery housing her collection of Nelson memorabilia was created in the north ground floor shell of No. II storehouse, with a new floor, contemporary display cabinets containing the memorabilia, a false ceiling and modern lighting. This encasing meant that the visitor lost most of the experience of an eighteenth century room: it concealed the fact that it was a 'converted' building and turned it into an anonymous modern space. After that there was little physical adaptation of the fabric of the storehouses apart from the creation of the colonnade until the creation of the King Alfred Library in 1989-90 on the first floor of No.11. This was undertaken in a much more conservation-conscious style than the McCarthy Gallery by the Portsmouth conservation architect. Terry Wren's design left the thick brick walls, lime-washed ceiling and wide-boarded timber floors intact, inserting carefully environmentally controlled heating and lighting and allowing natural ventilation. But there were problems from flaking lime-wash onto the books and floor. The collection the library houses predated its establishment; it has the plans from when the stores were first fitted out. The fittings dating from 1911 were said to have come from the Old Admiralty Library, which was in the Admiralty Arch, and later in the Empress State Building, though they may be more recent. Councillor Emery-Wallis opened the library in 1990.

The design chosen for the £4.9m adaptation of No. 11 Storehouse in 1988-1992 by the Fareham architects Hedley Greentree Partnership's Peter Allchurch - high tech - is an extreme contrast to the old fabric. A spectacular glass and steel lift soars through three floors in the central old goods lift opening. The bottom flight of wide, balustered stairs was reinstated and so was the timber wall lining. The MOD's concrete floor was removed in the south ground floor wing to reveal the original timber surface. The whole design is a combination of reinstatement of old fabric and dramatic modern intervention. Some may dislike this extreme contrast - between utilitarian eighteenth century brick, wood and stone and late twentieth century glass and steel. But it demonstrates the major change of use to museum and human occupation. According to the strictures of English Heritage the new work had to be reversible. The building contractors were a local firm, Henry Jones, part of Kier International. The new McCarthy Gallery redesigned by Robin Wade and Partners put right the failures of the original design while introducing other impressive exploitations of the space. The false ceiling and encasement of the columns were removed, while the windows were ingeniously covered with transparent reproductions of glass paintings of incidents in Nelson's life. Parts of the collection are displayed in the manner of sacred reliquaries - of a secular saint. Upstairs are highly serviced secure storage areas for the museum's collection as well as meeting rooms and offices.

Another problem in the profound change of use was the lack of a fire escape, since when the storehouses were built there was only one central stair. English Heritage decreed that these could not be added on the landward side of Storehouse 11 in order not to disrupt the fine elevations on the Victory approach, and if they were on the harbour side the exit would have been directly into the operational naval base, which would bring problems of security. The problem was solved in No. 11 by a new stair inserted within the end wall, invisible from the outside.

Historic ships and boats

HMS Victory from Storehouse 10

HMS Victory has been sitting in No. 2 Dock supported by 22 steel cradles positioned six metres apart since she was bought ashore in 1922 after a national appeal launched by the artist WL Wyllie in order to save her from decay. Her condition had deteriorated so much she could no longer safely remain afloat. During the initial restoration period from 1922 to 1929, a considerable amount of structural repair work was carried out above the waterline mainly above the middle deck. In 1928 King George V unveiled a tablet celebrating the completion of the work, although restoration and maintenance still continued under the supervision of the Society for Nautical Research. Restoration was suspended during the Second World War, and in 1941, *Victory* sustained further damage when a bomb dropped by the Luftwaffe destroyed one of the steel cradles and part of the foremast. On one occasion, German radio propaganda claimed that the ship had been destroyed by a bomb, and the Admiralty had to issue a denial.[18]

Over 30 million visitors have explored her since she was opened to the public in 1928. As the world's oldest commissioned warship, she is the Flag Officer Portsmouth's flagship. The MOD spent about £1.5m a year on her maintenance. From 2011 she has looked denuded when her three masts, 26 miles of rigging and bowsprit were dismantled. This was the first time since WWII that she had been seen without her topmasts. They had last been taken down in 1941 after damage in a Luftwaffe raid. Survey work revealed that the hull was leaking, suffering from dry rot and being pulled apart by its own weight. The restoration by BAE Systems, paid for by the Royal Navy, was said to take ten years, the biggest rebuild since she was repaired after the Battle of Trafalgar and the storm of 1805.

Only a fifth of the original ship remains. As the director of the Naval Museum said, "Preserving a wooden warship is a battle - a battle against nature, and, just as epic, in its way, as the Battle of Trafalgar."[19] In March 2012 responsibility for the ship was transferred from the MOD to the HMS Victory Preservation Trust. £30m was allocated for a major conservation programme that would take

15 years.[20] Former naval officer Sir Don Gosling who made his fortune as chairman of National Car Parks donated £25m to fund work on the ship. A detailed laser scan of 89.25 billion measurements and computer modelling revealed that the hull had moved 20 centimetres over the last 40 years. A new support system was designed to mimic how the ship would sit in water. Rather than the existing steel "blades" which were placing considerable strain on the hull structure, the 22 steel cradles installed in 1922 when Victory was drydocked were to be replaced by 134 15-foot adjustable steel props fitted over two levels. This aimed to completely revolutionise the support system and share the ship's 3,600 tonne load between them. Each prop is telescopic and features a cell monitoring the load around the clock and which is easily adjusted.[21] Although thirty million people visited the *Victory* over the years, they had not been able to get close to the hull. Visitors are now to be offered special access deep into Drydock No. 2 underneath the hull to see the elm keel that has supported her since it was laid down in 1759. A new gallery: the Nation's Flagship, documents her construction, service and ongoing conservation. Once the support system was installed 'the next 12 years will see a £35m programme of conservation work to ensure the longterm preservation including re-installation of her masts'.[22] In August 2020 Andrew Baines of the NMRN announced that the two-decade restoration programme was completed. A network of technology designed by BAE Systems (based in the dockyard) mimics the pressure of the sea and monitors the ship's weight distribution, identifying early warnings of faults or weaknesses in the hull so they can be identified and rectified. When the Historic Dockyard opened on 24 August 2020 visitors were be able to walk underneath the ship for the first time in 100 years.[23]

HMS Warrior 1860

HMS Warrior 1960 with Tall ships

HMS Warrior 1860 was a 40-gun steam-powered armoured frigate built for the Royal Navy in 1859–61, at the time the largest warship in the world. She and her sister ship *HMS Black Prince* were the first armour-plated, iron-hulled warships, built in response to France's launch in 1859 of the first ocean-going ironclad warship, the armoured steam-powered wooden-hulled *Gloire*, which started an invasion scare in Britain. By 1871 *HMS Warrior* was already obsolete when the mastless and more capable *HMS Devastation* was launched. In reserve from 1875, and decommissioned and offered for scrap in 1883, but luckily there were no takers. In 1904 she was assigned to the Royal Navy's torpedo training school in Portsmouth. In 1924 *HMS Vernon* took over her work and she was again put up for sale as scrap but - no takers! In 1927 she was converted into an oil pontoon in Pembroke Dock. In 1941 a bombing raid on the tank farm there narrowly missed her. Archivist Andrew Baines said "We were incredibly close to losing her – she had spent 50 years as a fuelling pontoon".[23]

As a pivotal design, she was the only surviving example of the 'Black Battlefleet' – the 45 iron hulls built for the Royal Navy between 1861 and 1877. A campaign to restore her began in 1967. Her saviour was Sir John Smith, MP for the Cities of London and Westminster, financier, landowner, philanthropist and founder of the Landmark Trust, which buys, restores and lets historic buildings as holiday homes. He raised the issue in the House of Commons: *Warrior* could serve as "a potent source of education and inspiration for our children".[24] He said "If every warship in the 19th century still existed and was available for preservation *Warrior* would still be my first choice". Smith's drive and persistence led in 1968 to formation of a committee chaired by the Duke of Edinburgh, to discuss *Warrior's* future. This became the Maritime Trust with the aim of raising money for the preservation of the country's naval heritage.

HMS Warrior moored to new jetty beyond Cell Block

When the oil depot closed in 1978 the ship was handed over to the Warrior Trust in 1979. Sir John Smith agreed that the Manifold Trust would underwrite the cost of restoration, estimated between £4-8 million, the world's largest maritime project ever undertaken. Her £9m restoration in Hartlepool took eight years. In 1983 ownership of the ship was transferred to the Ships Preservation Trust. Cllr. Freddie Emery-Wallis also played a role in securing *HMS Warrior* for Portsmouth. Sir John Smith was not keen on commercial development of the historic dockyard. He wanted *Warrior* to be anchored next to a working naval base.[25] The city council acquired the site just inside the Victory Gate where she is now moored and spent £1.5m dredging a new berth and constructed a new jetty in 1985 in preparation for her arrival. It was believed it was important that the jetty did not obscure the view of the ship, so it was constructed just high enough to be above the tides.[26] She was towed back to Portsmouth.

When she entered the Harbour in June 1987 thousands of people lined the city walls and shore, and over 90 boats and ships welcomed her. She opened as a museum on 27 July that year, but at first there was no interpretation of her history on board. Her role has evolved over the years: by 1995 she received over 280,000 visitors.

Preserving her historic fabric is a continuous task. In 2001-3 420 feet of weather deck teak was laid on the top deck by the Maritime Workshop based in Forton Lake on the other side of the harbour to replace the surfaces worn down by tourists' feet and years of exposure to the weather. In 2017-8 major reconstruction of her decayed metal frame was undertaken. Accoya wood was selected for its durability and ability to withstand extreme conditions. In 2011 a £60,000 restoration of the ship's sickbay lasting three years was completed. The restoration team used a detailed plan of the ship drawn by a 14-year old midshipman. A Gosport furniture maker recreated the wooden fittings, a local blacksmith made the cast-iron bunks.[27]

In 2015 the Heritage Lottery Fund awarded £2.6m to the Warrior Preservation Trust to make good the failed junction between the deck and the iron hull and to replace the bulwarks and water bar on both sides of the upper deck. These works are part of a wider £3.6m project 'HMS Warrior 1860 – Revealing the Secrets of Shipwrights and Sailors'. The project also digitised the archival collections and enhanced the visitor experience to the ship with new interpretation. The Warrior Trust had to raise the remaining £1m.[28]

Warrior was managed by the Warrior Preservation Trust whose purpose was to preserve and present her as the world's first modern battleship, to explain her influence on warship design and to demonstrate life at sea. She is used as a venue for weddings and functions to generate funds for her maintenance. The Warrior Trust also had a collection of material related to the ship. In April 2017 midway though the Heritage Lottery Funded project the Warrior Preservation Trust merged with the National Museum of the Royal Navy – thirty years after the ship arrived in Portsmouth. The ongoing conservation work on the upper deck, costing £4.2m was being completed. The museum's expertise and professional stewardship of historic ships was said to be one of the compelling reasons the Warrior Preservation Trust transferred ownership of HMS Warrior 1860 to the National Museum. Commander Tim Ash said 'the time is right for us to align our shared expertise more closely". During 2018 over £4m was invested in historic ship conservation.[29] In contrast to the formal guided tours aboard HMS Victory, sixteen expert guides in Victorian sailors' uniforms made visits an especially enjoyable experience: you could ask them detailed questions as you explored the ship. But after the NMRN took over, their numbers were reduced to six, and then cut out altogether.[30]

Mary Rose

After serving for 33 years, the *Mary Rose*, Henry VIII's flagship was leading the attack on the galleys of a French invasion fleet in 1545 when she keeled over and sank at Spithead drowning nearly all the crew and taking their tools, clothes, weapons, eating utensils, food, medical implements, musical instruments, games - and a pet dog - to the seabed. Professional divers John Deane and William Edwards had adapted breathing apparatus for firefighting into a diving helmet which they used to salvage some bronze and iron guns in 1836. But after that the hull lay undiscovered on the seabed until 1971 when Alexander McKee identified her position. Divers began exploring the surviving starboard section and salvaging its extraordinary time capsule of Tudor contents. More than 26,000 artefacts as well as the bodies of some of the people on board have been recovered and conserved. A selection of these was put on display in Boathouse 5.

To house the hull itself a site at Eastney was first considered but this was rejected because it was too far from the dockyard. Locating it next to the *Victory* was considered much more appropriate, but design of a permanent building to house the ship and artefacts together in such an important position was controversial. In January 1991 the Mary Rose Trust commissioned Prince Charles's favourite architect, the internationally renowned architect Christopher Alexander based at the Center for Environmental Structure to design a new museum for her. But his classical temple dubbed 'The Cathedral' was rejected: was it too expensive to build?[31]

The Mary Rose Trust raised the money to lift the hull from the seabed: a milestone in maritime archaeology, comparable in complexity and cost to the raising of the Swedish 17th century warship, the *Vasa* in 1961. The Royal Engineers at Marchwood in Southampton Water designed the steel cradle to enclose the remains of the hull. It was raised to the surface by the Norwegian crane *Tog Mor* on 11 November 1982, watched from the shore and on television by thousands. The wreck's slight shift in position was a heart-stopping moment, but there was no damage. The cradle was towed into No. 1 Basin, and then into No. 3 Dock, a Scheduled Monument, where a temporary cover enabled the long task of conserving the hull to begin. Until 2013 she was sprayed with filtered water then with polyethylene glycol, also used to conserve the *Vasa*, to replace the water in the cellular structure of the wood and strengthen the mechanical properties of the outer surface layers. The ship was then allowed to dry out.

Mary Rose Museum from No. 1 Basin

Instead of a classical design, Wilkinson Eyre's timber clad cocoon – or oyster shell – enclosing the hull like a precious pearl - was chosen. Its curved dimensions could only have been shaped with the aid of computer- aided design. The balconies echo the curve of the ship's decks. Built by Warings, the grant of £27m was the largest ever grant offered by the Heritage Lottery Fund. The design fulfils all the competition's criteria. Its complex functions are to protect and display the sixteenth century hull and its unique artefacts in a controlled environment, while maximizing visitor experience and learning. The construction was challenging because the museum was built inside the listed historic dry dock, which remains virtually intact and unaffected by the design of the museum's foundations.

The museum's design is utterly modern, but it is traditionally clad in black stained curved timbers with shipwright's marks - as a reference to traditional English tarred boatsheds. The view from the VIP balcony which resembles a partly open oyster shell is across the stern of *HMS Victory* towards modern warships and Portsmouth Harbour. The starboard section of the ship's hull and the many Tudor artefacts retrieved by divers are securely housed in a temperature-controlled chamber at the heart of the building. For many years the hull was only visible through a screen on three different storeys.

The interior by Pringle Brandon Perkins+Will was designed to recreate the low ceilings and dark and claustrophobic atmosphere found in lower ship's decks. Lighting is directed only onto exhibits and handrails so that visitors can find their way through the galleries. Two rectangular extensions branch out from the sides of the museum: a reception area, café and shop to the south, while the second contains an education centre and staff quarters.

Charles Haskell remembers the first day the public got to see the ship. "On the first day the

Mary Rose opened I was one of the first guides and started late morning. We'd been trained up as guides in sessions held in the central library by Sylvia Townson but each of us had our own individual self-written scripts. The ship was lying on its side in the cradle. We used a microphone on a very long lead that was in the middle of the aluminium bridge which spanned from side to side the whole of the dry dock. The bridge was positioned near the front of the ship so visitors were looking along most of its length. It was very high up and freezing cold in there because they were spraying the ship with refrigerated cold water and wax. The air was full of mist and the hissing sound of the sprays was continuous."

"On the very first day the ship opened we were told to take 40 people at a time from the expected queue. They went through an area with doors that kept the cold in and sunshine heat out. However, on that first day the queue was way bigger than expected. It stretched through the dockyard then a long way up Queen Street. We were supposed to whittle the queue down in groups of 40 but due to the size of the queue we were instructed to increase that to 80. With 80 on the aluminium bridge the structure moved a bit and we had to help people off if suffering from vertigo or freezing or generally scared. The first day was the only day that I am aware we took very big groups and the norm after that was about 20 to 40 visitors max. We could wear what we wanted. A lot of us wore waterproofs over our trousers, fingerless gloves and multiple layers. We kept warm when waddling about but the public in shorts and t-shirts were cold. We were allowed to say what we wanted as to the history of the ship while the group of public stood freezing on top of the bridge. After about 15 or 20 mins. max we shepherded the group out of the exit while the next guides outside got their group ready. I worked Sundays at the ship for about 2 or 3 years and met and talked with HRH Prince Charles on one occasion. The picture taken of us talking was called by the guides, "Charles meets Charles."[34]

In 2016 the museum was closed for nine months while a £5.4m funded the removal of the glass screen - so visitors can at last see the ship clearly from the curved shape of the enclosing galleries, echoing the ship's lines in all their glory – a visceral thrill. The story the contents tell are so engrossing that two or three hours pass before you long for daylight and a pause for reflection on what you've seen. Only the crew's quarters and the VIP balcony overlooking No. 1 Basin and Victory's stern offer daylight for those who work and entertain there, but not to the many thousands of visitors the museum attracts. Perhaps the ambient light levels could be increased as you rise. It's an irony of history that such a sudden horrific national and personal disaster, leaving the ship, crew and their possessions resting on the seabed for so many years offers us in the twenty-first century such an extraordinary face to face and minutely detailed experience of the Tudor navy's world. But the graphic video of the sinking and drowning fatally distances us from the imagined event, because it has no sound. Surely it is not to upset children that we hear no roar of the sea, no screams, no gasps for air? Children love being frightened! Celia Clark also missed the sensations of a sailing ship: timbers creaking, sails flapping, ropes snapping and the sounds of waves and wind - suddenly stilled. The top gallery has the ship's fittings: blocks, unique mast top, anchors and heavy coils of rope, but not the feel of a sailing ship. In 2013 80,000 people came to the historic dockyard in one month alone.[32] By 2018 visitor numbers reached over 850,000[33] The museum has won many awards including the Civic Trust Award and the Portsmouth Society's Best New Building Award in 2014.

HMS M.33

Launched in May 1915 *HMS M.33* in No. 1 Dock is a unique survivor, the sole remaining British veteran of the bloody Gallipoli Campaign and the only British warship from the First World War open to the public during the centenary year. A 'Monitor' of 568 tons it has a shallow draft that allowed the ship to get close in to shore to fire at targets on land. She carried two powerful and oversize 6" guns, but was basically a shallow metal box lacking in comforts. The 72 officers and men who sailed for the Gallipoli Campaign were crammed inside and away from home for over 3 years.

M33 in Drydock 2018

The National Museum of the Royal Navy (NMRN) and Hampshire County Council (HCC) worked as partners to develop the £2.4m project to conserve, restore and interpret her. Specialist metal preservation techniques were necessary since large sections of the hull had corroded and had disappeared over time. It is or was the largest metal object ever conserved. With the help of a grant of £1.8m from the Heritage Lottery Fund (HLF) the ship and interpretation were opened in 2015. Visitors start with a 6-metre descent into the bottom of the dock before stepping aboard. New interpretation, including a stunning immersive battle experience, brings alive its history and the stories of the men who served on board. The 'Commemorating Gallipoli Myth and Memory' exhibition was part of the NMRN's wider 'Great War At Sea 1914 – 1918' programme to mark the Royal Navy's First World War.

Preserved boats
PNBPT's Boat Collection

At the outset the Memorandum and Articles were drafted recognising that the term "property" in the context of a dockyard needed a broad definition and much wider scope than just buildings. Lock gates, caissons, penstocks, figureheads, pontoons, slipways, railway lines, weighbridges etc. would all require to be preserved under that broad definition of property. It was also understood that small boats were also a fundamental ingredient of the operation of any dockyard and that preservation of vessels should be specifically included within the objects of the trust. The Royal Navy first approached the PNBPT in the early 1990's seeking its agreement to take on the former barge of the Commander-in-Chief, Naval Home Command which was to be taken out of service. Soon after this was followed by requesting the trust to assume responsibility for Defence Harbour Launch 49, one of the last remaining workhorses of the Dockyard which ferried civilians and naval personnel from one part of the dockyard to the other. In parallel the trust itself had restored a naval cutter and whaler to display on the Mast Pond as part of its interpretation of how it was used historically. Brian Patterson recruited shipwright colleagues to work on the repair and maintenance of the PNBPT's boat collection. *MGB 81* and *HSL 102* were acquired through a purchase funded by the National Heritage Memorial Fund with additional grant assistance from the PRISM fund.

The PNBPT introduced the waterbus service following its acquisition of Priddy's Hard and did its utmost to promote the harbour, but without wholesale political will on the part of all the local authorities involved realising such an objective remains challenging. In Peter Goodship's view, encouraging all stakeholders to capitalise on its natural geography and illustrious history is vital.

In 2018 the PNBPT applied for planning permission to develop the waterfront space as a mini marina with £2.6m funding from the Chancellor's LIBOR Fund between *HMS Warrior* and No. 4 Boathouse by building bespoke pontoons for berthing a permanent display of some of their Memorial Fleet of small craft. These include the armed steam cutter *Falmouth* (the only surviving vessel from the Battle of Jutland other than *HMS Caroline*), a F8 landing craft used in the Falklands from *HMS Fearless* and a replica WW1 coastal motorboat. The property trust was also granted £2.4m LIBOR funding (from fines on banks) to create an operational Memorial Fleet of small craft that played a significant role in the country's defence in the twentieth century.

Thanks to this and the LIBOR Fund the PNBPT has the infrastructure in place to house the collection and undertake proper repair and maintenance of its collection. The trust is recognised by the maritime heritage fraternity as a responsible custodian of an important collection of historic craft; while at the same time preserving the heritage skills associated with its repair and maintenance.

Steam Pinnace 199

Steam Pinnace 199 off Haslar sea wall 2017   Martin Marks

This former 50-foot naval gunboat has a working steam plant and mounts a 3pdr Hotchkiss gun. Originally used to defend capital ships in undefended anchorages, she was sold off by the navy in 1949 and then bought privately and steamed to the Thames where she became a houseboat named "Treleague". In 1979 she was bought by the then Royal Naval Museum and restored in the Maritime Workshop Gosport by the late Peter Hollins MBE and other volunteers. Since then she has appeared under steam at a wide range of public events. In 2012 she went back to the Maritime Workshop for a five-year refit involving some 13,000 volunteer hours of work. This project was award three national prizes for restoration by the Institute of Conservation/ Institution of Mechanical Engineers, the Marsh Trust/National Historic Ships and the Transport Trust. She is maintained and operated by volunteers at the National Museum of the Royal Navy known as Group 199.

Originally incorrectly identified as Steam Pinnace 199 from a plate on her boiler circa 1979, recent research indicates her machinery was originally fitted in Pinnaces 208 and 224. The hull, stern cabin and funnel came from 224. The current boiler and engine donated by HMS Sultan when she was restored. She was built in 1909 by J Reid of Portsmouth. The NMRN, which owns her, has decided to retain the designation 199, as she has been known as this for so many years. She is now a hybrid anyway – an armed picket boat for'd and an admiral's barge aft.

In October 2014, the NMRN received funding to restore D-Day Landing Craft (Tank) *LCT 7074*. The craft was raised from where it had sunk at moorings in Birkenhead and was transported to Portsmouth for conservation. In 2018 the museum secured an HLF Round 1 pass to support a £5m project for its conservation and interpretation. In December the same year, the museum acquired *RML 497,* a WWII motor launch.

HMS Foudroyant/HMS Trincomalee – the one that got away…

Once afloat in Portsmouth Harbour, *HMS Trincomalee* was built in 1812 in Bombay by the Wadia family – in teak, because of shortages of oak in Britain as a result of shipbuilding drives for the Napoleonic Wars. Unlike the many demasted hulks moored off dockyards that served as prison ships, young people rather than prisoners slept in her hammocks. She was named after the battle which took place in 1732 off the Sri Lankan dockyard of the same name. She finished her Royal Navy service as a training ship and was sold for scrap two years later. George Wheatley Cobb bought and restored her in 1899 and renamed her *HMS Foudroyant*. She was used as a training ship, a holiday ship based in Falmouth, then moored in Portsmouth harbour until 1982. Celia Clark remembers her sons in the Sea Scouts enjoying learning to sail in classes run from the ship, and Martin Marks and his wife used to run inflatable boat handling courses from her for sub aqua divers.

In need of serious repair, she was towed to Hartlepool, where the skilled team of specialist craftspeople who also carried out the extensive work to bring the *Warrior* back to life made a splendid job of her. She reverted to her old name. She is managed by another outpost of the NMRN. She now sits in drydock surrounded by a replica enclave of shipbuilding workshops developed as a tourist attraction. But she cannot go to sea again because the dock entrance to the basin has been rebuilt to smaller dimensions.

Portsmouth Harbour tours

The busy launches of the Port Auxiliary Service carrying dockyard and service personnel across the harbour are seen no more, but tourists can explore the harbour by boat – the obvious way to do so, rather than circumnavigating the seventeen miles around the harbour by road. The trust leases its landing jetty alongside Boathouse 4 to Solent & Wightline Cruises who offer boat tours around the harbour. Seeing the operational naval base with modern naval ships including the giant *Queen Elizabeth II* supercarrier and her sister ship *HMS Prince of Wales* close up gives an idea of their immense dimensions. The commentary tells passengers how the harbour has developed since Roman times as well as details of naval ships currently in the dockyard. In summer there is also a boat service across the harbour to take visitors to Portsmouth Point, the Submarine Museum and Priddy's Hard.

Festivals

The managers of the heritage area work hard to attract the visitors needed to maintain the site by staging special events, whether they are the International Festivals of the Sea in 1998, 2001 and 2005, Trafalgar 200, and the Victorious pop festival which has grown every year since it was first held alongside *Victory*. It now takes place on Southsea Common. In 2015 £5.8m was spent by festival-goers; in 2018 120,000 people, 65% from Hampshire and a third from Portsmouth enjoyed Victorious, spent nearly £10m boosting the city economy and benefitted local charities.[34] The height of the summer tourist season is when the historic dockyard is really thronged. The Victorian Festivals of Christmas attracting winter visitors are also deservedly popular. In March 2020 the Covid-19 put a stop to all this activity.

Portsmouth's Historic Dockyard has an embarras de richesse: a national museum, historic ships, boats, docks and slips, fine dockyard architecture converted to new purposes, while the ambience of a sober working place still survives surrounding the ebb and flow of tourists. There are so many attractions visitors cannot tour them all in one day – but this is a gain to those who operate the site: repeat visits are the key to sustainable income. It is surprising to a local how many families can afford this considerable cost, but over 800,000 visitors a year enjoy the riches the historic dockyard has to offer.

The reinvention of the dockyard's historic core is jointly a major achievement of the Naval Base Property Trust, the trusts of the historic ships and the National Museum of the Royal Navy as well as of the many volunteers and specialists who work on the buildings and the collections.

Sources

Celia Clark 2008 'Adaptive re-use of the Georgian Storehouses of Portsmouth – naval storage to museum' *Transactions of the Naval Dockyards Society* Volume 4 October pp. 27-37

Jonathan Coad 1983 'History and Architecture of the Royal Navy' *Mariner's Mirror*

Jonathan Coad 1989 *The Royal Dockyards 1690-1850: Architecture and Engineering Works of the Sailing Navy.* Royal Commission of the Historical Monuments of England Scholar Press Aldershot pp 132-136

David Lloyd 1985 *The Buildings of England: Hampshire and the Isle of Wight*: Harmondsworth p. 410 http://www.pnbpropertytrust.org/index.asp?upid=11&msid=2 Accessed 3 July 2018

Portsmouth Society 2000 *Annual Report*

Ray Riley 1985 *The Evolution of the Docks and Industrial Buildings in Portsmouth* Portsmouth: The Portsmouth Papers pp. 7, 10

http://www.blueplaqueplaces.co.uk/thomas-telford-blue-plaque-in-portsmouth-10793#.W6I4Uy2ZNEI accessed 2019

The Mary Rose Museum Christopher Alexander, Gary Black, Miyoko Tsutsui *Oxford University Press*, 1995
Martin Marks 2018 'Steam Pinnace 199 - 107 Not out' in *Dockyards* The Naval Dockyards Society December 2018 Volume 23 Number 2

"The Wadias of India". zoroastrian.org.uk. Retrieved 25 July 2015. "Trincomalee Construction". The National Museum. Retrieved 25 July 2015

The Portsmouth Telegraph letter dated St. Helena Jan. 29, 1819

"HMS Trincomalee - Royal Navy Service". The National Museum. Retrieved 25 July 2015

HMS *Foudroyant* "HMS Trincomalee - Training days as TS Foudroyant". The National Museum. Retrieved 25 July 2015. "Restoration and the present day". The National Museum. http://www.nmrn-portsmouth.org.uk/online-information-bank Retrieved 25 July 2015

"HMS Trincomalee - About us". The National Museum. Retrieved 25 July 2019

https://www.bbc.com/news/uk-england-hampshire-14131279

https://www.hms-victory.com/restoration-log/world-first-hms-victory-ambitious-engineering-project-resupports-flagship August 21, 2017

http://www.buildingconstructiondesign.co.uk/projects/the-babcock-galleries-at-the-national-museum-of-the-royal-navy/

https://www.nmrn.org.uk/sites/default/files/impact_2018_web_0.pdf https://www.historicdockyard.co.uk/news/itemlist/date/2017/8?catid=6

References

1. No. 4 Boathouse Drawings held by Unicorn Consultancy, Establishment Works Consultants, Murray's Lane, HM Naval Base Portsmouth. Notes by Celia Clark 10 December 1996

2. Boathouse 4, Portsmouth' in Marcus Binney 2016 *BIG SAVES Heroic transformations of great landmarks.* SAVE Britain's Heritage

3. Walters & Cohen https://vimeo.com/192616620 November 2016 cited in 'A Brief History of Boathouse No. 4 Portsmouth Historic Dockyard' Steve Dawson Boathouse 4 Volunteer. August 2019

4. Portsmouth Society Design Awards report 2016

5. https://www.waltersandcohen.com/projects/boathouse-4

6. 'Court battle will be charity's final move' The News October 10 2018 p. 7

7. Ray Riley 1985 *The Evolution of the Docks and Industrial Buildings in Portsmouth*: The Portsmouth Papers Portsmouth 1985: 7, 10

8. Celia Clark 'The Re-use of the Georgian Storehouses of Portsmouth – naval storage to museum' *Transactions of the Naval Dockyards Society* Volume 4 October pp. 27-37

9. Jonathan Coad 1989 *The Royal Dockyards 1690-1850:* Aldershot: 1989: 132-136; David Lloyd 1967 *The Buildings of England: Hampshire and the Isle of Wight:* Harmondsworth: p. 410; O'Brien et al 2018 *The*

Buildings of England Hamhsire South p.488

10 Jonathan Coad 1989 *The Royal Dockyards 1690- 1850: Architecture and Engineering Works of the Sailing Navy.* Royal Commission of the Historical Monuments of England Scholar Press Aldershot

11 P J Larkham 1997 "Remaking Cities: Images, Control, and Postwar Replanning in the United Kingdom' *Environment and Planning B: Urban Analytics and City Science* https://doi.org/10.1068%2Fb240741 accessed 30 March 2020

12 Artists' Harbour Email to Celia Clark 18 September 2018 .1, 7

13 https://portsmouthdockyard.org.uk/about/history-of-the-trust accessed 23 June 2019

14 Portsmouth Society Annual Report 2000

15 http://www.constructionmanagermagazine.com/onsite/project-month-babcock-galleries-portsmouth/ 15. accessed 23 March 2020; http://www.buildingconstructiondesign.co.uk/projects/the- babcock-galleries-at-the- national-museum-of-the-royal-navy/

16 http://www.pnbpropertytrust.org/index.asp?upid=12&msid=2).

17 https://www.thehistorypress.co.uk/articles/hms-victory-saved-for-the-nation

18 BBC News 25 January 2011; BBC News 12 July 2011 'HMS Victory's masts removed as restoration starts'

19 Paul Brown 2018 *The Portsmouth Dockyard Story From 2012 to the Present Day* The History Press p.180 https://www.bbc.com/news/uk-england-hampshire-14131279;

20 https://www.hms-victory.com/ restoration-log/world-first-hms-victory-ambitious-engineering-project-resupports-flagship August 21, 2017

21 *The News* 2019 'Visitors to be able to go under HMS Victory as part of new experience' Simon Toft The News September 19 p.11

22 Millie Salkeld 2020 *The News* 11 August

23 https://www.bbc.co.uk/news/uk-england-hampshire-14556126

24 http://www.hmswarrior.org/history/restoration

25 Richard Trist 2019 interview with Celia Clark 9 May

26 Richard Trist op cit.

27 https://www.bbc.co.uk/news/uk-england-hampshire-14556126

28 https://www.historicdockyard.co.uk/news/item/261-lottery-cash-to-secure-future-of-historic-navy-warship-hms-warrior-1860

29 https://www.historicdockyard.co.uk/news/item/912-record-breaking-year-for-portsmouth-historic-dockyard

30 William Cushion to Celia Clark 2019 Portsmouth Society meeting 20 February

31 Christopher Alexander, Gary Black, Miyoko Tsutsui 1994 *Mary Rose Museum* Centre for Environmental Structure Oxford University Press

32 http://www.attractionsmanagement.com/index.cfm?pagetype=news&codeID=306012

33 Charles Haskell 2019 Interview with Celia Clark 13 March

34 The News 2018 June 18 pp.1.7

Chapter 8
"Make Art, Not War!" Museums and galleries in military spaces

In the wider harbour area museums and art galleries, both public and private have colonised former barracks, bastions, casemates and Southsea Castle. This chapter draws on the memories of those involved – in setting them up, designing and managing them. Inevitably Portsmouth and Gosport are particularly rich in museums focused on defence, and there are other defence oriented museums in southern Hampshire: Calshot Castle, Fort Nelson, Southampton Hall of Aviation. These are run by a variety of organisations: English Heritage, the Royal Armouries and by local authorities; while volunteers set up and run No. 6 Cinema in the dockyard's No. 6 Boathouse, the Hovercraft Museum at Lee on the Solent, the Diving Museum in Browndown, the WW1 Remembrance Centre at Hilsea Lines and the Farnborough Air Sciences Trust (FAST) next to the former Royal Aircraft Establishment. But all museums, whether national, local authority or voluntary, face continual fund-raising challenges and the need to produce new initiatives to maintain both their precious collections and visitor numbers.

As active participants we are aware of the sometimes fraught and sometimes dynamic interplay between the collections, the trustees or governing bodies and the many volunteers whose passion is the collection, services to visitors or the conservation of the collection, ship or site. While the governors or trustees hold the legal and financial responsibility for each museum, volunteers, with their investment of knowledge, expertise and time often feel they are the spirit of the collection, ship or place. In the best-run museums and galleries both parties have clearly defined and interlocking roles and work creatively together. But this is not always the case; volunteers' roles may be downgraded or even eliminated, sometimes justified by health and safety rules. In our view volunteers bring alive museum collections, particularly in times of staff shortage, so losing them may be both to the detriment of the attraction and to the visitor experience.

Portsmouth Museums

In Portsmouth the city museum in the High Street was destroyed in the 1941 Blitz. In the 1970s three Portsmouth military buildings were purchased from the Ministry of Defence. With key political support Southsea Castle and Clarence Barracks were converted into city museums and the NAAFI into the City Records Office. Ken Barton who was appointed museums Director by Councillor Freddie Emery-Wallis said: "When I became Director of Portsmouth Museums in 1967 I soon found how important Freddie was to their wellbeing. He was on hand with advice at the political level, never interfering but always helpful. With his guidance we took the concept of a comprehensive Museums service to the community further through firstly, opening Southsea Castle to the public and secondly, setting up a new Museum and Art Gallery in an imposing former Naval barracks building especially converted for this purpose and complete with its own bar. Freddie was also an early supporter of the *Mary Rose* project, which was to become so important to Portsmouth."[1]

Ken Barton "was a remarkable director, bullish, determined, iconoclastic, free-thinking, and probably the only person who could get up enough peoples' noses in Portsmouth to make this project work. With the political help of Freddie Emery-Wallis he had already fought to save and re-use the remaining barracks building, overcome the inadequacy of Cumberland House in Southsea, and balance the historical / military purpose of Southsea Castle."[2]

To carry out the conversions and other restoration work, Freddie appointed two historic buildings architects: Brian Young and Deane Clark, with David Lees and later Cyril Hathaway as clerks of works. Their task was top priority; specialist historic buildings input, missing previously, was required because before that repair work had unfortunately damaged the historic fabric of Southsea Castle. In addition Forts Purbrook and Widley were acquired by the city for leisure activities – as explored in our second forts chapter. Deane Clark said: "Ken Norrish, Bill Worden (City Architects) and Ken Barton interviewed me for the new post created for me - because I was a local. I had brought my drawings of the city down to the interview. They asked me where certain buildings were, and I replied 'You destroyed them!' Conservation and reuse instead of comprehensive redevelopment were new ideas then. Brian Young had a rare historic buildings diploma from the Architectural Association, so I was very privileged that they appointed both of us – and I learned about conservation on the job. We were moving away from using the direct works department and only employing small contractors whom we had trained up. Up to then trainee stonemasons had spent one day a week in Portland, but this had been cut back for lack of money. I noticed the rather harsh treatment of the historic brickwork inside the keep next to the recently added staircase. It had been cut back and smoothed over using a disc cutter..."[3]

Southsea Castle

Southsea Castle from the southeast

In 1960 the castle's military use ended. It was sold to Portsmouth City Council for £35,000. Restoration to its 1850 condition began about 1955. The sixteenth century part of the Castle and moat were restored. Much of the structural work had been completed by 1970, removing years of WWII military accretions, cable runs, external lights and blocks of concrete by the moat. "During [Freddie's] time as Chairman of the Development and Estates Committee in Portsmouth he ensured that good budgets were available for conservation work including purchase of historic buildings material and artefacts and ongoing maintenance."[4] "Money for the restoration only became available because the city treasurer had made other savings, and Freddie steered the funds towards this project. We liaised with the Department of the Environment and the Ancient Monuments Inspectorate. They were particularly concerned about the spending slopes on the seaward side. Each winter something like 250 tons of rapid setting concrete replaced the gaps scoured out of the slopes by heavy seas. This was very much the responsibility of the city engineers department. An interlocking concrete block was designed as a continuous shield. The city engineer John Easterling also rescued historic stonework including a large stone roundel base for a gun, which was incorporated into a new ramp in the sea defences adjacent to the promenade for access to sea level if emergency repairs were required. Our client, Ken Barton, was very demanding, determined to have his own ideas incorporated to make the fort a popular visitor attraction. For example, the well situated in the courtyard, just a hole in the ground, gained a stone parapet. He was so impatient he used to send the city architect up to two typed messages a day – for example on uncompleted details such as putting a non-slip tread on the spiral steps up to the roof of the keep."[5]

Southsea Castle entrance and lighthouse   Deane Clark 1985

An expert bricklayer, Toby, repaired the large expanses of brickwork on the exposed rooftop, working in horrible conditions to replace the top layer of bricks above the seaward casemates with no protection against the wind and rain, to complete the job before the end of the financial year. Len Kidd repaired the stone walls, "but unfortunately at that time we did not understand the use of lime mortar. While restoring the casemates on the southeast corner for the new museum layout later plaster was removed, exposing the original seaward facing Tudor gunport in the keep. This was a very exciting time…. As part of bringing the counterscarp gallery, which had no floor, back into use, a new brick floor was laid by a small local firm Purbrook Construction (Norrie Barnes) using Burseldon bricks. That was a grim job carried out in the winter months, done in the darkness because there were no lights. The east and west stairs to the top of the seaward wall were rediscovered, full of rubble, and restored. The complex brickwork of their tops is visible on the seaward walkway. A recently rediscovered Armstrong cannon was mounted on a specially designed brick and stone plinth outside the entrance to the fort. This was again a winter job. Len Kidd was on site making final alterations to the profile of the plinth as a sling was lowering the cannon onto the base. Some years later the Palmerston Forts Society were very keen to remove my design and replace it with the correct carriage of the Crimean period, but there was no money for this – which pleased me!"[5]

The raised banks of the west and east batteries had been remodelled and landscaped by the parks department in municipal style. They are a very special high-level vantage point for locals and visitors to enjoy views of the sea and the Isle of Wight, passing shipping in the Solent and special maritime events.

Archers fire longbows from Southsea Castle in celebration of *Mary Rose* May 2013

In 1967 the castle was opened to the public, "a most appropriate setting for an excellent military and maritime museum";[6] the lighthouse remained in operation. As a tourist attraction over 90,000 visitors visited it in the year 2011–12. The castle houses a collection of historic cannons brought into the castle for safe keeping from other locations. It is a scheduled monument, and its lighthouse topped by a fox wind vane in memory of Admiral Fox added in 1828 on the north face of the Western Gallery gun platform of the Castle is listed grade II. It's unusual to have a listed building inside a scheduled monument.

To raise revenue and increase footfall a temporary pavilion let out for weddings and special events occupies the area east of the keep. In 2015 the Courtyard Café was opened in the western seaward range, while a year later Southsea Brewing Company was established in an ammunition store. Its choice of ales includes 'Low Tide', 'Searchlight' and 'Six Wives'! In March 2017 the lighthouse was retired from active service when the new Portsmouth Approach Channel was deepened and realigned for the supercarriers, with large warning lights planted on high columns in the sea. In 2018 investigation of the ground conditions for the design of new sea defences in the face of rising sea levels and damaging storm surges produced worrying evidence of the weakness of the castle's foundations. As part of the protection of Southsea seafront engineers proposed to reinforce them on the seaward side via piles of rock armour.

Barracks to museum

Clarence Barracks demolition late 1960s. Left: City Museum and City Records Office converted 1970-73.
Maritime City Portsmouth 1945-2005 Portsmouth Society 2005

Victoria Barracks between Alexandra Road and Pembroke Road were built by the War Office from 1880 to 1886 to provide living accommodation for infantry units and artillery, partly over the site of the town's ramparts which were flattened in 1876.[7] They were a unique design unrelated to other English barracks, which were mostly standardised in their planning and architecture. In effect they were an extension to Clarence Barracks, sharing the same parade ground. The first unit to arrive was the 1st Battalion the South Lancashire Regiment. David Lloyd described them as an: "extensive large-scale group in red brick with heavy stone dressings….architecturally exuberant" with a "fantastically irregular outline with circular turrets, gables and sharp roofline seemingly influenced by the images of romantic German castles or Scottish baronial halls… Assertive Gothic with Butterfieldan or Teulonic derivation and Scots Baronial cross-reference. Skyline with variegated turrets, steeples, gables, looking weird from certain angles. Some of the blocks are bomb-damaged and half-ruined, others empty and derelict; this heightens their romantic appeal…. In terms of accommodation they were said to have been a great advance on previous barrack blocks."

In March 1948 the northernmost block which is now the city museum was converted for use by the Women's Royal Naval Service, renamed the Duchess of Kent Barracks after the Princess Marina of Greece and Denmark who was Commandant of the Service. It fronts the parade ground to its south; Lt. Col. R Dawson-Scott RE designed it for the War Office in 1893. In contrast to its fantastical skyline, its interior is plain with large open well stone stairs with cast iron balusters at either end of the block. Celia Clark remembers the bare barrack rooms, divided by curtains and niches for the slop buckets before it was converted.

Hardy House Officers' Mess Pembroke Road before demolition in 1970s   Demolition   Deane Clark

The central tower of Hardy House in Pembroke Road, the officers' mess, was hit by a bomb and seriously damaged during WWII. After the war the barracks were taken over by the Royal Navy. The complex was decommissioned in the early 1960s. Most of the blocks apart from the northernmost and the NAAFI – which was converted into the City Records Office - were demolished in 1967. They were so solidly built that demolition was a giant task. Michael Laird describes the fire in early December 1970 which destroyed Victoria Barracks.

"I had joined the staff of what was then Portsmouth Polytechnic in 1970, and in December of that year was sitting in my office across the road in the Kings Rooms, an extension of the students' hall of residence. So I had a good view of the Victoria Barracks on fire – and very dramatic it was, as the pepperpot turrets in the buildings flared up into the dark sky like giant candles! There was a rumour that it had been started deliberately to expedite the demolition! So it was empty, and I don't think anyone was hurt. I had an additional personal interest, as back in 1954 I had spent six weeks there doing basic training at the start of my National Service."[6]

The converted Portsmouth City Museum and former Records Office are the only survivors of the site's military past, although the latter faced demolition in 2019-20. David Lloyd's 1974 account was written just after the barrack block's conversion into the City Museum and Art Gallery when he said "The site adjoins Southsea Common and would offer a wonderful opportunity for imaginative redevelopment." 'Pembroke Park' was redeveloped on the site as a private estate, with ordinary Ministry of Defence houses at the western end, tower blocks and a hotel, not fulfilling the promise which Lloyd anticipated. The only reminders of the barracks are the roundel in the gable and a triangular stone set into the city museum's southern boundary wall, and the southern gate piers.

South Elevation and garden

The conversion was Deane Clark's first job as city Historic Buildings Architect working to a brief by Ken Barton, the director of museums.

Measured drawing of central façade 1972   Deane Clark

"My experience was limited in museum requirements. However, I had converted a Georgian house in Bembridge into the Isle of Wight Maritime Museum for a private syndicate. Converting a Victorian barrack block surrounded by asphalt parade grounds was a challenge. It was empty with a bleak interior, but weatherproof. I prepared several working drawings, traced off the original plans and elevations; a very large set, probably handed to the city on the transfer of ownership. The work was going to be done in phases. Phase 1 was the first floor and one ground floor gallery and the entrance hall."

"The biggest challenge was to unite the first floor with the side galleries, achieved by breaking through the walls and adding three steps, with a handrail balustrade at each side, reused from a demolished block. The steps were later replaced by ramps. The required schedule of works included the condition of all the windows and doors. Hardwood floors rescued from the demolished blocks on the site replaced softwood ones. More toilets were introduced and existing ones adapted to male and female – though still very basic. All the locks and keys and a safe were supplied by Mr. Buckle, the locksmith in Highland Road. New doors were contemporary designs. Interior colour design was by Terry Riggs including bright yellow on one wall in the main central gallery. This was repainted white the evening before the opening – on the city architect - Bill Worden's instructions. It was originally the Warrant Officers' mess. We retained the bar with its sliding sash window above the counter, and incorporated at Ken's request a Georgian external doorcase rescued from a prostitute's house in Portsea… Landscape design was by Ken Fieldhouse, including building a boundary wall to the south incorporating a roundel and a triangular pediment from Hardy House, the officers' block. The new flagpole opposite the entrance was shaped by Bernie Clark from a redundant telegraph pole – as an economy measure! I designed a brick surround for the external substation against the Alexandra Road boundary wall. Emery-Wallis objected to this and had the railings reinstated by the city engineer. The external downpipes were diamond shaped cast iron. It was impossible to find replacements so the alternative was to make them in glass fibre painted black. Over the years the complicated roof had to be replaced in phases. The fire brigade came each year at my request to clear out the gutters – as a training exercise. I also appeared at the magistrates' court on Ken's behalf to apply for a license - which was approved! The conversion took two years. The contractor was George Austin. Bill Worden's secretary was his daughter – both excellent at their jobs!"[5]

James Hamilton, who was appointed as Curator of Art in 1972 said that Barton, with Janet Chamberlain at his right hand, was the driving force behind the renovation of the remaining block. "Its long thin structure, perhaps about ten times as long as it is wide, was "ideal for soldiers' beds, but an odd, chancy choice for a major museum." Ken Barton was a demanding client, changing his mind on the design frequently.

James describes his interview and the fitting out of the museum in detail. "I was appointed in February 1972 after an interview in Southsea Castle during which the lights failed, so we carried on by candlelight. Then the Chief Attendant Harold Chennels came in to tell Ken that there was a contractor crisis somewhere on the barracks site, and could he please have some guidance. Ken took all this in his stride; he lit the candles, gave Harold some steer, and gave me the job!"

"I was daunted by the size of the building, the extent of the first floor galleries to be set up in the three months left before the already fixed opening date in May, and the monthly programme of temporary exhibitions planned and to be planned – two running at a time, one upstairs, one downstairs, and a change-over every month. 'I want exhibitions – bang, bang, bang,' said Ken."

"The museum's entrance was on the dull north front of the building, by the wrought-iron road gate. This was already buckled before we opened, not by a contractor's lorry backing into it, but by Ken driving through too fast one morning and breaking his bumper on it. The north front was gloomy, with drainpipes, lowering turrets and stone-dressed windows, but it was lit by a good nineteenth century iron lantern, an exhibit from, I think, a former church or school in the town. Past the entrance desk – I think you had to pay to get in even then – to the left was a generous-sized exhibition gallery, to the right, by 1973/74 when we got money from the Arts Council under their 'Housing the Arts' scheme, there was a long, neatly proportioned gallery for more exhibitions. This ran to the centre of the building."

"We all worked extremely hard getting the rooms laid out, the pictures hung on rods, the display cases erected, the objects set out in them. There were no designers involved, thank goodness – this was before the Age of the Designer in museums – so it probably all looked muddled and over-full. But at last it was done. White walls, grey paintwork with primary colours at the staircases to denote the various floors, crisp display cases, and hideous screens covered in mustard and olive green hessian, set on poles with wide circular feet."

"On the first floor, another temporary exhibition gallery on the left, echoing the one below it, and all along the line of the building, across the raised central hall, were the main galleries where we exhibited the City's collections. Ken's vision was to demonstrate through the collections 'The History of English Taste'. Ken was not an aesthete looking for the most breath-taking exhibits, nor a connoisseur searching only for the genuine (though he was very effective as a judge of fakes), but a curator searching for the typical, the fascinating, and the 'right': 'right as ninepence' was a favourite saying of his as we trawled together through Nesbits, the local auction rooms. (Ken taught me many things – for example in an auction room never look too closely at the objects you are interested in; pass by them once or twice, give them a glance. You do not want to alert other bidders to the presence of something the museum might be interested in, and remember, everybody on the auction room knows who you are, even if you may not know them)."

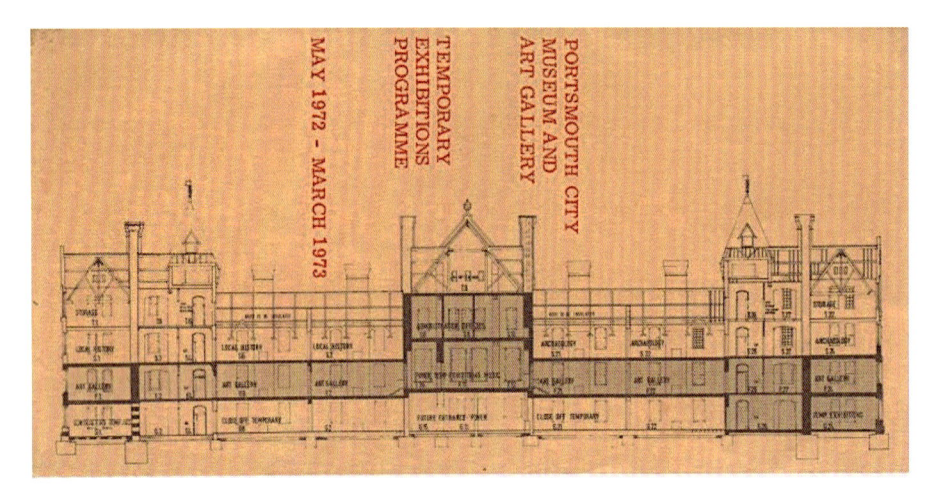

Portsmouth City Museum and Art Gallery Temporary Exhibitions Programme May 1972-March 1973

"The gallery was arranged chronologically backwards in terms of objects, with the Edwardian and Victorian material at the east end of the building, and Tudorish oak furniture, including a huge oak bed and an oak fire surround apparently from Chilham Castle in Kent. These dominated the end gallery, west. Large objects were on open display, smaller more fragile things in large Edmonds cases, with their unwieldy glass sides that needed two people and courage to open them. The general plan was 'muddle', wound into an underlying chronological skeleton. Thus, contemporary prints were hung with period furniture, and modern things like David Hamilton's ceramic cast of a boy sitting on a bench, and the occasional sculpture we bought, such as a tall glass piece, and an elegant smooth assemblage of polished granite by David Thompson. By the way, we had a lucky break with the prints – at Nesbits we bought prints by Richard Hamilton (*Kent State*) and Claes Oldenburg which were being sold by the GPO having been lost and damaged in the post; and at Percy Beer's two-day sale when his print dealership in Southsea closed, we bought many dozens of mixed folios of prints."

"It was opened by the Lord Mayor in chain of office, robes of state, and an entourage, all guarded by the Portsmouth Militia, a group of old men who liked dressing up as a Nelsonian fighting force carrying World War I rifles."

"My curator colleagues were not under quite so much time pressure to get the place opened, as only the Art Department was involved in the Alexandra Barracks building initially. Bill Corney was Curator of Local History. A tall, fastidious and kind man who Ken wanted to get rid of, and who got a hard time mainly because he was much older than everybody else, had been there longer, and had had it easy for years. His expression when thinking of an answer to a question stabbed at him by Ken was to look upwards, bend back slightly, say 'aaah –um', and scratch his bum. The glamorous Liz Lewis was Curator of Archaeology, who, when she was not out digging somewhere, had an office in Southsea Castle. This she shared with the Education Officer, Tony Higgott, who like Bill Corney was being ridden by Ken, and suffered greatly from his onslaughts. Tony's job inter alia was to create collections and displays to go to schools, and to circulate them in boxes made by the Museum technician. It was all neat and elegantly done, but took far too long to satisfy Ken's chronic impatience. Somewhere deep in the dungeons of Southsea Castle a craftsman lurked. He made Tony's boxes. His name if he had one escapes me, but he was reputed to have three wives and ten children down there."[8]

At the same time as the conversion a formal parterre garden within a semi-circular wall was laid out on the former parade ground, incorporating monumental stone features from the other demolished barrack blocks. The walls came from Crichel Down in Dorset, which had been compulsorily purchased in 1938 by the Air Ministry for use for bombing practice by the RAF. Cllr. Freddie Emery-Wallis instituted the practice of architectural salvage – stored at Fort Widley, which was the source of this stonework and of the classical temple re-erected in the grounds of Gatcombe House in 1973. But it was decided not to install the central fountain from Crichel Down behind the city museum.

Alexandra Road was renamed Museum Road. The museum and displays have evolved over the years. A lift was installed at the western end. In the 2000s the city's military history and the museum building itself was explained in the central gallery. Thanks to the Lancelyn Green Bequest Portsmouth also has the largest collection of its kind of memorabilia relating to Sir Arthur Conan Doyle, who set up his first medical practice in Elm Grove Southsea. The display features items relating to the iconic Consulting Detective from across the decades, as well as a few items that belonged to Doyle himself. 294,590 visits made to Portsmouth Museums in 2015/16.[9] Recent popular exhibitions on local themes include 'The Tricorn Controversy in Concrete' in 2014, ten years after its demolition, 'Edward King. A Life in Art' in 2015-16 and 'Game Over: The Future of the Past' the largest private collection of TV games, home computers and consoles in the UK – still available to play in Game On in Portsmouth High Street.

City Records Office

Next door to the east of the City Museum is the former City Records Office, once the Portsmouth NAAFI. A single story gabled red brick building set back and the barracks caretaker's house parallel to the street in a similar style and materials to the museum, it has been standing empty after temporary uses moved out once the archives were transferred to the city museum. It has severe structural problems because it was built over the town's ramparts and had cracked its foundations.

We must go back to the 1950s for the earlier stages of this story. Before Freddie Emery-Wallis set up these museums he said: "one of the first things I got involved in was the National Register of Archives, which - Mr Tuck, he was the Electoral Registration Officer of Portsmouth - the archives had survived under the steps of the Guildhall - and one day he said to me, 'would you like to be the convenor of the south east branch of the National Registry of Archives?' and foolishly, I said, 'yes'. And I have never left that path since that day and I campaigned for a Record Office in Portsmouth. That took me seven or eight years, I suppose, we got it in 1958. We had our Record Office and our Archivist and…. my report that I sent to the Portsmouth *News* established that you could open a Record Office in Portsmouth in 1952, 53 or 54 for about £750 a year. Then the job, of course, was to persuade the council, that one: they had a responsibility - and they had cared for the city records quite well - but they were packed away in boxes – and, they should in the reconstruction of the Guildhall, build a Record Office within, and eventually they agreed. Very largely through … a remarkable lady, Miss Toats, who then ran the National Register of Archives, who was very persuasive, and they had a remarkable tea party at Cumberland House, with an exhibition, in 1956, and that then did the trick."[10]

Rosemary Billett describes local people's efforts, which culminated in Freddie's election as a city councillor: "In the 1950s there was an informal group meeting weekly in a room over an antique shop in Southsea to catalogue and where necessary translate the written archives of the city, which Freddie had discovered, which had been saved from destruction in the devastating bombing of WWII. This group and others who were greatly concerned about the need to assess and conserve what was good about Portsmouth met informally at Cumberland House. They felt that this need was not being addressed by the council, and that their only hope was to get a member who had this interest at heart elected to it. As Freddie was the only self-employed person (i.e. able to attend council meetings) it was suggested that he was the ideal candidate, and with unanimous support, he agreed. The first urgent task was the appointment of a City Archivist, on which Freddie had already been engaged. This was achieved when Betty Masters took the post. A dinner was given in her honour, to mark this big step forward. It was a joyous occasion!"[11] Betty was succeeded by Michael Willis-Fear, Margaret Hoad, Sarah Quail and Dr John Stedman.

The NAAFI building proved difficult to convert. Sarah Quail said "Records offices are places of deposit and there are official standards set out for them. In those days it was BS5454. They have to have the right atmospheric conditions - temperature and humidity for storage of records. The choice of the NAAFI building for off-site storage in about 1970 was a short-term expediency, which became in due course a long-term necessity. Remember that the most frequently consulted material - and the Search Room - was still in the Guildhall when the NAAFI was acquired for additional storage. The City Records Office did not quit the Guildhall and consolidate on the one site in the old NAAFI building in Museum Road, until the late 1970s." "The NAAFI was built as a space for people – to drink cups of tea, eat a currant bun, drink warm beer and play billiards. It's not suitable for storing anything. The efforts to make it into a records office were very difficult to do. It was never constructed for storing documents, and we were always on a hiding to nothing trying to improve things. It was not ideal. The site did have some positive points though. Putting it into public use was a good idea: it's well placed. It is on a main thoroughfare with public transport and there is space for parking so you can certainly argue that it was accessible in all senses of the word."[12]

A new floor had to be laid to take the heavy rolling storage racks. John Parkinson established a bindery and repair workshop in the western end of the building. "The old record office building was never ideal as a store and its physical condition was deteriorating. It had grown as new rooms were built on to the original NAAFI and barracks caretaker's house when more space was needed. There were many flat roofs, which leaked, and it had been extended across the moat of the fortifications, without adequate foundations, so one wing was subsiding. The National Archives were questioning its fitness before the end of the 1990s. There were several possible schemes explored for providing alternative accommodation for the collections, but none came to fruition until the conversion of the second floor of the Southsea Library in 2013-14. In the meantime the former NAAFI building suffered 'planning blight' and its structure deteriorated without substantial maintenance, until c2011, when the roof of the subsiding wing was replaced. That resulted in further movement and that wing became unusable."

"The City Records Office was closed in towards the end of 1994 or start of 1995. The records office and the city museum service were merged on 1 April 1994. It made sense for the public searchroom to be moved across into Portsmouth City Museum (as it was then called) as that brought all the public services into one building on the site. A gallery was turned into a searchroom. Part of the museum was converted into a store for the archives most frequently called for. The former public and office areas of the records office building became stores for more archives (allowing expansion of the collections), the museum costume collection and the museum shops, which were being revamped. The stores were decanted to the first floor above Southsea Library in the former Woolworths shop in Palmerston Road and in the Portsmouth History Centre in the Norrish Library."[13] In 2020 a block of flats was proposed for the site.

Sarah Quail was impressed by the efforts that have gone into providing new and better accommodation for the Record Office collections. "The Local History Centre in the Central Library provides excellent space for people to work in and the most popular original records are available to consult there with of course the excellent secondary resources, the Local and Naval Library Collections. The new additional storage space for records above Southsea Library has been designated as a suitable place of deposit by the National Archives and provides good, secure accommodation. The temperature and relative humidity of this space is well within the required parameters. Staff should be congratulated for securing these improvements despite the very limited finance available for such enterprises today."

The D-Day Museum

D-Day Museum south elevation March 2020

Soon after Richard Trist took up the job of Chief Executive of Portsmouth city council in January 1978 he received a letter from Lord Dulverton who had commissioned the Overlord Embroidery in 1978, a 272-foot long (83m) textile as a tribute to the sacrifice and heroism of those men and women who took part in Operation Overlord, the Allies' operation to liberate Europe during the 1939-1945 war. Designed by the artist Sandra Laurence, it took 20 women of the Royal School of Needlework five years to complete. He thanked the city council for their offer to house the embroidery and explained that he thought a greater number of people would visit it in London. He had decided to display it in Whitbread's London Brewery.

Overlord panel

"Out of interest I glanced at the file where I read the correspondence involved in the City bid. I noticed a rough sketch, back of envelope style. It showed a building shaped like a conch shell, initialled by Ken Norrish, then the Deputy City Architect. I closed the file, sighing that you do not win every bid." [14]

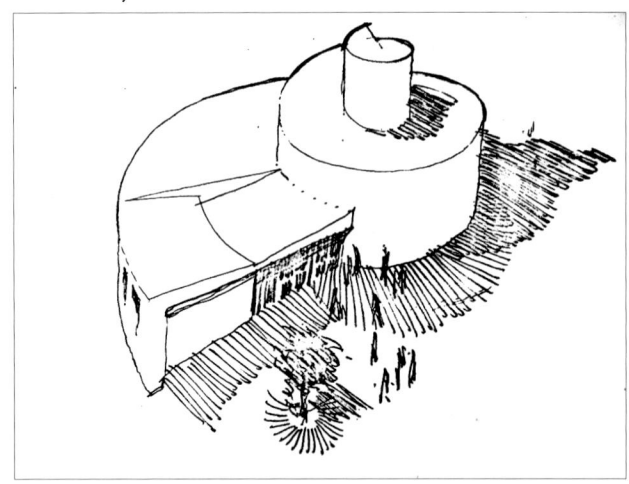

Initial design Ken Norrish Collection
Portsmouth History Centre

In 1983 the approaching 40th anniversary of D-Day – 6 June 1944 – and the key role played by Portsmouth – galvanised the city council into planning a dedicated museum. 156,000 American, British and Canadian forces had embarked from many southern departure points and landed on five beaches along a 50-mile stretch of the heavily fortified Normandy coast. Codenamed Operation Overlord, the invasion was one of the largest amphibious military assaults in history and required extensive planning. Southwick House behind Portsdown Hill still houses the D-Day map where the advanced planning of the invasion took place while the communications base was underground in Fort Southwick.

Fort Widley which overlooks the panorama of Spithead and the Solent was first considered as the location for the D-Day Museum, because Generals Eisenhower and Montgomery watched the invasion unfold from the top of the fort's seaward facing barracks. The naval base was also considered, but the trust believed that it would be a possible disadvantage to be surrounded closely by the *Victory*, *Mary Rose*, *Warrior* and the Royal Naval Museum. The other stimulus to building a new museum was that the Dulverton Trust needed to find a new and popular location for the Overlord Embroidery.

Richard Trist takes up the story. "One evening in July 1983 I was working a bit late in my office when a phone call came to me. The caller was the local admiral, Rear Admiral Tony Tippett, Flag Officer, Portsmouth and Naval Base Commander (Later Vice Admiral Sir Anthony Tippett KCB, Controller of the Navy). He asked whether I was aware that the Overlord Embroidery was on the move. He explained that Lord Dulverton had decided to move it from London. Quite simply, there were too many other attractions in the capital. The admiral had this information in confidence but was one of those asked to help in the relocation. Here was a great opportunity for the city."

"Next morning I quickly looked again at that obscure file. It was clear to me that Portsmouth had been the runner-up choice of location at the time of the original decision. I also noticed that the enthusiastic Chairman of the Libraries, Arts & Museums Committee back in 1978 (Councillor Leslie Kitchen) was now (1983), by good fortune the current Lord Mayor of Portsmouth. Meanwhile Ken Norrish had risen to be City Architect and his sketch was sill on the file. I found Ken Norrish at a drawing board (he was not a desk-bound man), and asked him to provide an estimate of cost to build the conch-shaped museum shown in his sketch. He asked for a week and I suggested an answer after lunch."

"After lunch and armed with the rough estimate given me by the City Architect I met the Leader of the City Council, Councillor Ian Gibson. I asked if he was willing to back a bid for the Overlord Embroidery at a likely capital cost of half a million pounds for a display museum. He did not hesitate. I easily convinced the Lord Mayor that it would be a good use of the prestige of his office to sign a letter to Lord Dulverton. I drafted such a letter, inviting Lord Dulverton to consider Portsmouth again, the Lord Mayor signed it and it was dispatched before tea time that day."

"A couple of days later Lord Dulverton telephoned me, said he was very interested in Portsmouth's bid and invited me to London. The Lord Mayor who as onetime Chairman of the relevant committee had a great deal of knowledge about museums came with me. We offered to erect a good building, at the City's expense, to display the Overlord Embroidery on a fine site near the Portsmouth seafront. Shortly after the meeting Lord Dulverton again telephoned, this time to say that he accepted the proposal. For Portsmouth a new and important chapter was opening."

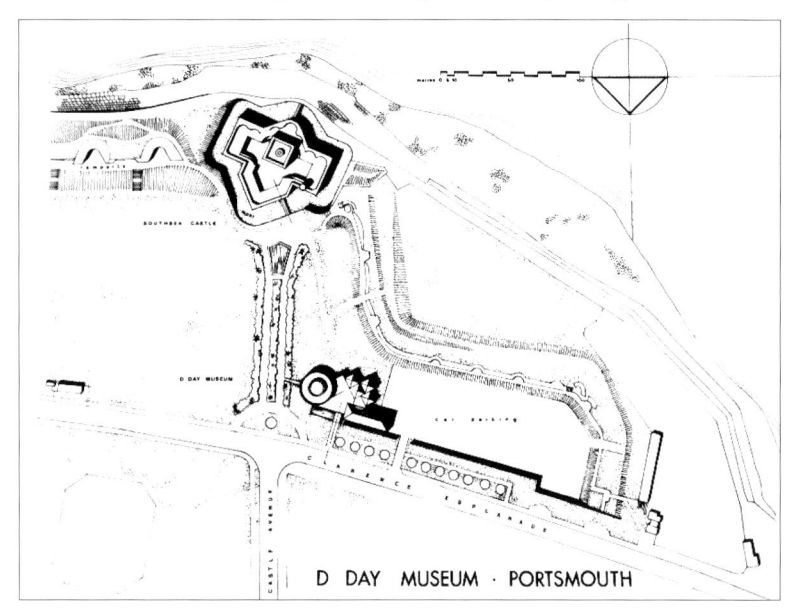

Southsea Castle and D-Day Museum Ken Norrish Collection Portsmouth History Centre

The Southsea Castle location appealed to the trust because of its connection with Henry VIII and its last military role in WWII, in particular the events leading up to D- Day. The Castle attracted over 200,000 people, who would also see the embroidery. The proposed location for the museum in Clarence Parade north of the castle was adjacent to the original War Department Transit Shed, ideal for displaying vehicles connected with D-Day, rather than at Fort Widley at the other end of the city, including a landing craft.

"The Leader of the Council then called together some of his senior colleagues from the Council majority party and asked for their support – it was happily given. There was however no real opportunity to seek formal approval from the Council or a Committee because the summer recess was upon us and many councillors were on holiday. I myself departed on holiday."

"Whilst I was away Ronald Tweed, the City Secretary and Solicitor took charge and the project moved ahead with consideration of the details – it remained pretty confidential. When the City Museums Officer, Tony Howarth looked at the project in outline he was reminded that there had always been a Portsmouth idea for a permanent exhibition to commemorate the departure of the Operation Overlord forces – largely from Portsmouth – and the D-Day landings in Normandy. He could see that here was the opportunity to realise this dream on a major scale. Ronald Tweed organised another meeting with the Leader and his senior political colleagues. An informal agreement was given to a bigger project now estimated at a million and a quarter pounds."

"I returned to find the project much grown and proceeding with enthusiasm. There was extra pressure on everyone because the June following, just ten months away, would see the fortieth anniversary of D-Day itself. It became a clear target date and an obvious occasion for a high profile opening ceremony."

"As autumn arrived we needed to address the formalities of a major project which had progressed a long way without formal approval. Committees were summoned and reports sent off towards the City Council. In the Council meeting some questions were raised, obviously, but in general everyone recognised that the D- Day Museum would be good for Portsmouth. The City Council adopted the project."

It was built in ten months. The challenge to Ken Norrish (City Architect 1970-1988) who designed the museum was that the 34 panels needed to be shown continuously. In London they were arranged around four walls of a huge rectangular room, which would be expensive to provide. Many different configurations were considered, to save both space and costs. Norrish's sketches from 1984 show the evolution of his design, from a hexagon to a coiled snail shell rising in height to a central tower.[15]

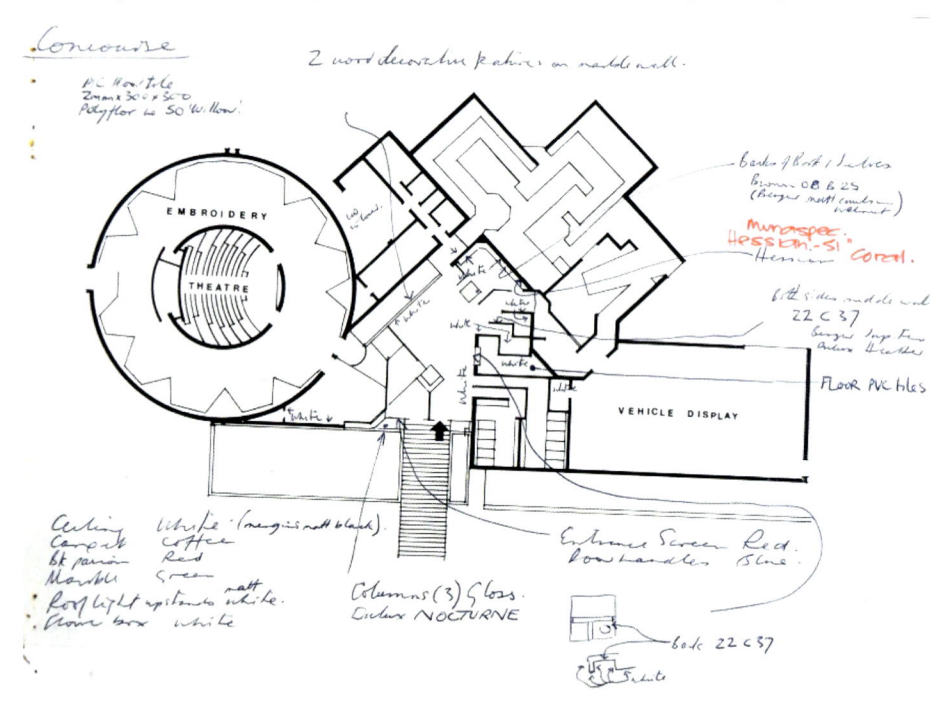

Plan with annotations by Ken Norrish.   Collection Portsmouth History Centre

"By now it was October and the target date was rushing towards us. The Council approved a construction contract with Warings, a respected local building firm. Teamwork and trust developed from the urgent nature of the project – the builder pointed out areas where the budget might be exceeded by design changes, which were inevitable. He remarked that many of the drawings were initialled by Ken Norrish himself and were dated on Sunday. Ken Norrish even designed a special tool for the bricklayers to ensure that the inset pointing was even throughout. Construction moved ahead at a pace and the target date (June 1984) remained firmly in sight."

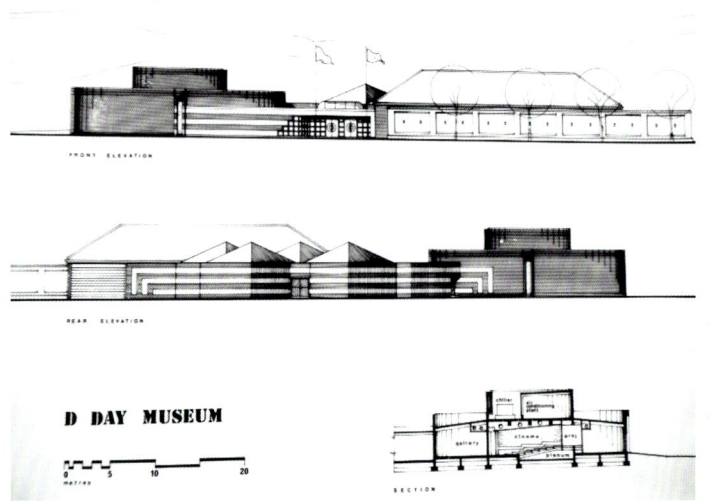

Elevations Ken Norrish Collection
Portsmouth History Centre

Eventually, by continual refinement, a circular building in striped red brick was built to display the embroidery. The entrance flanked by flagpoles flying the Allies' flags was approached by a wooden bridge over water - to symbolise the crossing of the Channel on D-Day, and to offer interesting reflections. Construction was completed in a remarkable ten months – in time for the 40th anniversary. To the left of the gift shop and bookstall and toilets, a gentle ramp took visitors to the oval auditorium – a drum within a drum, surmounted by a large plant room above, a silhouette to echo Southsea Castle to the seaward. The pairs of cranked concertina panels displaying the embroidery were intended to excite interest and gain by juxtaposition of the vivid imagery and colours of the textiles. Adjoining the drum the museum tells the D-Day story.

"Meanwhile the Museums staff worked on designs for the interior layout in conjunction with the architect. New showcases were carefully designed to allow maximum illumination for the public and minimum damage to the embroidery. An expensive mock-up display case was built to make sure it would work. A shed which had once had a military use was incorporated into the design of the museum. Vehicles and military hardware were gathered – some already owned by the city. Others helped too. One display showed Portsmouth dockyard machinery that was used in the preparations for D-Day. The Ministry of Defence insisted on them being paid for so Captain Colin Allen RN took out his chequebook. He was the Chief Staff Officer (Administration) for the Admiral and the official link with the City Council. Members of the public arrived with contributions. One such was a mattress handmade for a family Anderson shelter."

"It had, from the very start, been pretty obvious that the ideal person for the Opening Ceremony would be Queen Elizabeth the Queen Mother because she had, with her husband, King George VI, been a powerful leader to the nation through the war years. She accepted the proposal – the opening and the commemoration of D-Day were to be spread over several days."

"There was a rush at the last moment. In the final hours the City Museums Officer was seen painting a display himself and finishing touches were added through the night. At ten in the morning the building was cleared for a security check and brooms swept over the red carpet. The Leader spoke in welcome; Queen Elizabeth made a gracious speech, cut the ribbon and declared the museum open. Her Majesty then toured the display and was shown a portrait of her late husband the King within the embroidery itself."

"The Queen kindly lent the Royal Yacht *Britannia* to add glamour to the proceedings and the Queen Mother invited veterans of the D-Day landings and those concerned with the museum project to a party on board. An exciting Portsmouth project reached a happy conclusion."[14]

In giving the museum its Best New Building Award in 1985 the Portsmouth Society praised how well it fits into its special seafront setting, providing a round counterpoint to the square stone of Southsea Castle. "Its interior planning brilliantly manages the large numbers of visitors… The judges particularly commended the display of the D-Day embroidery, enclosing the oval auditorium, and the use of three dimensional displays and sound". Celia Clark was amused by the reactions of her family to the audio-visual commentary when the museum first opened – shushed by others in the auditorium. Her father, a D-Day veteran, said loudly: "Stuff and nonsense!" about the commentary… and her aunt, who married a GI, sitting next to her, dissolved into giggles at "Overpaid, Oversexed and Over here!"

On 5 June 1994 D-Day 50 was celebrated by heads of state including Bill Clinton, John Major and Lech Walesa, royalty and many veterans in ceremonies on Southsea Common nearby, followed by a fleet review. *HMY Britannia* sailed past the vessels anchored in the review on her way to Normandy.[16] The Queen was accompanied across the channel by veterans: a representative soldier, sailor and airman who described their experiences of fifty years before. Celia Clark's father, Sir David Willison Royal Engineers was the soldier. He had landed at Sword Beach to secure Pegasus Bridge on the Caen canal, where he was severely wounded and brought back to Haslar Hospital.

In preparation for the 75th anniversary in 2019, a £5m project was set in motion in 2015 to upgrade the museum. A key aim was "to build the profile and reputation of Portsmouth as a

Waterfront City by showcasing its national and international heritage and attracting new audiences". The project included the removal of the collection and equipment to enable specialist conservators to undertake remedial conservation work on the embroidery and vehicles, and to develop the storylines and content in new displays. The museum was closed to the public from summer 2017 to 2018 while it was completely refurbished as part of the city's £4.9 million contract, with funding from the Heritage Lottery Fund and Arts Council England[17] and a donation of £45,000 from the Victorious Festival in 2014. The museum reopened on 30 March 2018 with three new exhibition galleries, a café and shop. New landscaping replaced the pond and also the 1930s floral clock with hard paving.

LCT7074 under restoration in the dockyard shiphall @SeaSpitfires

The largest item outside is LCT 7074, the last surviving tank landing craft, which delivered troops to Gold Beach, Normandy at about midnight on D-Day. Jim Jarman's book *Those Wallowing Beauties The Story of Landing Barges in World War II* has eyewitness accounts of the severe difficulties of operating landing craft in the storms of 17-18 June 1944 on Gold Beach. LCT7074 was used to remove vehicles damaged or swamped in the surf blocking access to the beaches and to refloat other landing craft. After the war it was used as the floating headquarters of the Master Mariners Club. It subsequently became a nightclub ("Clubship Landfall") owned by George 'Jud' Evans. From the late 1960s to early 1970s it was moored in Canning Dock, next to the (then unrestored) Albert Dock in Liverpool. In 1972 it moved to Collingwood Dock and then to Birkenhead Docks where it later sank. In 2017 the Heritage Lottery Fund awarded the National Museum of the Royal Navy, which now owns it, a £4.7m grant to raise and to restore it in the naval base[18] and then to put it on display in front of the D-Day museum transit shed. The plan was to install it with the Sherman and Churchill tanks on board to coincide with VE Day in May 2020, but Covid 19, the need for further fundraising and storms delayed its move from the dockyard.

LCT7074 passing Portsmouth's War Memorial    LCT7074 approaches its new shelter

On 24 August 2020 it was finally transferred on heavy rotating wheels from a pontoon moored parallel to Southsea beach to trundle slowly along the seafront road, watched by a large crowd of interested onlookers: a surreal sight! It took a further week to lift the two tanks inside it and to slide it sideways to rest under its heavy wavy steel shelter designed by architect Giles Pritchard. As it's in the public domain outside the museum entry was to be free via its ramp, raised each morning and closed at night. Nearby are an anti-aircraft gun and the statues of Gen. Montgomery and an exhausted soldier.

The Submarine Museum, Gosport

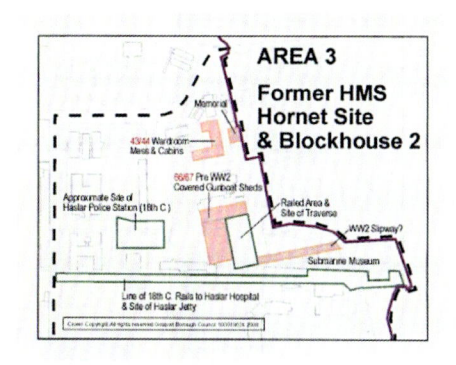

Blockhouse and Haslar Gunboat Yard form part of the wider Haslar Peninsula Regeneration Area that also includes the Royal Hospital Haslar site, plus Haslar Marine Technology Park that is home to QinetiQ. The site occupies four parcels of land known as Blockhouse 1, 2 and 3 and the Haslar Boat Yard. In December 2018 the three Blockhouse sites were owned by the MOD, whilst the Haslar Gunboat Yard was until 2020 owned by Starvale Developments. Blockhouse 2 includes the RN Submarine Museum and the Joint Services Adventurous Sail Training Centre (JSASTC). The museum is a popular tourist attraction.

Gosport Borough Council

The Submarine Museum started life as a small Submarine Branch Collection above St Ambrose Church in HMS Dolphin. Along with its sister museums: the Fleet Air Arm Museum and the Royal Marines Museum, it was officially recognised by the MOD in 1967 and the following year a curator was appointed. In 1970 it was registered as a charity and in 1978 it moved outside HMS Dolphin, opening access to the public. The submarine *HMS Alliance*, a 279-foot post-World War 2 submarine was given to the museum. Funds were raised for it to be put on display and the revised museum was opened in 1981. In 1983 a display building was opened and access to the inside of HMS Alliance and the much older submarine *Holland 1* were provided.

By the early 1990's *Holland 1* was showing signs of rusting badly and a four-year preservation exercise was undertaken. This stopped the deterioration and she was put back into view in a special climate-controlled building. The project won a national conservation award in 2002.

In 2008 a process was started to integrate the four of the four naval service museums: the National Museum of the Royal Navy: the Submarine Museum, the Fleet Air Arm Museum and the Royal Marines Museum under a single Director General. In 2011 the Royal Navy Submarine Museum was "vested" in a new company. It became a company limited by guarantee and all assets and liabilities were transferred to the new charitable company. The original charitable Trust was closed. The museum has archives holding over a million pages of documents, countless photographs, and 4,000 books – a resource eagerly sought after by researchers.[19]

Martin Marks

HMS Sultan Marine Engineering Museum

Within HMS Sultan in Gosport is a small museum: the Marine Engineering Museum. It is operated by a registered charity, the Royal Naval Engineers' Benevolent Society (RNEBS) Alston Memorial Historical Society. It is run by volunteers. The Society's aims are: "To provide and maintain an exhibition for members of the public including the engineering branch of the Royal Navy of documents, artefacts, memorabilia and other items connected with engineers of the said engineering branch for the purposes of educating the public and members of the said engineering branch in engineering and in its traditions, of furthering espirit de corps through knowledge of the past, and of promoting its efficiency." It achieves this over two floors of a small part of the building which a gives the history of marine engineering in the Royal Navy, displaying a collection of records, artefacts and memorabilia covering marine engineering training in HMS Sultan, Fisgard (at Torpoint, Cornwall) and Caledonia (Rosyth) from 1903 to present day. There are some interesting displays of the careers of naval artificers who achieved officer status at flag rank (admiral). The museum is usually open to the public during Gosport Heritage Open Days and by request to the museum for private visits.[20]

Aspex Gallery Gunwharf

Towards the end of the 1970s a small group of teachers and former students from the art department of Portsmouth Polytechnic (now University) had come together: "to create an artist-run space of studios and gallery. Artist-run spaces are now recognised as crucial to the artistic ecology of the country and are known for their exciting, subversive and inventive approaches to programming that tend to be focused on other artists rather than on art collectors or the general public, and these small organisations are often able to create small, yet flourishing, artistic networks."[21]

Art Space developed two floors of a converted chapel in Brougham Road Southsea as their studio space and Aspex art gallery. The first exhibition in 1981 was the work of Rachel Fenner, who designed the brick sculpture in front of Hedley Greentree's dramatic office block at the junction of Lake Road and Kingston Road. The Aspex Gallery's move into the Vulcan building in Gunwharf is explored here, while the site's redevelopment is set out in our chapter on the two Ordnance Yards.

Gunwharf, an ordnance depot developed over the town's northern fortifications from 1709 played a key part in two services' technological history - where innovation was a constant thread. It was used as the navy torpedo and mining school, and redesignated as HMS Vernon from 1920-1986. Conflicting values - in economics, including potential employment regeneration, heritage, and tourism - were aroused by debate about its future. The Portsmouth Society was concerned at the lack of any cultural dimension in Gunwharf's otherwise commercial redevelopment of housing and leisure shopping. They recognised the potential of the major surviving historic building, the Vulcan, an ordnance storehouse dating from 1814, as an art gallery. Celia Clark and Roger James got permission from the MOD to take Chris Carrell, the City Arts Administrator to see the uninterrupted three floors of space in the south wing. Chris exclaimed, "**This** is my art gallery!" The Society invited major national institutions including the Tate and Imperial War Museum to use the space, but they turned the opportunity down. But in 2006 the Aspex seized the chance.[22]

Far left: Vulcan Building converted for Aspex Gallery 2009.
Left: Inserted office above reception

In April 2006 they moved into the conversion of the Vulcan's southern ground floor wing. Designed by award-winning architects Glen Howells, funding came from Arts Council England and the city council amongst other sources. Joanne Bushnell, Director of the gallery, feels the building has just the right mix of new and old. "We wanted to retain its historic structure yet give it a contemporary layout."[23] Complex negotiations with English Heritage were involved. The height of the indoor racks once used for storing war materiel was lowered in order to insert a first floor range of offices inside part of the space. The surfaces of the massive brick walls were left unplastered and the massive doors were retained beside modern openings.

"The gallery, which is all on one level, consists of two exhibition spaces and an open plan shop and café selling locally produced organic food. In a bid to offer support to aspiring creative practitioners the gallery also has an artist resource centre, called ARC, which provides information and support for creative practitioners in Portsmouth and across the southeast."

Aspex Gallery
Winner of Portsmouth Society
Best Restoration Award 2007

Inaugural events celebrating the re-launch of the gallery included Greetings from Pompey by the internationally renowned and Portsmouth based Caravan Gallery. Their travelling exhibition of locally taken photographs in a reincarnated 1969 yellow caravan takes contemporary art to unexpected locations and to tens of thousands of people both in Britain and abroad. "Inspired by tourist information and the rose-tinted language of advertising, the Caravan Gallery invites the public to help them depict a more diverse view of their home towns by contributing artwork, ideas and artifacts to Pride of Place Projects – temporary information centres in empty shops."

The elegant conversion of the Vulcan storehouse won prizes for design from the RIBA and Portsmouth Society in 2007. Unfortunately the closure of the footpath access by the surrounding residents severely affected the gallery's visitor numbers; there is also little on the site to remind visitors of Gunwharf's important technological history.

Old Portsmouth

"For hundreds of years, Portsmouth's importance as a naval base meant that it was one of the most heavily defended cities in Europe. Recognising that whoever controlled the Point also controlled the Harbour - it was here that the early inhabitants erected the first permanent defences. At its peak, it was the cultural heart of the city, attracting thousands of sailors and visitors from around the world, and has been immortalised in works by artists such as cartoonist Thomas Rowlandson and the composer William Walton. " (Submission to the Heritage Lottery Fund by Portsmouth City Council).

Most of the elaborate defences of Portsmouth and Portsea were destroyed in the late nineteenth century, so these seaward survivors are all the more precious now. The final and most significant modifications took place following the Old Portsmouth fortifications' decommissioning and acquisition by the city in 1958. A popular top-level seaward walkway was added, which has marvellous views of the harbor entrance, out to Spithead, Haslar seawall and to the Isle of Wight. In 1962-3 the boundary wall and buildings associated with Point Barracks adjoining Broad Street were demolished. The rear walls and ammunition recesses of the main gun casemates were removed and the soldiers' barracks cut back to half their width. Their original extent is inscribed into the cobbled surfacing of the courtyard. This clearance ultimately opened up the space and adjacent foreshore to the public. In the early 1970s Deane Clark, city historic buildings architect supervised restoration of the brick and stonework of the casemates. As well as the award for the dockyard colonnade, Portsmouth City Council won a European Architectural Heritage Year award in 1975 for this work and for the restoration of Southsea Castle. On the other side of the return battery is a boatyard for Portsmouth Sailing Club, enclosed in a wooden wall incorporating rescued wooden stanchions from sailing ships.

Wedding in Square Tower　　　　　Steve Hender　　　Round Tower 2017

The fifteenth century Round and Square Towers in Broad Street Old Portsmouth both belong to Portsmouth City Council. Halfway along Eighteen Gun Battery only a masonry projection into Broad Street survives of King James's Gate – most of which was moved to Burnaby Road. The Grade 1 Square Tower has been used in various ways - as a museum and exhibition space. One especially vivid display in 1987 commemorated the bi-centenary of the First Fleet's sailing to Australia, when the continent's strange animals and landscape were evoked – on a bed of sand. Now leased to Stephen and Julia Hender, successful entrepreneurs, it plays host to many weddings, parties, concerts, plays and arts and crafts exhibitions.

The Round Tower's special rooftop viewpoint looking out to sea, across the harbour to Gosport and to the harbour is dear to locals and to sailors' families. We crowd onto the tower top to welcome or to wave farewell – to ships and their crews. The round rooftop marker shows the distances to places far and near. The interior has suffered from damp for many years which damages temporary exhibitions inside it. Recent repair work may tackle this problem, making it available for events and shows.

According to Richard Massey, Inspector of Ancient Monuments, Eighteen Gun Battery has "a powerful sense of place... With commanding views across the Solent and the mouth of Portsmouth Harbour, the battery also occupies a pivotal location within the historic defences of the city which, along with the adjoining Round and St Edward's Towers, is fundamental to any appreciation of Portsmouth as a naval and maritime centre". He identified it as "having the potential for re- defining this area as a cultural focus and public space."[24]

The Falklands memorial by the Sally Port alongside the Square Tower is often decorated with wreaths and flowers. The massive entwined shackle rings of the Settlers' Memorial has a sister sculpture in Sydney commemorating the First Fleet of convicts transported to Australia. The Pioneer statue of an emigrant family who left for a new life in America was presented to the city of Portsmouth in 2001.[25] The unusual tamarisk trees with tall trunks were propagated in Sweden. They enjoy the maritime environment. The casemates were occasionally used to sell their work by the Point Artists but over the past fifty years maintaining the battery in good condition become untenable without ways for it to generate its own revenue.

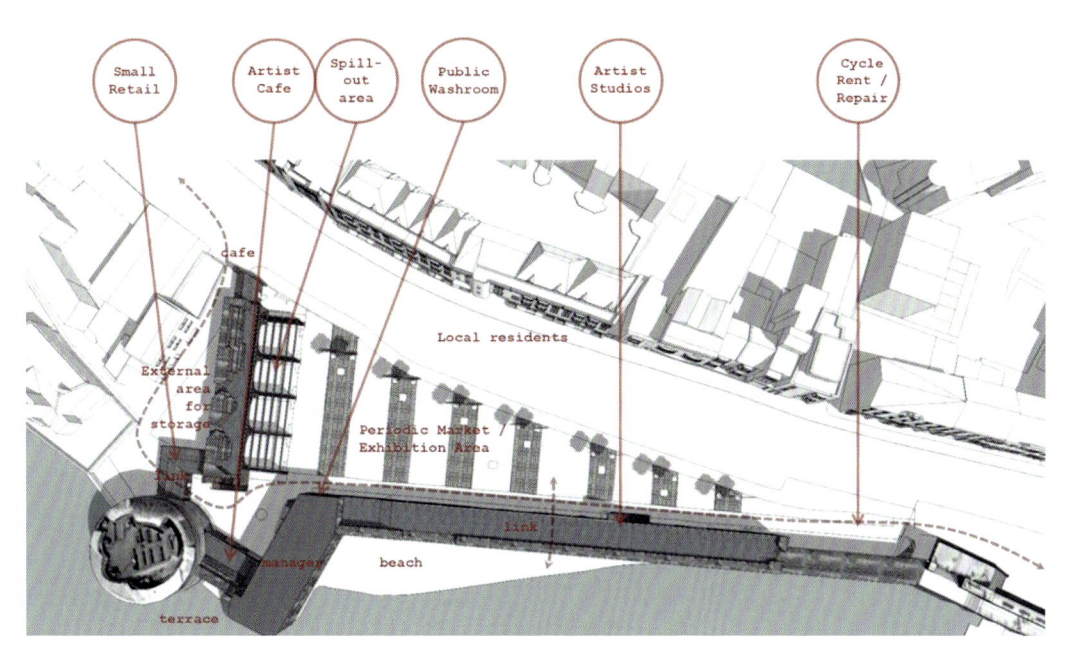

Hot Walls concept drawing before application for planning permission Deniz Beck RIBA

A most successful example of cultural reuse is the transformation of the casemates inside the defensive walls of Old Portsmouth by the insertion of new art studios and an atmospheric café. The origin of the ARTches bid for Heritage Lottery funding was the University of Portsmouth partnership 'Sea Change' consultation and report, which was related to the city council's Seafront Strategy. The project was outlined in the 2013 Seafront Masterplan Supplementary Planning Document which involved community engagement and considerable local consultation, as well as the city council's planning, asset management, seafront team, legal and property services. A councillor and the MP for Portsmouth South wrote to government ministers asking for financial support for the proposal. Portsmouth City Council was awarded £1.75 million from the government's Coastal Communities Fund to transform the historic and largely-disused Hot Walls arches into 13 creative studios: the 'ARTches', with the possibility of a brasserie, securing a sustainable future for them. The plan was to create 23 jobs directly, with a further 82 new jobs created indirectly. Apprentices would be taken on in the construction.

"It is incredible news for the city that we can make a cultural hub in Old Portsmouth a reality. The funding is not only a boost for our creative industries, but also secures the future for the local historic monuments which will be converted. They are vulnerable to damage and need ongoing investment. We are delighted this project has had so much support from residents, creative organisations and English Heritage. It was also the most popular suggestion in the seafront masterplan" according to Cllr. Linda Symes in 2014.

PLC Architects were instructed by Portsmouth City Council. Conversion and renovation of each space was designed to form a versatile working environment for its occupants and collectively act as a gallery, showcasing a variety of work and encouraging visitors to interact and engage with the artists. The new toughened glass frontages to the arches ensure that the visibility of the structure is not impeded. Secure shuttering is integrated within the framework, completely concealed when open, yet providing secure and attractive metal protective screens out-of-hours.

The new access terrace was designed to be simple and unobtrusive. The architect Deniz Beck said "Our ambition is to ultimately see the parade ground used as a regular, temporary marketplace, supporting activity associated with the arches and reinforcing its role as a hub for the city's considerable creative community. A new timber structure added to the return battery was designed to serve as a modern, three-dimensional interpretation of the original extent of the soldiers' accommodation (already demarcated in the paving) and, along with the new bench seating and foliage, extends the appeal of the arches into the adjoining open space. This encourages people to feel comfortable using the space for leisure, relaxation and socialising."

Layout for Canteen café with original position for kitchen   Deniz Beck RIBA

The project was granted £1.775m by the Heritage Lottery Fund, Portsmouth city council contributed £100,000 and the Partnership for Urban South Hampshire £40,000.[26] This paid for extensive brickwork cleaning, service installation for the toilets and the studios and restaurant, and for excavation and careful relaying of the Parade Ground. Mountjoy Building were the contractors. The popular Canteen café in the casemates between the Flanking Battery and the Round Tower has brilliantly opened up spaces which were hidden and inaccessible for many years. The kitchen has a fine view of the harbour entrance; the triple glazed ground floor of the searchlight battery makes a cosy room looking out to sea. The café also occupies rooms on the ground floor of the Flanking Battery. The landward-facing gunports were reopened and fitted with toughened glass - since they face a shingle beach. Especially dramatic is the corridor with ovoid arches and narrow gunslits towards Point Beach.

Hot Walls Studios – lateral view 2017   Karl Rudziak at work in his Hot Walls studio

The Hot Walls Studios and Canteen café were opened on 8 July 2016.[27] The project was shortlisted for an RIBA South Award, for another for Constructing Excellence, and an Architect of the Year Award 2017; it was also a Building Excellence Awards Regional Finalist in 2017. The new activities

animate Broad Street at ground level as never before, and the Canteen is so popular that chairs and tables now spread out into the open space. The design gives local people a whole new perspective on the city's maritime heritage, bringing us closer to the water, shipping and the life of the harbour. Historic England is now using this project as a 'good' example to demonstrate to other design teams in terms of how sensitive conversion can be done successfully.

The architect, Deniz Beck said that this most loved area of the city with the Scheduled Ancient Monument needed a particularly sensitive approach. It was one of the most difficult projects she had ever worked on, becoming an emotional roller-coaster, with the local residents polarised by the project into pro- and anti-camps during the public consultation process. However, many of those same anti- residents later said how pleased they were with the result after seeing the finished project.[28] "One of the most satisfying part of this whole process was that during the public consultation there was a petition against the project, then another one started supporting it. Once the development was completed we heard that the value of the houses in the area increased by 15 %; also seeing some of those angry residents regularly in the Canteen terrace makes my day."

Canopies under construction 2016

Hot Walls Artists' Studios 2016

As well as the studios where we can see the artists and craftspeople at work the architect played a master stroke: daringly, she proposed an outdoor platform for the café, connecting the searchlight battery to the Round Tower. When you open the flood resistant door you see what she achieved: a balcony flung out to the tower where you are nose to nose with passing ships: a timber deck and glass balustrade fixed to a steel frame structure. Despite fears that the heritage authorities wouldn't allow this, they did – so now you can enjoy a glass of rosé in the sun and wind, very close to passing ships entering or leaving Portsmouth harbour.

The Canteen café: seaward view: Round Tower, converted searchlight battery and balcony link

Royal Armouries in Fort Nelson

Fort Nelson restored access bridge

The Royal Armouries are custodians of the United Kingdom's national collection of arms and armour, national artillery and national firearms, one of the largest in the world. As it continued to expand the Tower of London became too small to house it all properly. As a result in 1988 the Royal Armouries took a lease on Fort Nelson, the large 19th- century artillery fort on Portsdown Hill to house their collection of artillery. The fort's history and restoration is described in our second forts chapter.

Fort Nelson Parade Ground Armstrong guns 2019

"The collection at Fort Nelson opened in 1995 as home to the national collection of artillery, with over 350 big guns and historic cannon on display. Covering nearly 19 acres and now fully restored, Fort Nelson sits majestically on top of Portsdown Hill, with amazing views of the Solent and the Meon Valley. It stands today as a monument to the skills and ingenuity of Victorian engineering and architecture."[31,32]

The fort itself is a great draw for visitors, who enjoy exploring its tunnels that run below the parade ground connecting the magazines with gun emplacements and the barrack rooms, parade ground and ramparts. Its panoramic views over Portsmouth, the Solent, Hayling Island, Portchester and Gosport, with the Isle of Wight beyond are splendid. In recent years visitor numbers have steadily increased from 106,000 in 1987-8. They doubled between 2000 and 2005. 232,000 visitors were predicted in 2018-9. This was attributed to the 'Poppy effect', people coming to see the travelling exhibition commemorating the dead of WWI originating from the Tower of London.

Part of Saddam Hussein's supergun ordered from a Sheffield ironworks Fort Nelson

In 2008 the Armouries gained a £2m grant to transform the site as a free heritage attraction from the Heritage Lottery Fund, added to the £1.5m matched funding.[33] The total cost of this work was £3-4million. From 2010-2011 further major redevelopment was again supported by the Heritage Lottery Fund. There are video presentations, and displays demonstrating the living and working conditions of the soldiers who manned the fort. These and the recreated officers' kitchen are amongst the most popular places in the fort. A large glazed gallery houses its most famous exhibits including the giant supergun ordered by Saddam Hussein. The bungalow to the northeast of the fort was demolished for construction of a new café and shop, which are buried under an earth cover. Live firing demonstrations were held; currently these do not happen but their resumption was planned. There used to be costumed guides, professional actors. It was hoped to provide more services in-house by training the existing staff. The glazed roof of the artillery gallery was damaged by high winds. £200,000 was needed to replace it with a double skin and insulation to manage the temperatures inside. The contents had to be moved for this work to be carried out. The Armouries acknowledge their debt to the Palmerston Forts Society. Until 2109 the society had received yearly donations towards its work, which includes the study of many other Victorian forts.

Recruiting office/art gallery and bookshop/office The Hard Portsea

Former Army and Navy Recruiting Office and Cell Bock

An early campaign by the Portsmouth Society succeeded in preventing the demolition and replacement of the small red brick Recruiting Office outside the Cell Block with a modern white building. The Society suggested that its location on the Hard right next to the historic dockyard was ideal for a tourist information office – which it became, following negotiations between its subsequent owners, the Portsmouth Naval Base Property Trust and the City Council. The Trust added a neatly detailed triangular glasswalled extension at the back and in the space adjoining the Cell bock. Following the City Council's closure of the most popular TIC in the city it was used as a gallery and maritime bookshop, but according to the trust, this use ceased as it was uneconomic; the tenant being unable to afford a meaningful rent necessary to keep the building in repair. Renamed Victory Gate Lodge in 2020 it was let as offices to the National Regional Property Group, which was working on restoration and conversion of other nearby former defence sites: Fraser Battery and Fort Gilkicker.

WWI Remembrance Centre Bastion 6 Hilsea Lines

No. 6 Bastion Hilsea Lines

Tucked away behind factories the WW1 Remembrance Centre has brought Bastion No. 6 in Hilsea Lines to new, appropriately military, life. From about 2013 the centre's private collection of artefacts, exhibits, a walk-through trench system and memorabilia was first housed in Fort Widley, but the way into it was down steep steps in the main barrack block not suitable for easy public access. In order to grow the collection to a national status as a museum, the owners then took on the decayed and leaking Bastion No. 6 in Hilsea Lines. It was in an uninhabitable state with a smashed up interior. Foliage was hanging down from the earth roof above and growing up from the ground. About five years ago work began to restore and open up the building, which is leased from Portsmouth city council. The Fort Cumberland Guard moved here from the fort when the rent there was raised too high, and their Monday evening parades, appropriately in 1860s uniforms, on the tarmac space outside encouraged the group to open up the bastion to the public. Volunteers have undertaken the considerable repair work and interpretation, many of whom are also enthusiastic collectors of WW1 memorabilia. They had to restore the floors, repair and repaint the windows and doors in green – based on surviving paint samples, working closely with Historic England. The hopperheads have the Royal Artillery emblem on the side. The armed forces, Skills Force and other interested parties have helped with such tasks as stripping lead paint from the vaulted ceilings. Students from the university's faculty of Creative and Cultural Industries produced design proposals and helped by their physical labour. In 2015 designer Sarah Roberts produced measured drawings and plans. English Heritage and Portsmouth City Council fully supported the venture. Regular events attract audiences such as the Army Cadet Band playing outside.

The centre's aim is to teach not just youngsters but everybody about WW1. The restored bastion is now filled with many genuine artefacts, a timeline of unusual stories from the war, a ship hall in the former battery room full of model warships including several models 10 feet long, and reconstruction of a trench system. The building is lit by solar panels since there is no mains electricity supply. It also has a big lecture area which is being developed. The National Army Museum is interested in working with the Centre. It organises battlefield tours to see the places where so many soldiers and civilians died between 1914 and 1918. An increasing flow of both British and foreign visitors enjoy and learn from the centre.[33]

Other museums, galleries, nightclub and gym

The huge span of the military Connaught Drill Hall in Stanhope Road has been adapted to new uses twice in recent years: first for the Liquid Envy nightclub with two dance floors and a high balcony overlooking Victoria Park, and then as a gym.

Historic England's excellent museum inside the Keep at Portchester Castle is not covered in this book, while the Diving Museum is described in our second forts chapter and the Explosion Museum in the chapter on Ordnance Yards.

Sources

A L Boxell 2010 *The Ordnance of Southsea Castle*. Portsmouth, UK: Tricorn Books. ISBN 978-0-9562498-4-5 Stephen Brooks 1996 *Southsea Castle*. Andover UK Pitkin Guides ISBN 0-85372-809-7

Stephen Brooks 1996 *Southsea Castle East Battery*

Southsea Castle West Battery Historic England, retrieved 7 May 2018

Freddie Emery-Wallis Interview: Portsmouth History Centre PORMG:2832A 11.4.1995

Welcome to Southsea Castle Portsmouth Museums and Records, retrieved 8 May 2016

Southsea Common Historic England, 2002, retrieved 7 May 2016

Arthur Corney 1965 *Fortifications in Old Portsmouth: A Guide*. Portsmouth, UK: Portsmouth City Museums. OCLC 24435157

Arthur Corney 1968 *Southsea Castle*. Portsmouth, UK: Portsmouth City Council. OCLC 504812915.

J. King, Cathcart 1991 *The Castle in England and Wales: An Interpretative History*. London, UK: Routledge Press. ISBN 9780415003506

Peter Harrington 2007 *The Castles of Henry VIII*. Oxford, UK: Osprey. ISBN 9781472803801. Laura Hodgetts 2017 Portsmouth Small Boat Channel Update *Practical Boat Owner* April 27 David Lloyd 1974 *Buildings of Portsmouth and Its Environs*. City of Portsmouth

Charles O'Brien, Bruce Bailey, Nikolaus Pevsner, David Lloyd 2018 *Buildings of England Hampshire South* Yale University Press

The News 2012 "Barracks tower hit". The News. 8 June 2012. Retrieved 23 September 2017.

The News 2017 "Fond memories for Barry on his return to barracks. *The News*. 23 June 2013. Retrieved 23 September 2017

Nikolaus Pevsner & David Lloyd 1967 *Hampshire and the Isle of Wight* The Buildings of England Penguin Books Ltd.

Andrew Saunders 1989 *Fortress Britain: Artillery Fortifications in the British Isles and Ireland*. Liphook, UK: Beaufort.ISBN 1855120003

Richard Trist 2009 *The Coming of the Overlord Embroidery to Portsmouth*

Stephen Weeks, Clark Celia et al edited by Grahame Soffe 2017 *FREDDIE EMERY-WALLIS – Tribute*

WD 'Jim" Jarman 1997 *Those Wallowing Beauties. The Story of Landing Barges in World WarII* The Book Guild Ltd. Sussex pp.159-162

References

1. Ken Barton Email to Grahame Soffe 2017
2. James Hamilton Curator of Art Portsmouth City Museums 1972 Email to Celia Clark 2019 3
3. Deane Clark Interview with Celia Clark 1 January 2019
4. Christopher Hyson in Weeks 2017
5. Deane Clark Interview January 1 2019
6. Michael Laird Emails to Celia Clark 18 & 19 March 2019
7. Duncan Williams 'The Development of the Fortified Lines of Portsmouth and Portsea in the Nineteenth Century' *The Redan* Palmerston Forts Society Number 79 2018
8. James Hamilton Portsmouth City Museum 1972-74 *Memoir by James Hamilton, Curator of Art,1972-74*
9. Portsmouth Museums Service *Unlocking Potential, Transforming Lives Portsmouth Museums Service Strategy 2015 – 2020*
10. Freddie Emery Wallis Interview with Dr. John Stedman 11.4.1995 PORMG: 2382A
11. Rosemary Billett Letter to Celia Clark in Weeks 2017

12 Sarah Quail Email to Celia Clark 2 August 2016

13 John Stedman Email to Celia Clark 12 December 2018

14 Richard Trist Portsmouth City Chief Executive 1978-1993 Note given to Celia Clark

15 Ken Norrish Collection Portsmouth History Centre

16 https://www.portsmouth-guide.co.uk/local/dday50.htm

17 Portsmouth Museums Service 'Unlocking Potential, Transforming Lives Portsmouth Museums Service Strategy 2015 – 2020' March 2015

18 BBC News 22 September 2017; https://theddaystory.com/plan-your-visit/what-to-see-at-the-d-day- story/d- day-landing-craft-tank-lct-7074/

19 https://www.nmrn.org.uk/submarine-museum; https://en.wikipedia.org/wiki/Royal_Navy_Submarine_Museum; Royal Navy Submarine Museum Account 2009-2010;

20 https://assets.publishing.service.gov.uk/government/uploads/system/uploads/attachment_data/file/2476 25/0392.pdf;.Royal Navy Submarine Museum Account 2010-2011; https://assets.publishing.service.gov.uk/government/ uploads/system/uploads/attachment_data/file/2472 79/1306.pdf
RNEBS Alston Memorial Historical Society– a registered charity dating from 1984 http://beta.charitycommission.gov.uk/charity- details?regid=289693&subid=0;. https://interests.me/org/portsmouth-info/story/28456

21 http://strongisland.co/2014/05/30/history-of-aspex-beginning/ accessed 10 December 2018

22 'Make Art Not War. Defence sites find new life as centres of creativity' in *Sustainable Regeneration of Former Military Sites* eds. Samer Bagaeen and Celia Clark Routledge 2016 pp. 52-72

23 https://www.culture24.org.uk/art/art41819 accessed 10 December 2018

24 Quoted by Deniz Beck RIBA Hot Walls Studios speaking at Hampshire Buildings Preservation Trust/RTPI seminar Sustainable Regeneration of Former Defence Sites October 2017

25 http://www.memorialsinportsmouth.co.uk/old-portsmouth/pioneers.htm

26 *The News* 15 September 2016 p.8

27 https://www.bbc.co.uk/news/uk-england-hampshire-36736540

28 http://shapingportsmouth.co.uk/conferences/shaping-awards 2019).

29 Celia Clark 'Make Art Not War' 2014; 2016 in Sustainable Regeneration of Former Military 30. Sites. Routledge 2016 p. 61https://www.youtube.com/watch?v=QqTLr_hviPs

31 https://royalarmouries.org/about-us/history-of-the-royal-armouries

32 http://www.portsmouth-guide.co.uk/local/frtnsn.htm;
The Royal Armouries at Fort Nelson – Official Site. Page retrieved at 01:53am 5 May 2008; https://www.dailyecho.co.uk/news/2130247.2m-lottery-grant-for-fort-nelson/

33 Emails from Charles Haskell to Celia Clark 20, 22 June and site visit 7 August 2019

Chapter 9
A tale of two hospitals: the Royal Naval Hospital Haslar and Queen Alexandra Hospital Portsdown Hill

Royal Naval Hospital Haslar 1746-62
Entrance and Pediment
Buildings of Portsmouth and Its Environs David W Lloyd 1974

Introduction

At Haslar five important elements in this complex transition have guided or impeded the process. These were: Royal involvement; Gosport's conservation officer's crucial anticipation that the MOD would release the hospital; the campaign against the hospital's closure; the Veterans' Village proposal and the problem of resupplying civilian services to a formerly independent defence site – all of which were major features influencing the outcomes. The Haslar Peninsula Conservation Area Appraisal which also covered Fort Blockhouse, the Gunboat Yard and HMS Hornet was prepared in co-operation with the MOD in March 2007.'

The Veterans' Village, an entirely appropriate proposal which would have linked the army veterans at the Royal Hospital Chelsea with naval pensioners and their wives/husbands in flats at Haslar convinced the MOD's Defence Estates to sell Haslar Hospital to developers at a low price on that basis - only for the backers of the Village to be rejected by the new owners.

Which hospitals?

The surviving naval and military hospitals serving the Portsmouth and Gosport communities were built to treat navy sailors and army veterans. This chapter explores the history and recent change of ownership at the impressive Royal Naval Hospital at Haslar of 1754, and Queen Alexandra Hospital on Portsdown Hill, which opened to treat veteran soldiers in 1908. The link between the two is the Queen Alexandra's Imperial Nursing Service which was established in 1902 and restructured in 1949 as the Queen Alexandra's Royal Army Nursing Corps. Its history is set out in the *Joint Services Guide to the Royal Hospital Haslar 1995*, when medical treatment for all three services was consolidated at Haslar.

The closure of the hospital at Haslar was strongly resisted in a public campaign to keep the hospital open, while at Queen Alexandra only a few peripheral buildings from the veteran's hospital survive. New buildings replaced the pavilions along the slope of Portsdown Hill in the 1970s and in 2009 a huge new western extension and other facilities were added. QA now serves a considerable area of southern Hampshire.

The Portsmouth Royal Hospital and Gosport Memorial Hospital are not discussed here. The Royal in Commercial Road is no more: closed and demolished in 1979; its site is occupied by Sainsbury's supermarket. The Memorial Hospital is still in use for local civilians, as is St. Mary's in Milton, which developed from the Portsea Island Union Workhouse designed by Thomas Ellis Owen and Augustus Livesay in 1843-6. The workhouse and St. Mary's Lunatic blocks of 1873 to the north of the present hospital have been converted to social housing and private flats. St. James's Hospital in Locksway Road Milton was empty of patients by 2018. It was designed by George Rake was purpose-built as the Portsmouth Borough Lunatic Asylum between 1875-1884 in farmland to the east of Milton to offer patients fresh air and therapeutic activity. Presumably, like Haslar, residential conversion will be the new use. There is a strong local campaign resisting housing development, which would obliterate the loved open space around it

Queen Alexandra Hospital

Portsmouth Hospitals NHS Trust

Unusually in the twenty-first century, much of the land north of Portsdown Hill and the village of Southwick still belongs to one individual, who owns Southwick Estates. From the 1860s onwards land was purchased from the estate by the Board of Ordnance to construct defences to the dockyard and in the case of Southwick House – requisitioned in the 1940s as the advanced planning base for the invasion of Normandy.

Queen Alexandra Hospital Western façade 2010

In the 1860s and 1870s Portsdown Hill was crowned with a series of forts a thousand yards apart. The northern slopes were cleared of trees to provide fields of fire for the forts' guns against a projected landward French invasion under Napoleon III. Southwick Estates' land also extended to the southern slopes. A deed was dawn up between Queen Victoria's Secretary of State for War and the Southwick Estate for the transfer of 29 acres of land from Wymering Farm to the War Office to build a hospital to care for sick members of the Portsmouth Garrison and other military personnel from Winchester, Southampton, Dorchester and Weymouth. It was named after Queen Alexandra of Denmark.

Building financed by the Admiralty began in 1904. On 10 March 1908 the first patients were admitted. Horse drawn ambulances supplemented by two of the latest motor ambulances were used to convey the sick from the former military hospital in Lion Terrace Portsea. The wards were built in red brick pavilions facing the fine southern view over the city, with linking blocks between. The Horndean tramway offered patients and staff transport to the hospital. Its embankments spanning the road to Southwick are still visible. The hospital was lit by gas although some specialist departments had electricity as an alternative light source. Initially four ward blocks were built - as shown on a plan currently displayed in a corridor signed by a captain and Lt. Col. Brandon of the Royal Engineers based in Hilsea in 1923.

As a result of the casualties of World War 1 more beds were needed. 500 were provided in hutted wards. In 1926 the hospital was taken over by the Ministry of Pensions to provide care for the disabled of the Great War. In January 1942 the first civilian patients were admitted after a bombing raid when the Royal Portsmouth Hospital was hit, and more were transferred the following April when the Royal was damaged again by a land mine. War casualties, including some injured after D Day continued to arrive in increasing numbers. To meet this need, bed numbers were increased to 640.

In 1948 the National Health Service was introduced, but it was not until 1951 that all but 100 of the beds – which were reserved for war pensioners who were the responsibility of the Ministry of Pensions - were transferred to the Ministry of Health. Only fine stands of Scots pines and two buildings survive from the first hospital at the eastern edge of the site: De La Court in red brick with crisp yellow terracotta dressings, which was the Sisters' Residence, next to the mortuary, and Southwick House which was the doctors' residence and accommodation for long term patients, converted in 1991 to a family centre.

In the early '60s planning began in earnest to transform QA into a general hospital with a major accident and emergency department. Building work began in 1968. In the first phase a new eye department and a nurse's training school were built. At the same time two 3-story blocks were built for staff accommodation, joined by two 9-story blocks in 1976. In 1979 patients were transferred from the Royal and in 1980 Princess Alexandra officially opened the new hospital.

Only nineteen years later it was decided to greatly extend it again. The Portsmouth Hospitals NHS Trust invited private finance initiative tenders for the new building from design/build/ operate consortia. At a public meeting in autumn 2003 members of the Portsmouth Society heard that two attractive designs for the mega-hospital had been rejected in favour of the winning bid by Carillion designed by BDP Architects, presumably chosen on cost grounds. Carillion bore the immediate costs of £236 million. It was the second largest construction site in the country at the time, after Terminal 5 at Heathrow. The contract was for 35 years; payments commenced after 3.5 years when construction of the new facilities was complete. The new hospital with large multicoloured façades and many internal lightwells was opened by the Princess Royal in 2009. The design incorporates airshafts using the stack effect for ventilation, but it did not include a green roof or solar panels, although these were later considered.

The Portsmouth Society judged the building for its Best New Building Award in September 2010, when they enjoyed the light airy feel of the huge new extension and the building's contribution to energy saving and innovation in clinical facilities.

Medical staff from all three armed services wearing the appropriate uniforms surprise patients who do not know the hospital's origins. The internal wards especially at lower levels facing deep light wells are dark and claustrophobic. Even climbing plants would provide a green stimulus - known to be therapeutic - to bedbound patients. Images of Portsmouth adorn the specially commissioned patients' curtains, and excellent photographs of local scenes by talented staff, which ascend in stages from underwater to the sky, adorn its corridors and staircases, which help in navigating the enormous building. Pride in the hospital's history is demonstrated by plans and photographs of the first building on display in its corridors. The large curved entrance wing faces westward. Staff and patients in its south facing wards enjoy fine views.

In July 2020 QA was awarded university hospital status. Patients would now have access to clinical trials and staff would be able to use cutting-edge technology for faster innovative treatment. Symbolising the continuing link with the armed services and their medical care of the community during the pandemic, a Spitfire with 'Thank U NHS' painted on its underside flew over the hospital on 1 August.

Haslar Hospital

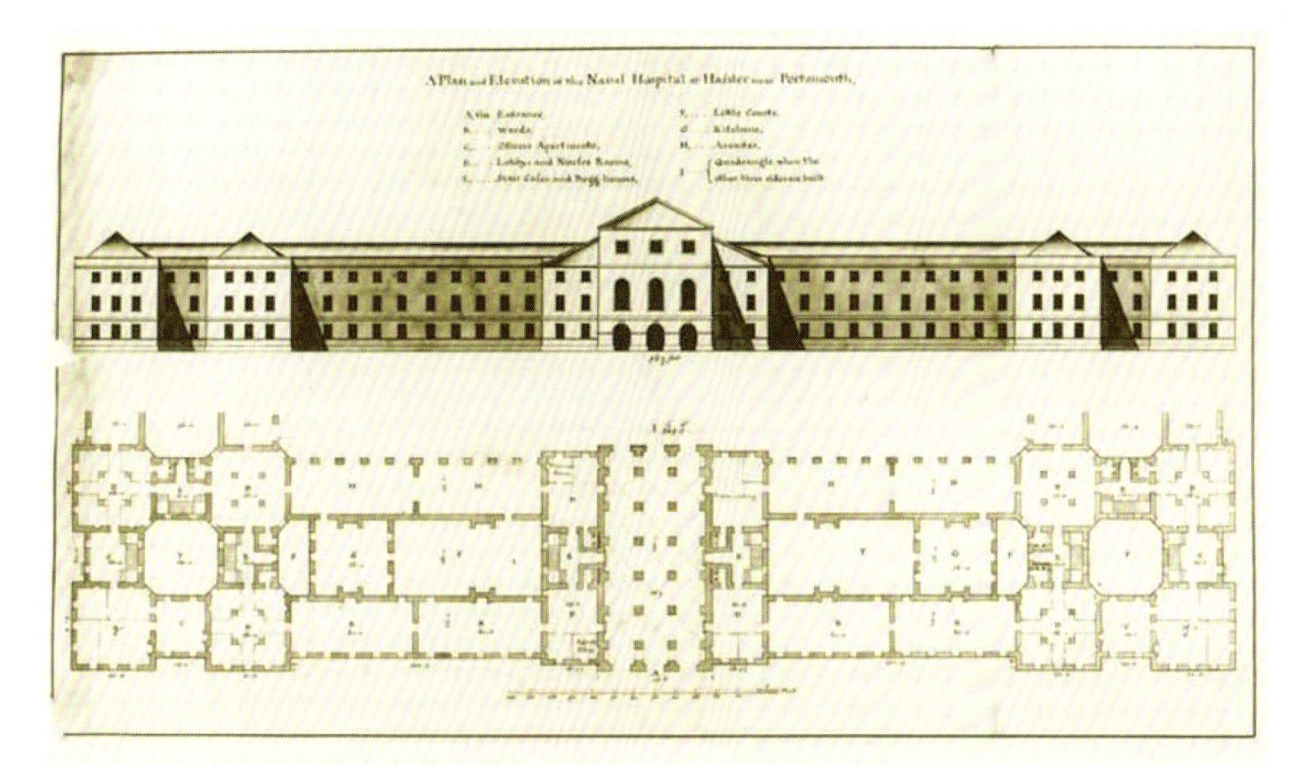

The magnificent Royal Naval Hospital at Haslar had a very long and distinguished history in the medical care of service personnel, treating many tens of thousands of patients both in peacetime and in war. Sailors' life on board national fleets was brutal and dangerous, their diet monotonous, hygiene and sanitation minimal. They suffered war wounds, burns, splinter injuries, tropical fevers, mental illness, 'the boody flux', the effects of malnutrition and venereal disease. In the eighteenth century the number of British sailors who died from disease or wounds far exceeded those killed by enemy action. At the beginning of the eighteenth century sick and wounded seamen were boarded out in hired accommodation which was far from conducive to recovery. The first purpose built British naval hospitals were built in Jamaica in 1704, in Lisbon in 1706 and Minorca two years later. The last was constructed on an island well away from the wineshops and people in the town of Port Mahon. In 1744 the Admiralty ordered a hospital to be built at Haslar on Gosport's isolated peninsula.

Designed for 1500 patients, it was a pioneer in hospital design, based on the quadrangular design first developed at Greenwich, double ranges of red brick embellished with Portland stone dressings. The architect was Theodore Jacobsen, who also designed the Foundling Hospital in Bloomsbury, London (1742-1752). Originally intended for civilians as well as naval seamen it was constructed between 1746 and 1753. It opened as the Royal Naval Hospital in 1754. On completion it was the largest brick building in Europe and the largest hospital in England. Building works cost more than £100,000, nearly double the cost of the Admiralty headquarters in London. Patients usually arrived by boat and were conveyed into the hospital from Haslar Creek. It was not until 1795 that a bridge was built over the Creek, providing a direct link to Gosport.

David Lloyd describes the building in his book *Buildings of Portsmouth and Its Environs* published as part of the local celebration of European Architectural Heritage Year 1975: "Grander in scale than any in the Dockyard… It was an enormous building by the standards of the time, with three ranges round a very large open courtyard. The architecture is domestic-institutional Georgian in deep red brick, embellished with white Portland stone dressings, having an affinity, on its much larger scale, with the later storehouses in the Dockyard. The main central block has a grand four-storeyed centrepiece containing the entrance, with a great stone pediment carved in symbolic sculpture representing Navigation and Commerce, on either side of the Royal Arms... Entrance is through a triple-arched passage, impressively vaulted in brick, leading into the courtyard on the axis of the chapel... This is a small elegant building of 1756, with a turret added a few years later (the clock is dated 1762). The effect of this austerely grand courtyard, opening at one end into the grounds with the chapel on the axis, is said to be reminiscent of monumental American colonial buildings of the same period. The hospital has been completely modernized internally, while carefully retaining the external appearance of the buildings: the only depredation has occurred, as a result of bombing, on part of the south range."

Free circulation of air to prevent disease occupied the minds of leading scientists, naval doctors and inventors, but cross ventilation was not achieved since no windows were provided on the ranges facing the light-wells. The chief disadvantage of the original design was that wards ran the full width, also having to serve as corridors, an inconvenient arrangement not in accordance with later medical practice. But Haslar was considered to be well ahead of its time, and its design in altered form was copied in naval hospitals in Plymouth and Chatham. By 1787 Haslar had 84 general, surgical, isolation and recovery wards, each with its own water closet.

An illustration from the Gentlemen's Magazine of September 1751 which shows clearly Jacobsen's original plan for a quadrangle building (Revell 1978 p.15)

Prince's Regeneration Trust Enquiry by Design report 2008

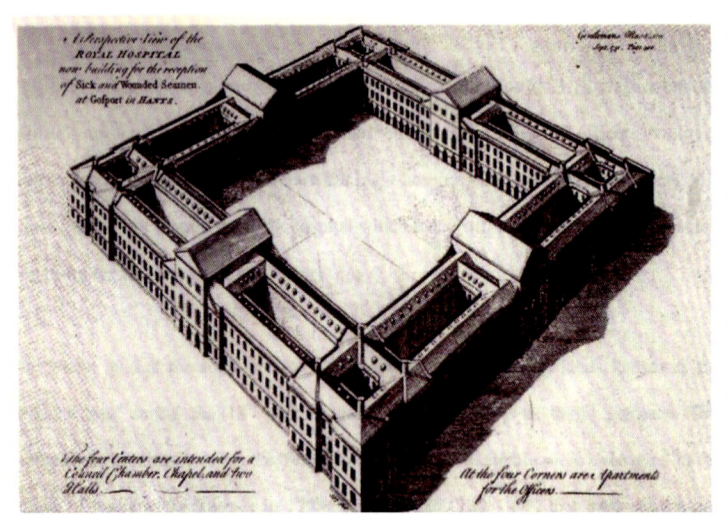

The original plan was a quadrangle but building of the fourth wing was cancelled in 1756 due to cost. Nothing changes with large government projects! Instead, the large open courtyard was closed by railings to improve security. Over 200 years later the concrete Brutalist 'Crosslink' building containing the operating theatres was built in the courtyard in 1984, connecting the two side wings.

The first physician was Doctor James Lind, appointed in 1758. During his long term of office, he did more than any other to raise the medical standards of the navy and to ensure for Haslar Hospital its pre-eminence in naval medicine, which it retained for 250 years.[12] The high brick walls and railings surrounding the site were designed to stop the patients from leaving: some compared it to a prison. They were "pressed men", who were escaping from Haslar in large numbers, so a guard of soldiers, stationed close by were instructed to patrol the perimeter wall to apprehend those attempting to escape. The buildings were overcrowded; discharged patients were taking up home in the attics and there were reports of drunkenness and petty theft among staff and patients. Female nurses allegedly imported rum from Gosport taverns in pig's bladders, suspended under their skirts and smuggled through the hospital sewers at low tide, despite the presence of the guards.

Railway lines were laid from the adjacent waterfront (now part of Fort Blockhouse) in through the original main gates of the hospital on the east side and then passing through an archway into the main block. Rails can still be seen there inside the building.

West elevation — Martin Marks

A second branch line went south to the seafront and then west to serve the zymotic blocks, described below. Small carriages were pushed on the lines by staff or sailors; there were no engines. It was the route by boat across Portsmouth harbour to Haslar Creek that generated the phrases 'up the creek' and later 'up the creek without a paddle' where injured and dying sailors from Napoleonic times would be sent. If they were sent to Haslar it was serious; many sailors were so severely injured they would soon die there.

In 1902[3] the hospital became known as the Royal Naval Hospital Haslar (abbreviated to RNH Haslar). Evidence of this change of name has been recently confirmed by the uncovering of a packing crate used as floor boarding in one of the residences. A section reused as floor boarding was inscribed 'Royal Hospital Haslar' in flowing Georgian script.[4] In 1954 the word Naval was formally incorporated into the title of the hospital.

Oval staircase in southeast corner
Haslar Heritage

Over the years the internal layout has been modified. Many new facilities were added. During the twentieth century much of the historic interior was swept away to create modern conditions. The hospital was a very busy place during both the First and Second World War. Celia Clark's father, David Willison RE, who was wounded taking Pegasus Bridge on the Caen Canal on D-Day was brought back to Haslar with a large lump of shrapnel in his neck, but when her mother came down to see him, crossing the harbour with her bicycle, he had already been transferred to the military hospital in Solihull where he recovered and six months later went back to fight. Following bomb damage in WWII an impressive oval staircase was inserted into the southeast corner.

Tri-service hospital

In 1966, the hospital's coverage expanded to serve all three services – the Royal Navy, Army and Royal Air Force, but it remained under the control of the Royal Navy. As a training hospital and because of the limited range of medical problems suffered by most relatively young and fit military personnel, it also took in civilian patients from the local community. Without this, the hospital would have lost its recognition from the various medical colleges.

Water

Unusually the hospital had its own water supply derived from two deep artesian wells in the Gunboat Yard, close to the hospital within a few feet of Haslar Road. The south well is the older, sunk at the time the hospital was built. It is on a much smaller scale than the north well. It goes down to a depth of 146 feet to the first water-bearing stratum of sand. The well is lined with a cast iron cylinder 6ft 10in in diameter. It still yields a fair supply of water. The north well was sunk in 1859 to a depth of 340 feet. It goes through the Bracklesham beds and other strata to the chalk. From the wells the water is pumped into two large iron tanks at the top of the water tower. Each tank holds 125 tons of water or 50,000 gallons. The water tower itself built in 1885 is 120 feet high, a conspicuous landmark for miles around. There is a story that the tower survived World War II bombing raids on Portsmouth Dockyard as it was using as a navigational mark by the Luftwaffe.

Water Tower   Martin Marks

There is also a large reservoir in the Gunboat Yard across the road, which was at one time supplied from Haslar Mains, but is now used in connection with the Gosport water supply. This reservoir, which is really a small artificial lake, is used as an experimental tank in which the models of newly designed ships are tested to ascertain their stability and sea worthiness. It contains 3,375 tons of water. This water could be pumped through the mains into the water tower and so become available for firefighting - although it is not up to normal potable water standards. The storage of water at Haslar exceeds 1,000 tons, and it is also in direct connection with the Gosport water supply. The Haslar well water was a good water for drinking and other domestic purposes, although it is very hard and contains much sodium chloride and ammonia. On one occasion in the mid-1990s, the Defence Evaluation and Research Agency (DERA) who then operated the Haslar test tank, were refilling the tank after maintenance work. In so doing they managed to drain down the water supply to the hospital, inadvertently putting the hospital's equipment disinfection unit out of action. A connecting valve was then fitted with a chain and padlock to prevent a repeat.

In 1996 a decision was made to staff the hospital on a tri-service basis, expand its bed numbers and it became known as the Royal Hospital Haslar again. An army medical unit, 33 Field Hospital, moved in next door in HMS Dolphin, a former submarine training base which later reverted to its original name of Fort Blockhouse. The field hospital staff trained and provided nurses in Haslar. Large numbers of RAF medical staff also arrived. This influx created a serious accommodation problem, as many were single and expected to live on or near the site. Earlier planning in MOD had assumed spare accommodation levels in HMS Dolphin that were not made available.

The author Martin Marks served as the Commander or Executive Officer (i.e. Second-in-Command) of the hospital from 1993 to 1998. When the tri-service expansion took place, he became the Administration Director and Officer Commanding the Support Unit. These were busy times, which threw up interesting new problems. The hospital was controlled at arm's length by the Defence Secondary Care Agency in London, which caused chaos in some areas. A remote, unannounced decision was made to retitle the Medical Officer in Command as "The Commander" (an Army title) without appreciating that it was used by the Navy to describe the second-in-command of a large ship or shore establishment, causing a flood of C.O.'s mail into the (naval) Commander's in-tray. The services all had different systems for travel expenses, so three people from different services in Haslar attending the same away event would receive different expenses. Senior, single Army and RAF officers turned up with pet dogs, expecting to live on site. In former times, dogs were an anathema to the naval way of life (apart, notably, from the dog on the *Mary Rose* and "Judy" on *HMS Gnat/HMS Grasshopper* circa 1936/42) and there were no facilities for them. There was no tri-service Discipline Act and these matters had to be operated for all in the hospital under naval rules, causing further concern and complications. A certain amount of local flexibility was needed and the guidance of "Best for Haslar" was invented by Martin, allowing for a choice from all three services when trying to decide on a single set of rules for all. It seemed to work well.

Medical-officer-in-command

The subtle change of title of the naval commanding officer from "Medical-officer-*in-charge*" (MOIC) to "*in-command*" is interesting. The former is the normal title for medical officers commanding a naval establishment, but it was changed at Haslar. The MOIC had been 'in charge' and did not have military powers of punishment; that resided with Chief of Staff to Flag Officer Portsmouth. When the latter post was abolished, the powers that be had to make MOIC 'in command' without recognising some of the implications!

When the commanding officer, (the late) Surgeon Captain Ian Jenkins was promoted to Surgeon Commodore, his deputy, co-author Martin Marks, recalled that commodores, on first taking up their command, were entitled to an 11-gun salute. A secret request from him to the local flag officers' staff for the salute initially produced just laughter and a "but he's a doctor" response. However, persistence and a long-standing naval reference won over. On the morning of his promotion, a completely unaware Ian Jenkins paraded for the morning colours ceremony to see his commodore's pennant hoisted and was advised to remain at the salute once that was done. "What for?" he asked. "Just trust me, Sir" Martin replied. The gun salute fired from the then HMS Dolphin, next door, was a spectacular sound. Ian, a gifted surgeon and astute administrator, went on to become the Medical Director General (Navy) and, later, as a Surgeon Vice Admiral, the tri-service Surgeon General. Sadly, he died suddenly in 2009 whilst serving in the unpaid position of Constable and Governor of Windsor Castle.

Closure – and resistance

"On the 14th December 1998, the MOD announced that the Royal Naval Hospital Haslar at Gosport was likely to close in 2002 under a sweeping review of defence medical services. It had a modern civilian role as well including accident and emergency and orthopaedic services for 110,000 patients every year. Around £25 million had recently been spent modernising the facilities there. The possible loss of this busy District General Hospital was a massive shock to local people and sparked probably one of the greatest and longest campaigns to save a hospital this country had ever seen. The campaign was led by the Gosport Borough Council sponsored 'SAVE HASLAR TASK FORCE " as Councillor Peter Edgar recalls.

"The Haslar Task Force arranged a march for Sunday, 24th January 1999 expecting 10,000 people to turn up but, on the day, over 20,000 people walked the mile from Walpole Park over Haslar Bridge to the hospital gates. More than 170 police plus 50 council stewards were on hand to direct traffic and some roads had to be closed to make room for the vast crowd. It took the protesters nearly an hour and a half to walk past the old Haslar gate to make their views known. Immediately after the march there were encouraging vibes from the government regarding the future of the hospital, especially after Health Minister, Frank Dobson, visited Haslar. However, after further meetings with him and the Defence Minister, George Robertson, it became clear that it would be unlikely that we would save the hospital but with a sustained campaign we could ensure it stayed open until new facilities were in place. These would be for the military in Birmingham and for civilians at Queen Alexandra Hospital at Cosham."

"The Save Haslar Task Force decided to keep up the pressure on the government to make sure the arbitrary and dangerous closing of the site in 2002, should not take place before adequate replacement facilities were provided for both the country's military, and also the local population across the south."

"The campaign continued until the final closure of the hospital in 2009, with numerous events keeping its profile extremely high both locally and nationally. It included a 220-person deputation that marched on London. Around a 100,000-signature petition was delivered to the door of 10 Downing Street. Some of the other many events that followed were ministerial meetings, evidence given to the Defence Select Committee by medical members of the Task Force, a torchlight vigil and enlisting and achieving the support of councils across Hampshire, including the County Council." "The pressure was kept on the Government by the Task Force by producing a number of booklets on the case for retaining the hospital, and there were regular public meetings to keep people informed. A Gosport adventurer, Mike Truman, even displayed a Save Haslar Hospital scarf and banner on the summit of Everest!"

"At a Freedom Parade through Gosport in March 2007, local people said a tearful farewell to the UK's last military hospital after more than 250 years. The hospital was handed over to the NHS who gradually ran it down to its complete closure in July 2009."

"The success of the Save Haslar Task Force has to be that, with the support of hundreds of thousands of people and the longest and most forceful campaign anyone can remember, this Task Force team of local politicians and medical practitioners managed to keep the hospital open until 2009. We should always remember that, if the closure of this excellent historic and caring life-saving facility had occurred any earlier, there is no doubt that its closure would have put people's lives at risk."

In 2003, local MP Peter Viggers spoke in the Commons: "The background to the Haslar hospital problem is that the Government decided in 1997–98 to investigate the clear deficiencies, particularly in respect of senior consultants, in the Defence Medical Services. Those deficiencies were well known for some time, and a committee comprising no medically trained staff decided that the way ahead was to build a new centre of medical excellence — along the lines, they hoped, of Guy's hospital in London or the John Radcliffe hospital in Oxford — and to close the only remaining military hospital in the United Kingdom, the Royal Hospital Haslar in my constituency."

"When they pressed ahead with the plan, they found that, apart from Birmingham, there were no takers for the new centre of medical excellence. I have to say that it is with some considerable reluctance that medical staff are now being moved to Birmingham as the plan proceeds. I have met all the senior people involved and I know that they are capable. They are not medical but military people; if they had been tasked to take Basra, no doubt they would have taken it. However, they were tasked to close Haslar hospital and to move defence medicine to Birmingham: they are getting on with it efficiently."[11]

The NHS takes over

To mark the handover of control to the civilian NHS trust, the military medical staff marched out of RH Haslar on 28th March 2007, exercising the unit's rights of the Freedom of Gosport led by a Royal Marine band. At a closing ceremony attended by Celia Clark the Mayor of Gosport, Councillor June Cully said: "It's the end of an era for Gosport, for just as servicemen and women have relied on Haslar's excellent staff and medical treatment during times of peace and war, we as a community have come to rely on them. We all knew we could rely on Haslar and her trustworthy staff."

In 2001 the provision of acute healthcare within Royal Hospital Haslar was transferred from the Defence Secondary Care Agency to the NHS Trust. The Royal Hospital was the last MOD-owned acute hospital in the UK. The decision to end the provision of bespoke hospital care for service personnel was taken prior to the UK's expeditionary campaigns in Iraq and Afghanistan, but was nevertheless followed through, largely on the grounds of cost. The change from military control to the NHS, and the complete closure of the hospital have remained the subject of considerable local controversy. The hospital formally closed in 2009 as noted above.

In 1986 the combined military and civilian staff had been 986 whilst in 2008 before its closure Haslar Hospital employed 662 people of which 357 lived in Gosport. Whilst many military staff would be redeployed this represented a significant loss of jobs and the site contributed some £3.5 million to the Gosport economy per annum.[10]

Heritage Protection

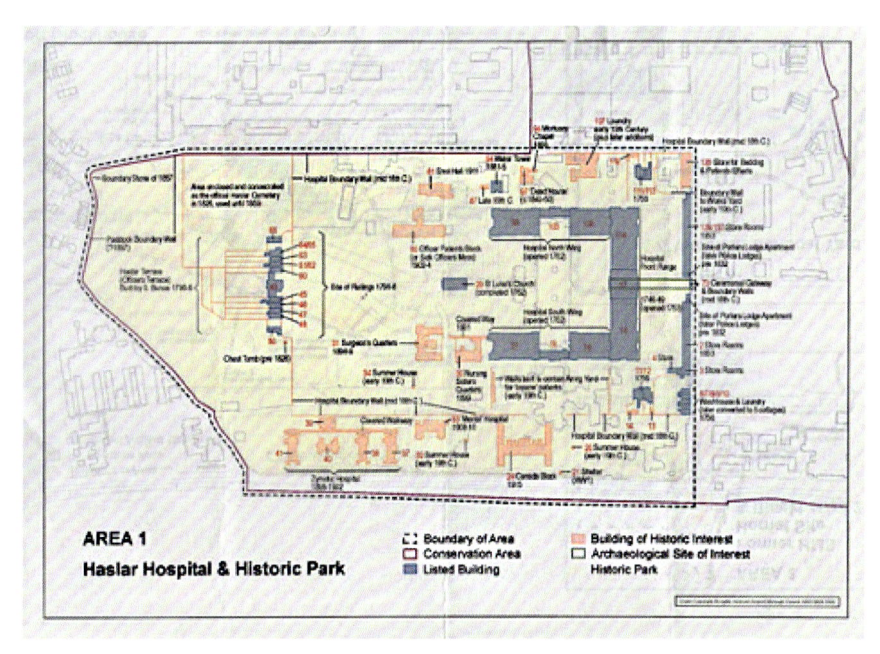

Haslar Hospital Conservation Area plan   Gosport Borough Council

In 2001 Haslar was designated a Grade II listed historic park. Several of the buildings are themselves listed. In addition to the main 18th-century main block (comprising Ward Blocks A, B, C, D, E, F and Centre - grade ll*), the grounds contain many interesting items.

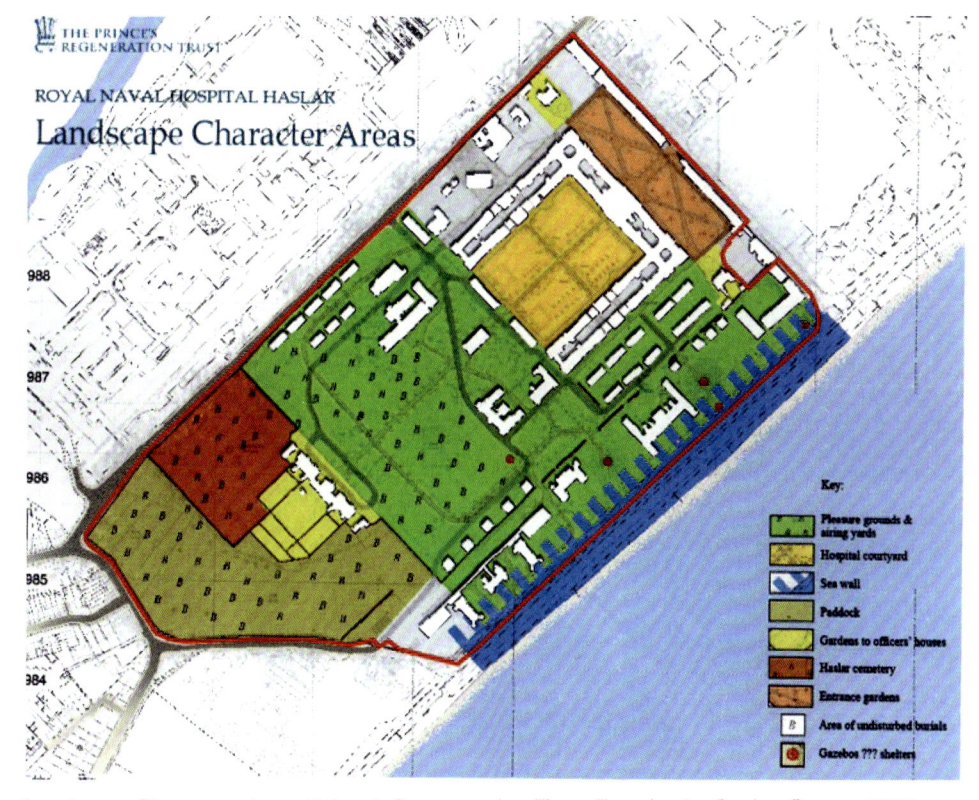

Landscape Character Areas Prince's Regeneration Trust Enquiry by Design Report 2008

- The original main entrance (1750s, grade II) is marked by a carriage gateway but formerly provided direct access from Haslar Jetty lying 300m north-east of the entrance, now part of the Submarine Museum. An avenue of lime trees flanks the approach from the jetty, which was the main arrival point for sailors approaching from their ships in the Solent before admission. The entrance is in turn flanked by long, narrow, single-storey medical stores buildings (1853, listed grade II)

Martin Marks

- The dead house (or mortuary, mid-C19 with an 1869 addition) and laundry c1854 (both grade ll)

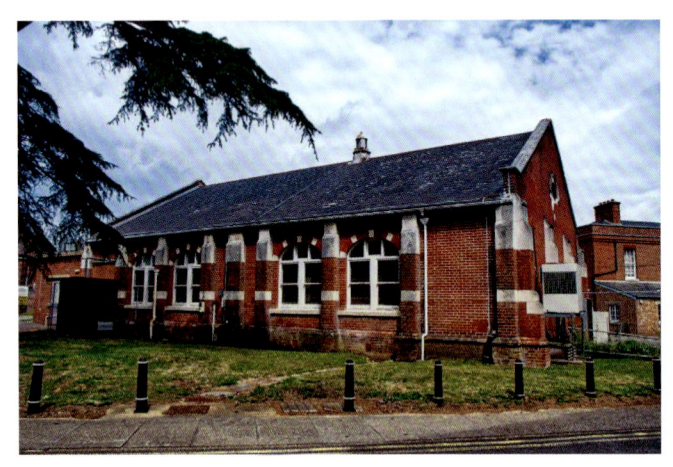

- A separate psychiatric ward on the seafront (complete with a padded cell pre- restoration) and a pathology lab on the north side. Its site arrangement was set by access to sunlight for microscope use.

- An isolation hospital in the form of detached "zymotic" (denoting a contagious disease regarded as developing after infection) blocks was built in 1898/1902.

Mortuary   Martin Marks

- It originally consisted of an administration block and four pavilion ward blocks. These were sited along the seafront, isolated from the main hospital by a brick wall with a "hole in the wall" for supplies and served by a rail line that ran from the pier in HMS Dolphin.

- A line also ran straight into the main hospital building through the east archway where rails can still be seen. In the late 1990's word was received that the wall was to be "listed" as a historic monument. Unfortunately, it had an asbestos roofed bicycle shed attached to it. Rapid informal arrangements were made for the shed to "fall down" safely one night before listing, and one ward block was demolished for safety reasons shortly afterwards.

Martin Marks

- Separate mess blocks to house medical officers (c1901), nursing sisters and staff; the latter funded by the women of Canada during the First World War when they became aware of the poor living conditions for the nurses. It was named Canada Block.

- Administration blocks.

- Errol Hall: opened in 1913 following a bequest from the widow of Commander George Errol RN it was a community hall providing entertainment and communal space for the patients and later a gymnasium and small theatre for the staff. Martin recalls a weekend spent in the Hall constructing the set for a production of "Wind in the Willows".

- To the northeast of the modern entrance on Haslar Road stands the massive brick water tower with stone dressings (1881 listed grade II) already described above.

- St Luke's Church: a rectangular building also by T Jacobsen (1762, listed grade II*) with a little turret containing a Harrison (of longitude fame) clock. The clock was fitted with a much kinder electric winding system in the mid 1990's. The church is still used for services once a month.

- Some historic items from the hospital were returned and placed in the chapel for safe keeping such as ornate boards of postholders over the years.

- Residences for senior hospital staff and officers. West of the chapel, along a short avenue, is the Terrace completed in 1798, consisting of the original Medical Officer in Charge's (MOIC) house on the main axis, stuccoed three-storeyed, with tall windows S Bunce 1796, listed grade II).

St. Lukes Interior   Martin Marks

Medical Officer in Charge house The Terrace   Martin Marks

203

They were not well built, and numerous problems resulted in several unsuccessful attempts at improvements until in 1800 when the governor reported that "the house is constantly wet, and the servants' rooms streaming with rain" he asked the Admiralty to stop their architect from making further changes. In more recent times it became the residence of the Medical Director General (Navy), the Royal Navy's senior medical officer. The hospital's Medical Officer in Command had been moved to a smaller residence (one of two pairs of residences that were built c1756, listed grade II) in the northeast corner of the site. In the Terrace on either side of the old MOIC's house are also two pairs of smaller officers' residences. All these residences have now been restored and sold as private homes.

The boundaries are largely marked by a high brick wall (north-east section mid C18, listed grade II) on all but the southeast side. Historic burial grounds: during the 18th Century, the whole area of land to the southwest of the hospital, including the grounds known as the Paddock, were used as burial grounds covering 9 acres and within a walled area. It has been estimated that between 1753 and 1859 13,000 sailors and soldiers who died at the hospital are buried there. The total number of graves may never be known.

At the northern end of the Terrace is a Memorial Garden, once a naval cemetery laid out in 1826 and enclosed by a brick wall. A gateway though a brick wall gives access. It is largely laid to lawn with scattered stone memorials, some dating from the Battle of Trafalgar. It closed in 1859 when a replacement cemetery was opened nearby in Clayhall Lane. There is a nursery garden with greenhouses against the northeast wall now run by Shore Leave Haslar as a project for ex-Service Veterans with mental health support needs. They offer horticultural therapy and related activities to make this garden a magical place for our veterans to recuperate. There is a beautiful parkland area, lovely flowerbeds and there is a shop selling vegetables and handcrafted items. Entry is through the green pedestrian gates on Haslar Road.

Redevelopment

Haslar Hospital Conservation Area, Listed Buildings, Archaeology. The Prince's Regeneration Trust

As mentioned at the beginning of this chapter, the disposal was anticipated, and conservation protection plans were put in place before it took place. In March 2007 the Gosport Conservation

Officer Rob Harper prepared and published the Haslar Peninsula Conservation Area Appraisal. The hospital was Area I, followed by Fort Blockhouse, the former HMS Hornet site and the rest of Blockhouse and the Victorian Gunboat Yard. These sites' recent experience is explored in other chapters. The plan was intended to act as a guide to the buildings and features that make a special contribution to the character of the area, assessing these areas' historic development, the scale and form of development and opportunities for continuing enhancement. Each section concludes with a Design Comment which offers guidance to any potential development in order to preserve and enhance the historic interest of key buildings in that area. Planning applications within these conservation areas were required to have a supporting Design and Access Statement to explain how the layout, scale and appearance of any planning proposal preserves or enhances the character of the area. This made it clear that a high quality of design was required in redevelopment of these special areas.

Plans for future use

Location of Haslar Hospital. The Prince's Regeneration Trust Enquiry by Design Report 2008

Planning the future of Haslar Hospital as a nationally important Georgian ensemble required particular care – and design flair - sparking the intervention of HRH Prince Charles, as a former naval officer whose first command was *HMS Bronnington* based at Gunwharf on the other side of the harbour. He and Prince William backed the Veterans' Village for naval veterans and their partners (the Boys in Blue), linked to the Royal Hospital in Chelsea (the Boys in Red). He also took an interest in future uses for Fort Gilkicker, the southernmost point of the Gosport peninsula.

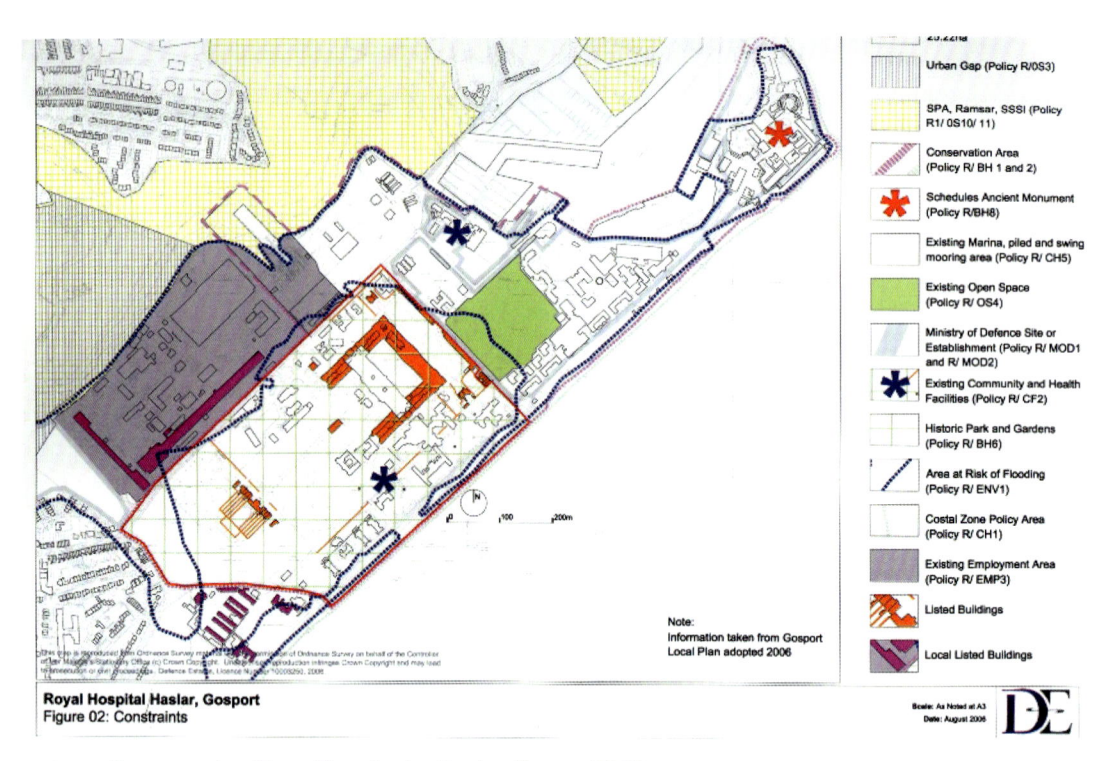

Prince's Regeneration Trust Enquiry by Design Report 2008

As one of their most important and complex historic properties, in November 2008 the Ministry of Defence funded a three-day workshop of local and professional interests to come up with a master plan for the hospital. The workshop was run by the Prince's Regeneration Trust, using the Enquiry by Design process. The Trust is a charity whose aim is to preserve buildings, monuments, structures or sites of particular beauty or historical and architectural interest, and to promote urban or rural regeneration in areas of social and economic deprivation. The Enquiry by Design process is a planning tool trademarked by the related Prince's Foundation for the Built Environment, an educational charity working to improve the quality of people's lives, by teaching and practising traditional and ecological ways of planning, designing and building. The process involves stakeholders and the local community working together to shape a place, testing every issue by drawing it. It differs fundamentally from the conventional planning process, in that data is built up on a plan-based system, a crucial element in achieving sustainable communities. It aims to deliver masterplans which are based on enduring design principles and to develop the place-making skills of the participants.

What happened at Haslar was quite different from the usual process of public consultation. At Gunwharf there was very little, but at Royal Clarence Yard across Portsmouth Harbour the same developers Berkeley Homes paid for a two-stage event where local people were encouraged to paste up their local knowledge and dreams on Post It notes. But there was no commitment to use their ideas to shape the final plan. The Enquiry by Design process claims to be 'not just a means of informing the community about a planned development but actively engages it in the planning and design of their community, helping build up the confidence and collective enthusiasm to allow the vision to be taken forward after the workshop has been completed.'

The event was held in the Medical Library of the Hospital, whose priceless historic contents have been transferred to the Royal Naval School of Physical Medicine in nearby Alverstoke. Lack of space limited the numbers taking part. The workshop was focused on the reuse of the many historic buildings within the hospital, many of them listed. Potential sites for new buildings were to be identified, the importance of features of the Registered Historic Landscape considered, types of uses identified, provision made for vehicular and pedestrian access in and around the local area all drawing on the experience and knowledge of the key stakeholders.

This was not the first time that the Ministry of Defence had used this particular planning tool, but they had not done so for a historic site before. They carried out major consultations in the urban extension of Aldershot and Catterick, and at RAF Cosford (internal to the MOD's competing operational uses). The gain to the MOD is establishing a site's value, and, of course, smoothing the site's future course through the planning system, as well as, hopefully, incorporating local people's knowledge and expertise into the plan. As the Defence Estates officers who commissioned the process said, this method goes further than the usual consultation, for example in the range of external stakeholders.

Significantly, the stimulus to greater consultation was the changing status of defence lands the loss of the Crown Exemption in 2006. Defence Estates now have to engage in statutory planning – whereas before abolition, they would bring forward substantial proposals (as they did at Gunwharf) and ask for responses from the planning authority. Not all the adjoining landowners such as QinetiQ or the Home Office, responsible for Haslar Prison, attended the Enquiry. The process took place over three intensive days, starting at 9am, with 70 participants. The aim was to establish a basis for putting value into the site. The first day concentrated on sustainable use, to ensure conformity with the local plan, with the aim of economic viability, which might involve some enabling development. Participants formed five specialised groups which developed final drawings of overall strategies for movement and connectivity, land use, heights and massing, landscape and ecology, heritage plan, phasing and mix and commercial viability.

The Prince's Trust undertook some documentary research. The whole of the Blockhouse peninsular used to be under water and is at flood risk; the seawall is particularly vulnerable. Access is the problem: Gosport's geography as a peninsular with restricted road access which no longer has a railway. In 2008 there was only half a job for everybody of working age. More employment was Gosport Borough Council's priority. Community facilities were needed at Haslar. Enabling development might be necessary – as at Royal Gunpowder Works Waltham Abbey – including breaches in the local plan for two housing developments. The challenge at Haslar was the maximum historic fabric, and its isolation – because in the eighteenth century hospitals were built away from the town. The aim of the masterplan was to reintegrate the site with the rest of Gosport. Defence Estates did not want to spend three or four years deciding on its future, but to identify beneficial uses more quickly. At that time no single entity was considered likely to buy the whole site. Defence Estates expected a consortium or a trust like the Greenwich Foundation to take it on. Improving the setting, perhaps by removing the 1970s and '80s medical buildings which was considered important, but reuse would have to be phased. Medical uses would have to be viable. Gosport wanted it to be a medical centre; the Gosport Society called it 'The People's Hospital' but the facilities were out of date. The medics' accommodation was probably to be reprovided at Southwick (the former HMS Dryad, now Southwick Park). One of the participants was Kit Martin who converted the Royal Naval Hospital in Yarmouth and was also involved in the potential development at Fort Gilkicker. Yarmouth has two exhibition spaces – the chapel and stairwell - but the wards on either side have been converted to apartments, a use which Gosport's local plan precluded. Subsequent history and the grant of residential development breached Gosport's Local Plan.

Not all the locals were happy with the process, which some dubbed the *'purchase by design'*. It was said that the Save Haslar Task Force would be disappointed if leaders approved anything other than the reinstatement of full district general hospital services - but these had already been reprovided in the huge rebuilding of Queen Alexandra Hospital on Portsdown Hill. The critics emphasised the importance of dedicated military hospitals for combat casualties.

Defence Estates and the Prince's Regeneration Trust collated the findings from the Enquiry which produced many good ideas for the future of the site for consideration by Gosport Borough Council. Defence Estates hoped that many of them would be incorporated into a masterplan that provides a long, sustainable and viable future for this most important site.

"This maximises the potential to produce a robust and deliverable vision for the site, where both opportunities and constraints have been fully challenged and a consensus agreed upon". This would not be an easy task for a group of listed buildings (13 structures at Grade II* or II plus other buildings within their curtilage, such as the wardroom) in a Registered Landscape with a Conservation Area (introduced to prevent unwanted commercial development, especially of the seafront areas) and an Archaeological Site of Interest. Mature trees within the site are covered by Tree Preservation Orders.

The outcomes of the workshop were as follows:

1. The shelter for the mental patients in the airing yard and the chapel required careful protection and security. Some monuments in the cemetery required restoration.

2. A range of modern buildings was recommended for demolition: some messes, the surgical block, squash courts, incinerator, boiler house, associated offices.

3. They identified distinct areas of landscape character: the paddock, cemetery, officers' houses gardens, the main hospital courtyard and entrance gardens, the airing yard, pleasure gardens, seawall and viewing shelters and pavilions.

4. Uses for listed and curtilage buildings should be those that require minimum alteration; retain significant spaces, interiors, fixtures and fittings; optimise the use of floor plates; reflect the fragility of the fabric.

5. The Georgian residential terraces to be retained as residential accommodation with as little alteration as possible. The chapel and all its fixtures and fittings to be retained and to be used for weddings, funerals and concerts.

6. Erroll Hall to be retained as an open community space.

7. The seafront Zymotic Hospital should be allowed some demolition to retain its symmetry, one wing having already been demolished. Further work into their importance was needed to decide on the justification for retention or full demolition.

8. The poor access to the site for non-car users was highlighted with the resulting requirement for adequate parking that did not conflict with the visual aspects. A study to compare traffic levels with a possible development layout against the amount when the hospital was in full operations showed no change. Suggestions were made for an internal, circular traffic route.

9. It was likely that future development would include a mix of residential and residential care space. The impact of providing affordable housing, often required for housing developments, needed to be considered. There was little justification for a hotel in the area. The Crosslink and adjoining building should be retained until the possibility of a private health contractor being involved had been explored.

10. A Masterplan with two options was generated with a series of design principles.

Closure - and the Veterans' Village proposal

In summer 2009 after the formal closure a programme of refurbishment of existing buildings and developing new buildings was proposed to create a residential Veterans' Village in the hospital buildings. This inspirational vision was to continue the strong historical connection and therapeutic use of the site for the armed services. "The Royal Haslar Veteran's Village, Promoting Quality of Life", the outline document was titled. Several well-meaning, knowledgeable and highly respected individuals were included in a steering group.

"The steering group have a good understanding of the site, local relationships and the market we are entering which will ensure optimum delivery to work with the board of the developers, Our Enterprise Haslar" said the brochure. A letter from the Veterans' Village Charity Consortium supporting the proposal was attached to it, along with others from the Gosport Borough Council, the Save Haslar Task Force and the Royal Naval Benevolent Trust. The Veterans' Village proposal was worked out in conjunction with Chelsea Hospital - so that exchanges between the boys in red and the girls and boys in blue could take place. It had City of London financing and backing from Prince Charles and Prince William. The local newspaper, *The News* reported that a hotel, a 160-bed elderly medicine infirmary, a heritage centre and 100 two-bedroomed apartments for veterans – 85 in Canada block and 15 in the former psychiatric block – were among proposals for the Gosport site. Developers also hoped to build a post office, hairdressers and shops in a dedicated 'veterans' village' on the historic site.

"The Veterans' Village Consortium's project is being bankrolled by wealthy expatriate construction magnate Paul Fraser who has committed to buying and redeveloping Haslar as his father was treated there." Veterans minister Kevan Jones announced on Tuesday that the consortium – represented by the firm Our Enterprise – had won the bid to develop Haslar. It was to undergo tests to check its plans were sound before a contract was signed that autumn. The consortium was also in talks with charities and NHS Hampshire about on-site services.

Roger Saunders, chairman of the Veterans' Village Consortium, said: 'The village will be open to all people who have served at sea – in the Royal Fleet Auxiliary, merchant navy and Royal Navy and also retired members of the RAF. A long-term aim is to make Haslar an example of how business and charities can work together to produce something of national importance." Peter Swan, co-founder of the Veterans' Village Consortium, said: "Our vision is a Chelsea Hospital with a military feel – so we would have Haslar pensioners, like we have Chelsea pensioners." Peter Viggers, Tory MP for Gosport and chairman of the Save Haslar Taskforce, said: 'I think the ideas are exactly what we had hoped for." "Gosport borough councillor and Save Haslar Taskforce spokesman Peter Edgar said: "Here we have a preferred bidder. We were all expecting a miracle, but this is unbelievable.'"

The plan which took two years to work out made use of every building. Commodore Peter Swan who supported the proposal pointed out that at the Enquiry By Design (EbD) it was concluded that because of the many restrictions currently in place on and around the site it was considered to be of little commercial value. It was recommended that the site house a closed community such as the Veterans Village and a proposal in this regard was submitted to the MOD in accordance with these conclusions and recommendations. This was accepted on the basis that the Veterans Village as outlined in that proposal would be implemented and this was reflected in the price that the MOD announced.

Even with the best intentions, excellent proposals do not always get realised. Haslar Hospital was sold in late 2009 for the low price of £3.5m - on the basis of the plan for this carefully worked out scheme for a Veterans' Village, dedicated to ex-naval pensioners and their husbands/wives. Unfortunately and inexplicably, no development conditions were placed on the sale to ensure that this actually materialised.

The hospital is sold

The new owner was not party to the recommendations of the Enquiry by Design workshop supported by Gosport Borough Council and English Heritage. They rejected the backers of the Village. Once Our Enterprise Haslar had approval as the developer, the Veterans' Village participants were quietly sidelined from the project.

Of course, this reversal aroused considerable concern. On 29 January 2010 Peter Viggers, MP for Gosport asked the Secretary of State for Defence "at what price was the Royal Hospital Haslar site sold and to whom; what the acreage of the site sold is; how many houses, flats and other buildings are on the sold site; what due diligence was undertaken prior to the sale; what undertakings were obtained from the prospective purchasers prior to the sale; what other offers and indications of interest were received for the site; what the monetary value of the site was on his Department's accounts prior to the sale; and what limitations and covenants were placed on the site at the time of sale?" Kevan Jones MP replied in a holding answer on 29 January 2010: "The Haslar Hospital site in Gosport, which extends to some 23 hectares and comprises around 75,000 square metres of buildings, was sold to Our Enterprise (Haslar) Ltd. in November 2009 for £3 million. The hon. Member will recall he met with me to discuss the proposals before the sale." "An Expressions of Interest campaign for the Haslar site was undertaken by Defence Estates in early 2009 and a number of bids were received by the closing date of 3 July, the details of which are confidential, for commercial reasons. The other bids were rejected either on grounds of insufficient detail regarding redevelopment plans for the site, the low level of the offer, or both. Prior to the sale the site was included on the Asset Registers of the Department at some £55 million based on the depreciated replacement cost of the site in its current usage. Such valuations are undertaken for accounting purposes and are not the same as a market valuation undertaken for disposal." "Additionally discussions took place with Our Enterprise, with regard to the consortium's financial status and intentions toward the site's heritage assets, memorial garden and future maintenance. No limitations or covenants were placed on the site at the time of sale."

As a member of the Hampshire Buildings Preservation Trust with a special focus on Gosport, Celia Clark was so concerned at the lack of any conditions or covenants being applied to the sale she asked Mike Hancock MP to ask another Parliamentary question about the conditions attached to the sale. She did not receive a reassuring reply.

The press release issued on 3 December 2009 by the new owners assured the public that they were committed to creating 'a veteran's village.' *The News* reported that "The Royal Haslar Hospital has been sold to developers who will build a 'veterans' village and are committed to keeping care and medical facilities at the site… The Ministry of Defence sold Haslar yesterday afternoon for an undisclosed fee to Our Enterprise - a company specialising in bringing charities, social investors and the public sector together for large developments. In a meeting with Gosport councillors last night, Our Enterprise pledged their commitment towards ambitious plans for the site. They include a partnership with veterans' charities to build a village for retired service personnel. There are also proposals for new education facilities for the University of Portsmouth."

"The company is drafting in two of the country's leading cancer and dementia specialists, Professor Karol Sikora and Professor Robin Jacoby, to develop a centre of excellence for rehabilitation and care at the hospital, which will include a dementia ward for 44 people. The announcement follows a high-profile 10-year campaign to save the 250-year-old military hospital, which was closed by the Ministry of Defence in July."

After a meeting with Our Enterprise the news that medical facilities were to remain at the site was welcomed by Councillor Peter Edgar who had fought to save the hospital: "It's a very encouraging project… The people of Gosport want jobs for local people and want people to go back to living and working on the site as people have done over the last 250 years… My big dream is that in 250 years' time, people will be celebrating 500 years of care on this site. Our Enterprise have definitely picked up on the affection local people hold this site in…"

"Andy Feculak, design director of Our Enterprise, who met with Gosport councillors last night, told *The News* plans for the site include a retirement community and village for veterans, specialist care facilities, accommodation and education buildings for university students and a four-star hotel with small conferencing capabilities. There are also plans for publicly accessible garden squares and a Haslar heritage centre displaying the history of the hospital. Mr. Feculak said: 'We strongly believe the site needs a vibrant mix of uses.' However, he warned plans are still at a 'formative stage'. "It's got to go through most of the planning processes with the local authority, but they committed today to make this move forward quite quickly. Hopefully, it will take six to eight months and we can move forward with the development plan thereafter. In the meantime, Our Enterprise will be renting out 13 properties on short leases in the listed buildings on the site."

"The redevelopment of The Royal Haslar Hospital is set to boost employment in the area. Our Enterprise… is working in partnership with national employment charity Tomorrow's People - which helps people out of long-term unemployment and welfare dependency and into jobs and self-sufficiency. The charity will be working to get unemployed people in Gosport back in to work through training and job opportunities at the new facilities planned for the site. Debbie Scott, chief executive of Tomorrow's People, said: 'We work with the most marginalised individuals helping them turn their lives around through work. The continuing care community for veterans and ex-services personnel offers a wonderful opportunity to make a real difference.'"

"Following the sale of Haslar, Our Enterprise's chairman Charles Green said: 'We have spent two years developing this scheme and I have to thank all of those individuals who have helped us to get where we are today. We also congratulate the MoD and the Secretary of State for making what we believe to be the right decision for the community, and now look forward to developing a scheme, a bright future for this historical site for the next 250 years.'"[22]

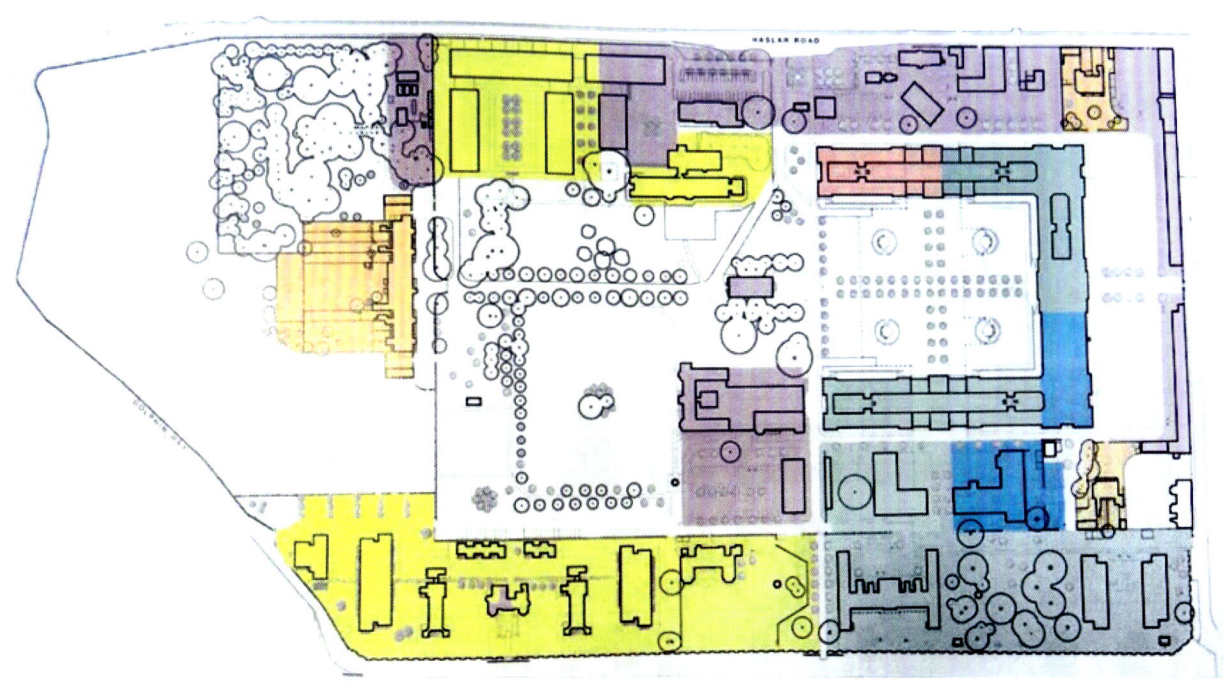

Planning application Use plan Blue: medical; mauve: commercial/social; grey: retirement village including social facilities; yellow private residential; pink affordable residential; orange; existing housing terraces.
By kind permission Gosport Borough Council

The developer's planning application ignored the EbD conclusions. "Delays were incurred because of the huge number of detailed planning applications required for such a large site subject to a range of restrictions arising from its heritage. There was an objection that the current owners obtained the site before the Veterans Village could be agreed and implemented by the various service charities."

One objector was Dr. Roger Saunders, a member of the Veterans Village steering committee. "The site was sold on the understanding that the site become a Veterans Village (not a Commercial Care Home) as this use was considered to be the best solution for the site and one that our Veterans deserve and expect…. Our Veterans deserve this site for their Village as it was originally planned but our Service Charities do not want another Care Home. They have these already! It seems that this and the EbD has been overlooked by the current owners who were not present at any time during the EbD and that they are proposing a number of Commercial activities on the site that would jeopardise not only the historical importance of the site but fly in the face of the EbD discussions, recommendations and conclusions and would not fit in with your strategic and transport plans."[23]

Archaeology

In May 2010, an investigation of the hospital's burial ground by archaeologists from Cranfield Forensic Institute was featured on Channel 4's television programme Time Team. It was estimated that up to 7,785 individuals had been buried there, although other estimates say there could be anything up to 20,000. Skeletons were regularly found in recent times all around the site when undertaking building work or drain repairs. Once established that they were over 100 years old, the problem went away, otherwise the police had to be involved. And as for the Irishman with no security clearance in IRA problem times who had been working digging holes around the site just before Prince Charles visited, this caused a serious panic and much extra digging until he was cleared.

Royal Haslar – planning permission

Haslar Hospital Planning application 12 February 2014. Conservation Management Plan. Trees retained light green; cut down dark green; blue: Listed buildings; grey buildings retained; red buildings: removed; red with black outline: subject to Listed Building Consent. By kind permission Gosport Borough Council

Negotiations on the complex planning applications were protracted. Royal Haslar, as it was decided to call the site, was to be redeveloped by Our Enterprise Haslar/Harcourt Developments. Erection of new buildings was to comprise 286 residential units; a continuing care retirement community containing a 60-bed care home, communal facilities and 244 self-contained retirement units; offices and business units; a health centre; hotel; tearooms and restaurant/bar); a convenience store; church, public hall and heritage centre together with alterations to existing vehicle and pedestrian access arrangements, open space provision and landscaping and parking. A programme of refurbishment of existing buildings and developing new buildings was planned to create a residential village.

In July 2014 five years after the purchase, outline planning permission was granted to the new owners for a £100m development: 15 exclusive listed residential properties; 271 private new-build residential properties; 244 retirement homes; a 60-bed care home; a community healthcare hub; a 78 room 4* hotel and spa; licensed and restaurant premises; over 50,000 sq. ft. of business space. The retirement accommodation was not specifically linked to naval veterans. The expectation that publicly owned land should be used to the public – and local – benefit was completely absent. There was no specific provision to continue to care for sailors and their partners. As we said earlier, Christopher's recently published book *The New Enclosure. The Appropriation of Public Land in Neoliberal Britain* makes the same point, that disposals of publicly owned land do not meet expectations of public benefit.

Sales

The historic Georgian houses for the senior medical officers at either end of the site, its most valuable assets, were sold first. Apart from these early sales, for several years there was little sign of development. The Officer's Terrace at the western end: residences 1,2,3 and 4 were sold for some £450,000 each; the old Admiral's residence No.5, sold for £975,000 in 2014. Caroline Dinenage MP and Mark Lancaster MP have the central, largest house, which they bought from the first private owner. Nos. 6-10 the Terrace were split into 10 flats that sold for £250,000 to £300,000 each. The interior of one of the flats seen by the authors was substantially and impressively remodelled by the new owner. Houses 11 & 12 and 13 & 14 - facing either end of the main east facade were also sold. A Section 106 agreement with Gosport Borough Council required the deposit of £6.5m from these proceeds towards restoration of the site's many historic buildings. The new owners of the houses and apartments are responsible for the development of their own properties and they have to apply for planning permission and listed building consent for alterations or demolitions.

As already mentioned, the developers paid about £3m for the whole site. They have had to pay £26m clawback (where the subsequent owner has gained more value than the sale price paid to the MOD) to the Treasury, according to Eric Birbeck of Haslar Heritage Group.

Haslar Heritage Group

In July 2017 the authors had a comprehensive tour of Haslar Hospital. Our guide was Eric Birbeck MVO, an ex-RN medical technician with 50 years continuous association with the hospital. In 2001 he was part of the founding team of historians: Haslar Heritage (see website). He leads the group of twelve experts (all past staff) who are documenting the hospital's history. The group has a business plan and is intending to seek charitable status. Their collection of objects, pictures and documents is currently in St. Luke's Church. They managed to get back the church fittings which had been both removed to the Royal Navy Trophy room in HMS Nelson and St Ann's Church in the Dockyard. The Heritage Group is involved with the ongoing use and care of the church for services, funerals, baptism and renewal of wedding vows.

It was also granted the use of the Old Medical Supplies Agency building (MSA) of 1756 for a peppercorn rent, eventually to become a Heritage Visitor's centre and archive. This building has an open space on the whole of the ground floor for a reception and historical display. The upper floor has offices, storage, meeting room and kitchen facilities. The intention was to reach agreement with the QARNNS Archive and RN Medical Branch Association for them to share office space with the Haslar Heritage Group. The Group continues to work with the owners' management team during the redevelopment of Haslar. They were in discussion with the National Museum of the Royal Navy for advice and support in planning the project and to seek grant funding for various stages of the refurbishment of the building. They intended to apply to the Heritage Lottery Fund to convert it to an archives store, museum and visitors' centre. The developer allowed them to name both the roads within the site and buildings.

Further Planning applications

- During 2015 and 2016 the developer submitted many final planning applications including:

- Demolition of kitchen block 42, zymotic block 39 and conversion of zymotic wards east and west and admin block 41 to provide 17 residential units and tearoom.

- Demolition of the galley, the general stores block and Albert block.

- Erection of new buildings to comprise 286 residential units; a continuing care retirement community containing a 60-bed care home, communal facilities and 244 self-contained retirement units; offices and business units; a health centre; hotel; tearooms and restaurant/bar); a convenience store; church, public hall and heritage centre together with alterations to existing vehicle and pedestrian access arrangements, open space provision and landscaping and parking.

- Demolition of 3 buildings and conversion of Canada block and the erection of 2 buildings to provide 93 retirement apartments, conversion of G block to provide 8 residential units

---

The maximum number of people living on the site was to be 1500 people – far fewer than the many more working there and the patients when it was a hospital.

In 2017 the developer became known as the Haslar Development Company Ltd. Delays were said to have been incurred because of the huge number of detailed planning applications required for such a large site subject to a range of restrictions arising from its heritage. The chairman of the board of management is Brigadier Parker Bowles.

Slow progress towards new life

In March 2017 the developers pulled down the ugly Brutalist Crosslink, which filled in the quadrangle. They aimed to retain the serving basement underneath which housed the hospital utilities as an underground carpark. The quadrangle was to be restored; it was once famed for its flowering cherry trees. The open landscape will be kept, and ultimately opened to the public.

In January 2019 the 1970's build senior rates' mess and the older galley and stores were taken to pieces. Ready for apartments to be installed, Canada Block was renamed Canada House and G Block (the old Psychiatric Department) was Goodrich House, named after the matron who was in charge during World War II. An American company from Florida was to run the care programme for residents. All the site roads were renamed and eventually so were all the blocks. The business suite for sales located in the Medical Mess was opened by early spring 2019.

In September 2020 50 people lived on the site including 15 families, a figure far from the target of 1,500. As well as the MPs, retired naval officers, medical and police officers and a professor of maths

1,500. As well as the MPs, retired naval officers, medical and police officers and a professor of maths from the university of Portsmouth were residents. A reflection of how far from the original vision to care for veterans it had become, 21 years after it was sold, five properties were reportedly recently sold for about £3m. The restored Canada House offers 36 'luxury one and two bedroom independent living apartments with parking.' In September 2020 a two-bedroom house was for sale for £520,000. A two-bedroom flat in a new build was £475,000.[26] Four flats in Goodrich House were already sold and four more were available. The majority of the spaces in the main hospital were labelled 'Senior Living', with 'a range of independent and assisted care packages' available. The business units and food and retail were on the north and east periphery. Two buildings were labelled 'Healthcare'.[25]

There is no bus service along Haslar Road, so owning a car is probably essential. Despite the promises of public access, in stark contrast to Royal Clarence Yard, it remains a completely gated community. The public can get no further than the security gate, but this makes it irresistible to urban explorers, who publish vivid photographs of the hospital's derelict interiors on their websites. There was no Heritage Open Day tour there in 2020.

Another challenge at Haslar is how the vital seawall will be raised and financed to combat rising sea levels and storm surges. In 2019 the public walkway along the shore was the only way to see into the site. Next door, another historic military site awaits its future: the early barracks, later Haslar Detention Centre owned by the Ministry of Justice, discussed in our barracks chapter.

The former Haslar Detention Centre from No.9 Officers' Terrace Haslar Hospital

Sources

A Brief History of Queen Alexandra Hospital, Cosham, Portsmouth Tim Lambert http://www.localhistories.org/qa.html

William G Gates *The Book of Jubilee* Evening News Portsmouth 1877-1927 page 48

J K Buchanan 2005 'An enlightened age: building the naval hospitals' *International Journal of Surgery* 3(3) pp. 221-8

Celia Clark 2010 'Naval Hospitals: History and Architectural Overview' *Transactions of the Naval Dockyards Society* Volume 6. June pp. 65-73

JG Coad 1989 *The Royal Dockyards 1690-1850 Architecture and Engineering Works of the Sailing Navy* Royal Commission of the Historical Monuments of England Scholar Press Aldershot p.293

Portsmouth Society News Autumn 2003 http://www.portsmouthsociety.org.uk/docs2003/nlsep03.pdf0

Nicholas Rodger 1986 *The Wooden World: An Anatomy of the Georgian Navy*. Annapolis, Maryland: Naval Institute Press. ISBN 0870219871 p. 110

David Lloyd 1974 *Buildings of Portsmouth and Its Environs* City of Portsmouth *pp. 67-68*

Memories of Councillor Peter Edgar, Spokesperson of the Save Haslar Task Force (1998-2009).

Millie Salked 2020 'New era for Portsmouth's Queen Alexandra Hospital as it is awarded university hospital status' *The News* 29 July

References

1. https://en.wikipedia.org/wiki/Royal_Hospital_Haslar
2. http://www.royalhaslar.com/Joe Hynes info@royalhaslar.com
3. http://www.qaranc.co.uk/haslarroyalnavalhospital.php
4. http://www.haslarheritagegroup.co.uk/ and thanks for information to Eric Birbeck of the Haslar Heritage Group
5. http://www.bbc.co.uk/hampshire/content/articles/2007/03/26/haslar_history_fe ature.shtml
6. http://www.mygosport.org/info_pages_htm/info_haslar.htm
7. https://www.gosport.gov.uk/sections/your-council/council-services/planning- section/local-development- framework/gosport-borough-local-plan-2029/gosport-borough-local-plan-2011-2029-adopted-october-2015/ evidence-studies/part-f- development-sites-allocated-and-or-suggested/ LP/F4/1 Royal Naval Hospital Haslar- EbD Workshop Report (Prince's Regeneration Trust January 2009) LP/F4/2 Royal Hospital Haslar Planning Application
8. "Haslar - The Royal Hospital" by A.L. Revell, a Gosport Society publication.
9. http://www.portsmouth.co.uk/news/government-snubs-haslar-handover-1- 1271560
10. Gosport Borough Council - Employment Background Paper www.gosport.gov.uk/EasySiteWeb/GatewayLink.aspx?alId=33212
11. Hansard 17 July 2003 Citation: HC Deb, 2 February 2010, c209W http://hansard.millbanksystems.com/ commons/2003/jul/17/adjournment-summer
12. Listed Buildings in Gosport
13. https://www.gosport.gov.uk/sections/your-council/council-services/planning-section/conservation/conservation-guide/list-of-listed-buildings-for-gosport/
14. David Lloyd 1967 *Hampshire and the IOW* Buildings of England Penguin Books (1967)
15. David Lloyd 1974 *Buildings of Portsmouth and Its Environs* pages 67 to 68: (1974)
16. Celia Clark the design of naval hospitals *Sustainable Development and Planning* by Dr C. M. Clark 2009
17. Thomas Dobson 1808 "Observations on Preserving the Health of Soldiers and Sailors; and on the Duties of the Medical Department of the Army and the Navy; with Remarks on Hospitals and their Internal Arrangements" Philadelphia
18. Crosslink demolition – The News https://www.portsmouth.co.uk/news/health/work-finally-starts-on-former- gosport-hospital-site-1-8416152

19 England and Wales Court of Appeal (Civil Division) Decisions >> Doherty v Fannigan Holdings Ltd [2018] EWCA Civ 1615 - 12 July 2018

20 https://historicengland.org.uk/listing/the-list/list-entry/1001558

21 Surgeon Rear Admiral Michael Farquharson-Roberts19-20 LP/F4/1 Royal Naval Hospital Haslar- EbD Workshop Report Prince's Regeneration Trust January 2009
LP/F4/2 Royal Hospital Haslar Planning Application
https://www.portsmouth.co.uk/news/health/work-finally-starts-on-former-gosport- hospital-site-1-8416152

22 *The News* 23 July 2009 http://www.portsmouth.co.uk/news/veterans-village-is-topping-list-of-plans-to-replace-haslar-1-1234927

23 'Haslar hospital sold to create a veterans' village' *The News* Thursday 03 December 2009 Read more at: http://www. portsmouth.co.uk/news/haslar-hospital-sold- to-create-a-veterans-village-1-1243934

24 Dr Roger Saunders 2014 Letter to Gosport Planning Department 22 May

25 Regeneration Investment Organisation 2015 https://assets.publishing.service.gov.uk/government/uploads/system/uploads/att achment_data/file/600615/RIO_Pitchbook_May_2016_withdrawn.pdf

26 https://www.nestoria.co.uk/detail-int/00000050000038586424368721/thumb/1/1-2?serpUid=0.5532818822151160009077 1&pt=1&ot=1&l=royal-hospital-haslar&did=90_high1&t_sec=1&t_or=2&t_pvid=c891f8e5-fc54-4131-b7a6-31db7d18fdf6
https://www.royalhaslar.com/property-search/
https://www.royalhaslar.com/property-search/goodrich-house/apartment-g205/ accessed 14 September 2020

Chapter 10
Ordnance Yards: HMS Vernon/Gunwharf and Priddy's Hard

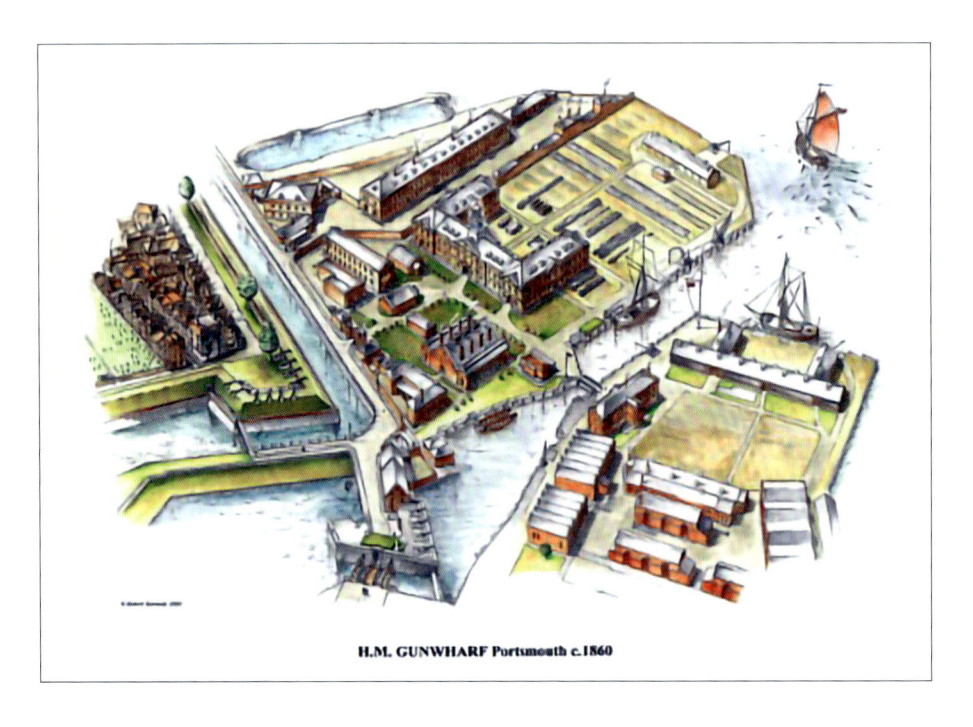

Gunwharf in about 1860   Robert Kennedy 2000

Gunwharf and Priddy's Hard on opposite sides of the harbour are linked by their shared function as places where the navy's ordnance was stored – and by their recent reinvention, linked to the Millennium. Gunwharf was one of Celia Clark's case studies for her PhD thesis. This was action research because at that time she was directly involved as a citizen in responding to the redevelopment.[1] Few people shopping, going to the cinema and gallery, bowling, eating and drinking or living in today's usually bustling Gunwharf are aware of its history. In the 1990s it was largely cleared of existing buildings and redeveloped.

The historic core of Priddy's Hard was acquired by the Portsmouth Naval Base Property Trust, which is developing it, despite its intractable contamination and conservation challenges. It contains the Explosion Museum of Naval Ordnance – part of the National Museum of the Royal Navy – and a nature reserve as well as proposed flats, houses, a pub and brewery and holiday and student accommodation.

History

Earlier storage of armaments around the harbour was problematical. From 1584 the Square Tower at the seaward end of Portsmouth High Street was adapted to serve as a gunpowder magazine, but the townspeople protested about the danger of this and requested that explosives be stored away from populated areas.

In 1662 the Board of Ordnance which supplied the navy with ships' armaments chose an area which had been fortified in the mid-sixteenth and seventeenth centuries between the town and the dockyard, well placed to receive the armaments of warships coming in for repairs or refits and to issue fresh supplies of weapons for those preparing for sea. Between 1709 and 1715 Gunwharf was constructed on land reclaimed from the harbour by contractor William Shakespeare, using fill from excavations associated with development of the nearby dockyard workforce town of Portsmouth Common, later Portsea, which was separated by the Mill Pond from Portsmouth proper. Guns, gun carriages, small arms, gunpowder and shot were removed from ships returning from a commission to be stored in the ordnance yard and wharfs of Gunwharf. Despite this new facility, until it was adapted as a meat store in 1779, Gunwharf also leased a small landing wharf for vessels to load up gunpowder kegs at Portsmouth Point at the north end of Broad Street.

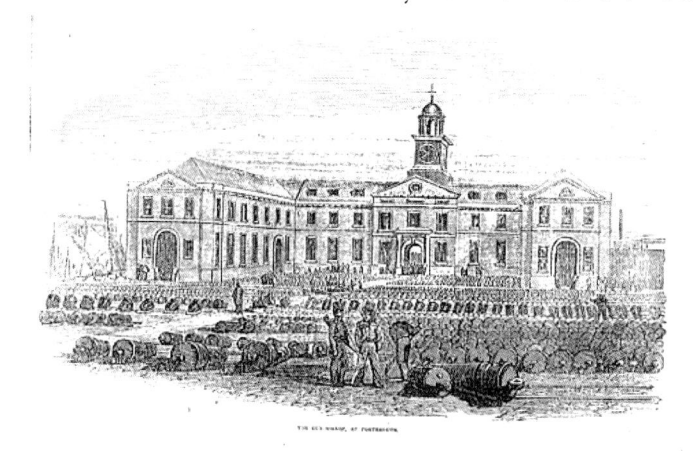

Gunwharf *London Illustrated News Supplement May 5* 1855

The *Illustrated London News* has an engraving of the Vulcan building of 1814 which describes the Gunwharf at Portsmouth as "one of our great reserves for the efficient progress of the war [Crimean]. The Wharf is a large area of ground, partly in Portsea and partly in Portsmouth; for it fronts the harbour opposite the junction of those towns. Here are the Ordnance Department of the Navy - the guns and other weapons offensive and defensive employed on shipboard. That the supply of guns for the Navy is an important matter may be made clear. The number of cannon and cannon balls on the Gunwharf is truly astonishing. These guns comprise, not only new ones for ships yet to be built, but the guns belonging to ships now laid up in ordinary [reserve]. In the latter case each ship's guns are ranged by themselves, with the name of the ship painted on the first gun of each parcel. Some of these guns are of such vast size and thickness as to weigh 60 hundred weight each. We acknowledge with much pleasure the courtesy of Mr. Steward, the head of the above department, in permitting our Artist to sketch this interesting scene."[2]

That last sentence encapsulates much that made the recent Gunwharf transition at once so difficult and so bound up with national and local pride: exclusion in the interests of security, and intensity of military value physically located, which in the context of the vulnerable local and national economy in the late twentieth century also raised expectations and arguments over appropriate futures for the site. Continuance of technical innovation was one such prospect.

New Gunwharf to the south of the Old separated by a camber spanned by a bridge was added in 1797-1808 over part of the Tudor and later Portsmouth Town fortifications. Between 1800 and 1820 an armoury, magazine for filled shells and Congreve rockets, workshops for armourers and smiths, a guardhouse and the Grand Storehouse or Vulcan were added. The armoury was designed in 1804 to accommodate 12,000 muskets, 5,000 pairs of pistols, 10,000 swords and was supplied with workshops and forges. The main smithery was completed in 1805. In 1799 the Old Gunwharf made up 1,000,000 rounds of musket ball cartridges, which were sent to Priddy's Hard. A dramatic 1896 photograph of piles of armaments piled up in the Shell Ground is in the collection of Portsmouth Museums and Records Service.[3]

Slight's description of Gunwharf buildings quoted in Chapter 1 as 'sacred to utility' eloquently describes their character. In old Gunwharf the Customs House, originally the Vernon Offices (c.1790) of sober Georgian design and cellular plan facing the Vernon Creek is one of the earliest examples of specifically designed office buildings, predating the North and South Office Blocks in the Dockyard (c. 1840). The Warrior block (c.1797) a long four storey flat roofed block shown as the Sea Service Store on a plan of 1859 was later gutted by fire and extensively rebuilt as Mens' Quarters in 1921, a function it retained until closure in 1991.

Gunwharf Gateway 1803

The 1803 Nelson gateway with stone mortars surmounting the pillars stands on the line of the original boundary of the New Gun Wharf.[4] Storehouses such as Building No. 47 now the Old Infirmary (c1810) and the splendid Vulcan building (1811-14) were built in red brick with grey headers. Henry Slight and the *Illustrated London News* described it as the most notable building on the New Gunwharf. The Duke of Clarence laid the foundation stone in 1811. Its total cost was £50,140 Is. 2d, of which over £16,000 was spent to provide solid foundations 18' deep on this unstable reclaimed land. It has an innovative composite construction of iron columns and timber joists for storage of heavy items.[5] It was one of the largest storehouses developed by the Board of Ordnance, its importance reflected in its embellishments - including pediments and the clocktower.

Vulcan shorn of clocktower by WWII damage

The Vulcan was partly used as offices for the Mining School, which came ashore in 1917 during the First World War. Cracks, evidence of settlement and of heavy loading bear witness to its many uses over the years until its closure in approximately 1955. The roof and top floor of the building were badly damaged by incendiaries on 10th/11th January 1941, when the clock tower was burnt out. The northern wing was pulled down in the 1950s to make way for the Creasy block; the position of the clocktower was roofed over with corrugated iron, so that it lost much of its ancient splendour. Even in its decayed state, before it was restored and converted in the redevelopment, the Vulcan Land Service Store was still the most notable building on New Gunwharf.

As already discussed in Chapter 2, in the 1870s most of the older landward defensive structures including moats and bastions were declared redundant and demolished to make way for the rapidly growing town, where once separate settlements were coalescing into one large urban area. The Mill Pond separating the towns of Portsmouth and Portsea was filled in. In 1876 the military authorities were finally persuaded to allow the fortifications to be pierced for the extension of the London and South Western Railway to construct a line from Town Station to the harbour. Tracks were laid over the northern edge of Old Gunwharf on a brick embankment with a siding to serve both the Old and New Gunwharfs. The site of the millpond was laid out as the United Services Sports Ground. All the landward fortifications were demolished to ground level or lower for the construction of Gunwharf Road.

A new perimeter wall built in 1876-1878 in yellow brick with red panels and corbelled stone castellated semi-circular 'pulpits' or machine gun points at regular intervals, with new gates and a porter's lodge, redefining the extension of the Gunwharf.

Gunwharf Gateway. Creasy block being demolished visible inside archway

Landward end of Gunwharf Creek   Private collection

Donegal Lodge 1897 Quartermaster's Quarters

The main entrance near Park Road and the railway line is marked by a miniature castellated tower with a pierced brick parapet and police cell, flanked across the archway by a smaller tower and the adjoining colonnaded gatehouse and police office. The cell still had the wooden pillow in situ. A steep gradient railway spur linked the site to the main line was used for goods traffic. It operated until about 1930 and was resurrected in 1937. Disconnected after the war, it was officially abandoned in 1962.

From 1855 to 1918 the area to the seaward side of Gunwharf was filled in with numerous buildings in a similar way to the cramped layout of the Dockyard and most naval establishments, where for many centuries land between secure boundaries was limited and buildings of various styles were fitted into any spaces available. New structures were added piecemeal as the site's function developed and changed over time. In 1897 Donegal House in dark red brick was built near the main gate as the quartermaster's quarters, later used as an officers' hostel. During the redevelopment it served as the developer's offices until it was demolished.

The Ordnance Board was disbanded in 1855 and as ships and armaments developed the requirement to offload armament diminished. The two Gunwharfs were amalgamated as the Naval Ordnance Department in 1891, and most of the site was subsequently adapted as the Naval Torpedo School. However, legal freehold of the 1709 wharf: Gunwharf Creek was retained by the Army for the balance of its entire long operational life. This portion of the site was marked on maps as War Department, and there was a strictly enforced boundary between this and the much larger naval area. It had its own road and rail access. In 1920 most of the military functions of Gunwharf ceased, except for its use by the Army's Maritime transport. By WWII Gunwharf was one of the Royal Army Service Corps bases, home to the Royal Corps of Transport 20 Maritime Regt.'s Landing Craft Logistics. Their duties included fast target towing; from 1957 they provided logistic support for the British Army of the Rhine, the firing range in the Hebrides established by the RAF in 1957 (transferred to the Royal Artillery in 1958), and range safety work. The history and evolution of the Army's maritime transport service are commemorated in a memorial stone in Gunwharf Quays, organised by Maj. David Glossop in the presence of Maj. Gen. Lyons, Commanding Officer of the Royal Logistic Corps. The RE Diving School was relocated to HMS Vernon when Marchwood port was rebuilt. In 1993 it was moved again to Horsea Island.[5] A sculpture of two divers with a mine between them by Mark Richards commemorates the mine warfare and diving heritage of HMS Vernon.

By tradition the Keys to the City of Portsmouth were held by the senior army officer in Portsmouth. After WWII to the early nineties this was an RASC officer and later an RCT officer. Sir Christopher Armstrong held the appointment between April 1981 and September 1983. "I presented the keys to the Lord Mayor at each annual Mayor Making. By tradition the holder could have the keys on his dinner table for special occasions and they were always placed at Mess Dinners in front of the CO. They were handed over to the OC of the RE Diving School in 1989 when the RCT gave up the Gunwharf and moved to Marchwood"[6] on Southampton Water. Now this duty is undertaken by 4PWRR, formerly the Royal Hampshire Regiment.

After starting his military career as a gunner in the army, Lionel Crabb joined the Royal Navy rising to the rank of Cdr. He worked as a diver clearing mines in WWII, later investigating sunken Royal Navy submarines and downed aircraft. Deane Clark's father Percy Clark, Senior Scientific Officer and Admiralty photographer maintained Crabb's cameras. On 19th April 1956 Cdr. Crabb dressed for his final dive in the boat pound at HMS Vernon with 'unofficial' help from a fellow officer, before entering the water at Kings Stairs Jetty to investigate the propeller of the Soviet ship *Ordzhonikidze*, a new design that naval intelligence wanted to examine. He was never seen again. Speculation about his disappearance, as well as novels and films ensued.[7]

There was a long-established public slip here which had a communal right of way, used by Southern Railway steamers. If the public rights of access to the adjacent area used by Southern Railway vessels had been upheld during the decision-making about the redevelopment, North Wall and Gunwharf Creek might not have been filled in and used for the vast two-storey carpark, the largest in the country.[8]

1709 wall visible in Gunwharf Creek south of the railway line
Private collection

Great storm damage to Vernon diver tender moored at North Wall 1987   Private collection

HMS Vernon

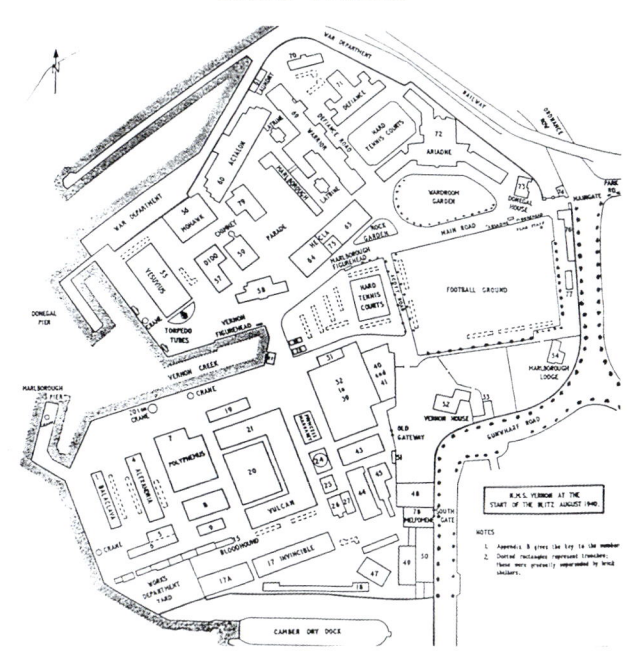

HMS Vernon at the start of the Blitz 1940 from *HMS Vernon A Short History from 1930 to 1955* E D Webb

Algerine class minesweeper *HMS Marvel* and another berthed on Maintenance Jetty. Mining trials tank astern of the ships; centre: south wing of Vulcan building and right Portsmouth power station. Date: Coronation Fleet Review by Queen Elizabeth II in 1953?   PJ Clark Admiralty Photographer

223

The major portion of Gunwharf was taken over by the Navy as HMS Vernon in 1923. *Vernon* was the name of a hulk where training in sea-based counter-measures had been developed. It specialised in mine warfare.

For over 110 years HMS Vernon was a centre of naval innovation and invention. From 1939 to 1945 Royal Naval trawlers manned by volunteers from deep-sea fishing fleets recovered German mines for analysis so that countermeasures could be devised. Vernon personnel prepared and deployed ships from Portsmouth for the evacuation of the British Expeditionary Force from Dunkirk. MX Wrens produced thousands of circuits for acoustic and magnetic mines. Vernon was integral to preparations for D-Day.[9] Commander Webb's *HMS Vernon A Short History from 1930 to 1955* published in 1956 described the evolution of the site and pioneering research there, including the defeat of the German magnetic mine and the technique of degaussing, one of a whole series of inventions.

Admiralty Experimental Diving Unit: 50 foot Tank 1921 and Vulcan Building
Deane Clark

Mine design, diving training and the army's water transport service were symbolised by two particular features: the tallest structure on the site: the Admiralty Experimental Diving Unit built in 1921 to train navy divers; and the military pier in the area of water to the north of the site bounded by the piled pier of Portsmouth Harbour Station and the 1709 wall to the south. This was the base of 46 Maritime Squadron, RASC/Royal Corps of Transport. North Wall and Gunwharf Creek were subsequently used as an operational base for revenue cutters. The Fleet Diving Unit and Management Systems Group buildings were added from 1936 to 1951.

The 94-foot Mining Tank, a column of seawater 55 feet high and 30 feet in diameter was originally supplied by the USA in WWI. It was used in mine design with observation ports on the various deck levels, until 1959 when underwater weapons work was concentrated at Portland. The tank was then refitted with a diving compression chamber, changing rooms and lighting for trials of underwater equipment. A moving platform, "the mat", could be moved to any depth. A 2-ton travelling crane raised heavy equipment from ground or lorry level. Stephen Payne said "I was taught to swim in the tank by a Navy diver. My father worked as a scientist (he designed the first underwater telephone & went on to design breathing apparatus for divers) in the Admiralty Experimental Diving Unit (AEDU) which was sited near the heli pad. I remember going up in an external lift and the guy that ran the Tank only had one arm. I went down the ladder into the tank on to a slatted wooden floor - at this time, the water was around my ankles. Then the floor started to descend & the diver said 'you've got to swim now!' I can't imagine this being allowed nowadays." [10]

It was also used to train aircrews to escape from downed aircraft. A helicopter body was secured in the middle of the tank; crew onboard. This was overturned and they had to swim out. Prince Charles apparently said how scaring he found the experience. The tank was demolished in 1988.

"Walker's Folly": Water tank for trainee divers to simulate working under water with mechanical equipment and repairs, observed by instructors. Base kitted out as a workshop with workbenches and vices etc.

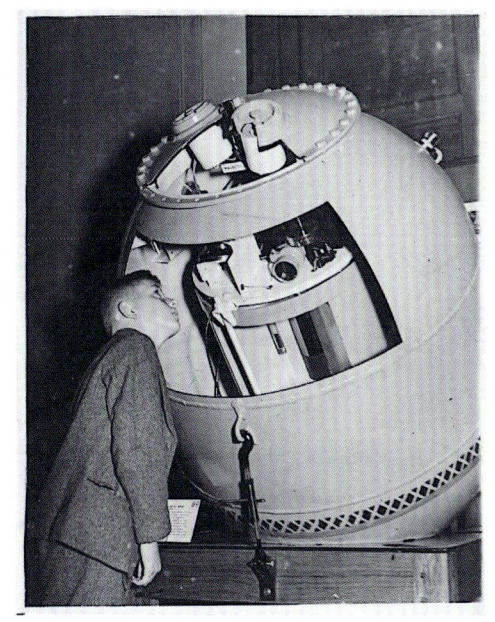

Deane Clark at the Science Museum peering into a magnetic mine
G H Goodman *Evening Standard* July 31 1946

In July 1946 the Science Museum displayed a mine from HMS Vernon and mine countermeasures from Leigh Park in an exhibition about British Mines and Firing Units. Aged 9 Deane Clark gatecrashed the launch: this photograph appeared in the *Evening Standard*. Some of the exhibits were hastily removed – for fear the Russians might see them….

Heritage protection

In February 1992 HMS Vernon – Gunwharf was designated by Portsmouth City Council as Conservation Area No. 25. Its purpose was "to conserve and enhance the architectural character of the designated area as a whole and to ensure that special regard is paid to the architectural and visual qualities of the part when proposals for development are under consideration…" Section 74 of the Act (Planning (Listed Buildings and Conservation Areas) Act 1990) provides that with certain exceptions, no building in a Conservation Area shall be demolished without the consent of the Local Planning Authority. These exceptions refer mainly to Listed Buildings which are already subject to stringent control, certain ecclesiastic building and Crown buildings…Development by the Crown does not require planning permission but special arrangements do operate whereby in general, proposals by Government Departments are notified to the Local Planning Authority. Publicity will be given to proposals on the same basis as publicity given to planning applications," as the letter from Mr. MP Kendall, Portsmouth City Secretary and Solicitor to the Secretary of State at the Department of the Environment South East Regional Office, London said in 1992. This last stipulation refers to the Crown Exemption, in place until 2006, which was perhaps why Defence Estates selected a developer and agreed on new land uses for HMS Vernon/Gunwharf without reference to the local planning authority.

The army and naval areas were briefly united into the site's final form, renamed 'Nelson-Gunwharf' and subsequently sold as Gunwharf Quay, the 's' being added later.[8]

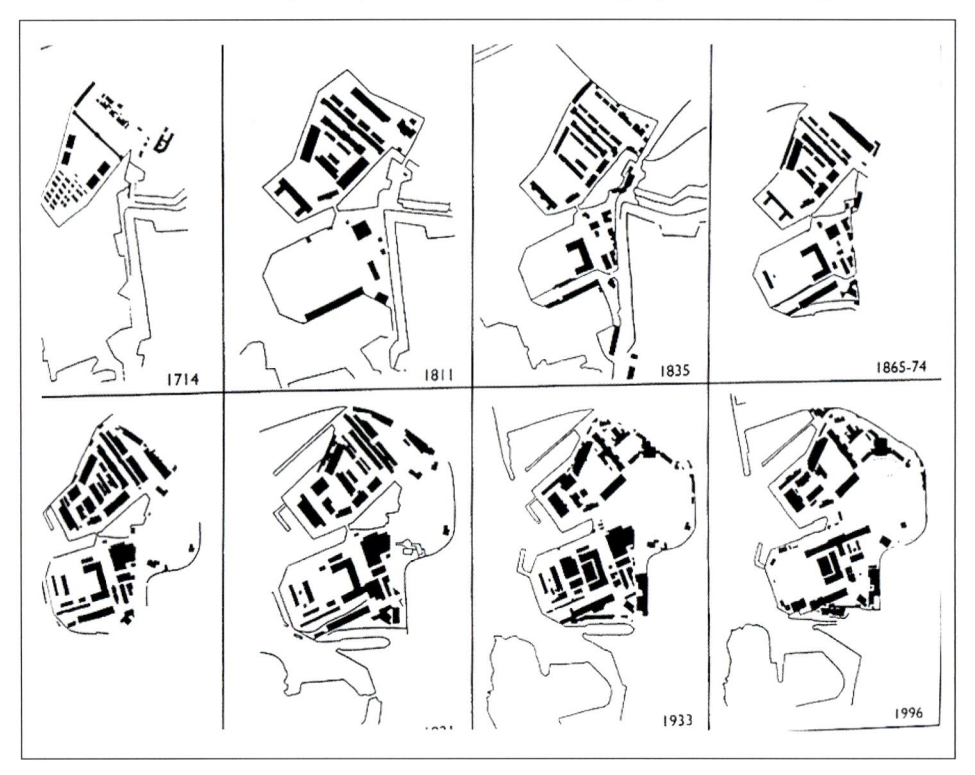

The evolution of Gunwharf's morphology.   Source: Gunwharf Quays (Lock 1997)

Until the redevelopment, a number of well-designed structures survived from this period. The neo-Georgian Wardroom, Ariadne - naval buildings are usually named after ships - was built in 1922-5 in brown brick with grey headers with Portland stone dressings on a stone plinth. The interior had a grandiloquent seventeenth century style staircase and two large halls at the back, incorporating the poop deck of *HMS Marlborouth*. Its wings were added in 1928. Its angled layout faced a garden and football ground where the foundations of the Mill Redoubt and traces of the eastern Mill Pond were still evident underground.[11]

Ariadne   Private collection

Across the parade ground behind Ariadne was the two-storey Defiance, the Non- Commissioned and Warrant Officers' Mess of 1922 in red brick and Portland stone quoins in an art nouveau style with a central cupola.

Defiance Non-Commissioned and Warrant Officers' Mess 1922

Behind it and close to the railway line was the Florence Nightingale Sick Bay also of 1922 of similar design

LCT Mk 8 Agheila moored on Military Pier. Warrior Block centre right with gunboat sheds lining the creek. From about 1977 Marchwood Military Port now stores Sea Mounting Centre archives

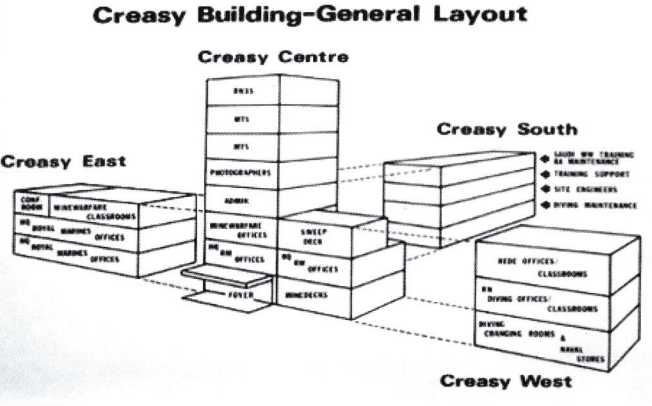

Creasy Block
Private collection

The nine-storey modernist Creasy office block was built in 1956-63 on the site of the demolished north wing of Vulcan. In 1966 a combined Mine Warfare and Clearance Diving force was established; in 1967 the Admiralty Experimental Diving Unit, a centre for deep sea diving, was set up to train divers in mine and torpedo clearance, increasingly necessary in the British offshore industry. They were later trained in a school at Fort William headed by a former Vernon officer.

An accretion of utilitarian structures as the site's uses changed produced a particular and recognisable character of functional but well designed buildings from the eighteenth and nineteenth centuries mostly in close juxtaposition, masking views of each other and generally not visible to the public from the land - and as already said, a clutter of more recent utilitarian structures. Apart from the 1920s Vernon buildings and the Creasy office block few were of architectural quality.

HMS Vernon finally closed its gates in 1986. Gunwharf/HMS Vernon had played a key part in two services' technical history, where innovation was a constant thread. This history was represented both by archaeology and by standing buildings.

HMS Vernon 1977 Reproduced by kind permission of the Minewarfare and Clearance Diving Officers' Association

This photograph shows HMS Vernon's heliport specially painted for the Queen's Silver Jubilee. She came to review the fleet at Spithead, as so many monarchs had done before. Many of the VVIPs and VIPS visiting Portsmouth for the Fleet Review on 28 June 1977 arrived and left by helicopter.

Aerial view of HMS Vernon 5 October 1981 　　　　　　　　　*The News* PP4255 JPIMedia

This photograph by the local newspaper shows the site as it still existed by the time of closure nine years later in 1986.

Disposal

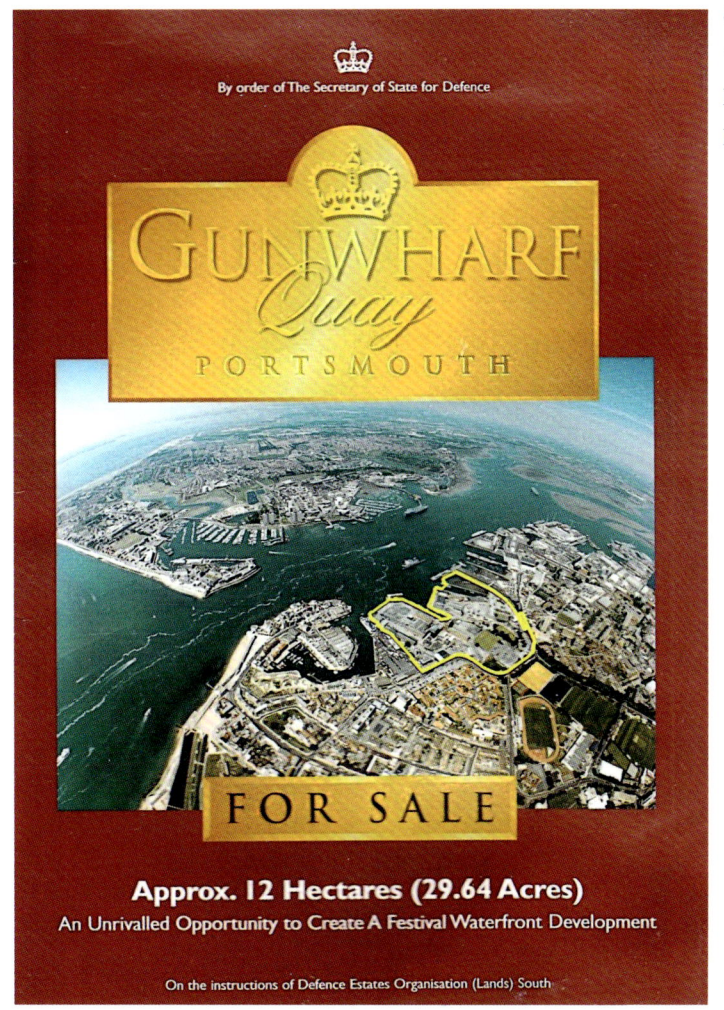

Gunwharf Quay Portsmouth
"For Sale An Unrivalled Opportunity to Create a Festival Waterfront Development"
Defence Estates Organisation (Lands) South

Vernon had been shown on the MOD's books as a potential alienation at least since 1986, after a rundown of activity of at least fifteen years before it had closed that year. Ten years later it was put up for sale by the MOD. With a frontage to Portsmouth Harbour of some 350 metres it had considerable potential development and amenity value. The Defence Land Agent Rod McDonald was responsible for its sale. He had an office in Brunel House on the Hard overlooking Gunwharf and the other redundant defence complexes round the Harbour. Celia Clark interviewed him there. The MOD sought to establish the site's value and to find a developer to buy it - without reference to the city council.

Richard Trist, Portsmouth City Council Chief Executive from 1978 to 1993 said "The big problem in Portsmouth is that the MOD decided secretly to release sites and didn't tell anybody, did a deal with developers, asking 'What can you do with this site?' before telling anybody. They said 'I can build this number of houses and make a lot of money...' but the community were never involved."[12]

As the largest redevelopment project Portsmouth harbour had ever seen, there was considerable local and national interest in what was proposed for its future. However, the land use mix – high end housing and discount shopping which was likely to bring in the highest financial return - had already been decided by the MOD in discussion with potential developers.

As a military and then a naval establishment from the early eighteenth to the late twentieth century, Gunwharf was generally forbidden and unknown territory for most people in the area, hidden behind a high wall on the landward side, although visible from across the harbour in Gosport and from passing ships. Expectations about its future were therefore almost without preconceptions as far as the majority of the population were concerned, except for ex-naval or military personnel who had a strong attachment to it and regretted its closure.

Many people who worked at Vernon or had associations with it were very upset when permission was given to demolish so much of it. From 1977 Dave Fricker obtained permission to record the buildings inside and out in photographs, which are contained in two large boxes of albums. In 2000 Robert Kennedy who worked there also researched the history of Gunwharf to produce the perspective aerial view of Gunwharf as it was in about 1860 reproduced at the beginning of this chapter, in order to record it as a historical site before demolition, and also to show via a Portsmouth *News* article, the potential loss of valuable Portsmouth history.

Until it became redundant, the city council also was unfamiliar with it - a factor that affected the drafting of the development brief, particularly as far as the few protected historic buildings were concerned. When the brief was written in 1995, no city officer had seen the interior of the Vulcan building, although its historical significance had been recognised in 1985. Knowledge of Gunwharf's potential as a new open space was not in the public domain until members of the Portsmouth Society were taken inside by the security guard, an ex-petty officer with considerable feelings about Vernon's closure. Emboldened by the grandeur of the uninterrupted space on the Vulcan's first and attic floors, as mentioned in the previous chapter, Chris Carrell was overwhelmed when he saw it. City officers began to investigate a property swap with the current city museum, which had been converted in 1970-2 from a barrack block of 1897, with the idea of transferring the city museum into the Vulcan with help from a lottery bid. They decided in July 1998 that Vulcan was in too poor a condition to convert, the price of £11 million was too high, and available space was too small without the rebuilt wing - which the developers Berkeleys had decided it should be flats over cafés and shops.

As Vernon/Gunwharf had been declared a conservation area, city conservation officers who were keen to see a conservation-led scheme for the whole site collaborated in the study by architect Alistair Hunt commissioned by the MOD's Defence Estates two years before disposal. In Hunt's opinion, the city council could have insisted upon a conservation-led scheme, which would have affected the sale price and survival rate of standing buildings and archaeology. "The MOD would not have liked this, but they would have had to comply. But once the property department got hold of it, that was the end of it." Their aim was "to release monies for reinvestment in other areas of defence expenditure", according to Gunwharf Quays' official website. "Portsmouth City Council were keen that the land was used to build and grow the prosperity of the city. Their goal being to regenerate a derelict City Centre site, to open up the waterfront for public use, to create jobs, to improve the attractiveness of the City and therefore increase visitor numbers, and to attract further inward investment."[14] By the time of the sale, putting maximum value and income-generating redevelopment above conservation or sustainability became the preferred option within both the MOD and the city council.[15]

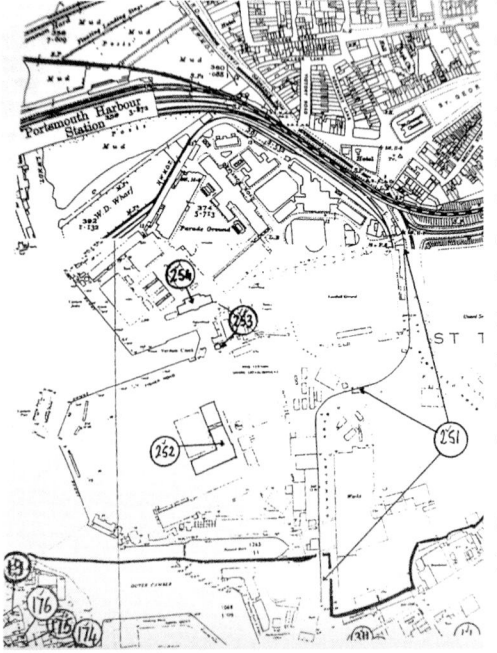

Scheduled buildings prior to redevelopment
Portsmouth City Council

At the time of the sale there were at least 130 standing buildings. After it was sold, every structure, listed below, except the four listed or scheduled ones, was demolished in the subsequent redevelopment.

Unlisted buildings demolished

1920s guardhouse near Main Gate
Store by Ariadne tennis courts
1935 workshop and rifle range near southern boundary
1935- 51 Invincible
1930 laundry on northwest boundary
Pump house by Vernon Creek
Sheet steel building in front of Vulcan
Store on railway track boundary
1930 store northwest corner
1950s and 1960s offices behind Vernon office
1922 workshops fronting quay
Small c.20 building and c.20 workshops around Vernon
1901 workshop east of Vulcan 1922 gymnasium against listed wall
1922 workshop South Quay
1897 workshop against listed wall
Modern shed garage block near south gate
1880 and 1922 store northwest of Vernon office
C.19 building abutting infirmary
1960 offices and garages southwest corner
1950 underground store west quay
Modern outbuildings on west and north quay

Retained
31 cannons, 5 bollards, stone paving near Infirmary and Vulcan
and any found under concrete, all trees, WWII pillbox on Marlboro
pier to be recorded before demolition.

Source: Portsmouth Society Executive Committee 19 August 1998

The Millennium: the Portsmouth Harbour Renaissance Project

The Renaissance of Portsmouth Harbour 2000
A Flagship Project for the Millennium
as the Gateway to Britain
Portsmouth City Council

A progenitor of Gunwharf as a destination was the architect Hedley Greentree, who made the decisive link between HMS Vernon and the celebration of the approaching millennium. His creative plans had been realised before, in the £220m Port Solent Marina on reclaimed land at the head of Portsmouth Harbour, the Town Quay redevelopment in Southampton and the Solent Business Park at Whiteley on the M27. In 1995 Greentree unveiled his vision of a Portsmouth waterfront so transformed it would vie with Sydney, Vancouver and Cape Town as a tourist attraction.

"His harbour dream started in November 1994 when he and property expert John Vail paid a visit to Portsmouth City Council's head of marketing Paul Spooner...'We wanted to persuade the council to buy the Gunwharf site when it became available. But on Paul's desk was a letter asking for bids to be submitted to the Millennium Commission. By the end of the meeting we put two and two together and came up with the idea of the Renaissance of Portsmouth Harbour millennium project.'" [15]

The Millennium Commission set up in 1993 offered funding to celebrate the Millennium to very large building projects. Those that succeeded in the first stages of the intense bidding process seem to have had an element of fantasy, so that they could in no way be accused of substituting for necessary infrastructure, but were most of all to be celebratory. This fund and the institution of the National Lottery in 1994 attracted cash strapped local authorities to bid for funding.

As the theme of the Portsmouth harbour bid was regeneration of the harbour communities, the promoters decided that the best way to revitalise the redundant MOD sites around Portsmouth Harbour was to link them together. Portsmouth and Gosport councils put forward proposals for the Explosion Museum at Priddy's Hard, the Spinnaker Tower, a giant water arc at the harbour entrance and a new millennium promenade. As partners in the bid, the Portsmouth Naval Base Property Trust's proposals in the naval base are described in our first dockyard chapter.

When it became available, Gunwharf/HMS Vernon in a commanding position on the harbour was identified as important to the Portsmouth Harbour Renaissance project. The partners applied to the Millennium Commission to fund the 150m tower and other elements of the project outside Vernon. The symbolic joining of Portsmouth and Gosport by a dramatic water arc across the harbour entrance was a leitmotif. This fitted the celebratory bill; perhaps this is why it was one of the first big five or six big schemes to be endorsed by the Commission. But the water arc was later opposed by captains of ships who did not want tons of salt water landing on their decks!

As described in Chapter 6 the Portsmouth Harbour millennium scheme's parent was the Portsmouth and South East Hampshire Partnership, a consortium formed in response to the government's pressure on local authorities which were directed to form partnerships with the local business community with offers of large funds. The partnership was chaired by Ben Stoneham, the Chief Executive of PPP Publishing, owners of *The News,* the local newspaper and the free *Journal*; other key members were two estate agents: John Vail and Jeremy Lear; Greentree, the local architect entrepreneur; the Tourism and Marketing Director for Portsmouth City Council: Paul Spooner and his opposite number, Ron Wilson in Gosport Borough Council.

In 1995 the Millennium Commission approved the Renaissance of Portsmouth Harbour project in principle and agreed to give up to £40m of lottery cash towards its cost of £90m. The city council claimed that with matching funding and local contributions this would add up to £112m, which would be injected into the local economy damaged by the rundown and closure of many Ministry of Defence establishments. Acting as a catalyst it would highlight the development potential for the creation of 3,500 new jobs on these hitherto closed sites. One creative way this would be done was by linking them via 5 km of new public promenades. Public walkways of 2000 metres along the Portsmouth waterfront and of 3000m in Gosport were funded by the Millennium project with the intention of "dramatically increasing public access… and acting as a catalyst to unlock the potential of further development sites". This was a visionary idea that continues to be much enjoyed.

This national endorsement enhanced government and press interest in the scheme, but the complexity of the decision-making bodies (the Partnership, MOD, local authorities, Portsmouth Harbour Renaissance Company, Naval Base Property Trust…) also distorted local input. The downside was that groups and individuals who wanted to influence the disposal process, particularly building conservationists who opposed the demolition of Gunwharf's historic structures in the redevelopment found this labyrinthine structure with its overlapping membership and responsibilities nearly impossible to clarify or influence. Even individual councillors or officers on the various millennial bodies often said that they could do little or nothing to affect the course of events. By the time of the disposal, both the MOD and Portsmouth City Council saw the site's potential in economic development terms, putting maximum value and income generating redevelopment above reuse of buildings or sustainability. The specialist conservation officers evidently had little status in local government hierarchies.

The Development Brief

A draft Development Brief by Portsmouth City Council planners was prepared in July and the final brief in November 1995 in consultation with Hampshire County Council as highway and strategic planning authority, the MOD and the Portsmouth and South-East Hampshire Partnership. It identified the Gunwharf site "with its water frontage it is a unique and very special development opportunity of national and conceivably international appeal…The City Council is very keen therefore to encourage an attractive mixed waterfront development of the highest quality that creates a vibrant festival atmosphere which will add to the attractiveness of the City, attract visitors, including residents, visitors to this area and expand employment opportunities in this part of the City."

"Water is a most powerful tourist magnet and the maximum level of access to the waterfront, together with leisure attractions e.g. tall ships, is fundamental to the vision. In addition, the existing tourist attractions in the Historic Dockyard and Old Portsmouth, and public transport services combine to make the area a natural meeting place. The City Council wish to build on these to assist the physical and economic regeneration of Portsmouth. The opportunities offered by the release of the Gunwharf site are considerable and, as a whole, [it] is a key site in the City Plan and the Economic Strategy for Portsmouth."[16]

The Millennium project's main elements were "the creation of a world class visitor attraction and magnificent public amenities by the expansion of the Historic Dockyard". As well as the new waterfront promenades, "a Harbour Tower about 150m high set in the Harbour off Gunwharf and illuminated water display features on either side of the navigation channel" were proposed. The redevelopment of the Gunwharf site "to include new public spaces linked directly to the City Centre by a new landscaped boulevard" was also included. On the Gosport side "the main heritage theme will be continued with the development of the Priddy's Hard Heritage Area and the enhancement and expansion of the Royal Naval Submarine Museum". "Linking the Harbour communities with a network of new environmentally friendly waterbus services" would also "open up fresh opportunities for tourism development."[17]

The City Council was expecting to see "a mixed development that included a wide range of leisure facilities, housing and job opportunities". Major retail development such as a food superstore, non-food retail park or shopping centre was not acceptable, but limited provision for leisure/specialist shopping was acceptable. While there was no restriction on the total amount of floorspace permissible, "individual units were to be limited in area and to sell a range of goods and services which were appropriate to the general location and development of the site and which would not by themselves, or cumulatively with other existing or proposed developments, adversely affect existing local or district shopping centres."

The aim of these constraints was said to be to protect the vitality and viability of Portsmouth's existing retail centres. The 3.2 hectares of water could, at the developer's discretion, be utilised and/or reclaimed in whole or in part from Portsmouth Harbour subject to obtaining all necessary consents under the Harbour Act 1964 and other legislation. Important parts of the Portsmouth Millennium project: the tower, boulevard from the city centre and harbourside promenade had to be included. The draft brief offered developers and architects: "the maximum flexibility in order to design a scheme that reflects the importance of this site, enhances its quality and reflects current and foreseeable market conditions. Townscape quality and the integration of the right mix of land uses are, in this instance, more important than rigid guidelines"[18]

It was this very open brief, together with the high sale price, which produced the combination of factory shopping and luxury housing with very large carparking provision which its critics identified as offering too open an opportunity: "in planning terms, you could drive a coach and horses through it", as Roger Brown, former County Planning Officer said on behalf of the Hampshire Buildings Preservation Trust.[19] A development brief generally applies planning and design principles to a site. However, as researcher Nancy Holman pointed out, the development brief was inadequate, particularly in its absence of a design guide, and the lack of advice on size, massing and materials giving no direction other than vague phrases about creating a vibrant atmosphere.[20]

There were very few objections to the draft brief. The City Planning Officer commented that "The Portsmouth Society failed to respond in writing during the consultation period". The Society did not realise that it could comment on the draft, but once it had done so, "the Secretary made his comments over the phone to Cliff Piper, after the closing date. A late letter was reported verbally at committee."[21]

This sequence of events may be contrasted with the later disposal and redevelopment of Royal Clarence Yard across the harbour in Gosport, discussed in our chapter on residential development. After extensive public consultation about that site's future, including two community planning events funded by the same developers as at Gunwharf, Berkeleys, Gosport Borough Council commissioned historic research which identified several more historically important buildings which the developers had to retain – a very different sequence of events from Gunwharf, which in contrast led to retention and reuse of more heritage buildings, less conflict and perhaps greater public satisfaction.

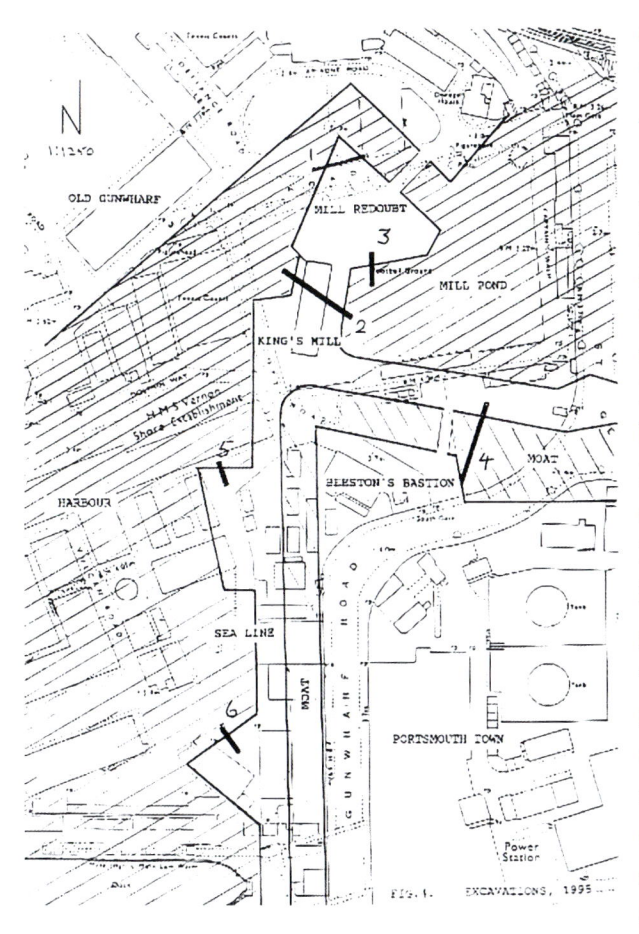

Gunwharf Archaeological investigations 1995
The Berkeley Group *Gunwharf Quays Environmental Analysis Addendum 1* Mouchel December 1996

Much more was learnt about the Gunwharf site in the historical, architectural, archaeological and environmental studies commissioned by the developer.[22] In her research Celia Clark had understood that the stonework of the North East Bastion was briefly exposed, but was then reduced by a metre in height in the redevelopment. Freddie Emery-Wallis, leader of Hampshire County Council and John Bingeman of the Society of Nautical Research were so concerned that they met in Freddie's house for Dave Fricker to hand over his photographs of the uncovered archaeology, which were then given to English Heritage. It is now clear from their eyewitness accounts and photographs that although six excavated trenches revealed the remains of Mill Redoubt, Beeston's Bastion and the King's Mill, some surviving archaeology was not preserved in situ. Only keystones from the sluice tunnels to the Mill Pond and timber from the wooden raft were recovered, but the bastion was broken up by a JCB. The excavation was so deep that the JCB seemed on the verge of falling into the hole. How did this happen in a conservation area of a city supposedly dedicated to conservation, particularly of its defence heritage, which earns a significant proportion of its living from defence heritage tourism? As a unitary authority determining the new land uses was the city council's responsibility.

JCB excavating bastion

Timber base of stonework

North East or Beeston's Bastion uncovered in cleared site. Simon Russell

Keystones from sluice in-out tunnels for Mill Pond

Timber said to be from wooden 'raft' structure below Beeston's Bastion

Dave Fricker whose father and uncle served at Vernon watched the destruction from where he lived overlooking Gunwharf. He described the redevelopment process as "scorched earth".... It was traumatic to witness such deliberate destruction of *Warrior* and *Acteon* buildings "with their magnificent Georgian and Victorian ancestry and the loss of the adjacent prized avenue of magnificent mature trees leading from the Gunwharf Gate." As he told Celia Clark in 2020 "The impact of the sight, sound, smell and even taste of the windblown dust of the smitten *Warrior* [block] I will simply never forget..."

The final brief differed little from the draft endorsed by the planning committee. In their urban design report for the developers, Lock and his partners recommended that in view of the site's physical boundedness and separation from surrounding areas, Gunwharf should become a new urban quarter, separate from its past as an area of exclusion, "deriving its character from its... waterfront location, new civic urban qualities, and its ability to draw in both tourists and the people of Portsmouth alike,"[23] thus fulfilling the requirements of the planning guidance and development brief. Because of the site's conservation area status, it was possible to accept the second or third highest bidder, but this made very little difference to the high return land uses proposed by all the bidders.

First public responses

Heritage Open Days in Portsmouth and Gosport provide the opportunity for local people to explore the closed military sites in their midst. As its contribution, in September 1996 the Portsmouth Society arranged with the Defence Land Agent and Portsmouth City Council for Gunwharf to be open for tours on 14/15 September 1996. At the suggestion of SAVE Britain's Heritage, the 150 people who came over the two days were offered a questionnaire. Its design was clearly related to the purposes of the initiating group, the local civic society, which had stated aims shared with its parent body, the Civic Trust, of "working towards high standards of architecture and planning" in its area of benefit: the city of Portsmouth.

Portsmouth Society visit to Gunwharf

All but one of the respondents to the questionnaire regarded the future of Gunwharf as very important to the future of Portsmouth, reflected in their taking the opportunity to visit it. As they filled in the questionnaire in the queue before they gained entry, there was a wide divergence of view of what should happen to it, though there was near unanimity about its importance to the future of Portsmouth. A large majority wished to see other historic buildings than the four identified as scheduled and listed retained. It was interesting to find that twice as many were against the proposed tower as an essential part of the Millennium Project as were for it. David Brock of English Heritage said that the survey was very useful in reflecting local opinion on the future of the site. Rod McDonald, the Defence Land Agent, said at the end "I think it was a very worthwhile exercise", despite the fact that the disposal and land uses proposed in the City Council's Development Brief for the site had already been determined: shopping, leisure facilities and luxury housing, allowing clearance of all but the few protected buildings.[27]

By the time of the disposal, both the MOD and Portsmouth City Council in its economic development and planning functions saw the site's potential in very different ways from the attachment to its past and remaining structures demonstrated by many of the visitors on the Open Day and subsequent objectors to the outline and detailed planning applications. Events proved that emotional attachments weighed little compared with the potential financial and economic value offered by a cleared site.

The first design

On 25 October 1996 *The News* reported that "Mr Greentree's firm, the Fareham-based Hedley Greentree Partnership (HGP), had backed the right horse: it was their designs which have been chosen for the consortium's £120m development". A consortium had been formed with Lordland Developments, a privately owned South African property and facilities management company which had already developed the Victoria and Alfred Waterfront in Cape Town, another British naval dockyard, to which 16 million visitors had been attracted in 1995. The Ministry of Defence announced that the international consortium of Berkeley/Lordland had won the race to redevelop the 30-acre Gunwharf site so crucial to Portsmouth and Gosport's millennium plan for the harbour. Perhaps in order to turn decisively away from southern Hampshire and Portsmouth's dependency upon defence, what was proposed for Gunwharf was modelled on the Cape Town redevelopment. The coincidental historical link between the two dockyards was explored in a 1998 BBC Southern Eye programme 'Out of Africa: Cape Town Lesson for Gunwharf'. Guns for relief of the siege of Mafeking (1899-1900) were shipped 6000 miles from Gunwharf to the Waterfront, and then dragged overland to supply the army fighting the Boers. The two sites' history was further entwined by the formation of the Berkeley Festival Waterfront Company, based on Lordlands Developments of Cape Town; Rob Tinckler moved from South Africa to supervise the Gunwharf redevelopment for Berkeley Homes.

Flagship Proposals for Gunwharf
Portsmouth City Council

However differences were soon apparent. In Cape Town, a much bigger site, existing buildings were converted to encourage start-up businesses and artists from the neighbouring impoverished area of Steamer Point; retail and educational uses were also catered for and new water links established, as explored in the BBC programme. Portsmouth University proposed to site their Business School in Gunwharf, and studios for local artists were to be provided on the upper floors above the shops. The first design showed a giant square shed of covered shopping, sweeping away all the existing buildings on the northern part of the site.

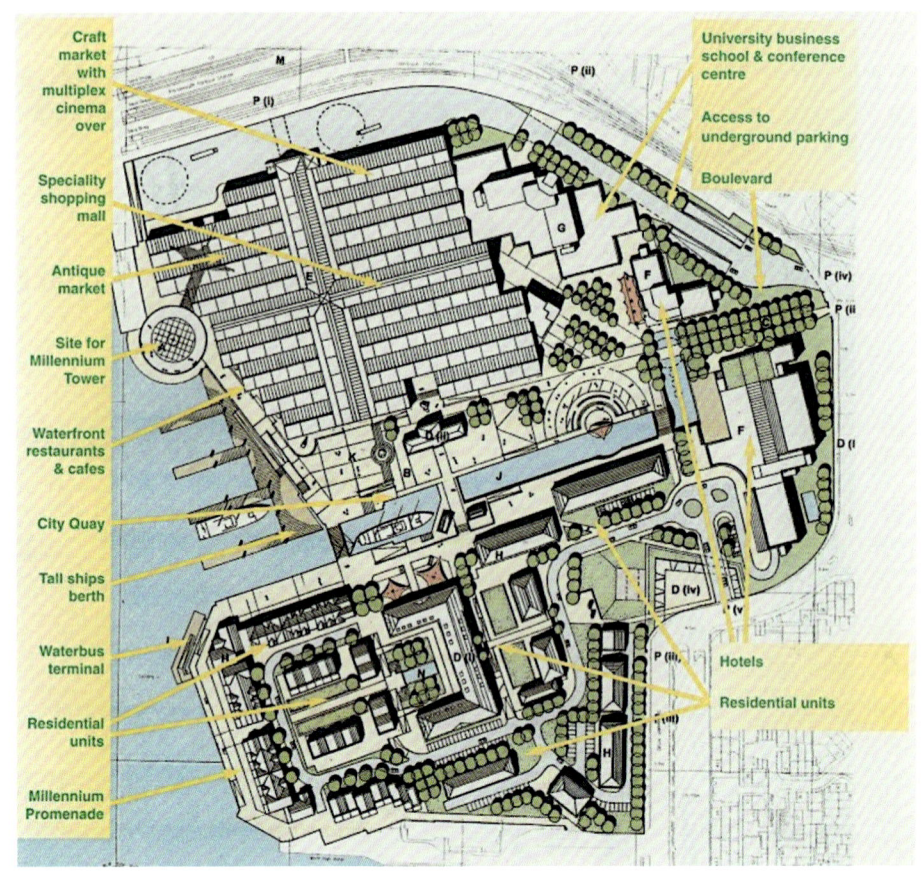

Flagship Special Issue
plan of development
Portsmouth City Council

Comparison between the site before redevelopment and the first design shows that very few existing structures were to survive, and that the northern area of water, Gunwharf Creek would disappear, while the central canal was extended inland. The site was split between two main land uses: retail and leisure to the north and residential south of the canal. Also missing was any indication that the huge roof of the retail shed would incorporate any green features or housing above the shops. When interviewed about this by Celia Clark the German architect for the shopping centre said that because British property tenures had different timescales for residential and commercial property, a retail/housing mix was not on the cards, but this would have been the norm in Germany.

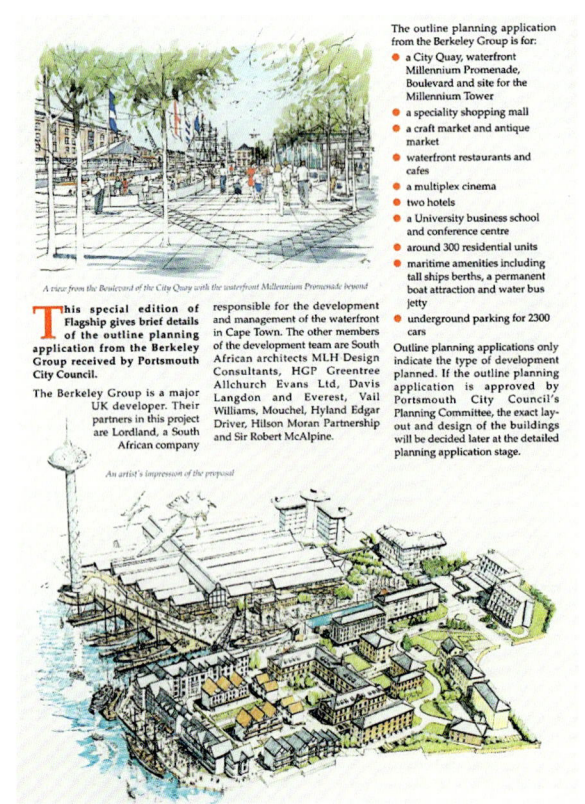

Flagship Special Issue
Berkeley/Lordland's first design for Gunwharf
Portsmouth City Council

Because Portsmouth Partnership's media element was promoting the Harbour Renaissance scheme, it was difficult for local objectors to gain public coverage of their views. HGP's first sketch designs for Gunwharf were published without any critical comment in the City Council's *Flagship*, circulated to 78,000 households. Responses were asked for, but the scheme was made to look inevitable, since as a result of the shortlisting process, there were no alternative schemes. Neither *Flagship* nor *The News* intimated that there might be opposition or alternative. *The News* printed overwhelmingly favourable comment about the Millennium project and about one bidder: Berkeley Homes/Lordland Development's redevelopment proposals for Gunwharf, often using copy supplied by Berkeley's PR.

"Key words in the newspaper such as 'gleaming', 'state of the art' and 'upmarket' also link to the themes of pride and cleanliness and the impression that Gunwharf was always planned to be 'better' than typical town centres" said Matthew Lambert, a student at Coventry University in 2020 from an analysis of the *News* coverage.[28]

In response to announcements by the developers, speakers from the Solent Protection Society, Gosport Society and Portsmouth Society were invited to put their views. In their Newsletter of April 1997 the Portsmouth Society said that what the city needed was social housing and executive housing "to help redress the balance in the population... [since] at present a disproportionate number of well-to-do people who work in the city or otherwise belong to it live in the peripheral towns... Substituting houses for most of the proposed shops would, by greatly reducing the car-parking needs, enable the abandonment of the vastly expensive proposal for subterranean car parking." As well as pubs, cafés and restaurants on the waterfront from which to enjoy the unparalleled 24-hour spectacle of boats and ships, the Society hoped for high-tech workshops, starter units or laboratories, to provide permanent jobs as opposed to leisure and shopping work, which would be merely transferred from elsewhere. The wardroom Ariadne with its very fine staircase was "eminently convertible into a hotel, while the Petty Officers' Mess behind it could be joined to it or used separately as a budget hotel or students' rooms. The gun-carriage sheds: Actaeon could be open plan retail, a covered market. "In advocating this we recall Jane Jacob's famous aphorism (in her seminal *The Death and Life of Great American Cities*): "Old ideas can sometimes use new buildings. New ideas must use old buildings." The tower would be better situated on the line of the 'canal' in the centre of the waterfront, to line up with the boulevard from the city centre, instead of being put in an obscure corner." Although they wanted to keep the site traffic-free, the Society wished to retain the open water between the station and Gunwharf as mooring for visiting yachts. If the southern wall of the station was glazed it would add to visual interest and the sense of arrival. These comments were unanimously approved by the Portsmouth Environmental Forum.

As the scheme progressed with few real improvements BBC staff made it clear privately that they shared the serious criticisms of it. In December 1996 'Millennium Man' by BBC South had focused on Paul Spooner, Portsmouth City Council's Director of Development and his efforts to attract inward investment via the Millennium project. Its uncritical approach was considered close to hagiography by those opposed to the quality of the Gunwharf redevelopment. As if to redress the balance, as already mentioned the BBC broadcast 'Out of Africa' in the same Southern Eye series, which compared the Cape Town renewal with Gunwharf. Cape Town's redevelopment mixed leisure, shopping, and museum elements with bustling shipping. The architect/city planner in Cape Town had demanded a high standard of building design. A variety of water transport was preserved - both points in contrast to Portsmouth.

More than four acres of public space were proposed at Gunwharf: the City Quay festival arena, a promenade and the Millennium Boulevard linking the site with the city centre. There would also be apartments and townhouses, shops, restaurants, cafés, pubs, offices, and two hotels. Restoration of the few scheduled or listed historic naval buildings such as the Vulcan building, the Customs House and the old infirmary was included. But even though Gunwharf was a conservation area with many reusable buildings, it was marketed by the MOD for clearance of all but the four listed or scheduled structures - in accordance with the very loosely worded local authority planning brief; this resulted in the complete loss of the site's historic grain and texture apart from the extended canal.

1790s office block, one of the few historic building survivors, now Customs House pub

The objectors' alliances

Since the Portsmouth Society had early on criticised the design and landuse, the secretary and chair were surprised and pleased to be approached by Berkeleys for a breakfast meeting and offers of talks to the committee. These encounters were intensely top-down: Berkeley's directors and managers explained what they were planning to build, and the Society committee asked questions, but there was never any real negotiation or change in the scheme. The Portsmouth Environmental Forum was promised a consultative role over design of the Millennium tower and boulevard by city officers (PEF 1998), but this did not happen.

Concerned that those likely to be most affected by the largest development project Portsmouth had ever seen and that public opinion should be as well informed as possible, the society proposed in February 1997 that the City Planning Officer hold 'focus' consultation meetings between various interest and residents' groups and the developers. Attendance was limited by invitation: "There never was a meeting where Joe Bloggs from Portsea could stand up and say 'What's in it for us?'" according to Roger James, the Society's secretary. The 1969 Skeffington report on public participation in planning had recommended that planning should be a genuinely democratic process involving mediation between a wide range of competing interests. One such mechanism would have been a community forum to encourage cross-fertilisation of ideas and greater understanding of local needs early on in the process,[24] but the MOD, city planners and city economic development officers had already decided on the content of the scheme.

On 6 April 1998 the chairs of the Portsmouth and Gosport Societies wrote jointly to the Government Office of the South East (GOSE) to express their extreme disquiet at the suppression of democratic debate about the demerits of the Berkeley Festival Waterfront Company's proposed redevelopment of Gunwharf on Portsmouth Harbour, and the speed at which it was being dealt with, which lead them to question the whole legitimacy of the planning process. They were concerned that the civil servants in Guildford (GOSE) might have perceived an apparent lack of criticism about the scheme in the local print media, and in the low number of letters of representation received by the city planning department. Many of the societies' members and other residents had written to *The News* and *Journal* (free paper owned by the *News*) raising issues on the scheme over the past two years; few of their letters were published. Reading the local press the civil servants might have believed that there was little local controversy about the proposals. This was very far from being the case. The councillors on Portsmouth City Council committees probably had the same impression; the Millennium timetable and the offer by Berkeleys to part fund the tower might well have blinded them to the scheme's considerable drawbacks. Roger Lewis, director of Berkeley Group, made this threat explicit in *Building Design* on April 3: "We must not allow a few dissenting voices, who do not represent the majority" to jeopardise the whole Millennium scheme.

Although the developers disliked public meetings they were prepared to attend semi-formal consultations with special interest groups and individuals. Three intense and sometimes heated encounters, chaired by the City Planning Officer, took place on the implications of the scheme for employment and retail impact, environment and transport, and design and conservation. The issues and snags emerged into sharper focus, a process that the Planning Officer regarded as thorough, constructive and therefore helpful. Other interests felt that from early on decision-making was invisible and impenetrable to ordinary citizens and bottom-up groups. It was dominated by tight-knit groups of the MOD, developers, Partnership members, city officers and councillors who did not appear to have any intention of involving the wider public, except during the ritual 'consultations' required in the planning process. Berkeley/Lordland arranged two short-lived public exhibitions at the outline planning stage in the Guildhall and Cathedral. In the objectors' view these were PR exercises. Planning officers attended, but members of the public were dominated by Berkeley directors such as Mr. Fleming, who told those who asked questions that there was no alternative to Berkeley's proposals and handed out a pre-printed letter of support for visitors to sign. Portsmouth Society members protested and tried to distribute their newsheet which was critical of the scheme. These were brusquely snatched away by Mr. Fleming. The tower design consultation, though longer, was similar.

Groups and individuals were activated by the Portsmouth Society and by Celia Clark as an individual. Alliances were formed locally with Portsmouth Environmental Forum and the Gosport Society, regionally with the Solent Protection Society and Hampshire Buildings Preservation Trust, and nationally with the Royal Fine Art Commission, Richard Rogers of the Urban Task Force, Lord Palumbo, SAVE Britain's Heritage and Prince Charles, patron of the Civic Trust, via his spokesman, Leon Krier. Prince Charles' first naval command was *HMS Bronnington*, stationed in HMS Vernon. Hampshire as national leader in conservation through the Historic Buildings Bureau in the county planning department had good connections with national groups, particularly over historic defence sites - where for example they had intervened together in the past. These groups and individuals contacted had a common focus: the conservation of buildings and knowledge of relevant planning law; the local pressure groups (Portsmouth Society, Gosport Society) have long experience of public participation as defined in the limited model offered by the planning system.

The conservation network used its connections to gain considerable coverage in the national professional planning and architectural press in the hope of influencing the dominant top-down groups and the local authorities. Reports about the disposal and redevelopment and the Millennium project appeared in *Building Design*, the *Architectural Review* and *The Times* - through articles by Marcus Binney, president of SAVE and others. Binney gave national press voice to such criticism.[25] Lord St John of Fawsley, Chairman of the [Royal Fine Art] commission said: "The current proposals totally lack a sense of place. They could be anywhere at all, in any suburban housing estate or shopping park. As an example of development at the time of the millennium, they are simply lamentable. In the commission's view, the architects have failed to rise to the challenge of designing proposals which pay due regard to the historic buildings that will be retained, exploit the magnificent site and link it to the rest of Portsmouth."[26] At the request of the Portsmouth Society, Prince Charles also intervened, via Leon Krier. "The sale of Gunwharf Quay and nearby Priddy's Hard - which is to be developed by Barratt Homes [sic] - has once again raised questions about the MoD's policy on disposal of surplus sites containing historic buildings. According to the latest annual report on the MoD historic estate: it still has around 800 listed properties to dispose of, including stable blocks, chapels and barracks. A conservation officer who worked closely with the MoD said: 'Each year a glossy annual report shows the marvellous work the MoD is doing, but the reality, in respect of the majority of the sites, is appalling."[25]

It was the Portsmouth Society and other objectors' contention that the redevelopment proposed by Berkeley/Lordland did not conform to Portsmouth City Council's Development Brief of November 1995. Far from being "in line" with the brief, the objectors pointed out that 17,000 m₂ of shopping was far larger than the "limited provision" sought in the brief, and that such a large shopping centre could not fail to have a severe impact on the existing shopping centres in Southsea, the city centre and Gosport.

Their case was also on aesthetic grounds. While this was not their only concern, a major preoccupation of the design and heritage interests was the poor quality of design and of sustainable land use represented by the proposals. As soon as the sketch designs were published, the Portsmouth Society and other local conservation groups such as the Hampshire Buildings Preservation Trust resolved to oppose the application, because far from enhancing Gunwharf's marvellous harbour-side setting and history, in the objectors' view the developers were proposing to strip it of its maritime associations and much of its history. Nearly all the buildings, many with historical significance and capable of sustainable reuse, were to be cleared. Marlborough Pier and its associated pontoon were ideally sited to use the tide to assist boats approaching from the south. Instead of retaining this usable area of water between the north wall and the harbour station it was to be filled in for a vast underground carpark. In fact, apart from the badly thought out proposal for jetties for tall ships, this whole development could, as the Royal Fine Arts Commission said, be anywhere, in any inland city.

As the result of pressure from English Heritage, able to intervene as a statutory authority because Gunwharf is a conservation area, the scheme's urban design was considerably improved. "Our concern is to make sure that the new architecture on the site properly reflects its historic purpose and pattern, and indeed the historic importance of Portsmouth harbour as a whole. The functions of Gunwharf required substantial buildings, enclosed against the elements, tough and reticent, with simple lines and substantial detail, and set off by large open areas which functioned as external storage. More generally, as you are aware, the waterfront lies the old town and the historic dockyard and forms part of a sequence of exciting and high quality views from the harbour; we all are conscious that this site is an exciting opportunity which demands a high quality response in design terms."[26]

They insisted that the diagonal path from under the railway line should lead directly to the harbour and that the housing blocks framing the view of the Vulcan building should be pulled further apart to allow a better view of the site's major historic building, the Vulcan from the harbour. These changes took place. But the land use and building design changed little.

There were also comments by voluntary groups and individuals: the Royal Fine Art Commission, the Civic Trust, Lords Palumbo and Rogers, as well as the Portsmouth and Gosport Societies, Environment Forum and individuals. These were brushed aside by the city council planning committee. Lord (Richard) Rogers, after a lecture at Portsmouth School of Architecture took a close personal interest, although he was not shown the final scheme. After the election when he was appointed by the government to the Urban Task Force to bring brown lands into productive use, he wrote to the Ministers at the DOE/DETR asking that the outline and detailed applications should be called in so that there was a chance of better proposals emerging. He was admonished by the Minister for becoming involved, the DETR preferring to leave determination of the applications to the local planning authority. Aesthetics in the members' and officers' view were not considered important in comparison with the hoped for economic gains. Indeed heritage interests' views on design are denigrated by both elements of the Portsmouth planning authority as trivial, "non-constructive" and unhelpful.

Objectors saw the proposal as a waste of a prime maritime site, where no real link was made with its history of defence research. The lack of sustainability: the waste of many reusable buildings, the land use mix of leisure/shopping/luxury housing were all identified as negative factors. Building a large traffic generating development in the south-east corner of the densely built up Portsea Island, the developers were asked to pay not for progressive sustainable transport measures, but only for redesign of road junctions for a project whose 1500 place underground carpark was to be next to the city's major train/bus/coach/ferry/pedestrian public transport interchange. According to the city planning officer the scheme was up to 70% of full parking standard, reflecting sustainable objectives, but volumes of traffic generated by the very large carparks were likely to be considerable. One gain was that at least parking would be underground so most of the site would be free for pedestrians to move around.

An outline planning application was to be submitted to the city council within eight weeks. As one of the largest construction and housebuilding firms in the country, *The News* said that the Berkeley Group had a strong track record in developing waterfronts. *The News* also attacked the motives and representativeness of those who oppose or criticised the Millennium designs or Berkeley's scheme. Ben Stoneham, Chief Executive of PPP: publishers of the News, chairman of the Portsmouth and South East Hampshire Partnership, and director of the Portsmouth Harbour Renaissance Company said to the chair and secretary of the Portsmouth Society at their meeting on 4 September 1996 that he had no control over editorial policy. His editor nevertheless many times attacked opponents of the scheme for being against regeneration of the harbour economy[28], which was not the case. Publicly conducted democratic debate that could have taken place in the local press did not happen.

As the councillors were keen to grant permission because of its perceived economic benefits, the Society asked the Secretary of State to 'call in' the outline application in November 1996 and the detailed one in February 1997 so that they could be examined at a public enquiry. Although these requests were supported at the highest level - and there were at least three reasons to do so: an outline application was submitted despite Gunwharf being a conservation area where most of the buildings were to be cleared; the shopping element was so large (17,000m^2) that referral to the ministry for possible call-in was triggered; and the architectural design had not improved during the course of planning determination. Nevertheless, the DOE/DETR did not use its powers of call in to hold a public inquiry at either stage.

Portsmouth City Chief Executive was so concerned about the poor quality of the proposed redevelopment that he wrote in a personal capacity to the director of the Defence Estates Organisation (Lands) in early 1997. The reply acknowledged his concern "on the way this prestigious waterfront site should be redeveloped, and also that you do not think the present proposals do the site justice". According to the DEO they had worked very closely with the city planners "since before the announcement of the proposed Millennium Scheme for Portsmouth Harbour… We have addressed the development potential for the Gunwharf Site and over the last twelve months have sought to bring the property to market as speedily as possible. As you will recall, a Development Brief for the site was agreed in November 1995 after proper public consultation. In preparing and submitting their bids and proposals for the site, prospective purchasers/developers had to consider the scale, location and density of the development. Eventually Berkeley Homes/Lordland were selected as the purchaser and I can tell you that their redevelopment scheme was the one most favoured and supported by the City. The way in which this waterfront site is developed into the future to include essential elements of the Millennium Scheme… ultimately rests with the community through the town planning mechanism."[31]

Demolition and carpark under construction in creek   Private Collection

The developers Berkeley Homes (Hampshire) Ltd, part of the Berkeley Group, moved the project into a Single Purpose Vehicle (SPV) company, Berkeley Gunwharf Ltd. Architectural and planning support was provided by the local firm HGP Architects. On the day that the outline planning permission was to be debated by Portsmouth planning committee, Celia Clark had a stark choice: as a member of the Hampshire Buildings Preservation Trust board she had an invitation to Highgrove to meet other trusts around the country. As speaking to the committee was unlikely to change minds, she went to Highgrove. She said to Prince Charles: "I'm sorry to have to say, Sir, but at this moment Portsmouth planning committee is granting permission for that development at HMS Vernon that we both object to…" He replied: "It makes one want to go and jump off a cliff….!"

There were twenty varied and carefully prepared deputations to the special meeting of the Planning Committee on June 4th, most of them against the scheme, which took one and a half hours to present. Letters in favour were mainly proformas arranged by the developers. Nevertheless, outline planning permission was granted that day in June 1997, and detailed permission in January 1998. The main elements were factory shopping, leisure facilities and luxury housing. The Millennium timetable had reduced the time and opportunity for creative input or opposition to the scheme. The imperative was to get the redevelopment and the Millennium elements built.

It might be argued that the Treasury did not obtain the best price for the site, said to be £2.7m. Tony Pidgley, founder, managing director and leading shareholder in the Berkeley Group said that his firm "specialises in urban regeneration, buying land at the bottom of the market and selling at the top". In 2001, the year Gunwharf Quays opened, he was worth £4.993 million and his son £2.26m.[32] The clawback provision where a proportion of post development profit made over the sale price is paid back to the MOD was activated in relation to Gunwharf at least twice. In 2012 the penthouse at the top of the tallest block, the 'Lipstick' was on sale for £2.5 million.

If a proposal is approved despite well founded opposition, bottom-up groups may feel that the effort of participation was not worth the time and money spent in responding to the invitation to make representations: that the process had been 'Consult and Ignore'.[33] Robinson and Shaw said that in practice, community consultation often means 'tokenism' - consulting with safe community groups and allowing the community to comment on schemes already accepted in principle.[34] But perhaps the 'creative delay' of six months did a little to redress the very unequal balance between the Ministry of Defence allied to powerful, well-funded developers on the one side and local interest groups and communities on the other.

Deep excavation   Garrick Palmer

HMS Vernon/Gunwharf redeveloped. Beeston Bastion lower centre.
Maritime City Portsmouth 1945-2005 Portsmouth Society 2005

Job Creation and Portsea

Many still believed that the project did little to meet local needs for permanent skilled work and social housing. A key factor influencing politicians was their hope for job creation. In April 1999 the *News* stated that the £200m project would help to create 4000 jobs around the harbour.[35]

Professor Harris, professor of economics at the University of Portsmouth was commissioned by Berkeleys to estimate the number of jobs resulting from the 'Gunwharf Quays' development.

He said that between 1100 and 2,500 jobs in construction, leisure, retail, restaurants, hotels, residential units and open space management were likely to be created.[36] Local politicians seized on the larger figure as a key benefit of the scheme. However, Harris's figures were subjected to scrutiny by Harvey Cole, a planning consultant member of the Hampshire Buildings Preservation Trust with a special interest in discount shopping. Cole considered them a considerable over-estimate: many jobs in retailing and leisure would be transfers from elsewhere such as the city centre and Southsea shops. The true gain in job creations was likely to be about 300.[37] Predictions were clearly sensitive: *A Tower too far?* a BA Business Studies dissertation of 1994 at the University of Portsmouth by R Compton was also critical of the developers' figures for employment generation and tourism income. Only local labour involved in construction and subcontracting and permanent skilled jobs as opposed to seasonal service work were likely to benefit local people. The city council and Berkeley Homes made a commitment to using local sub-contractors and to training local labour in construction, particularly young unemployed people in Portsea, Somers Town and the city centre where poverty and social deprivation are concentrated. As home to the premier naval base on the south coast of England, Portsmouth has traditionally been a low wage economy. Despite post-war diversification into electronics, education and high technology industry, it is still a comparative island of deprivation in the prosperous county of Hampshire. Although the Berkeley Group had a target of 20-25% local employment in the construction phase, they admitted that they did not have control over all their subcontractors. Furthermore, skilled labour was not necessarily available in Portsea.

In 2000 Geraldine Twaites, a geography student, examined the socio-economic consequences of the urban regeneration of Gunwharf for the historically and physically distinct community of Portsea, the dockyard town, sharply defined by the walls of Portsmouth Dockyard to the north, Victoria Park and the city centre to the east and the Hard to the west. The elevated Portsmouth Harbour railway line to the south separates it from Gunwharf, an area on a social fault line between social tenanted Portsea and owner occupied Old Portsmouth. Charles Dickens ward which includes Portsea had the highest level of deprivation in the city in 1998, the highest proportion of council rented accommodation (almost four times above the city average), the highest level of unemployment at 18%, the lowest average household income, with 45% of households below £10,000, while those with jobs are mainly semi-skilled and unskilled.[38]

It was the people in this area of high unemployment and social need who stood to gain or lose most by redevelopment of Gunwharf. The employment proposed was mostly in retail, catering or leisure, part-time and un- or semi-skilled. No social housing was proposed, although in the revised scheme a block of social housing was built near the railway line.

As many writers say, local residents continue to have little impact on development because the hierarchy of power relationships is still weighted against them. Attempts by the Portsmouth Society and local vicar to provoke responses to Berkeleys' proposals in Portsea were not successful, whether because of fatalism, lack of confidence, or interest, or the 'deference factor' identified in defence dominated areas by *New Society* (1977) is not clear. Some residents clearly felt excluded from the start: "Why bother about all this fuss that the land is going to be opened up to the public for the first time in years? They may just as well keep the signs 'Keep Out' up! If it is meant for us, why haven't they provided a more mixed development?"[39]

The *Estates Gazette* perceived that the emphasis at Gunwharf would be *decidedly upmarket* (March 1999). An earlier report (August 1997) said that although Portsmouth had a low average income, there was a wealthy market within a 90-minute drive. These reports seem to suggest that the scheme was to be exclusive, not geared to poorer people, including the residents of Portsea, a view reflected in the comment above. Dominic Blake recalls a section

106 agreement on the Lipstick building which required a certain amount to be 'affordable homes' to be included but that they proved to be out of the price range of key workers even with a discount, and that the council eventually agreed to the section 106 agreement being removed. In the era of rapidly rising house prices much of the residential property was bought off-plan and sold again at a profit once completed a year and a half later. Second homes and tenanted flats are part of the mix.

Paul Spooner, city chief marketing officer, spoke regularly to the Portsea Neighbourhood Forum a Portsmouth City Council initiated body - over a series of meetings about the developers' proposals for Gunwharf and the associated Millennium project for a tower and harbourside promenade. There was no real consultation - just exposition of what was going to happen. Councillor Peter Spencer, ward councillor and chair of the planning committee until May 1997 said in a private discussion with the Portsmouth Society, even before determination of the planning application, that the scheme offered Portsea "Not a lot". Despite this admission, he subsequently assured a well attended meeting of the Portsea Neighbourhood Forum that they should look forward to what Berkeley's wanted to build at Gunwharf, which would mean "Jobs, jobs, jobs" for them and their children. This view was challenged by a member of the Portsmouth Society at the end of the meeting, but those who expressed concern felt unable to counteract the prevailing optimism on what the scheme promised. The job creation potential of the scheme was stressed, but the councillor was silent on the likely environmental and economic outcomes of the development. Open debate was discouraged, even though, as shown, the estimate of 2000 new jobs had been subjected to professional criticism. Few responses to the scheme were received from Portsea by the City Planning Department.

As the scheme was completed, the economic reality was likely to be different from that predicted: contractors were often not local; the money was probably raised on the international market; profits by national and international concerns including retailers were unlikely to be spent or reinvested in Portsmouth. Elsewhere there are examples of job creation - in permanent, skilled categories - in waterfront renewal where buildings are retained and reused for education, leisure, housing and manufacturing - as documented in Celia Clark's research.[40] In Portsmouth's case Professor Pinder of Plymouth University pointed out that even when naval heritage is exploited in a highly orchestrated manner, the outcome in Portsmouth was likely to be only modest direct and indirect spending in the context of the entire city economy.[41] By the early 1990s the 600,000 visitors to Portsmouth per year were supporting the equivalent of less than 500 full-time jobs, compared with the loss of over 6000 civilian maintenance workers from the dockyard in the previous decade. The jobs were not comparable anyway. Only 10% of the city's total earnings from leisure and business tourism could be attributed to visitors to the naval waterfront. Although ambitious plans were being proposed to increase the attractions of the heritage area, on the most generous estimate these would have raised the total value of city output by only 1.3%.

As confirmation of this prediction, in July 2019 retail jobs advertised in Gunwharf were for sales advisors and part-time stockroom assistant, temporary part time sales associate, chefs, kitchen porter, full time supervisor, assistant manager, guest services host, bar tender, store manager...[42] Local children, realising that sea bass were breeding in the canal and that they sold for £17 per kilo to restaurants, began to catch and sell them, until purple dye was put in the water, which turned them an unappetising shade of mauve.[43]

Clearance of most of the site

Construction of first new building designed by Hedley Greentree Partnership June 1999   Garrick Palmer

Vulcan stands amongst rubble

Despite HMS Vernon's designation as a Conservation Area by Portsmouth City Council in 1991, perhaps because of its potential value next to the harbour and the main transport hub, most of the surviving buildings were swept away in summer 1998, except the few that were scheduled or listed. The Warrior block, Sick Bay, Wardroom and headquarters and the Gunboat Sheds were all demolished for the redevelopment. The northern seawall of 1709 was concealed inside the seabed carpark, and only the few listed or scheduled buildings of Old Gunwharf survived to find new uses. The Vulcan storehouse of 1811-14 was both Grade II listed and scheduled, as was the Customs & Excise House/Vernon office block. Nelson's Gate and the Infirmary were scheduled, and the 1870s walls and gateways are Grade II listed buildings. So these had to remain. The submarine escape tower also disappeared. But although Ariadne, the neo-classical wardroom of 1922 was listed following a request by the Portsmouth Society, it was subsequently delisted in 1996 when *HMS Marlborough's* poop deck in the principal hall was removed to HMS Sultan after closure of the site in 1986. Because of the Crown exemption by which the Ministry of Defence was exempt from civilian planning and conservation legislation, the planning authority were not informed of this major alteration to a listed building. The wardroom was demolished in 2000.

Reinstatement of the cupola on Vulcan by Michael Underwood of HGP

As so much of the historic character of Gunwharf has disappeared, it's apparent that consumers who may have travelled some distance may well have little or no interest in it – or realisation that it once was a historic site. [44]

The Millennium Tower

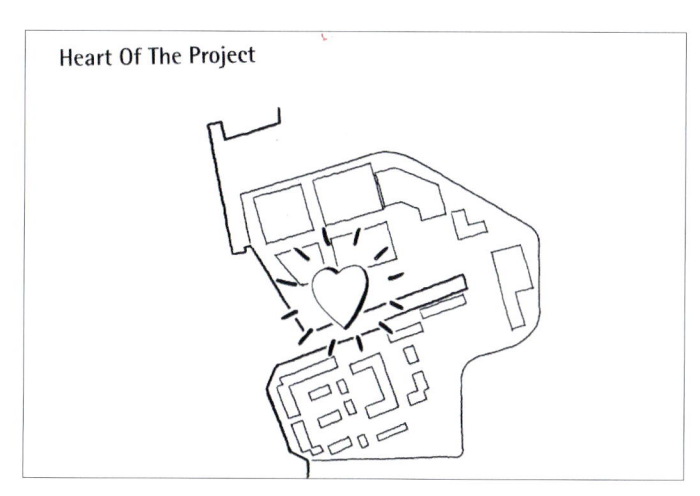

Robin Partington RIBA Apt
Location of Millennium Tower

The central feature of the Gunwharf redevelopment was to be a symbolic tower, a symbolic gateway to the city and celebration of the harbour. [45]

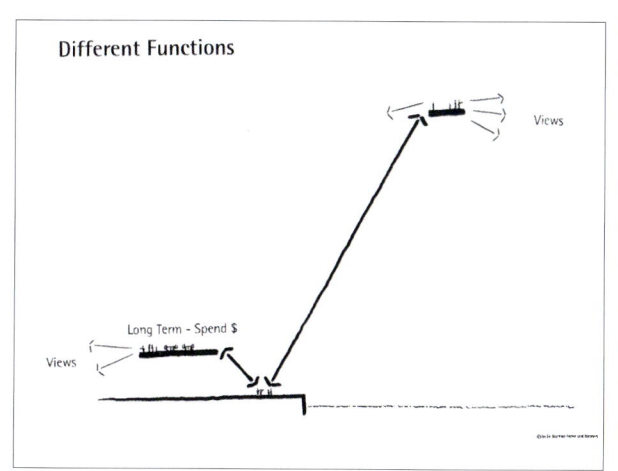

Millennium Tower Design by Robin Partington, APT, formerly of Forster and Partners

Robin Partington of the world famous architectural firm Forster and Partners designed a two-armed tower based on a dockyard crane, which fulfilled the original brief for a high level viewing platform offering long distance views, balanced by a shorter counterweight arm with low level views containing the revenue generating restaurant. Ove Arup provided the engineering design. They had to hire a fishing boat in order to assess the site. Robin

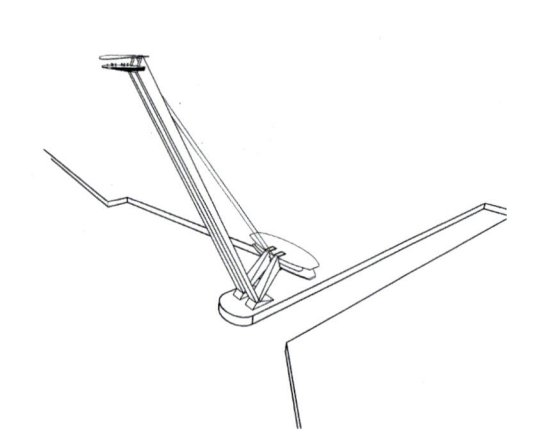

Partington said: "I have very fond memories of the work that we did on the Millennium Tower, and I am really proud of the solution that we prepared, as I genuinely feel that it was 'of its place': 100% engineering mixed in with 100% architecture, and a reflection of the very best engineering solutions that we find in our historic ports and shipyards... in keeping with the spirit, character and heritage of this amazing location."[46]

Robin Partington Design for Millennium Tower based on a dockyard crane

In February 1997 members of the Portsmouth Society went to Forsters' office in London to see it. As well as being a much more appropriate design for Portsmouth – since spinnakers are associated with yachting which is mainly based on the Gosport shore, Forster's tower was better sited than the Spinnaker – at the end of the canal instead of tucked away in a corner. But the contract was for Design, Build and Operate, which Forsters did not wish to enter, so this design did not proceed any further.

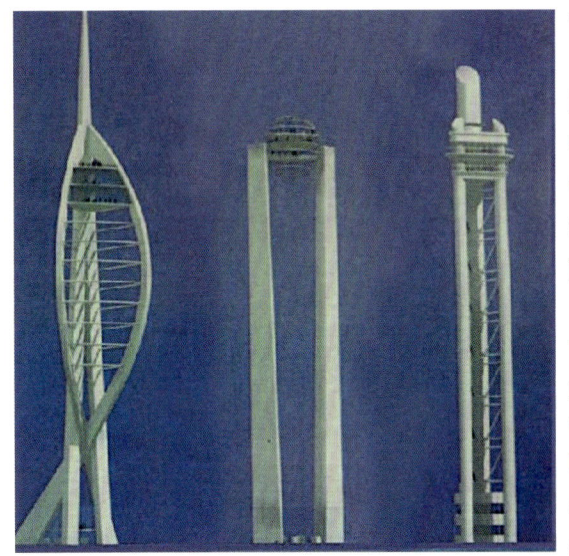

Three tower designs by Peter Warlow HGP Partnership

Instead, three different designs – all by Hedley Greentree's firm, were offered to a public vote. "All three stood at over 360ft tall, but the tallest was actually the Triple design, which had a viewing platform at 377ft... On Wednesday, February 11 1998 *The News* gave members of the public the opportunity to vote for their favourite design, in conjunction with polls from Portsmouth City Council and the Berkeley Group, who developed the tower. Voting closed on February 23, with 9,476 people voicing their thoughts on the outcome. Mr Greentree's Spinnaker design was announced as the winner, with a total of 6,137 votes (65 per cent)." Celia Clark did not like any of the designs, but there was no opportunity to register this. Paul Spooner, who probably did not want a contrary view broadcast, dragged her away from an interview with BBC Radio 4. The tower was repeatedly said by the Millennium Commission to be central to release of funds, but as the result of uncertainties in funding, guarantees against loss, delays in planning and Transport and Works Act permissions construction of the 170m tower did not begin until November 2001. Since it was not ready by the Millennium it was named the Spinnaker Tower.

The protracted negotiations and long delays before it was finally completed are not spelt out here... There were disputes between the contractors Mowlem and Berkeleys, owner of the site. Its eventual cost was £35.6m, to which the city had to contribute £11.1m. The council's handling of the project and its failure to exploit revenue opportunities such as the Millennium was heavily criticised in a report. Eventually the council took over ownership of it. BBC Radio Solent reporter Dominic Blake was covering its opening in October 2005 when the external lift with VIPs on board got stuck halfway up. The PR lady said desperately that it was designed to stop, but to her embarrassment the VIPs had to be rescued via escape doors; some even had to abseil down.[48] In 2015 Emirates agreed to sponsor the tower, rebranded as the Emirates Spinnaker Tower for five years for £3.5m. Protests at proposals to paint it red (the colour of rival football club Southampton) distracted attention from the terms of the sponsorship deal which was claimed to promote the city across the world through additional Emirates marketing, and according to the council leader, avoid the need for cuts to vital services.[49]

The tower became one of those projects, like the supercarriers, which would have cost more to cancel than to complete. Like the Sydney Opera House (also hugely over budget and way behind schedule) and the supercarriers, the Spinnaker became an iconic symbol of Portsmouth used in tourist promotion as far away as the railway line to St. Ives in Cornwall, on the London underground and to herald BBC South Today News and ITV Meridian News every day. It helped to attract big events such as Ocean Racing from Southampton and the Americas Cup to the city.

The planning decisions

There were many interested bodies and groups who put forward their views on the Gunwharf development but this story demonstrates that the Ministry of Defence retained its power over the first stages of the disposal and redevelopment. Despite the Defence Estate Organisation's letter to the contrary, the local planning authority and local residents had no input into the bidding process and little into the choice to sell to Berkeley/Lordland. Members of the Partnership and the Portsmouth Harbour Renaissance Board had a great deal of influence over what was decided and the manipulation of public opinion. The local print media censored criticism of the proposals. At the outline stage, the government's statutory conservation adviser English Heritage was apparently discouraged by political pressure from advising that the planning applications should be decided by the Department of the Environment, which might have provided them with an opportunity to participate. But because the site is a conservation area English Heritage was able to intervene positively by insisting that the axis path from under the railway should lead directly to the harbour. National voluntary groups in alliance with local ones were unable to persuade those with power over the site to listen to their case on the importance of excellence in design in attracting inward investment, the likely damage to city centre and Southsea shops and lack of benefit to local people. These arguments were judged irrelevant by both the local planning authority and the DETR.

Lessons

The Gunwharf story raises important questions about the MOD's power over the disposal and redevelopment of sites they no longer need. The development process was freed from constraints and it also prevented involvement of most of the people likely to be affected by it. Adverse national and professional criticism might have damaged Portsmouth's reputation but it had no impact on the local decision-makers. Portsmouth City Council was allowed by the Department of the Environment (DOE: later Department of the Environment, Transport and the Regions: DETR) to determine the outline and detailed applications, choosing to give prominence to the economic and commercial interests who had most to gain. The city council wished to see rapid development and substantial investment. Their planners attached little importance to objections based on the site's history represented by archaeology and standing buildings, or to arguments that the redevelopment was contrary to principles of sustainability.

The £100m investment into a complete change of use of the site was undoubtedly a considerable boost to the city economy. Nearly 30 years later, the Association of Hampshire and Isle of Wight Local Authorities believed the Gunwharf redevelopment was an example of 'good liaison with the

MoD on Land Disposals'. "Former HMS Vernon. A secure land based site of 12.14ha released in 1995. Development Brief published in Nov 1995, followed by marketing and an appraisal by MoD of all bids, shortlisting, and a final decision in mid 1996 behind closed doors by the MoD."

"This was potentially difficult because the MoD decision should have been made just on best offer. There was a real concern for the Local Planning Authority if the best offer was not the best scheme. However, all the bidders had the opportunity to make presentations to the Local Planning Authority and seek our advice. Most (not all) did, and as a result one or two entered a fairly detailed dialogue resulting in amendments to their final submissions. We commented to the MoD on the acceptability of each of the schemes in respect of adherence to the Brief and likelihood of planning permission being granted. This was important, because we understand (although we only suspected at the time, and it has never been officially acknowledged) that the selected scheme (Berkeley/Lordland) was not the best offer by some margin. It was certainly the scheme we felt had best potential for delivering the Brief for a mixed use waterfront development, as has subsequently proved to be the case. A deciding factor may have been the strong indications given about the relative prospects of planning permission being granted for the competing schemes."

"The process delivered the right scheme because the Local Planning Authority had a clear Brief, the developers talked to the Local Planning Authority, and the MoD listened. It could easily have failed to deliver because the MoD was not required to listen to our advice, and could merely have opted for best price. This would be a major concern where two schemes adhere to the letter of the Brief, and one is subjectively better, but not best bid. This is a real risk, particularly because it is difficult to write a very tight and prescriptive planning brief under such circumstances."[50]

This was not how local objectors felt at the time. Endorsement of the scheme also came from the property press. In the Centre Retail Autumn Issue in 2002-3 Gunwharf's success was highlighted: "Voted Britain's best mixed-use scheme, Gunwharf Quays has attracted brands".[51]

Old Gunwharf Gate (centre) surrounded by new development

Bringing the story up to date

Gunwharf Quays, now owned by Land Securities, opened in 2001 and today is, as envisioned, a new quarter of the city. It's a popular shopping, leisure and residential destination for locals and visitors alike. The outlet stores are global names – and so profits probably accrue to their owners, many in foreign countries. A few local enterprises – a newsagent, a souvenir shop and a Lebanese restaurant did open, but did not survive. According to Michael Underwood there are 8 million visits a year. The 14-screen Vue cinema, 26-lane bowling alley, Grosvenor casino, Holiday Inn, clubs and restaurants draw people in day and night. Luxury yachts up to 80 metres can moor in the small waterfront marina. Instead of one vast shopping shed originally proposed, the shopping arcade is partly open air, a welcome contrast to Southampton's closed box, West Quay. The traffic-free streets are called after famous admirals including the Japanese Admiral Togo, who trained with the British navy in Portsmouth. Views from two-storey row of chain-owned restaurants facing the harbour were much enjoyed.

View from BBC Office Gunwharf June 2020 — Dominic Blake

The BBC studio and gaming businesses occupied the top floor of the North Promenade Building above the restaurants. They shared wonderful views of the harbour entrance, Spithead, Ryde and the downs of the Isle of Wight. BBC Radio Solent reporter Dominic Blake joined the BBC in July 2001. He had previously worked for the MOD, where he was based in a building looking out at two brick walls. His first office in the Guildhall was a pair of stuffy dingy rooms with no exterior windows at all. So he relished the move to Gunwharf. It offered the best views of any BBC office in the country, with space for a sound proof studio, tv facilities for live tv broadcasts - and social space – which is where Celia Clark, Martin Marks and Deane Clark interviewed Dominic about the development of Gunwharf in August 2019. The D-Day and VE-Day commemorations, the arrival of the two supercarriers in the dockyard and local documentaries such as the Panorama investigation of deaths at Gosport Memorial hospital were all covered from these studios.[48]

Gunwharf Quays from Millennium Tower 2007

The promised artists' studios did not happen: high rents ruled out that possibility. Architect Robert Benn, whose firm was responsible for the repair and conversion of part of the dockyard foundry into offices for BAE Systems was based in Gunwharf for a time. The university Business School was built in Portsea, not in Gunwharf. The innovative slip-cast Spinnaker Tower designed by Peter Warlow of HGP won the Portsmouth Society's Best New Building award in 2006. As we said in the last chapter, the society played a part in the move of the award-winning Aspex Gallery to part of the ground floor of the Vulcan building, while the floors above were converted into flats. But footfall to the gallery was drastically reduced when the footpath past it was closed in response to demands by the residents.

A replica cupola for the Vulcan replacing the one lost in WWII was designed by Michael Underwood of HGP using photographic evidence - in the same way that Terry Wren reinstated the clocktower on Storehouse 10 in the dockyard. These clocks' prominence is a reminder of the importance of timekeeping in military and naval establishments. Michael Underwood also wrote a book on the history of Gunwharf. A replica new north wing to the Vulcan was built on the site of the Creasy Block. The Whale's Tail sculpture by Richard Farrington at the head of the canal was commissioned by Berkeley Homes as part of a section 106 agreement. Its skin is incised with imagery from 16th century charts of Ortelius and Apianus's Cosmographia, probably the world's first atlas. But when it was installed in 2008 it was shorn of its surrounding entry ripples amid fears that people might trip over them… A greater indignity is when it is sometimes surrounded by sand in a children's playground. The central square of the shopping centre was recently relandscaped to a higher standard than the first design. In 2018 the guardhouse was converted into a restaurant and tearoom.

In 2019 the reassembly of Isambard Kingdom Brunel's first surviving bridge across the railway line at Paddington, was discussed with Land Securities, Historic England and the city council. It had been in store in Fort Cumberland for sixteen years. Spanning the top end of the canal would offer reflections of its unique cast iron structure, as well as being close to his birthplace in Britain Street Portsea. To the promoters' disappointment this apposite site was rejected by Land Securities.

But HMS Vernon/Gunwharf's history has largely vanished from view, apart from the quayside crane which remains in situ by the lengthened canal and two figureheads from *HMS Marlborough* and *HMS Vernon*. The 1709 wall of Old Gunwharf is hidden in the huge underground carpark built on the seabed. The 'Customs House' pub's interior was once decorated with displays about the site's history, but a later licensee threw these away. There is scarcely any on-site interpretation of its historical importance. *The News's* nostalgia expert Robert Hind said that there is now a 'generation of shoppers who do not remember HMS Vernon'.[52]

Instead, to quote current publicity, it "is the UK's leading premium retail outlet, with big- name fashion brands. The focus on affordable luxury and the wonderful waterside location attracts a broad range of customers. It has its own on-site recycling centre, and over the last three years has increased its combined recycling and reuse performance from 48% to 80%. [It] has one of the largest arrays of solar panels on a shopping centre in Europe, which are used to power the car park. The site contains 4 Scheduled Ancient Monuments. Only 2% of listed buildings fall into this category, so to have 4 on site is exceptional. The canals on site are filled with salt water, and represent the last remaining piece of the marshlands from pre-1770. 76 unexploded ordnances were discovered when the site was being built. 7 acres of foreshore were reclaimed to build the centre. When it was first built, Gunwharf Quays was the largest man made marine deck in Europe. Rumour has it that Nelson marched through Vernon Gate on his way to the Battle of Trafalgar. The world's largest single masted yacht has stayed in our Leisure Marina. The site was one of several naval bases used for the D-Day preparations, with units based at Gunwharf responsible for carrying 3rd British Division to Sword Beach on D-Day."[53]

Gunwharf Quays has one of the largest arrays of solar panels on a shopping centre to date in Europe, which contributed to Land Securities' target to reduce energy intensity by 40% by 2030 from 2013/4. "In 2015 Gunwharf Quays won the National Recycling Awards 'National Retail Recycler of the Year' title for the second time, and an 'International Green Apple' award for our Waste Management Programme."[53]

When Covid-19 struck, the eerie silence of the usually thronged shopping malls and open spaces of the canal and harbour waterfront closed off to the public during the months of lockdown in spring and summer 2020 was only a demonstration of the immense employment and economic losses the owners of the shops, cafés and restaurants and those who work in them sustained during the pandemic. Gunwharf was closed again for the whole of November.

The BBC was forced to terminate the lease of the TV and Radio studio at Gunwharf Quays on 31 July 2020 as part of a plan to address a £125m shortfall in funding, caused by the pandemic.[54] How Gunwharf Quays might change in response to the closure remains to be seen. Will it presage a further paradigm shift in the way we shop towards ordering what we want to buy on line? Southsea's two department stores closed in 2019-20 – which was attributable to high rents but also of people moving to new housing developments along the south Hampshire coast close to the massively expanded Whiteley shopping centre – which is where the retail energy in the sub-region is now located. The Gunwharf shops opened again in mid-June 2020, with limited numbers allowed in and signage to manage the flow of people. Hopefully, Gunwharf Quays with its wonderful views of Portsmouth harbour will survive as a particularly special leisure destination.

Priddy's Hard

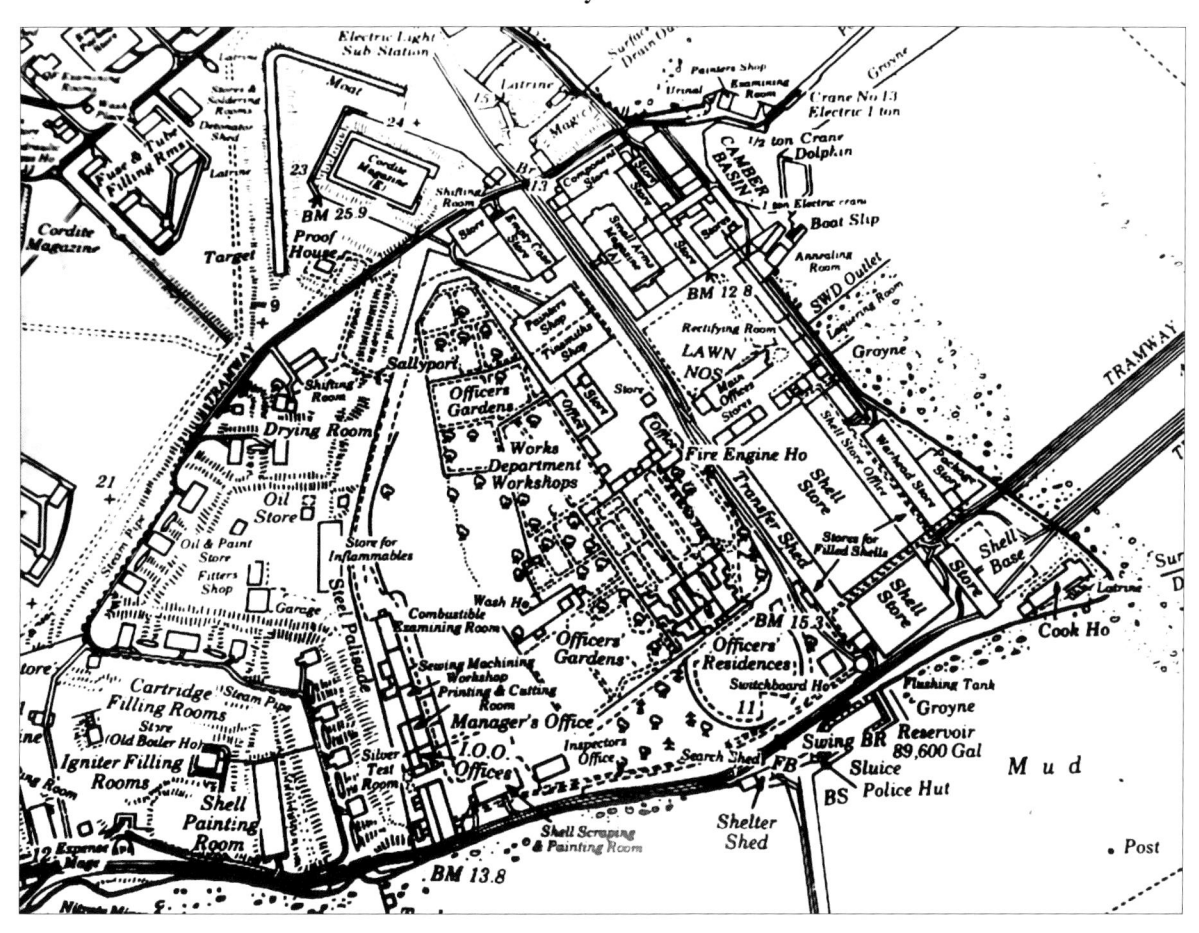

Priddy's Hard 1907 corrected to 1920

We now turn to what has happened to a very different but no less historically important ordnance yard across the harbour, which presented much greater environmental challenges, which have been met over a longer redevelopment timescale - as well as producing very different outcomes. It is perhaps the only example of free transfer of a MOD site to a local authority in the southeast. Gosport Borough Council was 'gifted' the 23 acres of highly contaminated and heritage-laden site of Priddy's Hard and the Naval Ordnance Museum Collection in 1994 55 – but without a dowry to pay for the high costs associated with regenerating such a complex and contaminated property. In 1998 an area of land around the ramparts, and north of Forton Lake, was protected as a Site of Importance for Nature Conservation (SINC) as a habitat for several protected species - to offset the 700 new homes to be built on the site, which instead of a dowry brought a financial return to the council.

History

Explosion Museum of Naval Firepower　　　　　　　　　　　NMRN

A "hard" is a Hampshire word that means firm beach or slope used to haul vessels out of the water. At various times, there were some ten hards on the Gosport side of Portsmouth Harbour and local historian Terry Hinkley offers an interesting guided walk "The Hards of Hardway" to explore them.[56] One of these hards was owned by a Jane Priddy: hence the name Priddy's Hard. In 1750 she and others sold the land to the Ordnance Board to build earthworks as part of the defences of Portsmouth and, especially, the naval dockyard.

Priddy's Hard ramparts　　Martin Marks

English Heritage note that: "The defences of Priddy's Hard, along with the adjacent length of fortifications behind Royal Clarence Yard, are the best surviving lengths of 18th century fortifications around Portsmouth Harbour… Whilst only the core buildings and earthworks are protected by listing and scheduling, the whole complex can be considered to be of national importance." It "is the most completely preserved ordnance depot from the era of gunpowder to TNT." [57] David Evans' *Arming the Fleet. The Development of the Royal Ordnance Yards 1770-1945* published in 2006 sets Priddy's Hard within its historical context.

Until well into the 19th century, naval and military armament needs were both met by a single authority, the Board of Ordnance. To meet the requirements of the fleet 10,000 to 12,000 barrels of gunpowder needed to be held at Portsmouth.[58] In about 1770 it decided that the gunpowder store in the Square Tower at Portsmouth (which only held 6,000 barrels) were too risky to be located there, and they were moved to Priddy's Hard. The siting of Priddy's provided an optimum location for access by water to the dockyard and warships anchored off the Gosport shore.

Jim Humberstone RCIS, a member of Hampshire County Council's Historic Buildings Bureau, who surveyed the site, comments that "the move to Priddy's was most likely prompted as a response to the petition presented by the mayor and citizenry to George III after a serious fire in the Square Tower, which took place in 1760... Perhaps Gosport residents were considered more expendable than those in Portsmouth?" [Few lived nearby.] "It may be that this development was linked back to the Civil War (1642-45) when Gosport and Portsmouth were on opposing sides![59]

Isaac Taylor's map of Hampshire from 1759 shows that the area around Priddy's Hard (marked as Prelley's Hard) was all farmland. In 1756 ramparts to protect the northern approaches to Gosport had been constructed on the land between the creek and Portsmouth Harbour. The ordnance yard was built inside these at an estimated cost of just over £6,921. 60.

Gunpowder Magazine in 2014

The most impressive building is the Grand Magazine of 1770-7 designed by Captain John Archer, Commanding Royal Engineer. Massively constructed in brick with 8 foot/2.5m thick double-skinned walls with ventilation slits, it has heavy supporting buttresses. Inside two aisles of beautiful brick vaults are supported on thick square pillars. Storage racks and airing passages were built into the thickness of the walls. In 1777 the first barrels of gunpowder were moved across the harbour by the Board of Ordnance. Gunpowder manufactured at Waltham Abbey was kept in these purpose built magazines at Priddy's Hard. The Cooperage and Shifting Room (for the workers to put on special clothing and shoes) were added in a two-storey range.

Priddy's Hard   Deane Clark

A two-storey office was added to the south in seven bays in chequered red and blue brick with a pedimented front and back. It had low two-storey wings; the eastern one was replaced in about 1815 for a six-bay extension. The imposing Officer's Residence was built facing south over Forton Creek by 1786.

257

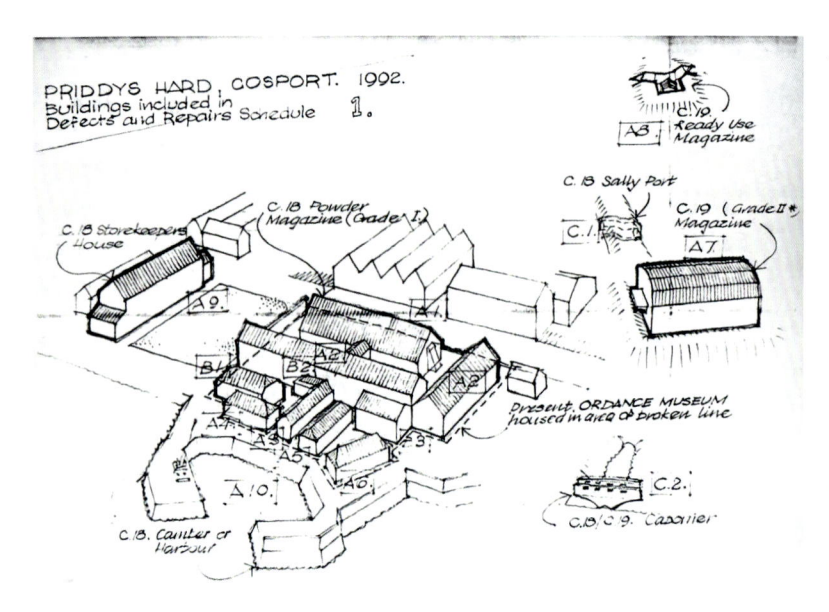

Jim Humberstone, commissioned by Hampshire County Council Historic Buildings Bureau

A small harbour, the Camber, was built at Priddy's, contemporary with the Grade I Magazine.

The covered 'rolling way' connected the harbour to the magazine – although powder barrels were never rolled, as this had a bad effect on the contents; they were wheeled in barrows or tumbrils.[58] The harbour consisted of two Portland stone moles, within which sailing hoys could safely load and unload their stores, presumably at most states of the tides. Interestingly, Priddy's Hard's little harbour is now badly silted up and dries out except at high water. It was proposed as a berth for the National Museum of the Royal Navy's Steam Pinnace 199 but rejected as impractical for this reason.

Aerial view
Explosion Museum of Naval Firepower

Artic II at Priddy's Hard. Last survivor of Army pre-WWII War Department Fleet built as static target tower with steam-powered winch for Shoeburyness Ranges. Post-war used as "floating barracks" at Yarmouth IOW
Dave Fricker

Outside the main site but within the defences the 'E' Magazine was added in 1878-9 for mass storage of gunpowder and cordite from 1913, one of the largest of its type. It has barrel vaulted chambers and a travelling overhead crane.[57] To the south near the Millennium Bridge were the remains of transit sheds and a control building at the southeast corner.

Priddy's Hard   Deane Clark

To meet increased demand from the Navy, additional powder storage was eventually added at Tipner on the Portsmouth side of the harbour with a magazine completed by 1800 and an additional magazine opened in 1856. This survives in mutilated form in Pounds scrapyard. The Army took over this complex in 1891. Priddy's Hard O/A Depot formed part of a chain of other locations, sequenced into the manufacture, supply and storage of armaments and ordnance and established throughout the 18th and 19th centuries. Associated facilities were the gunpowder works at Waltham Abbey in Essex and the Royal Military Depot at Weedon in Northamptonshire on the Grand Union Canal, while closer at hand, Marchwood on Southampton Water was important during the late Napoleonic period. Priddy's was also linked from 1915 onwards with the impressive complex of the Royal Naval Cordite Factories built at Holton Heath along the North West shoreline of Poole Harbour.

Over the years, the site expanded. The invention of breech loading guns generated a requirement for shell filling rooms, which were added. The invention of cordite made the separation of testing and filling from storage essential for safety. The size of the site necessitated a need for transport. In the mid-19th century an 18-inch gauge manually propelled narrow gauge railway, known as the Powder Line was constructed, including a link across the seabed to Burrow (or Rat) Island in the mouth of Forton Lake where unstable cordite was burnt. The tracks are reported visible at very low water. In 1904 an explosion in the shell stores raised concerns for the safety of the dockyard and a new store was built further north at Bedenham. Priddy's Hard's focus then moved from storage to shell filling. The railway system was connected to Elson and Hardway and by 1913 upgraded to 2 ft. 6 in. gauge as the "shell tramway". For safety, fireless steam engines were used which loaded up on high pressure steam and were fitted with brass allow wheels, all to avoid sparks or ignition sources.

The site was heavily involved in both World Wars as well as the Falklands conflict. There's a fascinating social history too, including the story of how 2,500 women worked on the site during its peak in World War II. The Royal Naval Armament Depot (R.N.A.D.) vacated Priddy's Hard in 1988 and transferred operations to Bedenham, Elson and Frater, under the new name of Defence Munitions Gosport. The last train left the depot in 1986 and it officially closed in 1989, with a volunteer caretaker under the supervision of RNAD Frater.[61] The Officer's Residence was demolished in 1952. A brick replica stands on the site.

Museum

The Priddy's Hard museum was established in the very early 1970s, formed from an original collection assembled by a former Superintendent for the interest and education of MOD staff, primarily of the Naval Armaments Service. In April 1994 the collection was acquired by Gosport Borough Council as part of the wider Priddy's Hard Heritage Area from the MOD, using a grant from the Millennium Commission. The oldest part of the site was opened to the public in 1991 as the Explosion! Museum of Naval Firepower centred around the original powder magazine of 1777. The museum traces the development of naval armaments from gunpowder to modern times and Exocet missiles. It uses audio-visual effects, interactive computer and hands-on displays. It also considers the historic aspects of its buildings.[62]

Explosion Museum of Naval Firepower 2014

Jim discovered an underground HQ on the site which he explored and photographed. He recalls that "the survival of the heritage at Priddy's, not least its unique ordnance collection, owes much to the curatorial enthusiasm and commitment of the staff working on the site. Obsolete or surplus equipment and materiel, passed to them for disposal was often discreetly retained as part of a growing collection." Martin Marks had such a move described to him when he met a former RNAD Gosport member - as 'Tuck it round the corner Ted, out of sight!' when confronted with say, a No Longer Required naval ordnance 12 pounder A/A gun. This led for instance to the retention of a complete Tribal Class turret and gun, together with smaller pieces. Other small items, including underwater weaponry of various kinds, were also taken in hand and stored. Care of the items in the collection continued after the vacation of the site by the Navy. During the period within which Jim undertook building surveys, the collection was administered by a retired Battery Sergeant Major Royal Artillery, detached from the Royal Armouries at Fort Nelson on Portsdown Hill.

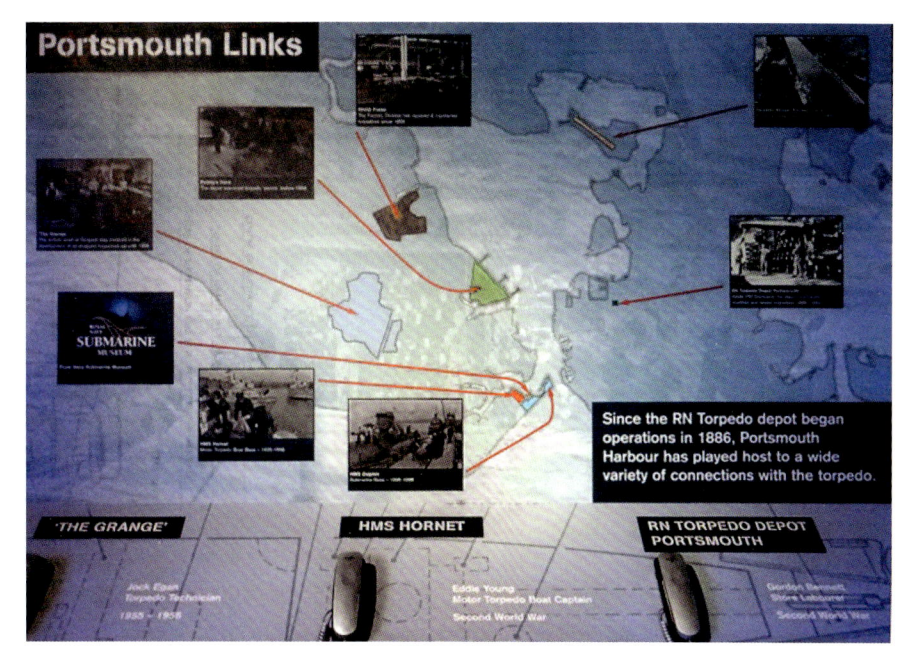

Associated facilities around Portsmouth Harbour   NMRN Explosion Museum

Millennium Bridge

Millennium Bridge with new block of flats on site
of Officer's Residence

Millennium Bridge and surviving buildings on site

In 2000 an opening bridge, the Millennium footbridge, was built across the mouth of Forton Lake where it meets Portsmouth harbour just south of Priddy's Hard to help with the development of the site.[63] The opening mechanism failed in 2016 and it was still not back in service in November 2018 – although repair work had commenced.[63] This trapped several yachts in the Forton Lake Marina, behind St Vincent College and west of the bridge, for an extended period.

New development

In 2015-18 nine new Modernist houses selling for £500,000 designed by John Pardy Architects were built between the blast walls of the shell filling stores for Dave Craddock of Elite Homes, who was also planning to build more within the ramparts. Decontamination, the proximity of the SSI and the complex permissions required make this intention particularly difficult. Other areas (including the former Royal Laboratory area) remained unoccupied and a programme of clearing the historic ramparts of vegetation was undertaken.

Blast walls of Shell filling stores before development
Forton Creek

New homes fitted between blast walls
Dave Craddock Elite Homes
Presentation to RTPI/HBPT seminar:
Future for Defence Heritage 2017

Portsmouth Naval Base Property Trust (PNBPT)

Map of Portsmouth Naval Base Property Trust holdings in the Naval Base and Priddy's Hard - from
Royal Clarence Yard Expression of Interest Portsmouth Naval Base Property Trust with Elite Homes Portsmouth

The museum struggled to operate and was supported financially by Gosport Borough Council. The scheduled ramparts, part of the ring defence of Portsmouth Harbour had become a wilderness and the museum was facing closure through falling visitor numbers and the inability of the local authority to invest in it. As a result it lost its museum accreditation. In 2009 freehold of the main part of the 23-acre heritage area of Priddy's Hard was acquired from the council by the Portsmouth Naval Base Property Trust - with a view to "developing the site, refurbishing the historic buildings and bringing them into new beneficial use". [64] Investment in infrastructure involved dredging a channel and the installation of a link span at the southern end of the site to bring visitors by ferry from the historic dockyard, where the Property Trust is the landlord. Investment in the museum led to the museum regaining its accreditation; the number of visitors quadrupled while commercial income cut its operating deficit of about £500,000 per annum to less than £150,000. In 2013 the Trust transferred ownership of the museum to the National Museum of the Royal Navy who were keen to operate it.[65]

In 2015 a Heritage Lottery Grant of £1.9 million was awarded to PNBPT under the Heritage Enterprise Programme which targets projects creating 'new sustainable economic uses for derelict historic buildings'.[66] This was to be used to restore seventeen listed buildings so that they could be used for residential, business including a micro-brewery operated by the local Oakleaf Brewing Company and a new waterside pub-restaurant where Oakleaf was to serve real ales and quality food made from local produce. But this didn't happen when the brewery failed.[67]

The trust planned to open up the ramparts, a scheduled monument, to the public. According to notices on site, they were working with Historic England to restore and reveal the historic fortifications to make them suitable for the public to enjoy through regular tours. Areas cleared of scrub vegetation and some trees would be replanted as wild meadows, enabling small mammals and insects to thrive. Reptiles, badgers and deer already live there. The trust aimed to establish a small visitor orientation centre by restoring one of the dilapidated buildings and to create new walkways though the ramparts for people to enjoy, with interpretation panels and seating on places offering the best views of Portsmouth harbour and the distant South Downs.

The trust's aim was to unlock £11m private sector investment in the site to create a new community and provide the financial means for it to maintain the 25-acre estate in perpetuity. 104 new homes were to be built by Linden Homes. PNBPT planned to use its income from the residential scheme to match-fund HLF's support for the project, establishing new community facilities and innovative sustainable uses for at-risk buildings on the site. Peter Goodship, PNBPT Consultant Chief Executive, said: "This is a great example of business-led regeneration and partnership with both public and private sectors. It has taken a while to reach this stage since we acquired the site from Gosport Borough Council at the end of 2010, but to bring 17 listed buildings back into sustainable use, help local businesses to expand, create almost 30 jobs, and build new homes with fabulous views of Portsmouth Harbour, will be a fantastic achievement and well worth the wait."

In 2018 a Heritage Lottery Stage 2 submission for some £1.9m included securing the future of 7 Grade II and Grade II* buildings and the scheduled ramparts.[64] The plan was to refurbish a further 10 listed buildings and to create a new Museum in P Building where visitors will be able to view stories of coastal forces reservists from the Royal Navy archives and two WWII vessels, *MTB 71* and *CMB331*, previously stored at Yeovilton. The total project cost was estimated at £3.8m with completion by summer 2021. The plans for the significant buildings were:

Building	Intended future use
E magazine	Distillery – the Chatham based Copper Rivet Distillery
Building PH – Proof House	Distillery Office
Building M – Case Store	Student Accommodation
Building U – Shifting Room	Holiday Accommodation
Building P – Mines and Countermines Store	Coastal Forces Exhibition
Building Q – Shell Store	Distillery Store
C Magazine	Volunteer Accommodation

In January 2020 Portsmouth Naval Base Property Trust was given the green light to proceed with the £30m first phase of its proposals to regenerate the historic Priddy's Hard former armaments depot in

Gosport. The regeneration included the refurbishment and conversion of several important historic buildings to provide a mid-size brewing facility - the Powder Monkey Brewery.[65] The initial phase also included the construction of 30 new homes on three sites on the western part of the estate to be developed by Elite Homes.

This presented many challenges, given its historic importance, the risk of rising sea levels and its ecology sandwiched between a site of special scientific interest and a RAMSAR site of international significance. In April 2020 Peter Goodship said that the trust was repairing E Magazine, C Magazine and four other listed buildings which will be converted into a brewery, pub/restaurant, Landmark trust style holiday accommodation with private foreshore, a new Coastal Forces Museum and accommodation for students attending courses at our International Boatbuilding Training College.[66]

The initial refurbishment of all the historic buildings was planned for completion by the end of July 2020, followed by the fit out over the following six months. The construction of the housing was planned to begin in May or June and be completed between April and September 2021. Delays meant that work was only starting on site in September 2020.

As Jim Humberstone says, "Three factors to bear in mind when looking back at the long saga of Priddy's and its significance: first the existence of the collection of various forms of ordnance and quite special in its own right, which helped to justify the retention of the eighteenth century buildings. Second, the survival of almost a complete Napoleonic period ordnance depot was something rather unusual. Once saved, the site and its buildings have contributed an important and vital element in the history of the Harbour and the Dockyard. Third, as experts like Andrew Saunders and Jonathan Coad pointed out at the time, the RNAD coupled with the Royal Clarence Yard and the still extant C17 and C18 fortifications and C19 Barracks, have as a complete historic complex, few if any Defence of the Realm equals elsewhere in the country."[67]

These two historic ordnance sites on opposite sides of the harbour presented different challenges to those who undertook to regenerate them. While Priddy's was mainly led by conservation of the standing buildings and nature reserve, Gunwharf was almost completely redeveloped. This difference was probably because of their locations and the times at which their redevelopment took place.

Sources

A S Blight, Superintending Engineer and Constructor of Shipping 1992 'Fleet Repair Through the Ages' in *Craftsmen of the Army The Story of the Royal Electrical and Mechanical Engineers* Vol. 11 by JM Kneen, D J Sutton Pen & Sword Books Ltd.

Dominic Blake BBC Radio Solent Interview with Celia and Deane Clark and Martin Marks 6 August 2019 at BBC Studio Somerville Suite 6A North Promenade Building Gunwharf Quays

Celia Clark 2002 "White Holes" Decision-making in Disposal of Ministry of Defence Heritage Sites PhD thesis University of Portsmouth

Portsmouth Society News April 1997 'Proposals for HMS Vernon/Gunwharf 'David Evans 2006 *Arming the Fleet The Development of the Royal Ordnance Yards 1770- 1945* Explosion Museum and English Heritage

Dave Marsden *The Hidden Railways of Portsmouth and Gosport* Kestrel Publishing

Michael Underwood, Julie Underwood ed. 2015 *Gunwharf Quays: The History, Architecture, Conservation and Development of a Remarkable Military Site*

HW Semark 1997 *The Royal Naval Armaments Depots of Priddy's Hard, Elson, Frater and Bedenham (Gosport, Hampshire) 1708 to 1977* Explosion Museum of Naval Firepower

E D Webb 1956 *HMS Vernon A Short History from 1930 to 1955* HMS Vernon Wardroom Mess Committee (Publisher)

Admiral E N Poland *The Torpedomen: HMS "Vernon's" Story, 1872-1968 - E.N. Poland* Kenneth Mason ISBN: 0859373967

References

1. Celia Clark 2002 "White Holes": decision-making in disposal of defence heritage sites PhD thesis University of Portsmouth

2. *The Illustrated London News* Supplement May 5 1855 London

3. David Evans 2006 *Arming the Fleet. The Development of the Royal Ordnance Yards 1770-1945* published by the Explosion! Museum of Naval Firepower in association with English Heritage) p.28

4. Historic England. "Details from listed building database (1378580) *National Heritage List for England*. Retrieved 1 October 2015

5. https://en.wikipedia.org/wiki/Royal_Corps_of_Transport0accessed025 May02020; http://www.memorialsinportsmouth.co.uk/others/gunwharf/hm-gunwharf.htm accessed 25 May 2020. These sources outline the evolution of the army's maritime expertise from a civilian fleet; Royal Army Service Corps; Royal Corps of Transport 1965; 20 Maritime Regiment. 20 Maritime Regt RCT was formed from 17 Port Regt RE and 46 Sqdn. RASC (Far East). 46 Sdn. RASC (FE) was originally 76 Sqn. RASC based in Gunwharf and then became 46 Sqn. RASC before deploying to the Far East. (Rob Hoole Emails to Celia Clark 2 June 2020). Dave Fricker: There were well over 1,500 Army vessels serving worldwide at the end of WWII. Army Water Transport was generically split into two main categories, the Blue Water component (ocean, and coastal seagoing, which was part of the Royal Army Service Corps, later Royal Corps of Transport, and now Royal Logistics Corps). Then there was IWT – Inland Water Transport – for the smaller craft supporting harbour/estuarial Military tasking.
These were manned by Royal Engineers and their vessels wore a different Defaced Blue Ensign. This unit was absorbed into the Royal Corps of Transport when it was formed in 1965. 20 Maritime Regt. left Gunwharf in 1988. It became the Royal Logistics Corps and later 17 Port and Maritime Regiment based at the Sea Mounting Centre Marchwood. This site was put up for sale by the MOD in October 2010 (https://www.dailyecho.co.uk/news/8471836.jobs-fears- as-marchwood-military-port-goes- up-for-sale/). Solent Gateway (David MacBayne Ltd. And GBA (Holdings) were the preferred bidders in 2015 (https://www.bbc.co.uk/news/uk-england-hampshire-31715480).

6. Sir Christopher Armstrong Emails to Celia Clark 2 June 2020

7. https://www.waterstones.com/book/crabbgate/john-bevan/9780950824284 Marshall Pugh 1956 Commander Crabb. MacMillan & Co. Ltd., London. "Mystery of missing frogman deepens". BBC. 9 May 1956. Dominic Blake 12 June 2006 "Secret spy diver report revealed" BBC
Don Hale 2 November 2007 *The Final Dive: The Life and Death of Buster Crabb*. NPI Media Group Ltd. p. 288. ISBN 978-0-7509-4574-5.Tim Binding 26 March 2006 "Buster Crabb was murdered – by MI5". The Mail on Sunday. Highbeam Encyclopedia.Tim Binding 2013 *Man Overboard* Pan Macmillan ISBN: 9781447219408.

Dr. John Bevan 2019 *Crabbgate* Submex Ltd. ISBN: 9780950824284

8. Dave Fricker Emails to Celia Clark 19, 21, 25 May 2020; Sir Christopher Armstrong Emails to Celia Clark 22, 26 May 2020

9. https://gunwharf-quays.com/history-gunwharf-quays accessed July11 2019 10 Stephen Payne Email to Celia Clark 14 May 2020

10. Michael Bateman and Ray Riley R eds. 1987 *The Geography of Defence* Croom Helm London and Sydney; Coad JG 1989 *The Royal Dockyards 1690-1850 Architecture and Engineering Works of the Sailing Navy* Royal Commission of the Historical Monuments of England Scholar Press Aldershot; Russ Fox 1995 *H M Gunwharf - Old Portsmouth Archaeological Excavation/Assessment* City of Portsmouth

12. Richard Trist Portsmouth City Council Chief Executive from 1978 to 1993 Interview with Celia Clark 23 July 2016

13. https://gunwharf-quays.com/history-gunwharf-quays accessed July 11 2019

14. Alistair Hunt July 2001 Interview with Celia Clark about his report for Defence Estates on the historic buildings in Gunwharf

15. Hedley Greentree in *The News* 1995

16. L C Weymes City Planning Officer. City of Portsmouth *Gunwharf/Millennium Development Brief* November 1995 p.1

17. Weymes 1995 p. 2 18 Weymes 1995 pp.13/14

19. Roger Brown 1996 Hampshire Buildings Preservation Trust meeting attended by Celia Clark

20. Nancy Holman 1999 *Networks, Design and Regeneration A Case Study of the Gunwharf Regeneration Project* PhD thesis University of Portsmouth p. 86

21. Paul Newbold 1998 City Planning Officer's letter to C Clark on draft thesis November 1998

22. The Berkeley Group *Gunwharf Quays Environmental Analysis Addendum 1* December 1996 2, 5

23. David Lock 1997 Associates with HGP Greentree Allchurch Evans Ltd 1997 *Gunwharf Qua Portsmouth Folio of Illustrations for Development Statement.* May p.9;

24. https://www.designingbuildings.co.uk/wiki/Skeffington_Report

25. Marcus Binney *The Times* 26 April 1998 *Architect's Journal* 7 November 1996 P7

26. David Brock 1997 English Heritage *Letter to City Planning Officer, Portsmouth* 5 February 28 Matthew Lambert 2020 "Critical analysis upon the regulation of contemporary urban space at Gunwharf Quays, Portsmouth" BA Dissertation Coventry University

27. A R Baillie Director DEO (Lands) 1997 Letter to R Trist Esq. 06 March 1997 30 *The News* 3 July 1998 p1,8 Richard Trist Portsmouth City Council Chief Executive 1997 Letter to the Director of Defence Estates Organisation (Lands)

28. *Sunday Times The Rich List November* 2001 pp.56, 86 33 Celia Clark 1997 Letter to the Editor *The News* 15 June

29. Fred Robinson and Keith Shaw 1991 'Urban Regeneration and Community Involvement' *Local Economy: the Journal of the Local Economy Policy Unit* May 1 p.66 Published online 2007

30. R. Bettsworth 1999 '£200m project helps to create 4,000 new jobs around harbour'. *The News* 10 April p. 21

31. Roger Lewis Letter to Celia Clarke [sic] 24 February 1997

32. Harvey Cole 1997 Hampshire Buildings Preservation Trust *Letter of 12th March* to Roger James, Portsmouth Society

33. Geraldine Twaites 2000 pp.22-28 derived from census data 39 Interview in Twaites 2000 p.34

34. Samer Bagaeen and Celia Clark 2016 *Sustainable Regeneration of Former Military Sites* Routledge 2016.

35. Pinder D Hoyle D & Husain eds. 1988 *Revitalising the Waterfront International Dimensions of Dockland Redevelopment* Belhaven Press 1988; Wiley & Sons Chichester 42 https://gunwharf-quays.com/jobs accessed July 2019

36. Dominic Blake 2019 Interview with Celia Clark, Martin Marks Deane Clark August 6

37. Matthew Lambert 2020 "*Critical analysis upon the regulation of contemporary urban space at Gunwharf Quays, Portsmouth*"

BA Dissertation Coventry University

38 *The News* 9 January 2017 'Nineteen years on – the story behind how the Spinnaker Tower design was chosen'

39 Robin Partington RIBA Emails to Celia Clark 20 May, 8 June 2020

40 Staff writers 18 October 2005. "Spinnaker opens five years late". *BBC News*. Retrieved 21 June 2013 Dominic Blake BBC Radio Solent Interview with Celia Clark, Martin Marks and Deane Clark 6 August 2019 https://onthewight.com/spinnaker-tower-details-of-the-five-year-sponsorship-deal- revealed/ accessed 2 June 2020

41 Association of Hampshire and Isle of Wight Local Authorities 2003 Report by Chief Officers' Group Paper 5A 31 January

42 'Gunwharf Quays Portsmouth' 2002 *Centre Retailing Autumn 2002-2003* 18 August, pp. 135-136

43 David George *The News* 2019 July 11 'Preserving heritage for our future generations' pp.10-11 https://landsec.com/properties/gunwharf-quays-portsmouth© L a n d Securities Group accessed 11 July 2019

44 *The News* Portsmouth 2020 'BBC to close city studio' 7 May p. 4; Dominic Blake BBC Radio Solent Emails to Celia Clark June 1, 2 2020

45 Letter from Ministry of Defence Secretariat (Naval Staff) Whitehall London SW!A 2HB 25 January 1995 to R Clayton, Head of Legal Services, Gosport PO12 !EB "Priddy's Hard Ordnance Museum Collection"

46 The Hards of Hardway - Guided Walk with Terry Hinkley in Gosport http://eventful.com/events/ hards-hardway-guided-walk-terry-hinkley-/E0-001-094928907-9 57; Charles O'Brien et al 2018 *The Buildings of England Hampshire South* Yale University Press p.300-1

47 David Evans 2006 *Arming the Fleet. The Development of the Royal Ordnance Yards 1770-1945* Explosion! Museum of Naval Firepower in association with English Heritage p. 19

48 Jim Humberstone RICS Email to Celia Clark 18 May 2017 60 O'Brien pp.300-1; http://democracy.gosport.gov.uk/documents/s837/Final%20RB%20agenda.pdf accessed 27 May 2020 includes a very detailed description of the ramparts

49 Fireless railway http://www.gosportheritage.co.uk/14-january/ 62 Explosion Museum www.explosion.org.uk/ 63 Gosport Cruising Club http://www.gosportcruising.club/documents/GOSPORT%20MILLENNIUM%20BRIDGE%20%20OPENING%20PROTOCOL%20revision%202%20April%202016.pdf; https://www.portsmouth.co.uk/business/bridge-gosport-will-be-closed-four-weeks-repairs- 86720#gsc.tab=0; Repairs to Millennium Bridge https://www.tenderlake.com/home/tender/6f49e667-ee6a-445c-9dc1-4ddb91c4acdb/gosport-millennium-bridge-drive-train-capital-repairs-20162017

50 PNBT www.pnbpropertytrust. org; www.pnbpropertytrust.org/propertymapgosport.asp

51 Royal Clarence Yard Expression of Interest Portsmouth Naval Base Property Trust with Elite Homes.pdf Friday 1st March 2019 accessed 22 September 2020

52. Heritage Lottery Fund;https://www.hlf.org.uk/about- us/media-centre/press- releases/regeneration-project-priddy%E2%80%99s-hard-secures-heritage-lottery;
David George *The News* 12 June 2020 'New Coastal Forces Museum 'on track' to open in summer 2021' p.6 Earthworks https://historicengland.org.uk/listing/the-list/list-entry/1010741PNBT

53 Email from Peter Goodship to Celia Clark 20 April 2020 Priddy's Hard Executive Summary Final

54 PNBT Priddy's Hard Statement of Change Final

55 PNBPT Priddy's Hard Development Appraisal with Appendices FINAL

56 Jim Humberstone RICS Correspondence with the authors

Chapter 11
Early flight to a civilian airfield

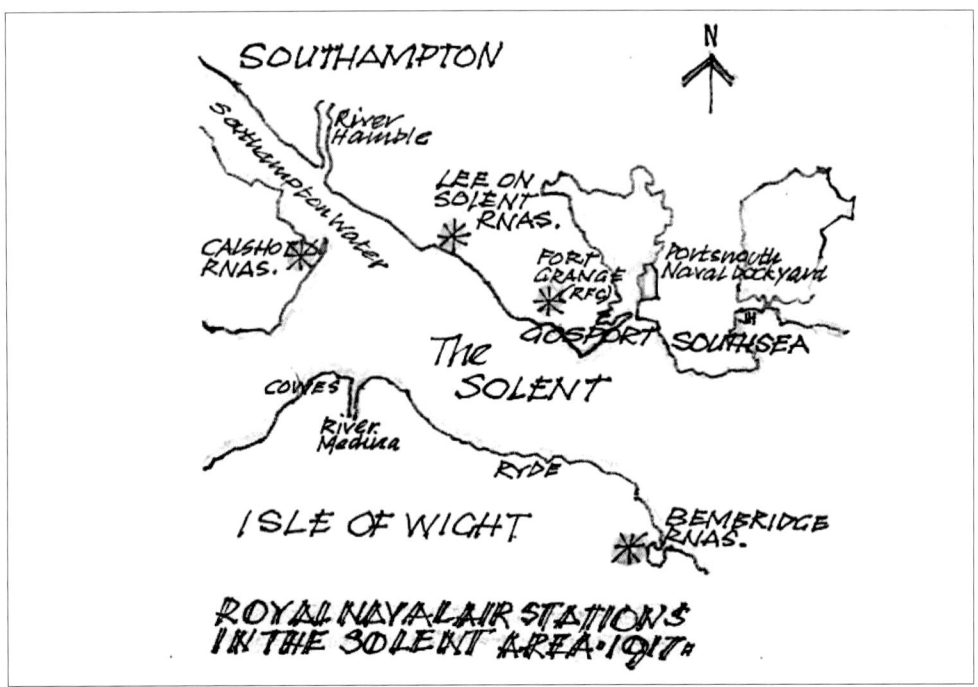

Royal Naval Air Stations in the Solent Area 1917   Jim Humberstone

There were two significant airfields near Gosport: Grange and Daedalus. The former was important in very early flying days whilst Daedalus had a long career as both a naval and RAF airfield throughout both World Wars and afterwards played a major role in naval flying and hovercraft development.

Grange Airfield, Gosport

In November of 1909, two Lieutenants from the Submarine Depot constructed a biplane which was taken to Gosport for trials. This was not a success; the machine was badly damaged when, in some way the starting tackle fouled it just as the machine was rising, and "it dropped heavily to earth". In April 1910, the Hampshire Aero Club successfully negotiated with the War Department for the use of site near Fort Grange for "for experiments in aeronautical science." In practice, this meant the use of a hand-towed glider constructed by one of the members and some model flying. In 1910 Grange Airfield, already in War Office ownership and manned by the Royal Artillery, was developed on flat land between Fort Rowner and Fort Grange for experiments in 'heavier than air flight', using a towed biplane built at the United Services College, Windsor. The wind tunnel in Fort Grange was used to test aircraft stability curves on aerofoil sections under the direction of Dr A. P. Thurston.[1] Louis Strange of 5 Squadron wrote about his experiences at Fort Grange in WW1; Albert Ball was an instructor there in February 1916. Hundreds of Royal Flying Corps personnel were accommodated in what BJ Hurren called a 'troglodyte existence' and Duncan Grinnell Milne 'dungeon-like' at Fort Rowner.

In 1914 the airship Gamma crewed by Royal Naval Air Service personnel moored in the shelter of Fort Grange before operating over the fleet at Spithead.

A reorganisation of the training units in the United Kingdom was in process from May 1917 and there was a proposal that a School of Special Flying should be formed at Gosport. Key developments in flying training were then devised by Lt. Col Robert Smith-Barry, Commanding Officer of 60 Squadron based at Grange. His system was founded not on avoiding potentially dangerous manoeuvres as had been the previous practice but on exposing the student to them in a controlled manner so that he could learn to recover from them, thereby gaining confidence and skill. This became known as the 'Gosport system.' Instructors were also trained to teach the principles of flying and pupils in the pilot seat at the controls communicating via the 'Gosport tube.'

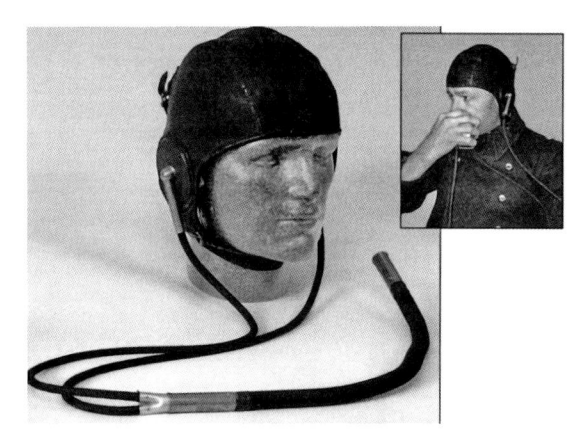

The Gosport Tube

Many trainee pilots died on their first flights because they could not communicate with their instructor. "Communications in any open-cockpit aircraft proved to be all but impossible as shouting over the combined din of the engine and the howl of the wind was simply inaudible through the thick layers of protection provided by headgear. One device developed by the Royal Flying Corps and soon adopted by the Royal Naval Air Service was the Gosport tube… a simple length of rubber tubing that was connected to the ears of the flying helmet, whilst the other end was connected to the face mask of the aircraft's other crew member. This reliable device was first designed for tuition so that instructors could shout commands to student pilots."[2]. In 1918 teams from Gosport travelled to America, France, Argentina and Chile to teach the Gosport method which became an accepted system worldwide. In 1919 the training role was transferred to the RAF's Central Flying School.

Throughout World War I the Aircraft Torpedo Development Unit (ATDU) was based at Grange and remained there until 1956. In the meantime, the Royal Naval Air Service and the Royal Flying Corps combined on 1 April 1918 to form the Royal Air Force. Were it not for the presence of the ATDU, this would have left RAF Gosport, as it was then known, redundant. In 1925 Captain Charles Lindbergh landed at Grange in the Spirit of St. Louis after his record-breaking flight to Paris. Between the wars Grange became a Royal Air Force Fleet Air Arm base. A dummy deck for aircraft carrier landing practise was added and it became a centre for deck landing trials. In 1937 the Royal Navy recovered control of its Fleet Air Arm and the 1940's map of the site shows the airfield between the two Victorian forts mentioned above: Fort Rowner and Fort Grange. The nearby Alverbank House was used as the officers' mess, familiar to locals today as the Alverbank Hotel. However, Grange remained in RAF hands until 1945. On 1st August Grange was renamed HMS Woodpecker but four days later this was changed to HMS Siskin as the former name had already been assigned elsewhere.

Also known as the Royal Naval Air Station Gosport, it became a centre for naval helicopters with the arrival of a squadron of Sikorsky R4 helicopters starting the Royal Navy's expertise in this field. By 1956, the rundown after the end of the Korean War led to closure and HMS Siskin became HMS Sultan.

HMS Daedalus – Royal Naval Air Station, Lee-on-the-Solent[3,4,5]

HMS Daedalus Lee on the Solent

This pioneering airfield had a distinguished career as a naval and air force base, and in its post-defence civilian history, it also experienced the vicissitudes the Hovercraft Museum went through in many changes of direction.

It dates back originally to 1912. The Royal Flying Corps took it over in 1914. The Royal Garrison Artillery, the corps responsible for static defences, armed usually with larger calibre ordnance, particularly significant for their role in manning coastal forts, were then moved up to the Portsdown Forts. In July 1917 it was opened as a temporary naval seaplane pilot training school to address the shortage of pilots for anti-submarine patrols in the face of the growing U-Boat threat. In both World Wars, coastal shipping routes, especially those down the east coast, Thames estuary and along the south coast, were very vulnerable to attack. The setting up of an extension training facility at Lee (Daedalus) to the one at Calshot, across the Solent in 1917 was a direct response to the problem of not enough RNAS trained aircrew coming through. Calshot where Celia's grandfather Leslie Bates RNAS was stationed was designed to protect shipping in the Solent and Southampton Water, but was heavily congested.

As 'Daedalus' was to be temporary, canvas hangars, a tented camp and some wooden storage facilities were established. The following November, a decision was made to transform it into a long-term seaplane base and orders were placed for Admiralty designed hangars and buildings. Work was started on a permanent slipway to avoid having to crane seaplanes into the water.

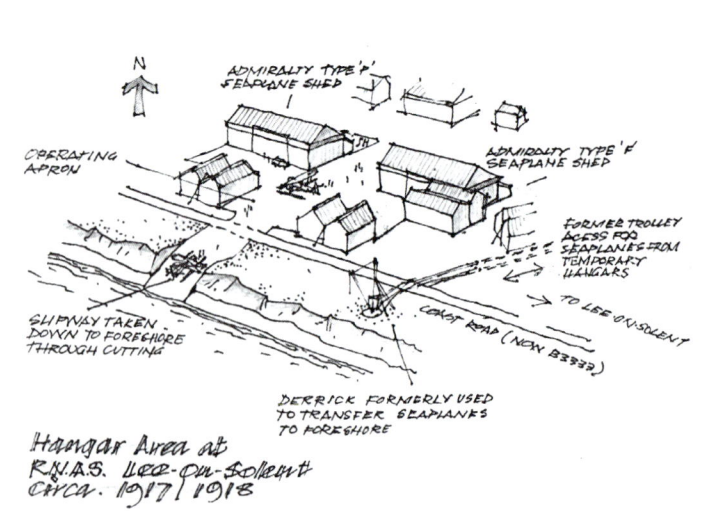

Hangar Area at RNAS Lee-on-Solent 1917/18   Jim Humberstone   Grade II Seaplane winch house   Martin Marks

270

Naval squadrons also used the airfield in the Great War. Expansion at Lee represented an extension of the Calshot seaplane base. Initially airmen were housed a few miles away at Warsash in the so-called "coastguard cottages". The airfield curtailed further development of Lee on the Solent as a seaside watering place. A decision was made to absorb the Royal Naval Air Service into the recently formed Royal Air Force. From April 1918 the airfield became No. 209 Training Station of the RAF. By 1918 there were just under 500 staff and 69 seaplanes and flying boats, but they were dispersed after the Armistice in November 1918 and facilities were gradually run down; in December 1919, the station was placed into care and maintenance.

Fleet Air Arm

In June 1920, following policy changes, the station re-activated for seaplane flying and aerial navigation training. A month later HQ 10 Group was established to control all air units working with the Royal Navy. In August 1920 the title School of Naval Co-operation and Air Navigation, RAF Lee-on-the-Solent was adopted moving again to the RAF Seaplane Training School, Lee-on-the-Solent in April 1921. In May 1923 it became simply the School of Naval Co-operation. In April 1924 the title of Fleet Air Arm (FAA) of the Royal Air Force – soon to be abbreviated to Fleet Air Arm - was adopted universally for naval flying elements of the RAF. Wykeham Hall which was built in 1895 served as the Fleet Air Arm heaquarters

In 1931 in furtherance of the RAF Expansion Scheme, a committee recommended the development of Lee involving the acquisition of 120 acres of land to the north of the existing site for use as an aerodrome. This land was bounded to the west by Stubbington Lane, to the east by Milvil Road which ran much further north than today to Peel Common and to the north by an east-west road joining these two. In January 1932, HQ 10 Group was disbanded on the arrival from London of HQ Coastal Command. Later that year a contract was awarded for the aerodrome and work began on erecting several aeroplane sheds as well as stores, barrack blocks, offices, galleys, sick quarters, dining halls etc.

In July 1936, the HQ of RAF Coastal Command was established on the site. In 1937 the Inskip Award recommended that the Royal Navy should no longer be an adjunct of the RAF but should have full control of its air assets. This led, some two years later to the official founding of the Fleet Air Arm. In December 1937, the School of Naval Co-operation moved to Ford, near Arundel and Lee became simply RAF Station Lee-on-Solent once again; the RAF were the first to use the reduced name. In 1938, Dunning Hangar was constructed and until closure in 1996, this hangar was reputed to be the largest ever built for the FAA. The following year the modern-day Fleet Air Arm was established, and HMS DAEDALUS was commissioned.

Lee-on-Solent was one of the first stations to have hard runways. The construction of concrete runways began in May 1939. Final runway lengths and orientation in 1942 were: 3,000 18-36 (which is the pilot's code for its $180_0/360_0$ compass headings in each direction), 4,290 (24-06) and 3,300 (11-29). To coincide with the construction of the runways in 1942, a new Admiralty designed control tower was built. During the early war years, some 23 hangars were added. Lee was at the very forefront of fighting activity. It was a key naval aviation base in WWII, especially for the D-Day landings. More airborne sorties were launched from here than from any other UK airfield on 6 June 1944. As a wartime Fleet Air Arm station Lee is notable for the fact that on 12 February 1942, Lt. Cdr. E. Esmonde led the Swordfish aircraft of 825 Squadron FAA from here to RAF Manston from where they attacked the German Battle Cruisers *Scharnhorst* and *Gniesenau*. All six aircraft were shot down with the loss of all but three of their aircrew. Cdr. Esmonde received a posthumous VC. On D-Day itself it was the busiest air station on the south coast: between 1939 and 1945 some 81 squadrons operating 21 aircraft types were based at Lee.

Sunderland Flying boat. Fleet Air Arm Airshow July 1952

Fairy Swordfish   Deane Clark

Post-war it was used for training, engineering and hovercraft testing. In 1946 the airfield was passed to the Navy as 'HMS Siskin'. Lee remained active on the fixed wing scene, albeit with mainly second line squadrons. It was heavily involved with helicopter work and trials. The establishment of the Air Engineering School brought a different emphasis to the base. It was later joined by the Central Air Medical School, the Naval Air Technical Evaluation Centre, the Naval Air Accident Investigation Unit and the Marine Aircraft Recovery, Transport and Salvage Unit. In 1959 the station was re-named HMS Ariel as this better reflected the transfer in of air electrical training. The name HMS DAEDALUS was re-adopted in 1965 in recognition of the fact that the station was considered the spiritual home of the FAA.

Attacker, Lee on the Solent   Deane Clark

During 1962 the slipway adjacent to the seaplane sheds was brought back into regular use upon the formation of the Interservice Hovercraft Trials Unit which primarily evaluated the amphibious hovercraft in military roles, initially amphibious landing roles and as fast response platforms. In 1963 the world's largest hovercraft SRN 3 was extensively trialled as a high speed, amphibious, load-bearing military vehicle and patrol boat by the British Inter-Service Hovercraft Team Unit based in Lee on Solent. On renaming as the Naval Hovercraft Trials Units in December 1975 much of the military interest in hovercraft interest was centred on its relative invulnerability to sea mines and thus its inherent advantage for use in mine-clearance roles.

Dragonfly   Deane Clark

In 1973, the closure of RAF Thorney Island saw responsibility for Search and Rescue transferred to Lee with 3 x Whirlwind Mk 9 helicopters. The RN retained this role at Lee until the award of the contract to a commercial company, Bristow's, in 1988.

The Hampshire Police Air Support Unit (HPASU) arrived at Lee in 1985. They were based in a now demolished complex of buildings on the eastern dispersal until 1996, when they took possession of the existing control tower.

Aerial view of Daedalus   Gosport Borough Council

In 1991 the decision was taken to de-commission the air station. Finally on 29th March 1996, HMS Daedalus closed after seventy-nine years of continuous operation. A whole range of possibilities for its development were voiced from housing and a business park, to gravel extraction, and even a horse racing course and a Hovercraft Museum.

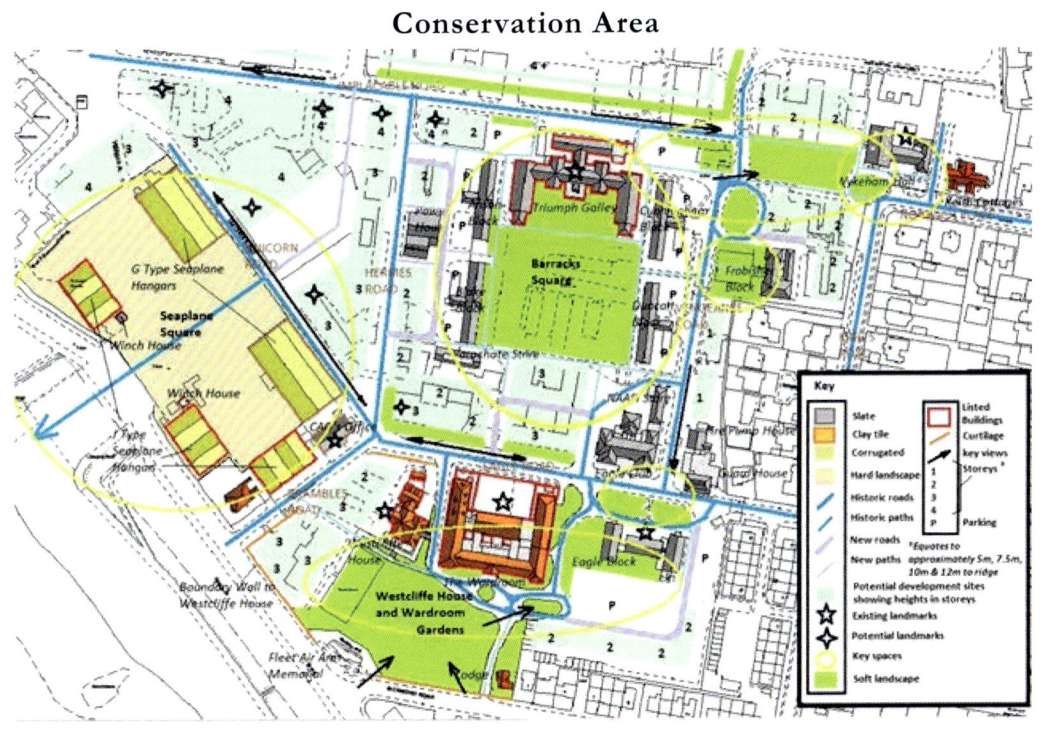

Plan of Daedalus   Gosport Borough Council

In 1999, anticipating that Daedalus would be closed and disposed of, Gosport Borough Council designated it a conservation area in order to protect the buildings and structures inside it.

Daedalus was later recognised as under-utilised for defence. The Ministry of Defence declared it surplus to requirements in 2004. In March 2006 the Maritime and Coastguard Agency (MCA) acquired the airfield which extended to approximately 106 hectares and the South-East England Development Agency (SEEDA)[9] acquired 82 hectares of land surrounding. SEEDA was one of several regional development agencies in England. It was set up as a non-departmental public body in 1999 to promote the region and to enable a few more difficult regeneration projects which otherwise might not take place. On acquiring their land SEEDA stated: "The lack of availability of serviced employment land and new business space has been identified as an important requirement in South Hampshire. Our intention is that development of the site will focus on new aviation and marine related businesses, exploiting access to the existing runways and the Solent. Plans are to create a quality business location that will attract inward investment and provide accommodation for start-up, growing and established businesses."

Aviation-related businesses, including an aircraft maintenance organisation, a microlight aircraft manufacturer and a flying school became tenants of SEEDA in 2006, as did the owners of around 50 aircraft based there. During its naval ownership, the airfield had been used for a variety of groups, including the Portsmouth Naval Gliding Club (PNGC) based at Lee since 1956. They occupy several buildings in the northern part of the aerodrome, with most of their gliders being accommodated in Bellman No 1 hangar.

The Lee Bees Model Flying Club, the Tigers Children's Motorcycle Display Team and two flying schools; several privately-owned aircraft were based at the airfield. When the RN moved out in 1996, operational management of the airfield was taken over by Hampshire Police Air Support Unit (HPASU). This airfield management continued until October 2010, when they became tenants of the Maritime and Coastguard Agency. In October 2010, the HPASU was closed and its tasks taken over by the newly formed South East Air Support Unit covering Hampshire and Sussex.

Hovercraft Museum

Hangar, Hovercraft Museum 2017   Martin Marks

In 1986, to try and save the last of the Hovertravel SRN-5's, Warwick Jacob approached the Hovercraft Society and successfully called in favours and sponsors to store the 38-foot hovercraft in various locations until a permanent site could be found. Formally registered in 1987 as a registered charity, the Hovercraft Museum Trust has concentrated on the restoration and collection of craft. The museum is run entirely by volunteers and houses the world's largest library of documents, some 15,000 books, publications, film, video, models, photographs and drawings on hovercraft, all of which are available for research. Some hovercraft manufacturers have deposited their complete archives with the museum for safekeeping.

Hovercraft Museum 2017   Martin Marks

In 2000 the Hovercraft Museum secured the safe storage of two of the enormous SRN-4's which had been plying their trade carrying motor vehicles across the channel for over 33 years. The hovercraft collection was spread over two WW1/WWII seaplane hangars with larger items outside. The main hangar contains an exhibition about hovercraft, with a range of examples.

The second hangar is used for maintenance.

The aerodrome was now strategically important. The growth of commercial air transport at Southampton Airport had left it with no capacity for general aviation (GA) aircraft. This left Lee-on-Solent as the only airfield in southern Hampshire with a hard runway available for general aviation. The slipway – which is not statutorily protected – could be essential if waterbombing aircraft are needed to put out fires in the New Forest.

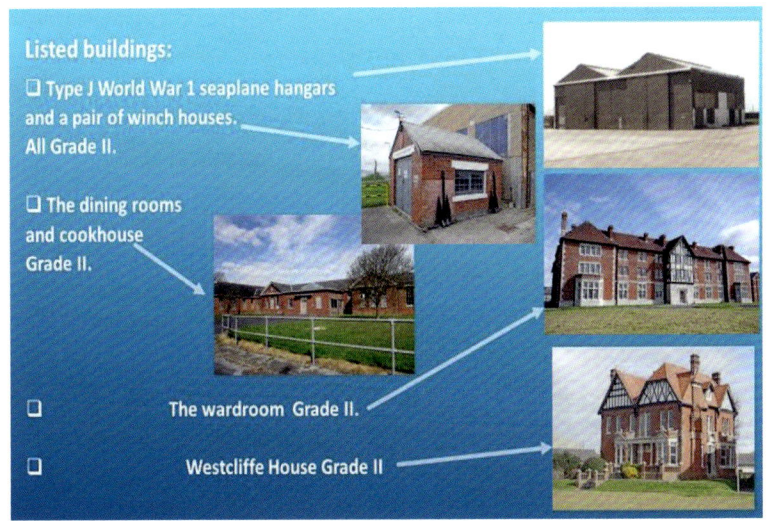

Listed buildings:
- Type J World War 1 seaplane hangars and a pair of winch houses. All Grade II.
- The dining rooms and cookhouse Grade II.
- The wardroom Grade II.
- Westcliffe House Grade II

Listed buildings Gosport Borough Council

In 2003 the government announced that they were considering using Daedalus as an asylum centre for some 400 male refugees. After a huge local protest and perceived difficulties converting the historic buildings as well as problems creating a suitable access point, it led to a decision to drop the proposal.

Bomb disposal

In early 2006, whilst conducting repairs to the runway, building crews discovered an unexploded pipe bomb, of over 60 feet long, placed underneath the runway by the military, designed to cripple the airfield's operational capabilities in the event of a German invasion. The bomb (along with 19 others subsequently discovered) was removed by October 2006. At one period, there were 265 pipe mines on the airfield, packed with a total of 2,400lb of gelignite. They were planted in the early 1940s so the airfield would be unusable if there was a German invasion. Documentary evidence had indicated that the site was free of pipe mines, but a review using modern detection equipment revealed that 20 mines were still buried in five areas of the airfield. Up to 3,000 people were asked to stay away from or stay in their Lee-on- the-Solent homes for a short period.[15]

The Princess Anne Hovercraft Museum 2017
Martin Marks

On 18 October 2007 airfield users were given 30 days' notice by the Airfield Manager that the aerodrome would be closed to all existing users except MCA, the police and PNGC. The closure was successfully challenged by Lee Flying Association which worked with other agencies such as the Civil Aviation Authority and AOPA to develop new operating procedures, an Airfield Manual and an air-to-ground service. The airfield then operated as a licensed general aviation airfield. In May 2008, the closure decision was reversed.

In 2007 SEEDA produced a development plan and organised a series of drop-ins and consultations.[10] From 1 April 2011 the airfield was leased to the British aircraft manufacturer, Britten-Norman under its airfield operations subsidiary, Fly BN Britten-Norman. It established corporate offices at the Daedalus Airfield site as well as a manufacturing base for its subsidiary Britten-Norman Aircraft and MRO facilities for two other subsidiaries, BN Defence and BN Aviation managed by Fly BN.

Solent Enterprise Zone

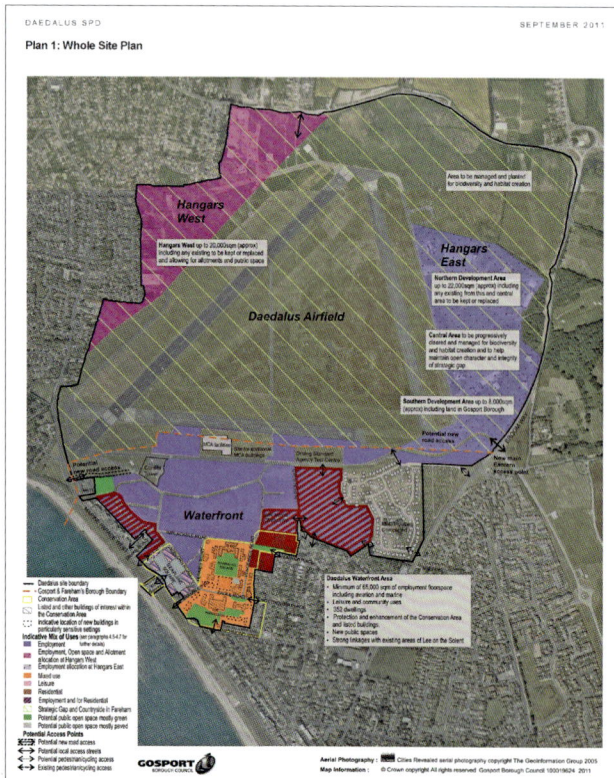

In August 2011 the government announced that the airfield would host an enterprise zone, the Solent Enterprise Zone,[6] one of 22 nationally. In 2011 Fareham Borough Council published its Vision and Outline Strategy for the section of Daedalus that fell within its boundaries, setting out a timetable from 2015 to 2020 onwards.[12] A month later Gosport Borough Council produced its Daedalus: Supplementary Planning Document to inform future planning decisions for the 38 hectares of the site within its area.

Daedalus Supplementary Planning Document September 2011 Whole Site Plan Gosport Borough Council

The intention was to transform Daedalus into a sustainable strategic business location, offering significant new job opportunities, particularly in aviation, high-tech manufacturing and marine industries as well as highly skilled jobs contributing to the area's economic growth and diversification. The design and use of existing and new buildings and spaces was to be of high quality to complement its identity as a business location, as well as preserving and enhancing the Daedalus conservation area and its listed buildings. The development was to be an identifiable place in its own right and to be of benefit to the wider community.[12]

In 2012 the coalition government abolished Regional Development Agencies. The South-East England Development Agency (SEEDA) closed at the end of March and the Daedalus site passed to another government regeneration body, the Homes and Communities Agency (HCA). On behalf of the airfield's new owner the airfield was prepared for licensed operations.

In March 2012 both Fareham and Gosport Borough Councils unanimously approved outline proposals for Daedalus. The development was expected to bring around 4,000 jobs to the area, mostly in the marine and aerospace fields. More than 200 homes were also planned, with shops, allotments and a 34-acre public open space. There were also plans to develop the airfield and improve the runway. The Solent Enterprise Zone was established[6]. Its main objectives were to:

- create 3,000 additional jobs on the EZ by 2026, contributing more than a third of the Solent local enterprise partnership (LEP)'s additional jobs target.
- promote an advanced manufacturing and technology cluster focused on marine, aviation and aerospace.
- utilise the incremental growth in business rates to unlock the full potential of this and other Solent employment sites.
- provide a catalyst for the regeneration of Gosport as it is the least economically viable area in South Hampshire.

Businesses starting up or relocating to an enterprise zone, qualified for enterprise zone relief. They had to start up or relocate by March 2018. They received up to 100% business rate relief for 5 years, up to a maximum of £275,000.

Hovercraft Museum at risk of closure

In 2014 the Hovercraft Museum was asked to close by the land agents for essential structural repairs to the listed hangars[13], reopening on 1st January 2016. In the same year there were fears that the last remaining cross-Channel SRN4 hovercrafts located at the museum site since 2000, but privately owned, were on the verge of being destroyed. Supporters started a petition to save them. The site's owner, the HCA[7] wanted to develop the land to create new homes. There was a protracted legal battle and more than 10,000 people signed the museum's petition in only two days, numbers later doubling. A deal was reached with HCA and the museum submitted a proposal to save one, "The Princess Anne" but scrap the other, "The Princess Margaret". By late 2016 the museum trustees had negotiated a three-year lease on the site, but their future will not be clear until the final phase of the Waterfront site has been negotiated. Grants are not easy to come by because the lease is so short.

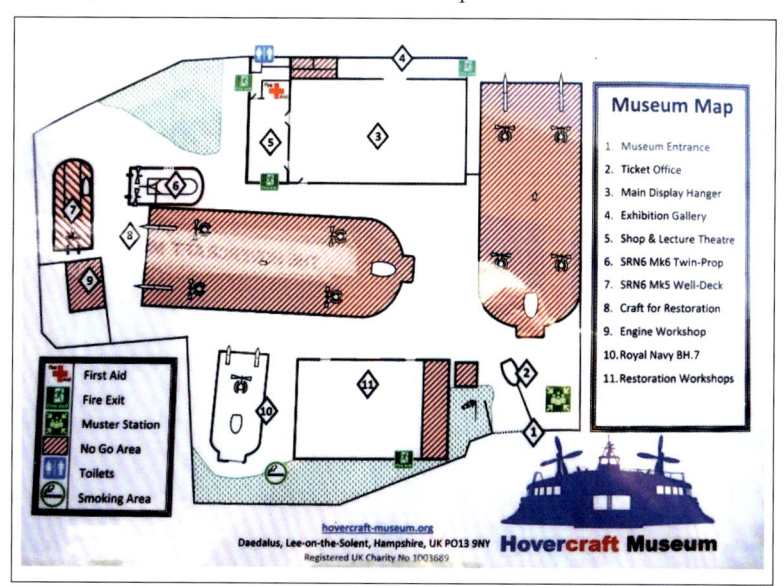

Hovercraft Museum map 2016

Warwick Jacobs BSc, FRGS and a founding trustee told Martin Marks: "Daedalus is the ideal historic home for The Hovercraft Museum as it saw 21 years of military trials and research. This in turn led to many forces using hovercraft worldwide. The Solent is the home of the hovercraft and still has the oldest commercial scheduled hovercraft service in the world. Lee's association with this great British invention spans 60 years!"[14]

The Gosport Society's publication to celebrate its half-century in 2020 included an account of the Heritage Action Zone with a reference to Daedalus Wardroom: "it is hoped that the evolving

Seaplane Square redevelopment could house a transport museum not only for hovercraft, but for other transport such as motor assets of the Provincial Bus Company." The article by Brian Mansbridge about the Hovercraft Museum echoes this aspiration: "a museum complex formed around Seaplane Square with common shared services and facilities based on a transport theme would be of real value, not just for the area, but for the nation..."

CEMAST

CEMAST, I Meteor Way, Broom Way Lane, Lee on Solent 2014

A Centre of Excellence for Engineering, Manufacturing and Advanced Skills Technology (CEMAST) was opened to students nearby in September 2014. It is the new home for Fareham College's Automotive, Aviation Engineering and Manufacturing courses. The site was planned to be the central training facility for more than 900 full time and part time students, and act as the main learning centre for students in apprenticeship programmes. Adjacent to it was the Daedalus Innovation Centre[8] which opened in April 2015 and offered office space and workshop facilities for new businesses, with a focus on the engineering, aerospace, aviation and marine industries. The project was aimed at creating some 150 new jobs. Within 12 months it reached 100% occupancy and in May 2016 proposals for a second centre were announced. It is managed by Oxford Innovation who also operate 17 other centres countrywide.

The ownership of the Airfield transferred to Fareham Borough Council (FBC) in April 2015. Regional City Airport Management took over as the airfield operators on behalf of FBC. It was given the name Solent Airport Daedalus.

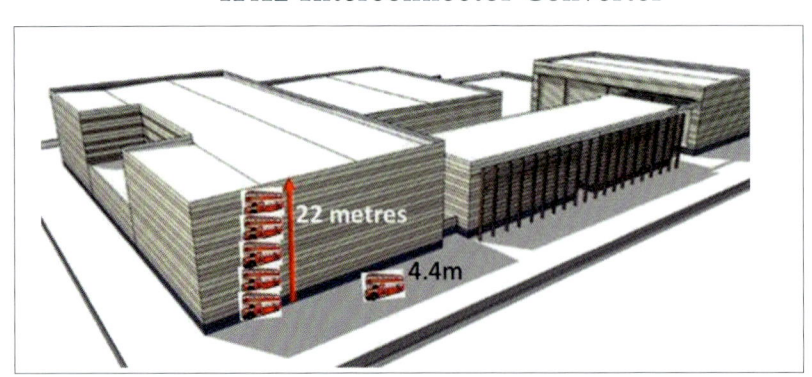

IFA2 Interconnector Converter

IFA2 Interconnector Converter

In November 2015 the National Grid announced proposals to build a massive IFA2 interconnector (Interconnexion France-Angleterre No.2) in the Fareham part of the site near Peel Common Roundabout. This was designed to supply electricity to/from France by means of cables under the channel. It required a transformer station at each end. This building was to be some 22m high and equal to four football pitches in ground plan, to house the huge inverters and transformers in a fully controlled environment. This caused much concern to residents over possible noise and electro-magnetic effects as well as the possible impact on the Enterprise Zone, airfield operations and its planned location in a Strategic Gap. An application for full planning permission for cabling and preliminary permission for the huge building was submitted in 2016 which received about 1,300 objections. The September date for the planning committee meeting was delayed to January 2017 following the issue of an independent report by Arcadis in October 2016 on the impact on airfield operations. This was not released to the public until December 2017, arousing concerns that it might have been adjusted to look more favourable. The final report admitted that there were areas of concern that would need to be addressed as the project progressed. The FBC Planning Committee waved through the application by 7 votes to 2, claiming that the breaches of its Local Plan and Strategic Gap rules was justified by the national importance of IFA2.[16] Four days later the same committee refused to accept two housing development applications as being "completely contrary to their Local Plan" because they were to be built in an area designated as countryside. The cable was energised for the first time on 15 October 2020 in preparation for testing before going live. At the same time another Interconnector Converter proposal by Aquind to come ashore near Fort Cumberland to tunnel on the west side of Langstone Harbour was strongly opposed by Portsmouth City Council and by local people. A public inquiry by the Planning Inspectorate was to determine the outcome.

Planning consents

Planning consents for the Daedalus site included:

- 300 dwellings as Married Quarters for the MOD. 148 of these were completed quickly.
- A new unit for the MCA as a helicopter base plus second phase as a rescue training facility.
- A new driving test centre
- Industrial Units on the East hangars site
- 3 New hangars adjacent to the old North South Runway
- A National Air Traffic Control Radar Site

Listed Buildings in the Daedalus Conservation Area

Gosport Borough Council owned the so-called Waterfront part of the site, which is included in a large Conservation Area, within which there are several listed buildings.[11] These are the Type J World War 1 seaplane hangars (Buildings Nos 31, 35 and 37) and a pair of winch houses – Grade II. Temporary repairs were carried out in 2014-16, though their ongoing state was not known. The hangers were partially waterproofed with to the extent of some re-skinning and putting polythene sheeting over the windows, making the sliding doors safe and working with replacement steelwork. They were made reasonably watertight and the roof panels were no longer blowing off. Since then the Hovercraft Museum made further improvements, although not having longterm tenure nor funds, they were limited in their efforts. The listed winch houses are intact; the westerly one is reasonably watertight and used for storage. The dining rooms and cookhouse (Building 91) – Grade II were externally not as bad as the seafront buildings and appeared weather tight.

Officers' mess and quarters (Building 118) – Grade II   2017   Westcliffe House   Martin Marks

Martin Marks was in the wardroom circa 2012 and found it in a very poor state – dilapidated, mould, wallpaper peeling, some minor structural damage. Contactors were seen in there and the open windows had been shut. HCA estimates it would take about £5m to restore it so that a potential developer would need to be offered something else with more potential for financial return to make it attractive. An obvious candidate is the MCA building in Marine Parade West, right on the seafront.

In the Gosport Borough Council document Daedalus Conservation Area Appraisal - March 2007 it mentions: "On the Daedalus site a few buildings of interest appeared, most notably Westcliffe House and its estate, Norbury House, Keith Cottages and Wykeham Hall." Norbury House was not listed but was in the Daedalus Conservation Area and a building of Historical Interest. It was not on the agreed demolition list for the site but was demolished without warning. Lee Residents' Association made a formal complaint, resulting in the HCA site manager apologising for the oversight and giving reassurances that the building was photographically recorded prior to demolition. At least HCA were made conscious that someone was looking over their shoulder in the future. The demolition was given retrospective planning permission in January 2015.

Westcliffe House (Building 119) – Grade II. the Residents' Association used it as a venue as part of Heritage Open Days some years ago – it looked better than the wardroom. Generally, it was not in bad condition as there has been some works internally and externally on the stonework. It was used for offices for a while.

The Gosport Borough Council/Fareham Borough Council boundary line runs close to the eastern edge of the airfield and along the southern perimeter road, recently improved to public highway standards and to be called "Daedalus Way". Interestingly, the local M.P.'s area extends from Gosport across the airfield west to Stubbington but not into Fareham.

The Future

Daedalus was split into several areas for development: Daedalus West - three triangles of land within the strategic gap - a strip of land separating Gosport and Fareham boroughs for a new access from Gosport Road and aviation and general employment use.

Daedalus East - an area off Broom Way & Gosport Road, a site for business units and possible aviation related industry. Following FBC's purchase of the airfield in 2015, its plans and developments progressed swiftly. It published a strategy for the overall development of "Solent Airport Daedalus" as it is now called. New-build industrial units on the east hangar's development site were built and in the same area new Airside Hangars were proposed adjacent to the old North-South runway.

The Waterfront - the area that is within Lee-on-the-Solent, thus Gosport Borough Council for planning consent, is the land south of the airfield, the hovercraft slipway and "Seaplane Square" including the Hovercraft Museum.

Proposals and outline planning permission were made for around 48,750 sq. m of new build employment space

- 21,240 sq. m of employment space in refurbished buildings
- 200 new homes and 36 refurbished homes (including starter homes)
- A residential care home with 32 units
- More than 1,000 sq. m of new retail
- More than 1,800 sq. m of community use
- Around 10,600 sq. m of leisure space
- Access to slipway and potential for a marina

The Homes and Communities Agency (HCA) planned to sell the whole site, but as this was not achieved the site was marketed and developed in phases. A separate and first phase was "Daedalus Park" a business and employment area. At the same time, work laying improved utilities and the upgraded East-West road, Daedalus Drive all contributed to significant and ongoing work areas. An early incomer in 2008 was a DSA Driving Test Centre.

DSA Driving Test Centre
Martin Marks

"Daedalus Park" phase 1 contains the business units next to the Barratt Estate. With new business already on site it looked busy as the developer progressed development on the western part, Phase 2. Once completed it was anticipated the whole site would provide about 250 jobs.

Plans for phase 2 units include bespoke facilities as well as several smaller business units. The laying of improved utilities and the upgrading the East-West access road, Daedalus Drive, was completed in 2017. This gave user and public access for both new and established business within Daedalus as well as a way into the next phases of development. Plans for the junction with Stubbington Lane on the seafront were revised to allow easier traffic flows, once funding issues were resolved. The junction included restrictions on the number of vehicles that can use the new access, at peak periods and for HGV's by laying a narrow road with tight curves. However, it did not state in the plans how any of these restrictions were to be enforced. 2017 heralded a move into the next major phase of development with further demolitions within the former base accommodation areas approved. There are, however historic buildings there such as Building 165 (Wykeham Hall – adjacent to the corner of King's Rd & Norwich Place) where the historic part was to be retained for redevelopment, but with the modern additions removed. Initially the work would be to provide internal roads and services for the planned 200 houses, about half to be the west of Drake Road and the other half towards King's Road. The land between was then to be developed as another business park.

Fleet Air Arm War Memorial

The final phase will be the most controversial, as it focuses on the remainder of the former Daedalus accommodation, offices, seafront hangars and 'Seaplane Square', including the Hovercraft Museum site. This development includes the seafront plots and all the listed buildings such as the old Wardroom and the land behind the Fleet Air Arm War Memorial. Until the developer announced final plans, the Hovercraft Museum was left in a state of suspended animation. Without certainty on the future of the site they were unable to apply for grants, putting a hold on any significant upgrades or development. In August 2019, Wates announced that Frobisher Block, Wykeham Hall and Keith Cottage, would be converted into 14 contemporary apartment apartments and six new houses, subject to planning permission.[16] There had been concerns that Keith Cottages would be demolished. The Conservation Area adjacent to the seafront is allocated to mixed use residential development. Remaining pockets of land at Manor Way / Stubbington Lane / Sea Lane are possible sites for residential use (except Manor Way land now owned by GBC for recreational use).

Round up

Grange airfield is no more having been swallowed up into what is now HMS Sultan and HMS Centurion. On the other hand, Daedalus, as described above, has seen exciting major development that has transformed it into a busy regional airport and rapidly expanding business development.

Sources

Fiona Callingham 2020 *The News* 'Portsmouth council spends £250,000 fighting Qauind plans' 7 October

Ben Fishwick 2020 *The News* 'Portsmouth residents set to stage protext against 'terrifying Aquind cable plans' 8 October

BBC News 10 October 'Portsmouth protest over Aquind Cross-Channel electricity link plan'

Gosport Society 2020 *Gosport 2020 For the Record: Essays in Celebration of the Half-Centenary of the Gosport Society*

Brian Mansbridge MBE MNI *Daedalus – the site* Lee Residents' Association http://www.leeresidents.org.uk/leeresidents_daedalus_site.html

Adrian B Rance editor 1981*Sea Places and Flying Boards of the Solent* Southampton University Industrial Archaeology Group with Southampton City Museums

Lesley Burton 2004 *Wings Over Gosport A Celebration of our Flying History* A Gosport Society Publication

LA Strange *2016 Recollections of an Airman Blackwells*

Ian Mackersley *2012 No Empty Chairs The Short and Heroic Lives of the Young Aviators Who Fought and Died in the First World War Weidenfeld & Nicholson*

Duncan Grinnell-Milne *1933 Wind in the Wires and An Escaper's Log John Lane*

BJ Hurren *1949 A Short History of Naval Aviation Ivor Nicholson & Watson*

References

1. Vivian Gibson 1956 *The History of Gosport Airfield*
2. Mark Barber *Royal Naval Air Service Pilot 1914-18* Osprey Publishing
3. Paul Francis *A History of HMS Daedalus* (RNAS Lee-on-the-Solent) Lee Residents Association
4. *The History of Lee on Solent Aerodrome* by Lee Flying Association http://www.eghf.co.uk/index.php/history-of-daedalus
5. Fleet Air Arm Archives http://www.fleetairarmarchive.net/daedalus/history_ww1.html
6. The Solent Enterprise Zone http://www.solentez.co.uk/
7. Homes and Communities Agency https://www.gov.uk/government/news/government- publishes-list-of-rda-assets-transferring-to-hca
8. Fareham Innovation Centre at Daedalus http://fareham-ic.co.uk
9. SEEDA plans for Daedalus airfield http://www.eghf.co.uk/index.php/aerodrome- news/171-seeda- planning-applications-for-daedalus
10. SEEDA Daedalus consultation 2007 http://www.secouncils.gov.uk/wp- content/uploads/pdfs/_publications/SinglePageStyle.pdf
11. Historic England - Daedalus listed buildings https://historicengland.org.uk/listing/the- list/results?q=daedalus&searchtype=nhle
12. Daedalus – A Vision and Outline Strategy visionhttp://www.fareham.gov.uk/PDF/business/daedalus/vision.pdf; https://www.gosport.gov.uk/media/677/Daedalus-Supplementary-Planning-Document/pdf/DAEDALUS_SPD_document_Pub_rsion_better_quality.pdf?m=637267929715670000
13. The Charity Commissionhttp://apps.charitycommission.gov.uk/Accounts/Ends89/0001003689_AC_20130630_E_ C.pdf
14. British Hovercraft Company http://britishhovercraft.com/blog/post/2016/10/19/Warwick-Jacobs-resigns-from-the- Hovercraft-Museum-Trust.aspx
15. Mine clearance http://www.dailyecho.co.uk/news/917379.display/
16. The many objections to the Interconnector carried no weight. The proposal for this huge building received accelerated planning permission from Fareham Borough Council, perhaps in response to anticipated site rental fees. The project conflicted with at least three elments of the Local Plan. Residents and other councils sent an appeal to the relevant minister. They were told that finalising planning approval would be held back until this had been resolved, but Fareham went ahead and granted permission, effectively bypassing the minister. Both organisations claimed there had been no such agreement. This led to a hardening of the government's rules on this appeal procedure to prevent a repeat but this was too late for this project.
17. *The News* Portsmouth https://www.portsmouth.co.uk/news/defence/abandoned-royal- navy-buildings- in-lee-on-the-solent-set-to-be-transformed-into-20-new-homes-1- 9026983

Chapter 12
What can you do with the harbour's legacy of military lines, batteries and forts?

Gosport

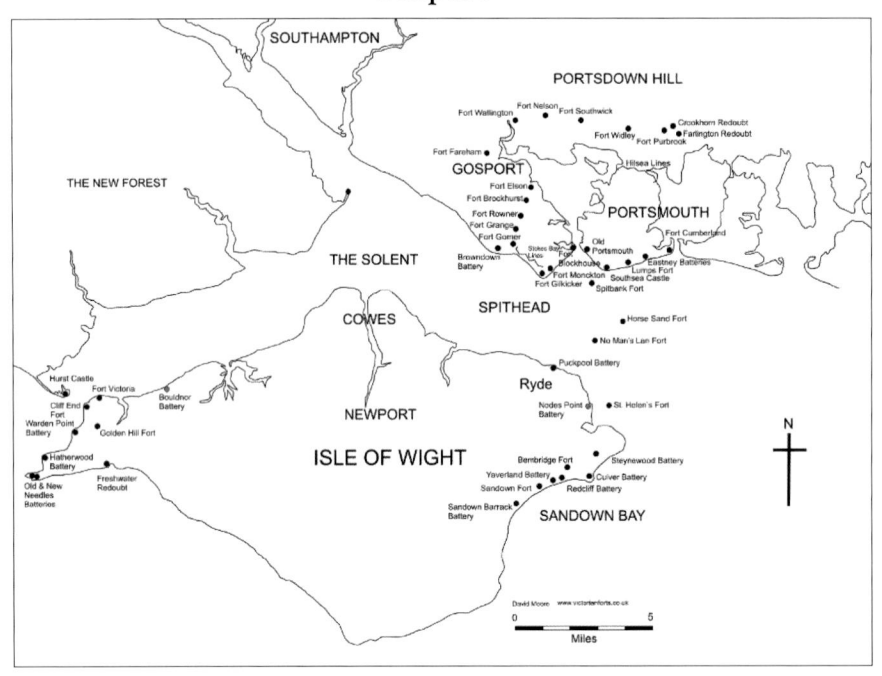

Plan of the Victorian defences of Portsmouth   Copyright David Moore

When and why were these many fortifications built around the harbour? The earlier defences are described in our second chapter. In these two we explore the rich legacy of forts, batteries and military lines built over the centuries to defend Portsea Island and the dockyard from the French – as well as their contemporary status. Several were the setting for important national events and technological advances, such as ship-to-ship communication and communications headquarters in WWII. We document their history, construction and their current uses and condition - in a west to eastern arc, starting with Fort Blockhouse opposite Portsmouth's Round Tower, followed by the Stokes Bay and Gosport defences on the west side of the harbour. The second forts chapter describes those on Portsdown Hill, the Southsea forts, Fort Cumberland built to defend the entrance to Langstone Harbour and the four seaforts. Between them these two chapters exemplify both the challenge of how to maintain these massive structures or whether to leave them to decay – and also the search to find appropriate new uses for the ones that are retained. It's surprising that more forts, batteries and lines have not been put to alternative uses, as they are fascinating structures, representing considerable embodied energy in their millions of bricks. However in view of their sheer numbers and the complexity of reusing them, perhaps it's to be expected that some - in Gosport and Fareham and along Portsdown Hill have disappeared. A few are in a state of 'controlled [or uncontrolled] ruination'.

Fort Monckton is still in active defence use - as No. 1 Military Training Establishment, operated by the army, the only fort around the harbour to be retained by the Ministry of Defence. Others have found new life in a variety of creative ways. No. 2 Battery in Stokes Bay Gosport is an acclaimed Diving Museum; Forts Widley and Purbrook offer young people climbing walls, horse-riding, athletics - as well as a Cold War command post; Spitbank Fort and No Man's Land in the sea at Spithead are luxury event venues… and Lumps Fort in Southsea encloses a lovely rose garden.

The importance of Gosport in national defence

In the 2018 conservation area appraisal for Haslar Barracks, Rob Harper, Gosport Borough Council's Conservation Officer described how important the western side of the harbour was to defence of the realm:[1] "Gosport, on the western side of Portsmouth Harbour, was a highly strategic area and recognised as such by the development of fortifications from as early as the 15th Century, with extensive defences added in the 17th to 20th Century in the form of earthworks, redoubts, forts, moats, coastal fortifications and anti-aircraft defences."

"By the time of the Napoleonic Wars it was a thriving settlement with many active military sites supplying the navy with food and drink, transporting and encamping troops, caring for sick and wounded soldiers and sailors, constructing small to medium sized naval and private vessels, supplying rope and sails, developing naval ordnance and being the last resting place for many military personnel. The extensive fortifications meant the town had a permanent garrison and its strategic location led to it being the home (for long or short periods) to many branches of the military, including: artificers, engineers, sailors, artillery personnel and many infantry regiments."

"The Haslar peninsula is extremely rich in military heritage including: Fort Blockhouse, Haslar Hospital, Haslar Gunboat Yard, the former HMS Hornet and the Admiralty Experimental Works (QinetiQ). The significance of the area was such that the entire coastline was remodelled and reinforced with stone facing walls at the end of the 18th Century to protect it from coastal erosion."

"To the immediate north of Haslar Barracks is the settlement of Clayhall, which in its current form developed during the 19th Century but which replaced an earlier medieval settlement known as 'Haselworth'. Further south-west are the extensive fortifications relating to Fort Monckton, Fort Gilkicker and Stokes Bay: all reminders of the strategic significance attached to this stretch of coast and the vulnerability posed by the potential to land large invasion forces along its beach."

"The Napoleonic Wars was a pivotal moment in the historic development of Gosport and resulted in extensive redevelopment of the town centre and the expansion and in the mid-nineteenth century the harbour's defences were greatly reinforced."

Bricks for the land forts' construction were generally manufactured in temporary brickworks as close to the point of use as possible and transported to the sites by temporary railway lines. The area has a wealth of good brickmaking clay. As early as 1785 land at Stokes Bay was acquired by the Royal Ordnance and used to produce bricks.

Fort Blockhouse, formerly HMS Dolphin

In March 1338, a large French fleet of small coastal ships crossed the Channel into the Solent where they landed and burnt Portsmouth. The town was undefended; the French attack was not suspected as they sailed towards the town with English flags flying. They attacked Portsmouth again in 1369, Gosport in 1372 and 1377. In response to the French blockade of Southampton Water in 1416 a chain was laid across the Portsmouth harbour entrance between the Blockhouse and the Round Tower. The blockhouse was probably built or rebuilt in stone in about 1495 to help protect the investment made in the building of a new dock at Portsmouth. It was upgraded and later replaced by an 8-gun bulwark in 1539 during the reign of Henry VIII. It remained in use until at least 1603.

As explored in Chapter 2, early in the Civil War, shots fired by Parliamentarians from Gosport damaged the tower of St. Thomas's church and Southsea Castle, and days later the Royalists surrendered Portsmouth. This action was a pivotal point in development of the Gosport defences. It was realized that Portsmouth could not be held unless Gosport was held as well. For 150 years Blockhouse had just a palisade fence facing Portsmouth. Leake's Battery was constructed on Portsmouth Point in the 1680s, consisting of massive 42 pdr. guns which, with a small traverse, could cover Blockhouse Fort, Haslar Lake, or even Gosport itself. The 21 gun battery formed part of the Queen Anne rebuild and part of those fortifications and the adjacent moat, seawall and guard house are still in place today – as are the 1790s building housing the furnace for heating shot and the reformed right demi-bastion from the French Wars. Most of the rest is 1840s and later. The use of casemates here is recorded as the first instance of their use in England since the time of Henry VIII, during the French War rebuild when main battery was reduced to 13 guns - also still visible although modified when a second tier battery was constructed over the top in the 1840s rebuilding.

Blockhouse was considered obsolete by the 1859 Royal Commission on the Defence of the United Kingdom. In the 1870s it became the headquarters of the Portsmouth Company of the Submarine Mining Engineers. It was turned over to the Royal Navy in 1905 and developed as HMS Dolphin, the submarine base. As Dolphin expanded beyond the lines of the original fort, a ferro-concrete jetty, a memorial chapel probably built in 1917 and a wardroom, partly of 1847 but mostly in place by the end of WWI were added. Paintings by Harold Wyllie of the fleet at sea were specifically designed for the mess as a war memorial to those lost at sea in Thames block on the other side of the parade ground. In WWII they were considered of such significance that they were taken down, transported across the Atlantic in the battleship *Resolution* and given into the safekeeping of United States submariners at New London. The prominent Submarine Escape Training Tank described below was built from 1950-54.

The submarines had left by 1994, although the training school remained on site for a further three years. HMS Dolphin was decommissioned in 1998 and reverted to its old name of Fort Blockhouse when it became the home of 33 Field Hospital. It is currently manned by personnel from the Royal Army Medical Corps, with Queen Alexandra Royal Army Nursing Corps, Royal Army Dental Corps, Royal Electrical Mechanical Engineers, Royal Logistic Corps and Adjutant Generals Corps within its ranks. The Portsmouth Harbour Entrance Range (HER) is located in Fort Blockhouse. This unmanned range automatically gathers magnetic signatures of vessels entering and leaving Portsmouth Harbour, used by QinetiQ, the privatised research agency.

A side effect of the dredging for the supercarriers is erosion of the Hamilton Bank. At Blockhouse, the consequent foreshore erosion and storm damage has caused partial collapse of the seawall and potential undermining of the fort itself. For three hundred years it has been recognised that a breach of the wall here, while immediately leading to flood risk for Gosport, would ultimately lead to loss of the harbour, through loss of the essential scouring effect of the ebb tide. Undermining of the Fort itself, built as it is on sand and shingle, would be an immediate and potentially catastrophic risk to the harbour. To date, rock armour is being used to provide emergency repairs, and the Defence Infrastructure Organisation [DIO] expected a full assessment to completed by the spring of 2021. No permanent repairs were planned nor were any efforts known to quantify the risk from Hamilton Bank erosion. These were urgently required and were likely to be very expensive. Meanwhile the planned sale of the establishment was unlikely while such uncertainty prevailed.[8] The DIO's review of Fort Blockhouse's assets was complete, but plans to release the site were put on hold in 2020, apparently due to shortage of shore-side crew accommodation, for example for the crews of the Portsmouth-based supercarriers *HMS Queen Elizabeth* and *HMS Prince of Wales*.

In preparation for the sale of Fort Blockhouse Fort Blockhouse/Haslar Gunboat Yard [2] formed part of the wider Haslar Peninsula Regeneration Area that also includes the 23 Royal Hospital Haslar site, plus Haslar Marine Technology Park that is home to QinetiQ. These four land areas are now known as Blockhouse 1, 2 and 3, owned by the MOD; the Haslar Boat Yard is owned by Starvale Developments. We now explore their current uses.

The SETT   Martin Marks   Submarine Escape Tower inside the tank   Mark Marks

Blockhouse 1 (former HMS Dolphin) includes the Submarine Escape Training Tank (SETT), now a Grade II Listed Building. Following several submarine accidents and the Ruck-Keene report of 1946, the Admiralty Board decided that safety training should be given to all submariners and the SETT was constructed as part of the facilities for this purpose. As built, the tower had a different arrangement of windows from that of today, indicating that there have been alterations believed to have been in circa 1995 when the tower was also re-clad. The SETT was a training facility to teach submariners how to escape from the pressurised environment of a submarine in a safe manner. Students would be sealed in a watertight compartment at various depths in the tank (30, 60 and 100 ft), the compartment would then be flooded, and the student would make his escape. Training allowed the student to build up tolerance, repress any instinct to panic and to instinctively, in an emergency, respond to commands with appropriate actions. Air filled observation ports in the tower allowed instructors to monitor the students.

This course was a requirement for all personnel before boarding a submarine. Co-author Martin Marks went through in about 1963. The SETT was used not only for Royal Navy training but also for Special Forces, submariners and SETT staff from other countries. Arising from changing submarine operations where escape of this nature would not be possible as well as health and safety concerns, Royal Navy training ceased in 2009, although staff maintained the capability to demonstrate using the tank. The SETT continued to operate as a classroom training facility and as a wet facility for sub aqua clubs until late 2012 and still had a very occasional use.

Design drawings and photographs of the building under construction indicate that piles support a concrete base slab and that it is of steel-framed construction. The circular tank is freestanding within the tower, supported by its own weight and that of the warm and chlorinated water it contains. In this last sentence lies a potential problem as draining of the tower could lead to structural problems. It is understood that the MOD intended to withdraw from the facility by 2019. Potential for reuse is restricted by its listed building status, its significant day-to-day operating costs and limited scope for alternative use. There are commercial interests who would like to take it on. It is identified by Historic England as a Building at Risk. Suggestions to make use of the site have included a public trust set up for a redeveloped visitor centre, use as a commercial diving training facility and use for sub-aqua club leisure diving. It would have to comply with current safety building regulations before it could be used by the general public, which might prove expensive to achieve.

Blockhouse 2 includes the Royal Navy Submarine Museum and the Joint Services Adventurous Sail Training Centre (JSASTC). The museum is a popular tourist attraction now part of the National Museum of the Royal Navy. A waterbus from the dockyard brings tourists across the harbour to see the museum and to explore the conserved *HMS Alliance* submarine, hoisted into the air on blocks instead of being immersed in its usual element, the sea. Exploring its interior crammed with machinery reveals the cramped conditions experienced by submariners.

HMS Hornet

Haslar Gunboat Yard entrance and perimeter wall 2019

Blockhouse 3 (former HMS Hornet, no longer active militarily from 1974) is separated from Blockhouse 1 and 2 by Haslar Road. As a highly valued site, it is enclosed on two sides within a high red brick wall with a corner sentry house and parapet for patrols and lookout posts. Inside is Haslar Gunboat Yard which contains the historic gunboat sheds constructed in 1856 for care and maintenance of the new type of gunboats, a term 'most often applied to a class of steam-powered ship, modified during the Crimean War, for navigating in shallow waters close to shore.[4]

Gunboat sheds   Martin Marks

288

"Royal dockyards have been credited with pioneering the use of large span metal roofs",[6] including the very large covered slip of 1847-8 designed by Col. Green in Portsmouth dockyard, now sadly lost. The columns of the sheds were slightly raised in 1873. The boats were winched from Haslar Creek on cradles and set on railed steam-driven traversers to move them in and out of their berths.

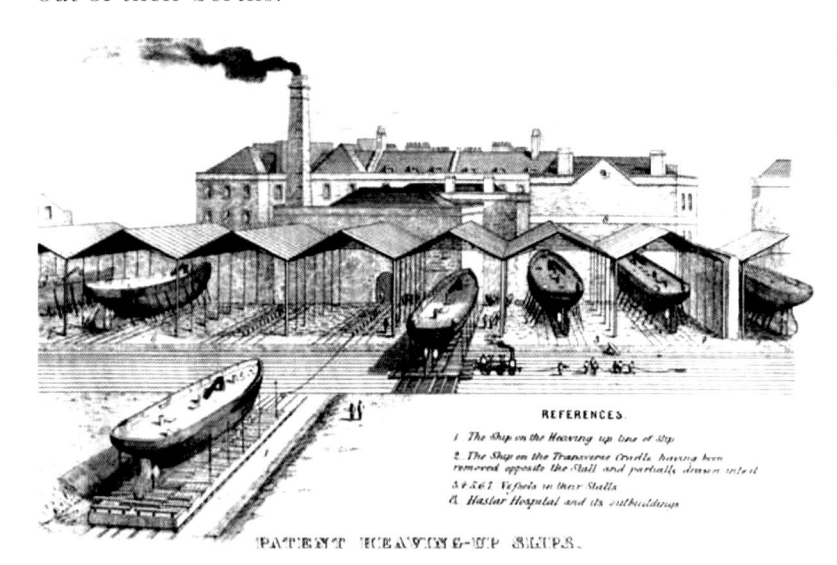

Patent Heaving-Up Slips' at Haslar Gunboat Yard (*Mechanics' Magazine*, 3 January 1857)

This amazing facility can best be described as dry-stack boat storage some one hundred and sixty years before its time! Its origins can be traced back to the Crimean War (1853-56) when Britain found a use for small, shallow draft but heavily gunned boats that could take on coastal forts and hence the description "gunboat". Large numbers were constructed but the end of the war brought this expansion to a halt. Thus, Britain had a large fleet of these vessels with considerable potential for protecting her coasts but their construction did not make them suitable for storage afloat and dry docks were not cost effective to store them in, especially as there were few and they were needed for larger warships. After some consideration, it was decided that a dedicated shed storage facility was needed and a site just to the north of Haslar Hospital was chosen. It was close to the naval dockyard at Portsmouth and abutted an existing creek connected to the sea.

A large number of specialist companies were employed to construct the sheds and boat lifts including Thomas White Jnr (1796-1863)), father of the Isle of Wight boat builder John Samuel White. Thomas had installed a slipway in his Forton Lake boatyard nearby and later specialised in building patent slips. The Haslar design included a steam powered, railed, traverser system using wheeled platforms carrying boat cradles enabling the vessels to be moved from the creek into the sheds. Overall control of the project was given to Mr. William Scamp, Deputy Director of Engineering and Architectural Works at Somerset House, who also designed the Naval Bakery in Malta. It is a unique construction. Ten parallel open-ended sheds survive (of the original 40) of wrought- and cast-iron construction with thin cruciform columns and slender roof trusses.

Trials in 1856 were not successful as there was not enough power available but in 1859, after a series of modifications, *HMS Cherokee* was successfully moved from her shed to the water in about 38 minutes. By 1860 there were 40 gunboat sheds on the site and the yard, an important military facility, was surrounded by a brick security wall with lookout towers, a guardhouse, a police barracks and a house for the Master Shipwright who oversaw the yard.

During the 1860's some of the sheds were transferred to Portsmouth Dockyard and there were reports that the Haslar yard would be closed. In the 1880's funds were allocated to build more sheds but by 1906 all the gunboats were scrapped as obsolete. It became a maintenance yard for Coastal Forces Patrol craft. During both world wars the yard was used to service motor torpedo boats and motor gunboats. In 1939 to 1956 this was done in association with the adjacent shore base HMS Hornet, which included part of the old yard. HMS Hornet closed in 1974 as a direct result of construction of the new Haslar bridge which was too low for craft to pass underneath. The traverser system was removed and the Master Shipwright's house demolished. The Yard has been disused since 1974. Some more sheds were removed in the 1980s.

Gunboat Yard Shed
Martin Marks

What has happened to the eastern, harbourward area is explored below. The section of the Gunboat Yard closest to QinetiQ was controlled by them from 2001. There are ten sheds remaining. The site is listed as Grade 1 for its special historic interest as well as being a scheduled ancient monument.

In 2003, 2004 and 2009 BBC Television ran the Restoration series of programmes where viewers decided which listed buildings in immediate need of remedial works was to win a grant from the Heritage Lottery Fund. At the same time as the first series, BBC South commissioned eight local programmes. As presenters Celia and Deane Clark visited and commented on key decayed Hampshire and West Sussex sites, including Haslar Gunboat Yard. "We were filmed shuffling up and down Brunel's shunting lines which were used to move gunboats in and out of the sheds. We were shocked to see the roofless state of the two lodges on either side of the entrance. We interviewed two WRNS who had grown up there, as we all balanced on exposed joists."

In 2015 Historic England, concerned at the underuse and decay of important local defence sites, commissioned the Portsmouth Harbour Hinterland Project from Museum of London Archaeology, which reported in February 2016 and presented its findings in Priddy's Hard in May 2017. Its aim was "to enhance understanding and heighten awareness of how the Portsmouth hinterland has developed as a result of the presence of the naval base (including the docks and the associated establishments of Royal Clarence Yard, Priddy's Hard and Gun Wharf), in order to assist local decision making, planning, development and management of the historic environment."

Historic England followed this up with a further project whose aim was to help particular historic sites towards achieve sustainable futures. One of these was Haslar Gunboat Yard. In 2014 the local Starvale Developments bought the western part of yard from QinetiQ to develop for business and residential use. They were hoping to have the site delisted to a Grade II status to simplify development. Advertising for developers to consider the way ahead for the area around the yard resulted in a firm of architects proposing: "…to create a centre for the historical preservation of the region's naval heritage. The site will be composed not simply of passive museums with 'static' exhibits, but include premises for the maintenance and reproduction of historic vessels; an 'active' history – an ongoing process of historic discovery and preservation, that will in itself be the cultural attraction."

The boat sheds were to be converted for business use. Up to 10 new homes and several business units would be built on the 5.5-hectare site. They were reported to be asking English Heritage to reclassify the gunboat sheds, downgrading their listed status. This seemed unlikely.

There are precedents for schools of traditional boatbuildng and repair in former naval establishments in Helsinki and Stockholm. However, the proposers for Haslar do not seem to have done much local research, as the nearby Maritime Workshop on the south shore of Forton Lake closed down its main workshop for the restoration of historic vessels through lack of work in 2016. It had operated there as the Steam Launch Restoration Group from 1973 and from 1983 as the Maritime Workshop. No progress was apparent by October 2020.

In March 2017 there appeared to be a move to reduce the QinetiQ occupancy of the site with Buildings 10, 12 and 23 available to let. In October the same year concern about deterioration led to Haslar Gunboat Yard being added to Historic England's Heritage at Risk Register. In February 2018 Starvale proposed 17 new homes, an art gallery and café in the gunboat sheds, a flood defence seawall and a new public slipway designed by their architects Knight Architectural Design.

Gosport's Modified Sustainability Appraisal to support the Gosport Local Plan (2011-2019) has a useful map which identifies it as a potential site for employment, with a public access path along the waterfront. But it was unlikely to contribute affordable housing. There was concern that the reuse of the sheds for housing by breaking them into separate units and the associated changes that would be needed could harm the historic building. Ideally maintaining it in its current form or as part of a marine employment use (such as boat sheds), or to display gunboats or other similar features was seen as preferable by Gosport Borough Council. The property was advertised for sale freehold in spring 2020.

Haslar Creek
June 2019

Police Barracks temporary roof June 2019

The poor condition of the Police Barracks was to be tackled by covering the building to dry it out. A Heritage Lottery bid was to be made, with a viability appraisal and advice from Gosport Borough Council's Business Development department and opportunities to develop skills with Fareham College. The Historical Diving Society, established in 1990 as an educational charity with research and consultancy capability run by volunteers and associated with the Diving Museum is also based there. The police barracks, gatehouses. Watchtower and perimeter wall were in a bad state when Deane and Celia visited the site in June 2019. Reroofing of the two lodge houses and repair of the police barracks was under way, but no work was being done on the eastern perimeter wall.

Northern gate lodge Haslar Gunboat Yard 2019

Watchtower June 2019

We now explore the more positive history of the eastern, harbour-wards section of Hornet. After the Royal Navy disbanded coastal forces in 1956 HMS Hornet was left to deteriorate. With the tacit approval of the Queen's Harbour Master naval personnel began to keep their yachts there. Keen yachtsmen on the Flag Officer Naval Air Command at Lee-on-the-Solent proposed to the Commander-in-Chief Portsmouth that a naval yacht club be formed for serving and retired naval personnel. There were few facilities, but the Hornet quartermaster's hut, the slip and workshop – one with its engine lift gantry, ideal for lifting masts – remained. A clubhouse was established in the old wardroom and the Captain's House was made into a flat for a barman and his wife. By

Building 98

1970 there were about 120 yachts berthed in Hornet.

In 1972 as facilities for servicemen overseas were closed, the MOD decided that the three services would each run an adventurous training facility in the UK. The Navy became responsible for sail training. The Joint Services Sailing Centre moved into Hornet. Nine Nicholson 55s, nine Contessa 32s and nine Halcyon 27s were berthed alongside newly piled pontoons. These were gradually replaced by rejects from the dockyard. When the club's future tenure of the site was under threat in 2011, the DIO recognised the role the club played in supporting servicemen and their families and renewed their lease in 2015 for 25 years, greatly increasing the area available, including the whole area 'over the road' as far as the QinetiQ boundary.

The club was renamed Hornet Services Sailing Club and increased its membership to include personnel from all three services, their veterans and families. It provides practical qualifications training to members of the armed forces as well as shore-based courses, and enjoyed 'close engagement' with Historic England, according to an on site notice seen in 2017. For the Development Phase – the Conservation Management Plan a tender process and appointment of a lead professional/Conservation Accredited Architect, structural engineer, QS, timber consultant, ecologist and builder were appointed. Building 98, an old boathouse which had been part of the Sailing Club lease has been returned to the MOD.

The Joint Services Adventurous Sail Training Centre (JSASTC) and the Hornet Sailing Club now use it jointly primarily as a boatyard. The sailing club has a 25-year lease until 2040 with the MOD, plus an option to renew for a further 25 years, for the exclusive use of Blockhouse 3. Thankfully in 2019 the decayed buildings were the subject of a Conservation Management Plan. The guardhouses and barracks were covered in a temporary roof to dry out the structure. A Heritage Lottery Bid was in hand to finance their use to accommodate the Gosport Sea Cadets and provide facilities for JSASTC staff and other groups.

In 2019 the MOD notified Gosport Council that it might release the Blockhouse sites, although timescales and details were not currently known. These special areas have scope for sympathetic development, although listed building status, traffic restrictions from the single track Haslar Bridge, possible flooding risks and the existing site users' rights outlined above to some extent limited the options. On a more positive note there is deep and sheltered water close to the Solent; a cluster of marina and other technology businesses close by as well as on-site assets which may be suitable for reuse – for example the SETT and unwanted MOD accommodation.

Stokes Bay Lines and batteries

Now we explore the defences of the vulnerable Gosport shore. Walking along the paths thickly wooded with evergreen oaks, you are aware that the slopes and ditches have been deliberately shaped. On-site signs give you an idea of how important this coastline was to the country's defence. The wide-open space to seaward is as impressive as the other former open military area, now just as valued: Southsea Common. To commemorate D-Day 75 in June 2019, the field was filled with a replica Spitfire, Hurricane, an American field kitchen, WWII vehicles… manned by re-enactors: soldiers demonstrating contemporary armament, WRVS… actors jitterbugging to Glen Miller, historic boats speeding or chugging along the Solent – and local people enjoying this evocation of a key international event that began on these shores.

The earliest defences of Stokes Bay are attributed to King Henry VIII, who built Haselworth Castle, probably on the site of what is now Fort Monckton. The need for a more robust defensive system in this area was recognised as long ago as 1587 when the threat from Spain encouraged the Earl of Sussex to request defences centred on the area of Haselworth Castle including damming the River Alver.

It was not until 1779 however, when the American War of Independence provided the spur to addressing the defences of Stokes Bay. Fearing that the war would provoke an attack on England by France or Spain, the first Fort Gilkicker was constructed. Redoubts were also built along the coast of Stokes Bay in 1782-3. Fort Gilkicker was rebuilt in 1789-90 on a new site and renamed Fort Monckton, just before the French Revolutionary War of 1793.

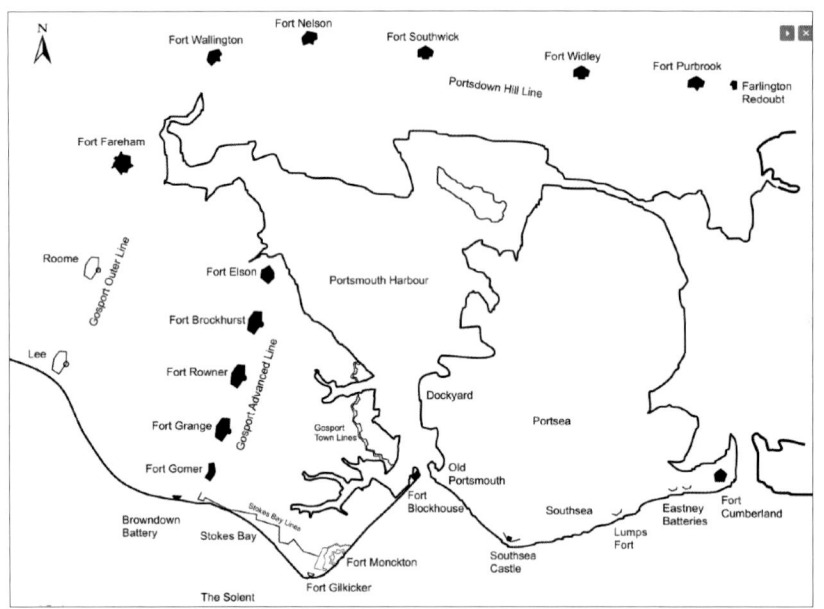

Gosport Lines map

In the 1840s and 1850s the first Browndown Batteries were constructed; two at Browndown Point and one at Gilkicker Point, prompted by the possible warlike intentions of France. As mentioned earlier, in the 1860s it was recognised that the guns of shore defences would have to be upgraded to deal with the new advances in shipbuilding and gun design. The Royal Commission of 1860 set up to consider the Defences of the United Kingdom in relation to the new rifled guns recognised that Stokes Bay was an ideal beach for possible invasion landings.

In 1857 Major Jervois had proposed a system of ramparts, moats and batteries to close the gap between a new fort inland at Gomer and Fort Monckton, which were to become the Stokes Bay Lines. The work was started in 1859 and the Lines ran from the rear of the Browndown Batteries in the west to Fort Monckton in the east. They consisted of a rampart with a road to its rear, a wet ditch or moat to the front 60ft in width and 9ft deep at high water of spring tides. Fort Gilkicker and the batteries at Browndown were to help in the defence with a series of five-gun batteries with overlapping fields of fire.

No. 1 Battery, Stokes Bay[6,7]

No 1 Battery guarded the Browndown Road, the only road to Browndown Camp along the coast. It covered the rear of a dam constructed to flood the Gomer marshes in time of attack by closing off the River Alver. The battery was first armed with 8-inch Smooth Bore guns firing through embrasures. By 1891 these were replaced by two 7-inch 92 cwt. Rifled Breech-loading guns. There were two ready-to-use magazines, one for shells and one for cartridges. There was a tunnel connecting No 1 Battery to the parade of No 2 Battery via a bridge across the moat.

294

The battery was damaged in the 1930s when the east end was removed during widening of Browndown Road. The northern part of the battery was also removed at the same time. It does not appear to have been reused during WWII. Little trace remains apart from the tunnel and short section of rampart with concrete revetments. In 2010, a set of granite blocks that held the gun racers for one of the gun emplacements were uncovered and were re-sited within the park. In 2012, the battery was listed. The southern part of the site is now a council owned mobile home park which abuts No. 2 Battery.

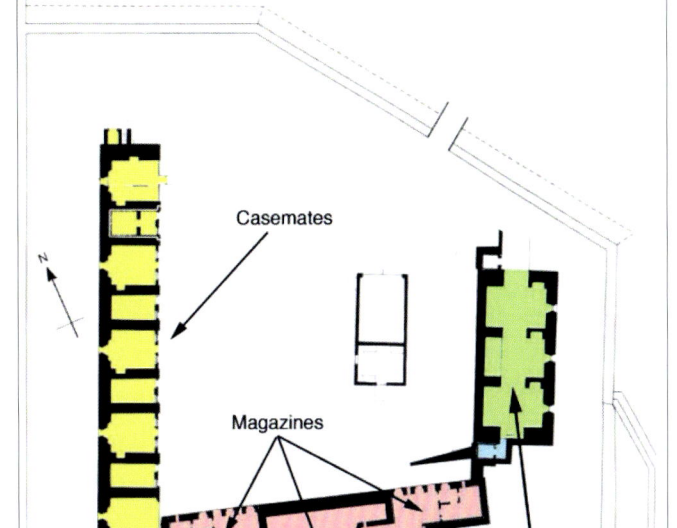

No. 2 Battery, Stokes Bay[7]

No. 2 Battery Layout in 1900

No. 2 Battery was a Grade II Listed Building that was upgraded to II* in 2017. It is the best preserved of the original batteries in the Stokes Bay Lines. This list is an account of changes to it over time.

1860: No. 2 Battery completed. It mounted 15 64 pounders on traversing carriages trained both eastwards across Stokes Bay, and westwards across Browndown. It never fired a shot in anger.

1891: The two sea facing emplacements built higher up. 1860 – 1902: Various guns installed/replaced.

1902: All guns replaced by 19 Maxim machine guns. 1907: All guns removed. **1932**: Gosport Borough Council (GBC) purchased the Battery.

1939: GBC moved all records from Town Hall into the Battery.

WWII: Requisitioned by the military. Used as a barrage balloon centre and for plane spotting. Provided a defence for the Canadian Forces who left Stokes Bay on D-Day for the "Juno" beaches of Normandy.

1947: Used by Special Armament Development Establishment (SADE). Testing of tanks and amphibious vehicles – there is a concrete ramp into Stokes Bay on the adjacent beach.

1950: Used by 7th Royal Tank Regiment Amphibious Wing

1951: Released back to Gosport Borough Council

1956: The moat surrounding the Battery was to be filled in.

The minutes of Gosport Borough Council (GBC) dated 1st April 1949[9] included a report from the Borough Engineer, who stated that as artificial waterways "they had no natural beauty, are not suitable for boating or fishing and the water looked dirty and uninviting". Appreciation of our military legacy has definitely changed for the better since then! When filled an area of nearly 17 acres would be added to the area and the intention was to fill the moats with refuse so that the cost will be small in proportion to the advantages derived. As to No. 2 Battery he advocated the filling of the moats around the battery and the piecemeal demolition of the battery itself, when an area of almost seven acres would become available for any scheme that appealed to the council. Fortunately, the latter never happened.

In 1982 No. 2 Battery was transformed into GBC's nuclear bunker and a Civil Defence Command Post. A new passage was cut through between casemates and lined with concrete. Several blast proof doors and an air filtration system were installed. In 1990 it was decommissioned as a nuclear bunker. The original entrance doors on the north side were bricked up. The inner blast proof door still exists as do two other doors and one air filtration system.

In 1994 an entrance door and stairs were installed to provide a summer Exhibition Centre, opened on 25th May 2001 by Admiral Bell-Davis and attended by the Mayors of Gosport and Fareham. The aviation history of Lee-on-the-Solent from 1917 to 1996 was the subject organised by Gosport Aviation Society. As well as charting the development of flying along the Solent's northern shores, the comprehensive display of photographs illustrated the development of the Lee-on-the-Solent and Hill Head areas particularly in the pre-World War II years. The exhibition consisted of 50 panels including photos, recollections by former servicemen and the Search and Rescue team with details of different aspects of the airfield's history, as well as a model of the airfield.

In 2010 the Historical Diving Society was granted a lease by GBC to open a Diving Museum. Gosport Borough Council enthusiastically seized upon their historic claim to diving fame through the residency of diver John Deane in the town for some eleven years (1835 to 1845) during which time he discovered the *Mary Rose* and, with his brother, Charles, the inventor of the first diving helmet, although originally a smoke helmet, launched the diving industry. No 2 Battery lends itself perfectly to a diving museum and Gosport provides a natural choice of location. The Museum is an outreach project of The Historical Diving Society and is operated and maintained entirely by volunteers.

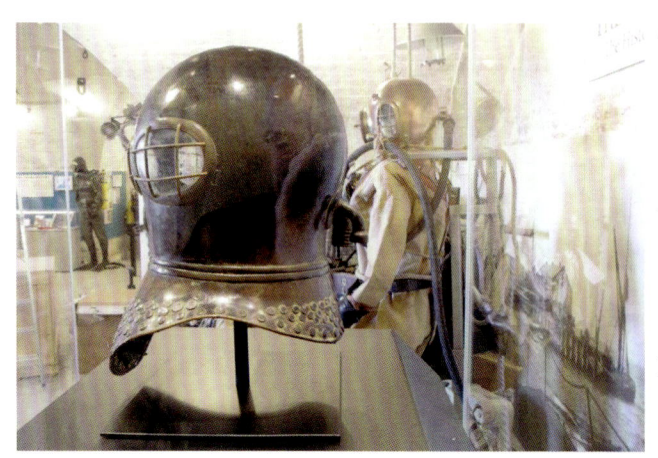

The Deane Helmet c.1828   Martin Marks

In 2015 the Diving Museum reached the top five out of 850 entries nationally in the *Daily Telegraph's* Family Friendly Museum competition, even beating off entries from the Science Museum and Hampton Court. The rear section of the battery is now a mobile home park and some areas are used by the residents as storage.

In 2017 the battery was upgraded to Grade II* status and a £20,000 grant was awarded to upgrade the fire detection and security systems to meet conditions imposed by the Science Museum to borrow their Deane helmet, the first ever and the Holy Grail of diving helmets, noted above. In 2019 the Diving Museum gained accreditation from the Arts Council which would formally recognise it as a museum, easing the loan of items from other museums. As a result of this, the Science Museum permitted the loan of the Deane helmet, the first commercial diving helmet. The museum was also bidding for a very large lottery grant to waterproof the whole battery which was damp inside and relied on dustbins to contain drips in some areas. A starter grant was obtained for a civil engineer and a business consultant to ensure that both the structure and the museum business are sound. This project was progressing on to Phase II.

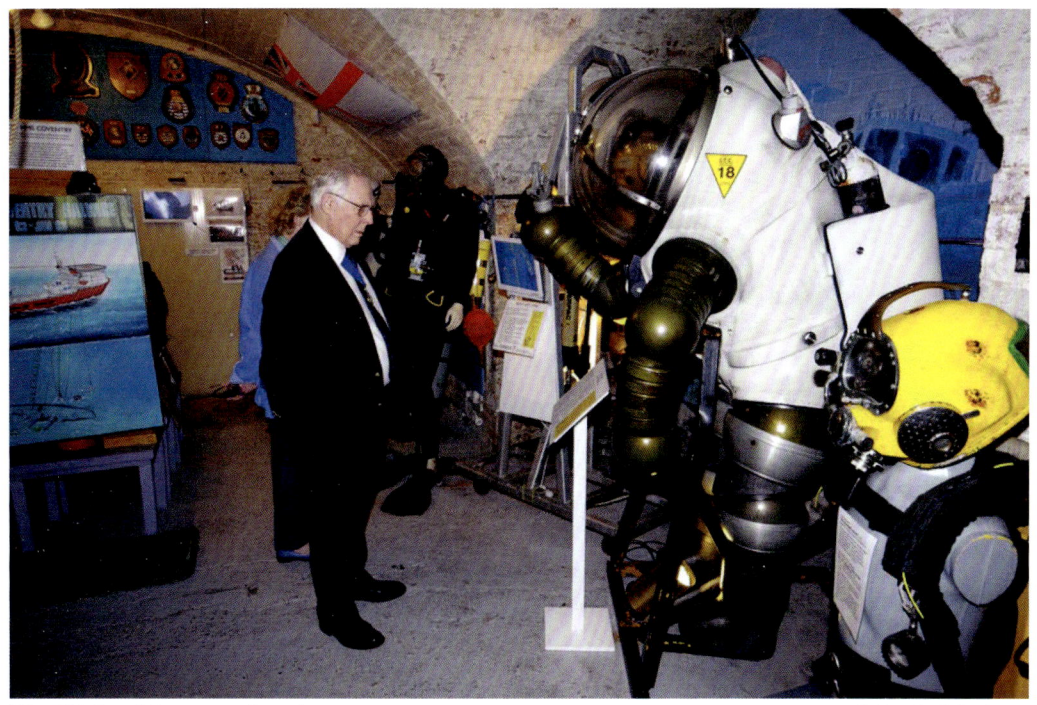

The Diving Museum – interior   Martin Marks

No. 3 Battery, Stokes Bay[7]

This was also armed with 8-inch Smooth Bore guns and four 7-inch Rifled Breech-loading guns in place by 1886. Two fired along the ditch towards No. 2 Battery and two protected the glacis in front of the ditch. The No.1 gun was removed by 1891. At No.3 battery the moat turned south towards the sea for a short section before continuing eastward to the lake at Gilkicker. Nothing of it remains.

No. 4 Battery, Stokes Bay[7]

This was designed for four 8-inch smooth bore guns mounted in pairs to fire down each branch of the moat. The 1886 armament list shows that these had been replaced with four 7-inch rifled breech-loading guns. The moat continued eastward to the lake at Gilkicker. This section of moat was flanked by No.4 battery. Nothing of the battery remains. The site is a stable and paddock for horses.

Diving Museum model

No. 5 Battery, Stokes Bay[7,10]

This was built in the early 1860's. At one time it had a mixed armament of 68 pounder and 8-inch smooth bore guns. The 1872 proposals called for replacements of nine 7-inch breech-loading guns but by 1886 only four remained in place. Another section of moat ran southwards from No. 4 Battery and branched east and then north to re-join the main moat west of No. 5 Battery. A small section of moat continued from the north end of Gilkicker lake along the rear of Fort Monckton towards Haslar sea wall. The battery is now an ancient monument within the former site of QinetiQ, Alverstoke. By 1900 it was used as a barracks. It was not brought back into service during the First World War. The 1932 Ordnance Survey Map shows several buildings occupying the yard behind the ramparts and it is marked as a "Royal Naval Camp" and from 1947 to about 1950 as an accommodation camp: HMS Vernon II. During World War 2 an AA battery was located there. By 1950 a road was cut through the battery site and in 1962 the Royal Naval Physiological Laboratory moved in from Peel Cottage and continued their work in diving research.

In September 2010, some of the post-1900 buildings on-site were due to be demolished when the current owners QinetiQ vacated the site. In 2014, it was reported that the site had planning permission for light industrial use. Gosport Borough Council however, wanted luxury homes that could draw in captains of industry for people leading companies at Daedalus Enterprise Zone in Lee-on-the-Solent. The Council leader Mark Hook said: "My idea to develop them is a very small number of million-pound plus houses." It is said that the home plan failed because there was a stipulation in the planning permission to improve the road and build a pavement from the development to the bottom of Anglesey Road, making it uneonomical. In 2019 the site remained empty.

No. 6 Battery, Stokes Bay

Located South East of No .5 Battery overlooking the Golf Course. Stokes Bay once had its own railway branch line from a junction on the Fareham- Gosport line to a pier for a steamer ferry to the Isle of Wight.[11] It opened in 1863 but the steamer service was suspended at the start of the First World War. The line was sold to the Admiralty in 1922 for transportation of munitions. Most of the route is now a public footpath[5] and parts of the moat survive.

The Gosport Advanced Line

We now explore the impressive ring of forts that protected Gosport from the west. The Line was made up of Fort Gilkicker, Fort Monckton, Gomer, Grange, Rowner, Brockhurst, Elson, Fareham and Wallington. Not all have found beneficial new uses.

Fort Gilkicker

Continual setbacks in attempts to achieve a new use has been the recent story of this fort. Until Gilkicker Point was reclaimed in about the 1830s or 40s as a redoubt for Fort Monkton, the site was the outlet of the river Alver into the sea. The redoubt collapsed and in 1860 the Defence Committee decided to replace it with a fort to be named Fort Gilkicker. A semi-circular casemated design with 26 rifled muzzle loading (RML) guns was chosen with five local supporting batteries to be called the Stokes Bay Lines. The group was intended to cover the local beaches as well as the deep-water anchorages off Browndown and the western approaches to Portsmouth Harbour. It was mainly completed by 1871, progress having been delayed by a change of contractor. A brick barracks protected the rear of the fort. The armament was switched to breech-loading guns on the roof for a short period from 1902 to 1906. During this time, the casemates below were covered up with earth banks to give greater

Fort Gilkicker   Bob Hunt

protection to the masonry and granite-faced front wall, the ammunition stores and to add camouflage to the fort. These were built following experience gained in the bombardment of Alexandria in 1882 which showed that such slopes would cause shells to ricochet over the works. Their design is generally attributed to General Clark. It also made the gun positions much more difficult to spot from seaward. At Gilkicker that might have been an added incentive. Because extra guns had been mounted on the roof, the walls were starting to tilt outwards. These guns were removed and the movement of the walls ceased when the slopes were finally completed in 1906.

Fort Gilkicker parade ground
Martin Marks

From 1908 to 1911 families of the Royal Engineers at Fort Monckton used the fort for accommodation, and again from 1939 to 1946. In 1916 the fort was armed with an early type of anti-aircraft gun. In World War II the fort was briefly armed with a 40mm Bofors gun and gun-laying radar was fitted outside the fort to direct the guns at the nearby Gilkicker anti-aircraft gun site. It was used as a Communications Centre during the D-Day preparations at Stokes Bay, especially during the Normandy landings when it was very busy. By 1956 all coast defences were abolished. The fort was taken over by the Ministry of Public Buildings and Works and was used for workshops and as a store.

Fort Gilkicker gun bay. Sea view simulated for potential buyers of flats　　　Martin Marks

In 1986 Gilkicker was bought by Hampshire County Council when the MOD declared it redundant, "to ensure a restored and continued life within Stokes Bay for future generations' according to a notice on site...Removal of the earth embankment to the front elevation will restore the original appearance of the Fort (pre-1906). The elevation with its exposed gun embrasures framed in Cornish stone will once again be seen from land and sea, making a stunning contribution to the Stokes Bay area…It is hoped that the property can provide homes for current and future occupiers without major structural alterations... It is the only surviving coastal battery of its type that has not been drastically altered during its life. This makes it unique and highly worthy of restoration and re-use" according to a notice seen on site in 2015. It is classified as a Scheduled Ancient Monument and is Grade II* listed to preserve it from unsympathetic development, but it is also on the Buildings at Risk register. English Heritage said that "without a bold and inventive solution for its continued use the fort faces a future of increasingly serious structural difficulties and dilapidation." Two successive development schemes were proposed but did not proceed. A third scheme was raised in 2010 by Askett Hawk under the name of Fort Gilkicker Developments to convert the fort to luxury apartments. This was given planning permission following a public enquiry. During 2016 there was a small fire on the site that was classified as arson, whilst on the other side of Portsmouth Harbour in early 2017 Fraser Gunnery Range at Fort Cumberland was experiencing similar vandalism.

In November 2016 planning permission for the Gilkicker conversion was extended for a further three years. In December 2016 it was announced that a developer, Wild Boar Properties Ltd. had finally agreed a deal with the former landowners to transform Fort Gilkicker into 22 townhouses in the fort walls and four luxury flats in the barracks section. This would be a £20m project, suggesting circa £1m per house or flat. It was hoped to complete the project in three years. There would be a museum and two viewing platforms at either end of the development. The site owners have purchased the road through the golf course, William Jervis Way, for access.

The Ecological Impact Assessment, carried out in 2012 would have to be updated and approved and then an Archaeological Methodology Statement and supporting Archaeological Impact Assessment had to be also be carried out and approved. There was a plan to remove the earth covering the seaward walls and to add a new top level with a green roof. The earth bank and between 7,000 and 7,500 tons of concrete would be removed so that the apartments could be opened up with picture windows. There were concerns that this unloading of the structure after over 100 years which might cause the casemate walls to distort, and that some of the required drilling and demolition to open the spaces might prove difficult.

There is also concern for the rare Gilkicker weevil found in the area, an issue that had been raised with earlier developments. In 2000 Hampshire Wildlife Trust with volunteers from the Hampshire Network for Invertebrate Conservation (HNIC) undertook a survey of sites around Gilkicker Point. Only a single specimen was found, but this was 250m west of the fort, indicating that the embankment at the fort was not the only breeding site left in Britain. Another survey commissioned by English Nature took place later in 2000. The objectives were to search for likely sites in Hampshire and to provide locations of all colonies found, as well as to record any useful ecological information on the species. The report concluded that although the distribution of the species at Gilkicker Point and Browndown was wider than was originally thought, the species was still extremely local, and protection of its fragile habitat should be a prime objective of any future conservation work. Part of the conditions for the current development was that when the earth wall was removed the soil would be taken eastwards, just off the beach, to provide a new home for the weevil.

According to the developer: "The rejuvenation of such a prominent and historic building will create a luxurious living opportunity in one of the UK's most desirable coastal hotspots. Fort Gilkicker sits at the heart of the UK's premier sailing arena." Gosport MP Caroline Dinenage said: "The new owners have a fantastic opportunity to restore these amazing buildings and bring them back into use. I hope they are able to do so, as this site has the potential to be a great boost for Gosport." In 2012 the fort was put back on the market after the Malaysian financial backers pulled out. In 2019 Gilkicker Developments Ltd. stated "As regards longer-term plans, the project has been mothballed due to problems encountered with Historic England over progressing implementation. We have no plans to re-start at present." They were seeking a buyer for the fort after the previous financial backer pulled out due to the project taking too long.

Today the fort's future remains unloved, and vandalism is starting again. A local resident says: "The problem, this time, is that the site is in private ownership. It might have been better to heed Dr Wright's advice a few years ago - to fill the fort up with silt from the dredging and form a mound over the top which would at least preserve what we have until a suitable idea, or the technology allows something more positive in the future…"

Fort Monckton

Fort Monckton was built overlying the site of the 1545 Haselworth Castle in 1783-90, following the 1779 invasion scare during the American War of Independence. It has a star-shaped plan of five bastions and demi-bastions characteristic of the late C18 like Fort Cumberland[12] with the casemates still prominent facing out to sea as you go past in the hovercraft between Southsea and Ryde. Much of the original fort still exists including the drawbridge, bastions, sea facing casemates, guardroom, one of the caponiers and the ditch.

It was considered too small and old to dominate the key anchorage of Spithead by the 1860 Royal Commission. Instead they opted to build the new Fort Gilkicker and give Fort Monckton a minor defensive role in the defence of Portsmouth harbour, rebuilding it in the 1880s. *The Illustrated London News* of 1856 and 1858 showed the fort being used as a viewing platform for the Fleet Reviews. In 1875 experiments were carried out at Stokes Bay with Wilde's Electric Light; the light was positioned on the southwest bastion of the fort. More experiments continued with Lime Light and signalling lamps for shore-to-ship communication in July 1875.[13] Mr Leather's yard and jetty in Stokes Bay was subsequently taken over by the Submarine Mining Engineers and eventually became the School of Electric Lighting. The platform on which it sat is still visible and the track of the light railway [18" gauge] which ran from there to the depot at Blockhouse is relatively easily traced - not to be confused with the later 20" track which ran into Monckton where there were at least two turntables.

As an adjunct to the Stokes Bay Lines the Royal Engineers moved into the fort in 1878 to train in the use of naval mines and later searchlights. In 1879 torpedo experiments were carried out at Stokes Bay, part of which consisted of a mock attack on Fort Monckton, also recorded in *The Illustrated London News*.[14] In 1880 the 4th Company of Submarine Miners of the Royal Engineers occupied the fort. They moved out in 1884 to Fort Blockhouse, leaving Monckton as accommodation for the RE Militia during their annual training. Anti-aircraft searchlights were located there during World War 1, and an anti-aircraft artillery unit was quartered in the fort during World War 2. After the war it was virtually abandoned, but retained by the Army.[15]

Now referred to as No.1 Military Training Establishment, it has the honour of being the only fort of the Portsmouth defences still in Military (as opposed to Naval) hands. In his book *The Big Breach From Top Secret to Maximum Security*, former officer Richard Tomlinson alleges that Fort Monckton is now the Secret Intelligence Service's field operations training centre, where both basic and advanced field training is given to SIS personnel[16], as well as providing liaison training with others services including

Fort Monckton   Bob Hunt

the Special Air Service and Special Boat Service. Protected by modern razor wire security fencing, CCTV cameras and high intensity lighting, the site has been heavily modified with modern offices and accommodation added in and around it.

Browndown

Browndown, located between Lee-on-the-Solent and Stokes Bay has been a military range since the early 19th century when French prisoners of war were used to build butts for a musket range. From around 1815 it was used to house German soldiers whose national links had been severed when they fought alongside the British against Napoleon. After a huge breakout from the camp for the delights of Gosport by more German troops in 1856 reminiscent of a Wild West brawl, they were sent off to South Africa.[17,18,19]

In 1845 Browndown was the site of the last fatal duel, with pistols, between a Royal Marine officer and a member of the 11th Dragoons. The dragoon later died of his wounds and the marine was acquitted at his court martial where poor medical care by the doctors was blamed for his death. In 1894 Browndown Halt on the Brockhurst to Lee-on-the-Solent railway was added directly opposite the ranges to provide an easy way to move troops between Fort Gomer to the ranges.

Browndown was then employed as a range for training the Royal Marine Light Infantry based at Forton Barracks in Gosport. Accommodation for infantrymen as well as their instructors and their families were built on site. The Musketry Camp was completed before 1900. After WWI the ranges were realigned following complaints of stray bullets near new houses as Lee-on-the-Solent expanded eastwards. In 1922 the site became the HQ of the Royal Marines Small Arms Corps.

In 1942 the Royal Marines Commandos were formed, and the R.M. Snipers' Courses took place at Browndown. The school continued at Browndown until 1957, when it closed. During WWII Browndown was used as a Royal Marine camp and for D-Day it was designated as Marshalling Area Camp A19.

In 1961 the site became the South-East District Minor Training Facility. In 1966 'Hovershow 66' was held at Browndown, during which the new Vickers hovercraft was displayed to the public. Over 3,000 official visitors from about 100 different countries attended. The Browndown Battery and the area adjacent to it were used by the Inter-Service Hovercraft Trials Unit from 1966 to 1974 when it was disbanded. 200 Hovercraft Trials Squadron (later 200 Hovercraft Squadron R.C.T.) was based there, using the SRN6-617 hovercraft. They added a hangar and some ancillary buildings to the north east of Browndown Battery and an observation post on top of Browndown Battery itself. The SRN3 built by Saunders Roe was tested at Browndown by the Inter-Service Hovercraft Trials Unit, but was never adopted for service.

The area was classified as a Site of Special Scientific Interest in 1987, which causes complications for managing what was still an active military training area. In 1990 it was included in the MOD list of "managed conservation areas". In 1999 it became HQ Army Training Estate Home Counties, which has since changed to the Defence Infrastructure Organisation (South East).

In 2002 it was used as the setting for the television series Lads' Army as a mock 1950's British Army National Service training course. It was used again twice in 2005 by the series renamed Bad Lads' Army as a mock Officer Cadet School.

As part of 'rationalisation measures' the Browndown camp became superfluous in 2009 and the MOD put the Browndown Camp up for sale in 2011.' It was purchased by a developer, Jumbuck Ltd. for £754,000. The site is restricted to recreational development only, which may explain why nothing more has happened to date. MOD retained the camping ground and ranges for military use. The ranges area is open to the public when not in use by the military for various day and night exercises, which include amphibious beach landings. Closure details are displayed on the Lee Residents'

Association website.[20] In December 2017 the CEO of Gosport Borough Council reported that the site owner had died, and the new managing agent seemed interested in creating a marina.

Browndown Battery

Browndown Battery was built on the site of Browndown Battery West, one of two earlier 1852 batteries – the other was Browndown Battery East. It was equipped with two 12.5" RML guns until 1904 when these were replaced by two 9.2" guns. The battery was closed in 1905, but retained by the army as part of the Browndown Training facility and Browndown Ranges. It saw a brief new lease of life when a lookout post was constructed on top of the east gun position as part of the training area used by the combined services hovercraft unit already mentioned. Browndown Battery is a Grade II listed building and according to the military authorities it is in excellent condition. Though modified and strengthened many times during its history much of the original building is still identifiable.

Fort Gomer

Fort Gomer was the first and most southerly in the line of five forts which formed part of the Sea Front and Spithead Defences. With Forts Gomer, Grange, Rowner and Elson the 'Gosport Advanced Lines' main purpose was to guard against a landing on the south coast for a flanking attack on Portsmouth to bombard the dockyard from the west. The five forts were in firing range of their neighbours to provide additional covering fire against a direct attack on one of them. They were constructed between 1853 and 1858; in 1869 the total cost was estimated at £92,000. In Fort Gomer a barrack block was later added with accommodation for some 350 officers and men. It was protected from a seaward attack by a moat using the River Alver which was diverted into the moat system at No.1 Battery and flowed out of the moat to the sea at various sluices along its length and from another at Gilkicker lake, thus maintaining a constant water level in the moat. (See No. 2 Battery above for later developments).

In the Victorian period the fort was used mainly for accommodation and training. In the latter half of the 19th century the gun armament was changed several times. It then saw more use as a training establishment for the Boer War and World War I. In World War II it became a headquarters for amphibious assault training playing a part in preparations for the D-Day landings. Work on amphibious fighting vehicles, using nearby Stokes Bay went on until at least 1946.

After the war the 7th Royal Tank Regiment was based at the fort in the 1950s, leaving in 1953. Briefly in the same year the fort was occupied by the Royal Army Ordnance Corps Boys Training School. It was eventually released in 1964 and sold at auction for £169,000. A Fareham firm demolished it to provide land for housing. The site is now bounded by Moat Drive and Broadsands Drive, west of Gomer Lane.[22]

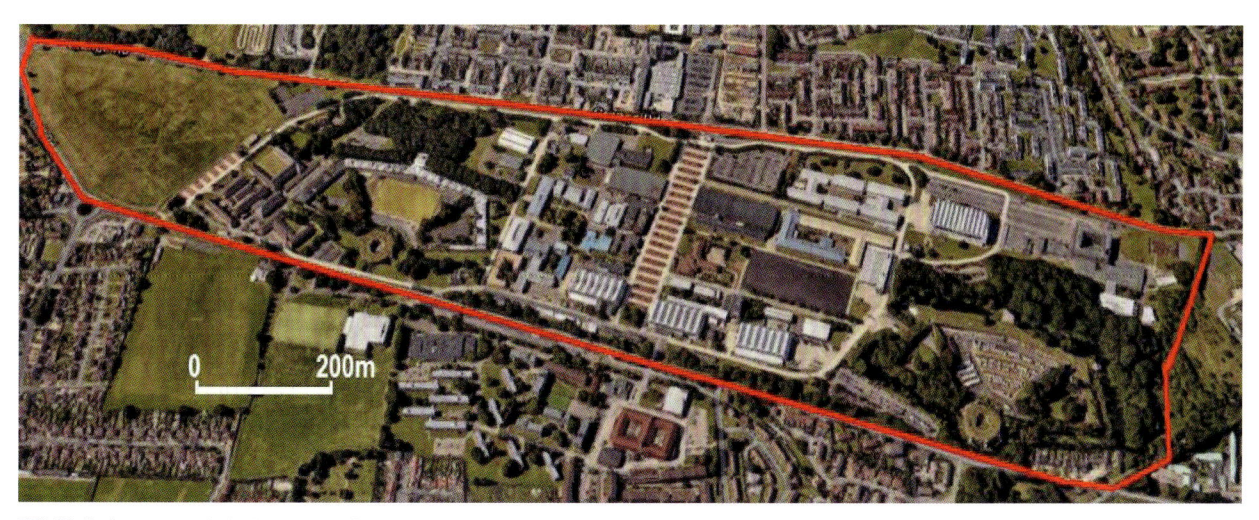

HMS Sultan – aerial view. Fort Gomer left, Fort Grange right   Gosport Borough Council

Fort Grange[22]

After the approval to build Fort Gomer and Fort Elson in 1852 a decision was made to fill the gap between them with three more forts. Fort Grange was at the southerly end of the gap and work began in 1858 and was completed in 1863. It is Grade II Listed and is virtually identical to Forts Brockhurst and Rowner. In 1916 it was armed with a 1-pounder heavy anti-aircraft gun on a travelling carriage. As we described in the previous chapter, in 1917 the Fort Grange Aerodrome became the home to Robert Smith-Barry's School of Special Flying. There are many surviving details and the main structure is intact; however the earthworks have been much reduced and much of the moat is infilled. The fort was used as a military headquarters from 1910 and was derelict by 1983. In 1984 Fort Grange became the home of HMS Sultan Royal Naval Volunteer Cadet Corps and continues to fulfil this role today. Jackie Percival reports that it and Fort Gomer, both in plain view on Military Road, one of Gosport's busiest roads used by hundreds of people every day, look neglected and overgrown, as though they were abandoned years ago.

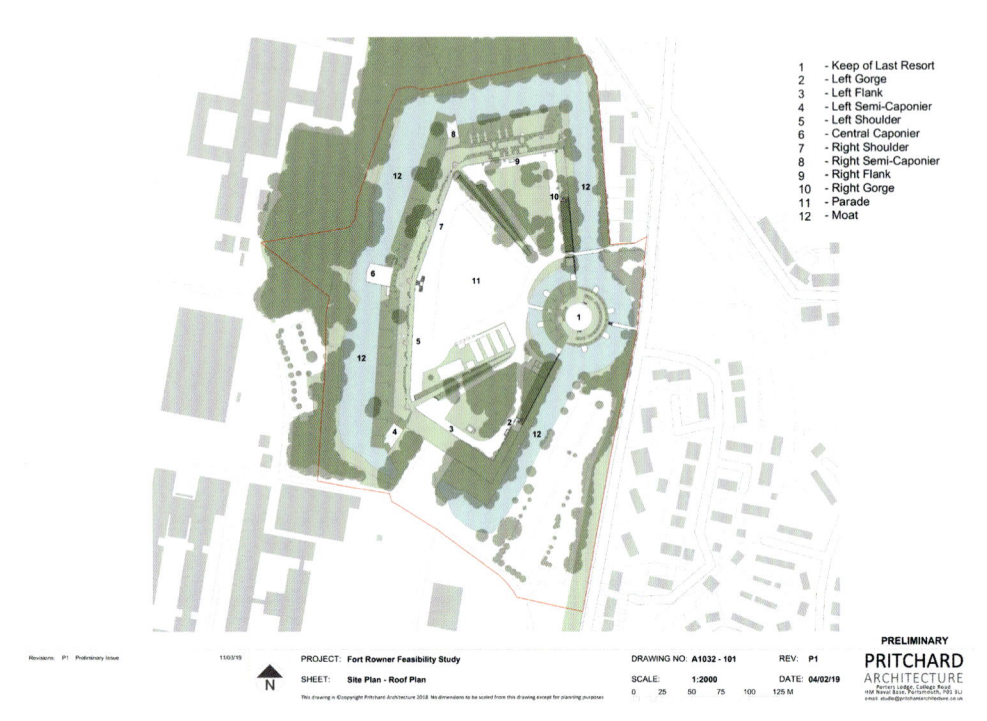

Fort Rowner Feasibility Study Site Plan 2019 Pritchard Architecture

Fort Rowner

Fort Rowner was built in 1859 as part of the Gosport Advanced Line. Forts Brockhurst, Rowner and Grange were all the same design, each a polygonal shape, a central 'keep of last resort', and caponier protecting the surrounding ditch. Around the central parade ground is a seemingly endless row of barrack casemates, above which are Haxo Casemates for the guns. In the late 60s naval officers in training slept in the fort's dorms, but some years later perhaps they were just used as storage. The pathway from the base to the wardroom passed through the fort in the early days but was subsequently bypassed by a direct road. The fishermen believed that there was a monster pike in the moat so big that it looked like a nuclear submarine! Half the barrack rooms in Fort Rowner remain in good condition and are used, the other half are derelict, some even lacking their front walls. The fort remains today within the HMS Sultan Naval Base but is mainly left to nature and a group of goats that graze the ramparts. A feasibility study of options for its future – commercial or residential conversion or newbuild housing was commissioned from architect Giles Pritchard as part of the Heritage Action Zone Zone funded by One Public Estate.

It began by preparing an understanding of the historic development and significance of the Fort and its context. A broad fabric and structural condition survey and as well as an ecological survey was undertaken and a costed repair strategy including localised investigations produced to inform the viability assessment for the site. This included a cost plan for mothballing the site for the next 5 years. A detailed assessment was then undertaken to consider the potential impact of the conversion of the Fort on its significance and setting.

Giles Pritchard explains that "One of the greatest challenges was the conversion of the casemates into residential use. To provide a suitable, dry environment inside the buildings, we looked at a number of options for waterproofing including the removal of the earthworks to allow a complete waterproof membrane to be installed, and reinstatement of the banks. Options were considered to add an upper floor to sections of the fort to increase residential floorspace." The study concluded that "residential alone is not viable, and this generally appears to be the case nationally according to Historic England's new study completed by Peter Kendall (now retired). Options focusing on commercial or business uses seem to have more potential, working on a long term phased programme of repairs as funds, and grants, allow."[24]

Fort Brockhurst

Fort Brockhurst   Martin Marks

Construction of Fort Brockhurst began in 1858 spurred by the threat of a French invasion. It was designed by William Crossman to house 83 guns and a 300-man garrison. Over thirteen million bricks were used, most of them produced in Gosport and Fareham, leaving many acres of excavated claypits. Fort Brockhurst was given a novel hexagon outline but with a more traditional circular keep and flooded moat. It has a low profile disguised by grass and earth banks. Three caponiers project into the moat to provide raking fire ensuring covering fire could be delivered around the entire perimeter.

The keep could only be reached by a drawbridge across the moat and then a sliding bridge across the interior. Under the ramparts are a series of casemates, chambers for storage, living quarters, and workshops. It was completed in 1862 and considered by some to be the ultimate in Victorian military architecture. However, it was overtaken by events, and only lack of finance prevented the Gosport Advanced Line being superseded by a line of forts even further west. As a result, it was never fully armed and for most of its military life was used for accommodation and storage.

As part of Kitchener's New Army, the 9th (Heavy) Battery, Royal Garrison Artillery was formed at the fort in 1914. It was an arm of the Royal Artillery originally tasked with manning the guns of the British Empire's forts and fortresses, including coastal artillery batteries, the heavy gun batteries attached to each infantry division as well as the guns of the siege artillery. In World War II bombing destroyed three casemates on the north flank, but the fort never saw action and was released by the military in the 1960s. It was handed over to Hampshire County Council and then to English Heritage.

In 1993 English Heritage opened a training centre which used replica ruins, known as "ruinettes", and halfbuilt walls to train its own staff and others in bricklaying and related techniques for restoration. However it failed to earn its keep through lack of paying customers and it closed in

1996. As the training need was still there, English Heritage negotiated with West Dean College to run courses instead. Today West Dean runs a wide range of courses relating to conservation.

In the past, there were a variety of events in the fort: classic car rallies, war-gaming weekends, concerts and plays held at the fort during the summer season. In 1995, there was a pop concert headlining James Brown, the Stranglers and Stiff Little Fingers.

In 2004 the vehicular bridge over the moat at Fort Brockhurst was restored. It had been supported with scaffolding for some ten years. It was produced as standard pattern for the military as a kit of parts and dated from 1862. Similar bridges can be seen at Fort Rowner and Fort Grange. English Heritage gained funding for the restoration of the project and Eura Conservation Ltd, a specialist restoration company, dismantled and restored the bridge at a cost of some £300,000.

In 2008 artist Caroline Cardus had a 'Go Make! residency at the fort in an ongoing series of bursaries for disabled and deaf artists, awarded by Dada-South, in partnership with English Heritage. Afterwards she presented a multi-sensory installation piece in response to her residency.

In 2013 the Gosport Shed, a branch of Men's Sheds which provides workshop facilities and a friendly meeting place for older men opened in the fort. They occupied two of the casemates which were fitted out as workshops. Sadly, after about twelve months, English Heritage decided that they were no longer able to use the site and they moved over to St Vincent College grounds, the old HMS St Vincent site.

In 2015, linked to Gosport heritage Open days, "Space Interrupted" artist Helene Kazanone, was one of six artists commissioned to make a new site-responsive work, re- enacted a routine second world war military test within the bounds of the fort.

English Heritage uses the fort as a regional store for archaeological, architectural and historical items for the southeast. John Rennie's impressive largescale model of Sheerness Dockyard which requires 1600 ft. to lay it out to scale was stored there. It was created as an aid to the builders, and updated and amended as changes were made to the dockyard. Carved stones that were excavated at Battle Abbey by Sir Harold Brakspear in the 1930s and John Hare in 1978-80 including fragments of Romanesque sculpture were also stored there. The Corpus of Romanesque Sculpture in Britain and Ireland is a complete online record of all the surviving such sculpture in Britain and Ireland at more than 5000 sites, also distributed to Fort Cumberland and Dover Castle.

Model of John Rennie's design for Sheerness Dockyard stored in Fort Brockhurst

In a building on the parade ground there is a superb, large, three-dimensional model illustrating the history of Portsmouth's defences circa 1870. One of the casemates houses an interesting armourer's workshop using old items dating around 1900 from a commercial company that closed down in the 1980's.

Fort Brockhurst – Armourer's Workshop   Martin Marks

The fort remains largely unaltered. The site was regularly open to the public under the auspices of the owners, English Heritage, but this was happening less and less. It is a scheduled monument, but it appears that English Heritage view it as very low down on their priority list. As a result, recent maintenance has been minimal, leading to an English Heritage view that "the quality of visitors' experience is not in line with their expectations". Up until the mid-1990s the fort was open 7 days a week then this was reduced to weekends and bank holidays. In 2004, it was closed completely for building works. It later reopened from April to September on the second Saturday of the month, plus Gosport Heritage Open Days (3 days in September). Since May 2016 it has been closed to the public except for the Heritage Open Days.

Fort Elson

Built between 1858 and 1860 Fort Elson is the most northern fort in the Gosport defence line and the oldest remaining. It was built using a method called 'escarp en décharge' – arches like a railway bridge were built and then the gaps filled in with earth. It had a dry ditch that could be flooded with seawater via a sluice from Elson Creek if necessary.[24,25] Over the years a range of various guns from 13-inch mortars, through 10, 8 and 7-inch RBLs (rifled breech loading) guns and machine guns were fitted, including 7-inch guns on the novel Moncrieff Disappearing mountings. By 1901 all armament had been removed.[25,26]

It is a scheduled monument but because it is located within an active naval armament depot at Bedenham, Gosport, which has a predicted long-term requirement by M.O.D, its future was difficult to decide. In 1994 a detailed Strategic Proposal for Fort Elson was submitted, outlining various courses of action regarding the future of the structure, ranging from a very expensive complete renovation through to "controlled ruination". The latter has been adopted and is to allow the fort to deteriorate under a controlled and planned manner ensuring adequate records and safety aspects.[27] Currently the fort can only be described as mostly derelict with areas of heavy ground subsidence, extensive tree and vegetation growth and many areas of dangerous/collapsed brickwork.

The next chapter takes the story on along Portsdown Hill, to the Eastney forts and Fort Cumberland and the seaforts at Spithead.

Source

Gosport Society 2020 *Gosport 2020 For the Record: Essays in Celebration of the Half-Century of the Gosport Society*

O Bayer, Mike Williams 2015 Historic England *Stokes Bay, Gosport: Five Centuries of Coastal Defences*

References

1. Gosport Borough Council - Haslar Barracks Conservation Area Appraisal - March 2018 Rob Harper
2. https://en.wikipedia.org/wiki/Fort_Blockhouse
3. Sarah P C Hendriks 2014 Haslar Gunboat Yard, Gosport, Hampshire Historic Buildings Report Research Report Series n. 15-2014 TheGunboatYardHaslarGosport_HistoricBuildingsAssessment.pdf
4. Jonathan Coad 1983 *Historic Architecture of the Royal Navy.* London: Gollancz p. 15
5. Charles O'Brien et al 2018 *Hampshire South Buildings of England* Yale University Press pp. 291, 301
6. List entry Number: 1405953 scheduled under the Ancient Monuments and Archaeological Areas Act 1979, No 1 Battery, Stokes Bay Lines
6. Stokes Bay Lines Fact Sheet https://www.victorianforts.co.uk/pdf/datasheets/stokesbaylines.pdf
7. Richard Tomlinson *The Big Breach: From Top Secret to Maximum Security.* Mainstream Publishing. ISBN 1-903813-01-8; https://web.archive.org/web/20111204102502/http://www.fas.org/irp/world/uk/mi6/
8. Christopher Donnithorne 2020 The threat to Portsmouth Harbour Unpublished paper
9. http://www.friendsofstokesbay.co.uk/stokes-bay-lines/
10. No 5 battery - https://historicengland.org.uk/listing/the-list/list-entry/1001829
11. http://www.disused-stations.org.uk/s/stokes_bay/index.shtml
12. Charles O'Brien et al 2018 *Hampshire South Buildings of England* Yale University Press pp. 291
13. *The Portsmouth News,* July 1875
14. *The Illustrated London News* 25 October 1879
15. G H Williams *The Western Defences of Portsmouth Harbour 1400-1800* Portsmouth Papers No.3 p. 47.
16. Richard Tomlinson *The Big Breach: From Top Secret to Maximum Security.* Mainstream Publishing. ISBN 1-903813-01-8; https://web.archive.org/web/20111204102502/http://www.fas.org/irp/world/uk/mi6/
17. Friends of Stokes Bay http://www.friendsofstokesbay.co.uk/browndown-camp- and-ranges/
18. BBC News http://www.bbc.co.uk/news/uk-england-hampshire-26471922
19. "Soldiers of the Queen" by Lesley Burton & Beryl Peacey
20. Browndown Range closures http://www.leeresidents.org.uk/Browndown%2019- 8to15-9.pdf
21. http://www.wikiwand.com/en/Fort_Gomer
22. https://en.wikipedia.org/wiki/Fort_Grange#cite_note-3
23. Jackie Percival Letter to Celia Clark 1 August 2019
24. Giles Pritchard Architect Email to Celia Clark 11 October 2020
25. https://wikivisually.com/wiki/Fort_Elson
26. https://www.routeyou.com/en-gb/location/view/48043326/fort-elson
27. http://www.mygosport.org/fort_elson.htm

Chapter 13
Forts, batteries and military lines: Portsdown Hill to the forts in the sea

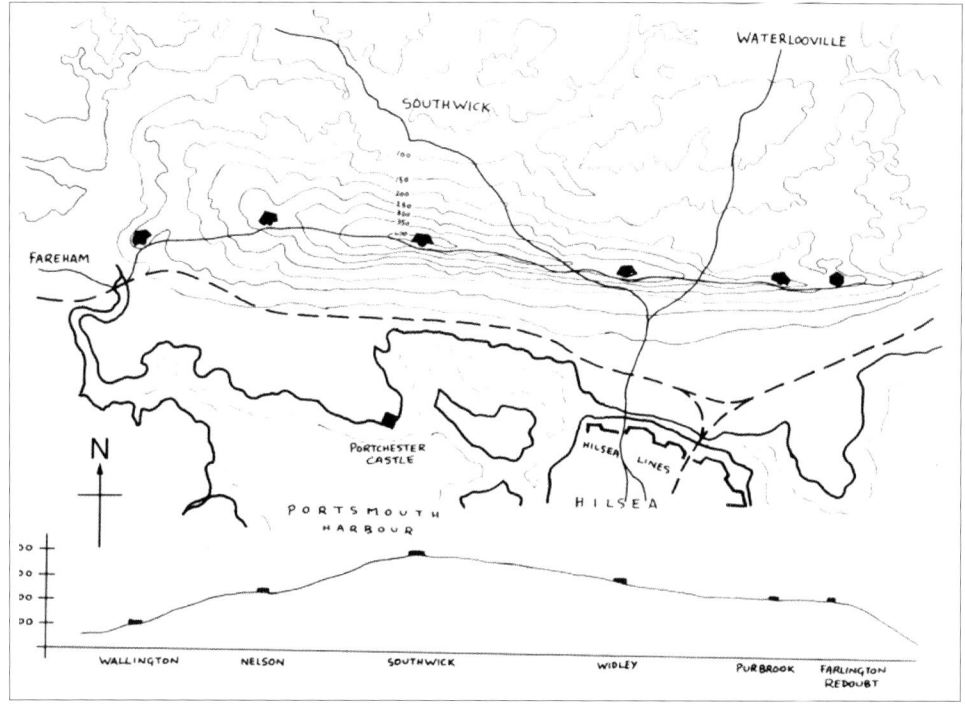

Local Studies Resource Guide to the Defence of Portsdown Harbour Part 1 The Eastern Defences. Portsmouth Teachers' Resource Centre 1980

The view from the long chalk ridge of Portsdown Hill, 131m above sea level at its highest point, with its commanding position above Portsmouth and its two harbours, Gosport and the Isle of Wight is justly celebrated in historic paintings. As you crest the hill on the A3 from London by the George Inn, the panorama that opens before you is breathtaking. To see a whole city in its maritime setting like this is rare. The carparks on the top of the hill are favourite places to enjoy the spectacular panoramic view of Portsmouth, Gosport, the Solent and the Isle of Wight. It's also an overview of all the places in this book, with the dockyard at its centre and the spike of the Spinnaker tower in Gunwharf.

Not surprisingly, Portsdown Hill is still strategically important to the nation's defence. It continues to host several active military research establishments, while the nineteenth century defences of the dockyard are also very evident. Before the forts were built, Portsmouth's defence relied upon smoothbore guns positioned close to the shore. The older Hilsea Lines which protected the northern shore of Portsea Island at Portscreek became obsolete at the same time and were remodelled. We describe them in our chapter on military ground as public open space.

The hill's silhouette was dramatically transformed from the 1860s by construction of six huge red brick forts: Forts Fareham, Wallington, Nelson, Southwick, Widley and Purbrook, built a thousand yards apart along the crest of the hill. The Portsdown forts were designed by the Royal Engineers and built by civilian contractors who competed for the work. Fort Southwick played a key role in the planning and operation of D-Day, remaining in operation as a centre of naval communications until 2001.

In the 1940s the impressive research building of the Admiralty Surface Weapons Establishment was added to the summit. Its history, demolition in 2012 and current defence research on Portsdown are described in the chapter on research.

Over fifty hectares of the south face of the hill are a Site of Special Scientific Interest as wildlife rich chalk grassland. In the early 1950s when the slopes were no longer grazed, they were gradually invaded by scrub: mostly hawthorn, dogwood, and wild privet. An intensive restoration programme funded by the Countryside Commission and Portsmouth City Council was initiated in 1995. Large areas of scrub were machine cleared and flowers and grasses allowed to regenerate naturally. Scrub re-encroachment is now controlled by cattle and horses grazing over winter. Most of the southern flank of the ridge is designated as access land under the Countryside and Rights of Way Act 2000, so it is freely available for the public to enjoy on foot. The footpath along the north side of the top of the hill has recently been improved with new gates installed by the Ramblers. The road running along Portsdown, formerly called Military Road was adopted by Portsmouth City Council in February 1997 and renamed James Callaghan Drive in honour of the former prime minister Lord Callaghan of Cardiff KG PC who was born in the city when his father was in the navy. The boundary line of Winchester City Council to the north and Portsmouth City Council to the south runs along the top of the hill.

Fort Fareham

Meticulous research by David Moore, Garry Mitchell, Geoffrey Salvetti and others published by the Palmerston Forts Society is the source for our accounts of the forts in both these chapters. To continue the sequence from west to east ending with the seaforts, we first consider the two within the Fareham Borough Council boundary. We begin with Fort Fareham, which despite being in council ownership has been partially demolished by its commercial tenants and works for the M27. It was built in 1861 on land immediately to the west of Salterns Creek, adjacent to the railway line (which it also protected) from Fareham to Gosport. It is 3,500 yards from Fort Elson, the most northerly fort of the Gosport Advanced Lines. It acted as a link between them and the Portsdown Hill forts, also protecting the road from Gosport to them. It was visible from Fort Wallington which is 2,500 yards to the north and east. Because of its isolated position it is a polygonal design: the only one of the Palmerston Forts of that shape, with a concentration of Haxo casemates. It was designed to defend itself from being overrun and to make an attack on the Gosport Lines hazardous without it first being captured or disabled.

In 1905 forty four men of infantry and artillery units occupied it as part of the Portsmouth Fortress Defence Scheme[1] as a base to meet an attack from the west, assuming that the enemy had captured or marched around Southampton. Sometime after 1900 stables, a harness room and vehicle shed were added to the parade ground adjacent to the west gun ramp. From 1928 to 1945 the Officers' Mess served as the HQ for 35th Anti-Aircraft Brigade, Royal Artillery and Gun Operation Room. In 1941 it was used a headquarters for several anti-aircraft brigades.

Fort Fareham became surplus to requirements in 1965 and was sold by the Ministry of Defence to Fareham Urban District Council. In 1974 a contractor acting for a property company which had leased the fort from Fareham cleared the ramparts of vegetation by bulldozing them flat. All the earth was stripped from the Haxo casemates, together with earth forming the parapets and merlons, leaving the gun emplacements and expense magazines devoid of cover. The guardroom and gateway were demolished along with the stables, sling wagon shed, movable armament and vehicle sheds - before an injunction prevented further damage. The parade ground of the fort is now entirely covered with Fort Fareham Business Park. The casemates are intact but have been converted to modern industrial units.

During the Cold War the main magazine and the adjacent south west caponier served for a brief period as Fareham Borough Emergency Centre, to shelter the council's chosen few in the event of a nuclear attack. It also served as part of 2 Group 10 Brigade Anti Aircraft Operations Room for the

Portsmouth and Southampton Gun Defended Area.

Today the ramparts are overgrown and inaccessible; the parapets and gun positions have been removed. The five Moncrieff pits are still extant. The range of Haxo casemates on the southwest and south ramparts are intact but bricked-up. The mortar battery and the caponiers are mostly derelict and overgrown with extensive damage to the brickwork.

The ditch was partially cleared in the late 1980s by Fareham Council with volunteer help from people of the adjoining housing estate (Fort Fareham Conservationists, now disbanded)[2]. In 1989 a serious fire caused considerable spalling to the interior roof arches. The gun embrasures, loopholes and sally ports have been blocked up to prevent access by vandals. The iron railings of the access bridge to the fort are still in place although the access road has been widened to two lanes by filling in the ditch parallel to the bridge. The surrounding wooded scarp, covered way and glacis now forms an area of wild ground which is apparently managed by Fareham council. The section of ditch from the West Caponier to the gorge is partially filled with water and the volunteer group stocked this with mallards, moorhens and kingfishers in 1981. A series of paths with seats at intervals provides a continuous walkway around the outer perimeter of the ditch. Fareham Borough Council sold the fort to businessman Graham Moyse. It is on Historic England's At Risk register in the Priority A category: 'immediate risk of further rapid deterioration or loss of fabric; no solution agreed'. The fort and access bridge were first listed in 1976. [3] The exterior is still owned by Fareham.

Fort Wallington

Building of Fort Wallington in Military Road began in 1861 and was completed in 1874. Constructed within the hillside with an angled projection in the centre it was faced with red brick with stone coping. The flanking portions had a castellated parapet. Beyond are plain blank sections with round-headed doorways, stone columns and stone soffits and dripstones above with flat arched and segmental arched windows. Like Fort Fareham it has partly disappeared. It was sold to Mr J. Sullivan of the Southern Counties Trading Company who used it as an industrial depot. By 1974 the eastern part and all the other earthworks except the south facing curtain wall were levelled by road works for the M27 motorway. What remains is now listed Grade II.[4]

The Portsdown Hill Forts

In response to the Royal Commission of 1960 the government acquired the northern slopes of Portsdown Hill. Farmland was cleared of trees to provide a clear field of fire, because the forts' main armament was directed at destroying an enemy attacking from the north. The same kind of gun was used both on warships and in the forts. Other guns covered the spaces between them. Hundreds of navvies worked on their construction. They lived in wooden huts nearby. Local birth registers reflect their getting together with the area's girls. An experienced navvy could shift about 20 tonnes of earth daily using a pick, shovel and wheelbarrow. It is a surprise to discover how completely most of the forts are almost invisible from the north – a tribute to the Royal Engineers who designed them. The whole of the crest of Portsdown Hill, including the forts and their connecting hilltop road built by prisoners from the Napoleonic wars was War Department property. Much still remains in MOD hands.

Ironically, the WW1 brick crisis put the forts' survival in jeopardy, when on 2 April 1919 Sir Bertram Falle MP for Portsmouth asked Mr. Winston Churchill, Secretary of State for War "whether use could be made of the obsolete brick forts by utilising the bricks for building purposes? Mr. Winston Churchill responded that as the bricks were set in cement it was felt that too few bricks suitable for new building could be recovered to make the effort financially viable". So the forts remained in position.[6]

Fort Nelson

Fort Nelson. Deane Clark on the restored access bridge

When the decision was taken to build the Portsdown forts it was decided that the one nearest the Nelson Monument of 1807-8 would also be named after him. In 1861 William Tredwell won the contract to build it and in the following year he gained permission to construct a railway line for steam locomotives to pull wagons with building supplies including brick and stone to the site from the main line near Fareham. About 4 million bricks were used to build Fort Nelson, made from clay at brickworks in Wallington, Funtley and in the surrounding fields. Constructed between 1861 and 1870 Fort Nelson covers 19 acres (77,000m2).

It is six-sided, with a deep ditch protected by three caponiers (projecting structures into the moat which provide covering fire along the ditch) accessed via long tunnels through the virgin chalk. Above each caponier is a well-protected emplacement for 13-inch mortars. The original entrance to the fort was via two Guthrie rolling bridges: retractable bridges commonly installed across the narrow steep sided ditches of these polygonal forts.[7] There is a barrack block for 180 officers and men and a hospital ward for 16 patients and stables for two horses, protected by a projecting V-shaped redan. A large open parade ground gives access via circular staircases to the magazines 40 feet underneath it. On the ramparts there are open emplacements for 64 pounder rifled muzzle-loading guns and 6.6-inch howitzers. There are also three Haxo double casemates for 7 inch rifled breech-loaders. Construction was protracted and Fort Nelson wasn't fully armed until the 1890s. A garrison of about 200 volunteers accompanied by regular army officers manned the fort in time of war.

In 1902 Nelson's garrison was stood down and the armaments were removed by 1903. The fort continued in use as a military depot and between 1914 and 1918 it served as a transit camp for soldiers and extensive stables on the parade ground for horses en route to France and Flanders. In 1938 it was converted to the Regional Armaments Depot for the Royal Army Ordnance Corps to store anti-aircraft ammunition for Portsmouth and Southampton. In 1939 ten large brick and concrete parallel chevron or arrow-shaped magazines were built on the parade ground by Fareham firm Crosby & Co. under the watchful eye of the Royal Engineers. The western entrance was greatly widened to allow access to articulated vehicles to collect the ammunition.

Access gate cut through Fort Nelson wall in WWII

Although the fort never saw action against the French it did against the Germans who bombed it in 1941. During the blitz on Portsmouth, on the night of 9th January 1941 despite being under heavy fire, Fort Nelson was commended for supplying a further 1,220 rounds of ammunition to the guns. As it was being used as an ammunition depot its own anti-aircraft guns were placed outside its walls. The fort continued to serve as a naval depot until the 1950s, but was abandoned to decay in the 1970s. A fire started by vandalism damaged it further.

As we explored in Chapter 3, led by the vision of County Councillor Freddie Emery-Wallis, Hampshire County Council anticipated and lead all the work on protecting and reusing historic defence structures around the harbour that took place. In 1979 the county established 'Defence of the Realm' as a tourism theme linked to restoration of key defence sites. Fort Nelson was sold to Hampshire County Council in 1988 for £50,000. What happened there was a catalyst and exemplar for the restoration and new life that developed in many of the other historic defence places around the harbour.

Fort Nelson ditch 2019

When the county took over Fort Nelson it was very overgrown and dilapidated. Very soon after it was acquired, Phil Turner of the County Planning Department and chair of the County Officers' Working Party and John Reynolds, Senior County Architect suggested that a working party of military heritage enthusiasts should work with Cyril Hathaway, Clerk of Works and builders Mick and Bernie Clark of B & M Clark of Copnor to restore the fort.[8] On 21 March 1984 a meeting took place in the Dolphin pub in Old Portsmouth to discuss ordnance being installed in Fort Nelson, the quality of work being undertaken by the county, clearance of trees and undergrowth, and the first Open Day. Present were shipwright Brian Patterson of the Fort Cumberland and Portsmouth Militaria Society who had already set up the Portsmouth Royal Dockyard Historical Society in 1982 to preserve the dockyard's history. He ran an adult education course on 'The defence of Portsmouth Dockyard' in 1982. Also attending were Cyril Hathaway, Peter Cobb, expert on fortifications, John Reynolds, Deane Clark of the Hampshire County Historic Buildings Bureau and Nicholas Hall, a former member of staff at the Tower Armouries and Keeper of Metalwork at the county.

Geoffrey Salvetti, a soldier and former member of the Commando Logistic Regiment Royal Marines with a lifelong love of military history offered to chair the Friends of Fort Nelson, which was formed there in 1984. He identified the link between Fort Nelson and his own corps, the RAOC, and drafted the Society's first constitution, creating a registered educational charity.[8] Guy Wilson, Master of the Royal Armouries gave a supportive address at the inaugural general meeting of the Society on 3 March 1986. Discussions between the Royal Armouries and the county about transferring the collection of artillery following the fort's reconstruction were already advanced. The long relationship between the Royal Armouries and the Palmerston Forts Society is documented in Geoffrey Salvetti's history of the Society. Its members still meet there every month in the atmospheric barrack block with its v-shaped courtyard. First Portsmouth Sea Scouts Celia and Deane's sons Adam and Tom raised the flag there in the early 1980s.

Over many years the society's members restored the fort with the help of working parties of unemployed people, teenagers on work experience and members of the Territorial Army and Army Cadet Force under the supervision of Cyril Hathaway. County Historic Buildings Architect Deane Clark remembers that it was so overgrown that "When the volunteers first arrived, you could hardly see the fort from the road. The bushes and undergrowth were so thick... we didn't even realise there was a moat to the fort. It was discovered by Cyril Hathaway. At the eastern entrance of the fort facing the main road there was an area of asphalt facing south, and when he probed it we found it was the moat – which had been filled in. In the rubble that was removed by volunteers working at weekends, Cyril found the buried bridge. To the northwest behind Fort Nelson there are vast oil tanks connected down to Gosport which are still there."

The Saturday and occasionally Sunday working parties gradually cut down the trees growing into the brickwork and cleared the scrub which filled the moats, assisted by regular small amounts of money from the county council. Cyril taught local builders the skills of knapping flint. He is commemorated by a plaque in the fort, and his ashes were scattered on the northern slopes. When Geoff Salter (PFS chairman from 1987 to 1993) died his ashes were fired out of a cannon by the society's uniformed group! The Society holds monthly meetings in the fort. Their focus is its history, design and restoration and its condition as a structure.

Fort Nelson is a Grade I Listed Building, the highest level of designation. As described in our museums and galleries chapter Fort Nelson was leased to the Royal Armouries in 1995 to house their collection of artillery, opened to the public in 1995.

Fort Nelson North Mortar Gallery

Fort Southwick

Fort Southwick

Fort Southwick (1861-1970) is the highest fort on the hill. It holds the water storage tanks for the other forts, supplying them via a brick lined aqueduct. It was designed to house a large complement of men (about 220) in a crescent-shaped barrack block. Its north projection has one full caponier to defend the dry ditch, and two smaller demi-caponiers at the corners. A small musketry gallery crosses the ditch at the southwest angle to cover a minor branch of it. Mortar batteries were set into the rampart behind the demi-caponiers. In 1893 the fort was armed with a total of 23 guns: nine 64 lb (29 kg) rifled muzzleloaders, eight 7 in (175 mm) rifled breechloaders, and six 8 in (20 cm) rifled muzzle loaded howitzers. Three of these were on ground platforms; the other three on travelling carriages. A central spiral staircase from the surface gives access to four main tunnels cut through the chalk, running radially outward from it at uneven angles past the magazine leading to the barrack block and caponiers.

Although disarmed in 1906, the fort was retained by the military as a barracks, also used to train soldiers how to capture and hold a fort. It became a demobilisation centre for three years after WWI. During WWII the fort housed the Underground Headquarters or UGHQ, the communications "nerve centre" for Operation Overlord in a secret network of tunnels excavated between February and December 1942 by 172 Tunnelling Company of the Royal Engineers 100 ft. (30 m) underneath the fort, well out of reach of any bombs of the era. The call sign of this base was 'MIN'. As part of the UGHQ construction project, a secure underground radio station was also built in the Paulsgrove Chalk Pit. It was once thought that a tunnel was also constructed linking the UGHQ to the Oil Fuel Reservoir, which lies to the west, but there is no evidence of this. The necessity of storing large quantities of fuel for the fleet at Portsmouth in a site less vulnerable than the depot at Gosport had led to the building of a large underground oil fuel tank to the west of Fort Southwick.

During the 1950's, between the Korean War (1950-52) and the Suez Crisis (1956), the Commander in Chief Portsmouth, who, on 21 February 1952 was appointed to the NATO post of Allied Commander in Chief Channel (CINCHAN), decided that to meet his national and NATO responsibilities he would require a Headquarters which could provide better facilities than those available in HM Dockyard. The answer was to reopen the WWII UGHQ at Fort Southwick.

In the Suez Crisis (1956) the entire UGHQ was refurbished to an operational state. On 21st January 1966 the CINCHAN appointment was amalgamated with that of CINCEASTLANT(Commander in Chief Eastern Atlantic) and in consequence Channel staff moved to Northwood during 1968/69. The underground headquarters remained operational as a communications centre (COMMCEN) for the proposed new Naval Home Command Organisation (CINHOME later CINCNAVHOME) but it became evident that it was too big and required modernisation as the communications equipment was long overdue for updating. There was also a problem of water seepage: the tunnels were becoming extremely damp and it was considered unwise to install new communications equipment in damp underground conditions. A further consideration was that after investigation of a serious fire in the underground operations room at RAF Neatishead, the Ministry of Defence fire authorities submitted a report that the UGHQ at Fort Southwick was a fire hazard. When it became clear that the cost of fully modernising the underground headquarters was out of the question on financial grounds it was decided to build a new COMMCEN in the parade ground above.

The fort was retained by the MOD as the headquarters of the Commander in Chief, Naval Home Command until the late 1990's and it formed part of an Admiralty Research Establishment. It was also used by the DCSA (Defence Communications Services Agency) as a communications centre for the Royal Navy. The COMMCEN remained in use until 2001 when its function was transferred to Portsmouth Naval Base.

Fort Southwick has been a Grade I Listed Building since 1987. In the late 1990s the underground complex was strengthened and stripped bare. Hampshire County Council expressed interest in purchasing it from the MOD because of its uniqueness and importance during WWII. It remained part of the Admiralty Research Establishment until 2002, when all operational use of the fort ceased. However because of budget cuts the county did not bid for the property when it came up for sale in 2003. About ten bids were received. It was eventually sold by the MOD to the Fort Southwick Company Limited in July 2003. This has local trustees associated with the Pounds and New family, who also own the Tipner Regeneration Company – another defence site beside the M 275. Their stated intention was to convert the barrack block into luxury apartments but this did not as far as we know, happen. In order to gain a return on their considerable investment, existing post-war buildings are let to other companies, including car sales.

They also intended to open the caponier as a museum depicting the fort's history and exhibiting their collection of military vehicles in the parade ground. In 2006 UCAP Airsoft leased the UGHQ, installed lighting and removed all the rubbish. A year later they opened the tunnels and woodland for combat games. The manager Andy Stevens feels privileged to work there, where world-changing events took place. It was listed Grade I in 1987. [9] It was also on Historic England's Heritage at Risk register in 2016.[10]

Forts Widley and Purbrook

Fort Widley 2009
Bob Hunt Portsdown Tunnels

Forts Widley and Purbrook are the easternmost of the forts along the seven mile crest of Portsdown Hill three to five miles from the dockyard. They were originally supported by two small redoubts known as Farlington and Crookhorn, now demolished. The Crookhorn redoubt suffered from subsidence, and was demolished by 1876. The Farlington redoubt had only the ditches and gun positions dug, and was finally demolished in the 1970s. The tunnel between Crookhorn and Purbrook was rediscovered.

Widley and Purbook were designed by Royal Engineer William Crossman who was on the staff of the Inspector General of Fortifications at the War Office. Their polygonal form has a deep ditch on the outer northern face lined with brick walls. Brick lined tunnels and staircases cut through the chalk give access to rifle parapets, gun emplacements, ammunition stores, and the central assembly ground, an earthen platform. On the south side facing the city are massive arrow-shaped red brick barrack blocks, including two tiers of casemates on the outer faces flanked by plain walling with entrances on each side. The gateways have Norman arches with roll mouldings and two recessed columns with cushion caps, all in stone. [11]

Barrack accommodation was built for both officers and other ranks in a massive red brick range overlooking the town. Fort Widley and the other Portsdown Forts accommodated various infantry units in the barracks. In 1875 the 30th Regiment of Foot were quartered there. By 1886 the South Lancashire Regiment was stationed there

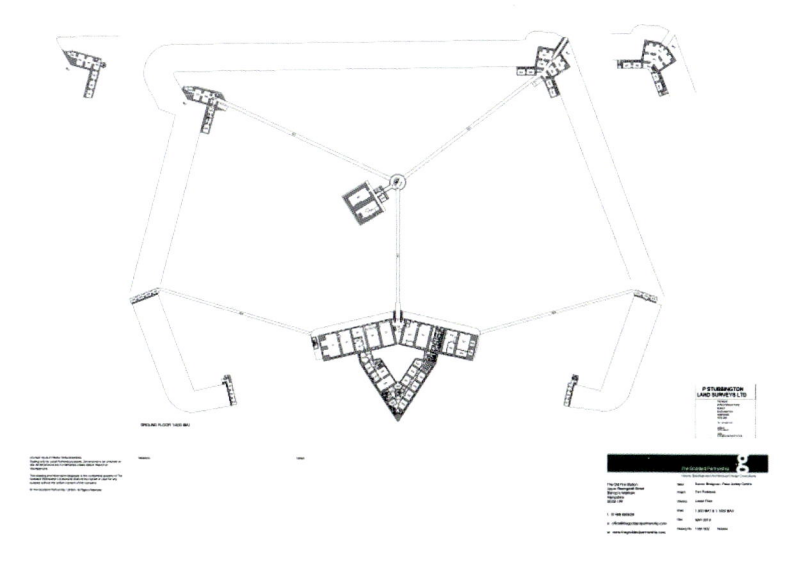

Fort Widley Conservation Management Plan 2020

Peter Ashley Activity Centres

The forts' main armament was mounted on a semi-circular rampart. 13-inch mortars in two protected mortar batteries provided high angle armament; close range armament was in one full and two demi-caponiers. Even before they were completed questions arose over their effectiveness in the face of improving weapons technology and the caponiers were not fitted with guns until some years after they had been completed. This delay resulted in changes to the fort's planned armament.

Interior of barrack block 2017

In WWI Widley was used as a transit depot. During WWII it was modified to provide more accommodation, used by a number of units before housing members of the Royal Corps of Signals and Auxiliary Territorial Service supporting the navy command at Fort Southwick. There was a brick lookout tower when General Montgomery and Eisenhower watched the flotilla of boats in the harbour assembling for D-Day. Their Advance Command Headquarters was at Southwick House a mile to the north.

Cold War Civil Control Centre Fort Widley

In 1952 the fort housed a bomb disposal squadron; a year later an emergency civil control centre for Portsmouth was constructed in the magazine. Celia Clark remembers exploring the fort's many tunnels and ditches when Deane was restoring it. She wondered how the city's elite would reach the fort in time after the two-minute warning of a nuclear attack. She also noted the lack of effective sealed air filtration in the event of contamination or radiation. As far as she knew there were no nuclear shelters for the people of the city, though we were told about the siren in the dockyard - which we would hear in the event of an attack. She was far more worried about an accident at the nuclear plant at Cap de la Hague.

Widley was let to Portsmouth Corporation in 1961 and sold outright to the council in 1972 - on the understanding that it could be used for practice by the fire brigade. It had a very large underground tank next to the parade ground. City historic buildings architect Deane Clark worked on restoration of Forts Widley and Purbook – also in city ownership. His first experience of Fort Widley was seeing smoke pouring out of the underground tank and the firemen escaping – in an exercise. He had a small budget of £25,000 at first, working low key because large sums would have been politically unacceptable - or the city treasurer would have vetoed them. To try and cut down the undergrowth inside the fort sheep were introduced. The ones born in the fort were much more aware of the dangers including steep drops. Rabbits were encouraged on top of the casemates, but they were eaten by foxes. Deane worked with Purbrook Construction on the priority areas: the flint moat walls and the west mortar battery. There was no specification, but he and his colleagues trusted the builder to do a good job. Mr. Pratt was the caretaker who lived in a flat by the entrance, taken on from the MOD by the city.

Cupola of Arundel Street School stored in Fort Widley
Deane Clark

The main eastern entrance to the fort had a series of artillery sheds: these and the parade ground were convenient for storing rescued historic building materials, particularly artefacts from the Grange, a historic Hampshire house which had fallen into ruin, as well as the ceramic pub front from the Hearts of Oak demolished in the widening of Commercial Road and the cupola from Arundel Street School. Rescuing historic building materials was Cllr. Freddie Emery-Wallis's idea. The plan was to reuse them: the city architect suggested that the gates from Knightsbridge Barracks were installed behind the Guildhall, and the lamp standards in Guildhall Square were adapted from the Army & Navy store in London's Victoria Street.

Widley is currently used in two different ways. In 1990 the fort was developed as an activity centre for young people offering sports including archery and an equestrian centre. "A wide selection of horses and ponies range in size & ability ensuring that riders of all abilities can be accommodated. Only BHS qualified instructors are permitted to take riding lessons, ensuring that every riding lesson is a safe & pleasurable experience. We have a purpose built indoor riding school with a heated viewing gallery that was opened in 1995 by HRH Princess Anne. This ensures a great riding experience whatever the weather and adjacent to the indoor school is a large outdoor floodlit manège" according to Darren Bridgman, the Chief Executive.

In September 2010 Widley was used to host a search and rescue exercise based around a simulated earthquake, Exercise Orion. The event involved teams from seven countries dealing with 35 scenarios over two days.

The fort and a gun in front of it are listed Grades II* and II respectively, but Widley and Purbrook are also on the At Risk Register. From about 2013 until it moved, the World War 1 Remembrance Centre occupied the lower ground floor of Widley's barrack block run by a small dedicated team of volunteers. On display were their WW1 artefacts and memorabilia, the All Quiet on the Western Front walk-through reconstruction trench system, memorials and temporary exhibitions covering aspects of the Great War. There were lectures on Ypres and the Somme and the chance to book battlefield tours. Although the centre was appropriately housed in a military building, access was only via steep stairs, so some years later the centre moved to Bastion 6 in Hilsea Lines. However, Pompey Pals still has a research centre and museum based at the fort. A charity originally set up in partnership with Portsmouth Football Club to produce a lasting memorial to the Portsmouth Battalions, it now aims to commemorate all those men and women from the Portsmouth region who served in any capacity during the Great War, whether at home or overseas.[12]

Fort Purbrook 2009　　　　　　　　Bob Hunt Portsdown Tunnels

Fort Purbook, completed in 1870, also run by the Peter Ashley Activity Centres, has many surviving original features: stencilled signs, metal shutter covers to gun ports and historic cannons. James Thomson describes his training as a RADAR technician and his time operating the signalling systems in the fort in the 1950s. "In September 3rd. 1957 I joined the RN at HMS Ganges in the Seaman's Branch. Also training in basic Radar and Plotting, because I had been told it was the cream branch to be in (I hated it). We learned to write upside down for plotting tables, so as the senior rate could read it opposite side, on the Plot. Also writing Back to Front for Air Plots. After Ganges, those trained went to HMS Dryad at Southwick for further training in more advanced operating systems, including reading and writing codes (CRYPTO)."

"The operating RADAR systems were housed in Fort Pubrook, where we learned how to switch on and operate the sets. There were a substantial array of aerials on the top: the 293P, (20 mile range if aerial at 15ft. above sea level); the 960 air search radar which looked like two old fashioned TV aerials opposite each other (200 mile range if the aircraft was flying high enough); the 277Q, nodding radar for gauging air craft height; the 974 navigational radar top of the Bridge. These were the main radars in use at the time. They were quite an impressive sight. (if you like aerials)…"

"The time in Purbrook was quite laid back. Ship's company duties included cleaning, dishing out food in the galley for training classes - general dogsbody really. Lunch times, (I think there were about 8 of us), we used to watch Second World War archive films which were kept there. Also [we] went down into the tunnels to explore, right down to the strong points to prevent enemy attacks in the ditch. The passage from the main accommodation block to the tunnels has firing slots; we were told this was for: THE LAST DITCH STAND DAN DIDDYY DAN!"

"For a walk to the nearest pub or bus stop was quite a trek down Portsdown Hill; we didn't go into Waterlooville, all the senior rates lived there (mixing problem). If a run ashore was needed in Portsmouth centre, it meant staying in 'Dame Aggie Westons', cost 2 shillings and sixpence (12.5 pence). Not including breakfast, for a cubicle, where the ceiling would spin, after 'The Fleece'. After walking up and down Portsdown Hill for 4 months, I had extended calf muscles."

"Having found that it was preferable to have my head under the water, due to working next to the Diving Store on HMS Gambia, (the two standard divers gave me an aptitude test), I had the temerity to request a change of SQ Branch to Clearance Diver, so was banished from the main Establishment to Fort Purbrook, for four months before going to Vernon (Diving Training). *HMS Gambia* was known by the fleet as 'THE GLAM GAM', so spotless, due to the amount of time cleaning. It shows the types of RADAR on the masts which were at Fort Purbrook."[13]

Keith Moon deified outside Fort Purbrook in the film Tommy 1975   Deane Clark

Extras in 1940s costume in Fort Purbrook taking a break   Deane Clark

In the 1970s, following his appointment as Historic Buildings Architect at Portsmouth City Council Deane Clark organised a working party from HMS Collingwood led by a naval officer to spend a Sunday up there hacking back the undergrowth. He supplied the beer! Ken Russell used the fort as a location for his rock opera Tommy (1975) about a psychosomatically deaf, dumb and blind boy who becomes a master pinball player and the object of a religious cult. The sound track was by The Who and the stars were Roger Daltrey, Ann-Margret, Oliver Reed. Deane won a *Portsmouth News* photographic competition prize of a trip to Venice in 1971 with his pictures of the film extras at Purbrook dressed in 1940s clothes. The Portsmouth Youth Activities Committee, a registered charity formed by representatives of most of the sports clubs in the city coming together to share experiences and funding was looking for a base. Peter Ashley who was Chairman of the charity as well as a local businessman and Cosham Councillor made enquiries about Fort Purbrook. "In 1980 the Portsmouth Youth Activities Centre signed a 21 year lease with Portsmouth City Council to open up Purbrook as an activity centre. It was in a very derelict state, windows and doors rotten as well as the place having been vandalised. So the first caretaker was moved in – Jack Tremlett and his family – to look after the security. The trust set about employing the Government scheme of youth apprentices to make repairs and restoration. This was over a 10-year period with over 30 youngsters being trained in the various skills of carpentry and painting that was overseen by two full time skilled foremen. Activity areas were made out of the old barrack rooms and a climbing wall was built in the old underground magazine. Internal showers, toilets and bedrooms were built in the former officers quarters." [14]

Compared with Widley there is currently far more outdoor and indoor activity at Fort Purbrook with many different facilities: climbing wall, archery range, air rifle shooting, assault courses, initiative tests, holiday activities and team building. The barrack block is used as a hostel for language schools and other groups. The centres are open all year round to welcome schools, colleges, clubs, youth groups, individuals of all ages as well as for weddings and private hire.

In October 2010 the trustees signed a new 40-year lease with the city council for the two forts. Both previous leases had run out several years before whilst the new lease was being negotiated. In 2015 the trustees decided to change the charity from being unincorporated to a new Community Interest Organisation that would give the trustees better protection and more leverage for external funding. The new charity was renamed the Peter Ashley Activity Centres; the Youth Activities Center was wound down. In 2017 the fabric of both forts needed considerable restoration work. The charity cannot apply for Heritage Lottery funding in its own right because it does not have a full repairing lease of the two forts of more than 25 years. In 2019-20 the trustees were considering applying to own the fort via an Asset Transfer from the city council – in the same way that Wymering Manor nearby was transferred to trustees in 2013. But this would be a considerable undertaking and responsibility, given the size and poor condition of both forts.

We now describe the forts at the eastern end of Southsea seafront and the seaforts.

Lumps Fort Southsea

In 1545 a circular redoubt was built of stakes and earthworks to prevent a beach landing to the east of Southsea Castle, called after the owner of Lumps Farm to the north. It is mentioned in the Board of Ordnance records of 1805. In 1822 it was in use as a semaphore station on the line from London to Portsmouth, to avoid the smog of the town of Portsmouth which impeded direct optical communications from Portsdown Hill. The line was replaced by the electric telegraph in 1847.

As part of the response to the Royal Commission the fort was reconstructed between 1859 and 1861, armed with 17 guns to command the entrance to the deep-water channel into Portsmouth Harbour. The barrack block to the rear accommodated 40 men and officers. The entrance gate into the fort today forms the north entrance to Southsea Rose Garden: the huge gate hinges are still there on the gate pillars. The fort was surrounded by a deep wide moat which was filled at spring tide low water to a depth of seven feet. By 1861 it had received its armament of thirteen 95 cwt 68 pounders and four 8-inch shell guns. The east and west cavaliers were armed with a 100 pounder Armstrong gun. It was not usually manned, with only a Master Gunner normally in residence. In 1885 the Master Gunner was H Orchard. Units of the Volunteer Artillery also used it - the City of London Artillery Volunteers undertook gun practice on the exterior of the fort in 1889. In 1889 a live shell fell into the ditch, which had not been revetted, and had to be retrieved manually – as dramatically illustrated in *The Graphic* of 27 April 1889. It was re-armed in the 1890s with three 6-inch rifled breech-loader Mk. IV guns on hydropneumatic carriages. At the outbreak of World War 1 the fort was rearmed as a beach defence battery, armed with a naval 6-pounder Hotchkiss anti-aircraft gun.

Obsolete in its defence role, Portsmouth City Council bought the 14 acres of Lumps fort from the War Office in 1932 for £25,000. The sale's terms stipulated that the area could only be used for recreation, parks, pleasure grounds, winter-gardens and public swimming baths. Plans to develop the site were interrupted by World War II.

Lumps Fort seen from the air in 1945

The fort had one last defensive role to play - in July 1942, when a small special Royal Marines unit, the Royal Marines Boom Defence Patrol Detachment was set up for commando raids into Europe, based in two Nissen huts in the fort. The 'Cockleshell Heroes', so called because their canoes were given the codename Cockle exercised in Portsmouth Harbour and patrolled the harbour boom at nights. They carried out the daring raid Operation Frankton on German shipping in Bordeaux harbour. In 1955 the film *The Cockleshell Heroes*, a fictional version of the story was made by Warwick Films, starring Anthony Newley, Trevor Howard, Christopher Lee, David Lodge and Jose Ferrer, who was also the director. A box office hit in 1956, it was quickly followed by Brigadier C. E. Lucas Phillips' book.[15] The fort's transformation into the popular Southsea Rose Garden is described in our chapter on open space.

Submarine Barrier   Dan Bernard

During WWII extensive submarine defences were built: large concrete blocks running about 1.8 metres (5.9 ft) below sea level from Horse Sand Fort to the shore at Southsea where they are still visible seaward of Lumps Fort. With only a single narrow gap to allow small craft to pass through, this barrier (and a much shorter one running south from No Man's Land Fort towards Ryde Sands) remains, as the cost of demolition is deemed too high.

Artillery Volunteers at Lumps Fort
Southsea Model Village

Eastney Forts West and East

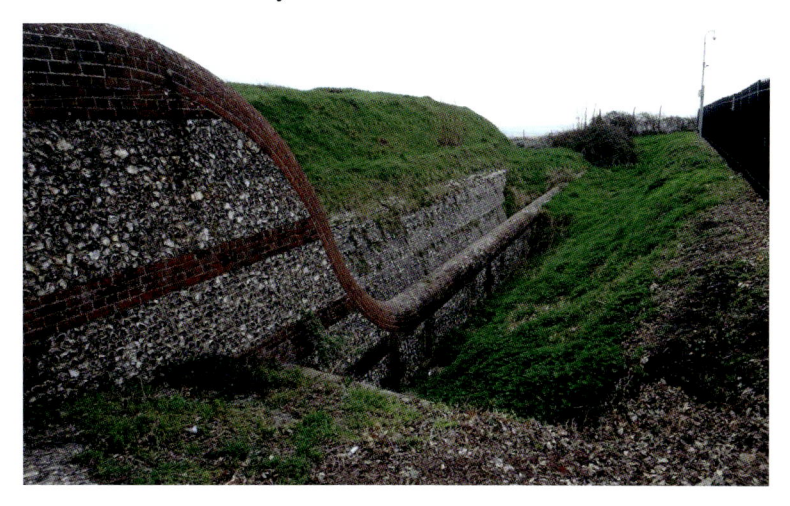

Eastney Fort East

In response to the Royal Commission of 1859 two forts: Eastney Fort East and West were constructed in 1861 in brick with galleted flint panels, completed in March 1863 at a cost of £17,435. They replaced the original Eastney Fort of 1747 designed by Major Archer. They were designed to protect the approaches to Langstone Harbour and to cover the water between Portsea Island and Horse Sand Fort. Each fort has a brick lined and vaulted magazine below ground. Their construction pre-dated Eastney Barracks by a couple of years but the land on which all three structures stood had been owned by the Ministry of Defence from 1860, as an extant boundary stone confirms. The fort and barracks are now bounded by Eastney Esplanade which was later constructed along the seafront.

Eastney Fort West, built in flint, red brick and ashlar, dates from 1862-67 with later alterations. It is a wedge-shaped enclosure facing the sea, with earth embankments, flint-faced internally on the west, south and east sides with a ditch along the front south side, now partly filled. The enclosing wall was built of flint rubble with redbrick bands, quoined pilasters and rounded coping. There are horizontal gun apertures at approximately one-metre intervals and some larger apertures with granite surrounds. The entrance on the north side has large square piers of banded brick and flint with ashlar quoins and capstones. Later brick infill forms a smaller entrance. Inside there are underground casements with arched entrances to lookout positions and steps up to the parapet. It is a Scheduled Monument and is now a private walled garden for the enjoyment of the residents of Eastney Barracks.

Almost hidden behind high ramparts, Eastney Fort East is not open to the public and as a result it has largely passed from the memory of most Portmuthians. Although far smaller than the Portsdown Hill forts, it is nevertheless a fully fortified unit; the only major difference from its larger cousins being the absence of any sleeping accommodation for the soldiers manning it.[16] On the east and west sides there are deep dry ditches, whilst the rear was closed by a loopholed wall three feet thick. A covered way ran between the two forts, but much of this was removed during World War II. The remaining section can be seen at the back of the museum car park, whilst on the side facing the barracks a firing step is still evident.

The earliest record of the armaments at the forts dates from 1869 when five 7" breech loading guns and two guns en-barbette were in place. These would not have been large enough to deter large ships in deep water off Spithead but were more than adequate for defending the coast against small ships and raiding parties. By the 1890s the fort had three 64 pounder, rifled muzzle loading guns with ranges up to 3,500 yards, but these were removed around 1900 and replaced by 6" breech loading guns on hydro-pneumatic carriages which had a range of 8,800 yards. These were the so-called Moncrieff "disappearing guns" also installed at Hilsea Lines, which used the recoil to propel the gun barrel well below the parapet so that it could be reloaded in safety.[17]

The Artillery Volunteers manned the forts from an early date. By 1889 they had been replaced by the 1st City of London Artillery Volunteers. This is perhaps surprising given that the Royal Marine Artillery were housed in the Barracks a few yards away. In an emergency the Marines could have manned the guns and in any case would have been ready to defend the covered way. Of course, neither the Marines nor the Volunteers were ever called upon to put their expertise to the test, as tension with France quickly subsided. There is no evidence that the forts were used at all during the First World War and by 1936 they were handed over to the Royal Navy.[18]

Both forts have been heavily altered, particularly Eastney Fort East which contained several brick buildings from the 1930s and 1940s when the fort was used by the Signal School for the development of radar, and during World War II by the Admiralty Signals Establishment – also on Portsdown Hill - for experimental naval radar. [19] These pioneer inventions took place in utilitarian buildings, similar to those surviving at Fraser Range. Those in the fort had already been demolished.

In 2002 the Royal Marines Museum Director, Chris Newbury, who was a member of the Palmerston Forts Society commissioned its members to restore Eastney Fort East to its first appearance. He was keen to see the fort developed, conserved and used. Potential uses included car boot sales, tournaments, conducted tours, some of which were to be money producing. "This was to be the start of a long term relationship including archaeological work and reconstruction of parts of the fortification including gun ramps, the 64-pounder blocked up carriage position on the southwest salient and the discovery of various features including firing steps and the granite platform blocks, hidden amongst the altered earthworks."

"Despite very bad weather, in a period of some eighteen months, the working party had rebuilt the ramparts, identified various features including musket loopholes in the buried carnot wall, excavated the caponier, and dug trenches to assess appropriate depths and original walkway levels. In October 2004 131 Independent Commando Squadron Royal Engineers (V) provided a group of sappers to clear and excavate the fort under PFS supervision. Prior to this, some 275 tons of spoil had been moved by hand. Their heavy machines achieved an enormous amount, but because of their commitments in Iraq and Afghanistan they were not able to return." [20]

In his interview with Celia Clark in March 2019, Geoffrey Salvetti, whose house overlooked it, described the many years of intensive work on the fort and the discoveries that the working party made. Fortunately most of the fort's original structure remained. Their programme of restoration one day a week for six months of the year began by clearing the structure of vegetation. The Royal Marines Museum gave them a few tools. The work was financed by goodwill. There was an old wooden Nissen hut, and the working party found the solid concrete bases of the three radar masts. In their trial trenches and pits 1m square they tried to identify the steps leading down; they hoped to find the staircase. In a trial trench to find the northeast passageway to the chemin de ronde, they found a ramp with a fillet down the side and massive debris that had survived: radio equipment, valves… These and other finds were put into trays and stored in the back guard post inside the main gate. The Victorian lintels, walkways and paths were preserved because they were covered by a layer of gravel and pea shingle. They found the original door still in situ. The southeast corner was cut off by Southsea Esplanade.

The western side of the fort was filled with debris. It was excavated by hand and with a JCB. The ditch debris had bomb proofing tiles and gravel, behind it earth and clay from the top of the ramparts filled in to create vegetable gardens in WWII– a flat growing area. The material was heaved down the bank. They excavated the southwestern open bastionette and discovered the edges of it.

Under the 1950s vehicle shed was a doorway underground. Its gravel walk was revealed under tons of earth. Adjacent to a 36 pounders' blocked up carriage they found a friction tube for a gun, a Victorian invention using ignition with formulite of mercury that sets off the gun. It's 2 inches long – it was the only evidence of a gun mounted and fired. They gave it to the Director of the museum, but it is no longer to be found. They also rescued a length of timber from the demolished Drill Shed, which had a timber floor. The timbers were lettered and numbered in sequence - which is how the army used materials.

By 2009 two-gun embrasures on the east side had been returned to a condition that would have been recognised by the artillerymen of the 1860s. They had plans to install some contemporary guns, including a disappearing gun repatriated from Hong Kong. It was hoped that when restoration was sufficiently advanced that the fort might be let out for local events. The Society's aim was to advise, help and interpret and interpret the restoration after the archaeology was complete.

The work was halted when a Heritage Lottery bid for funding work on the fort was submitted by the Royal Marines Museum, citing the Society and army's work as matched funding. The PFS was asked by Chris Newbury to stop work temporarily in about 2006/7 because of difficulties with the HLF bid of £650,000 to £750,000 to repair and reinstate the fort. A small Interpretation centre was planned, and finds were to be displayed in a museum. By 2009, however, one HLF application had been refused and two others were deferred. The PFS work on the fort has not been resumed.

"We handed two volumes of photos to the previous director Chris Newbury. The curator was Ian Main, a member of PFS. So the PFS left their tools and equipment in the guard hut for four years." Most of these records, photographs and files cannot now be found. The store where the tools and carefully sorted archaeological finds had been broken into and the finds scattered, losing their significance.[21] Geoffrey asked the Operations Director of the National Museum of the Royal Navy, into which the Royal Marines Museum had been subsumed in 2008, what the plans for the museum were. As we detail in the dockyard chapters the HLF bid to install the Royal Marines Museum into Boathouse 6 had failed, so in 2019 the collection and its archives were inaccessible. Part was moved into the historic dockyard; in September 2019 the uniforms and art were moved into other NMRN storage, including the Fleet Air Arm Museum in Yeovil.

In 2019 the former museum building and the fort were for sale. Ownership of the site is complex: part of the north wing is owned by the Royal Marines Museum, now part of the National Museum of the Royal Navy, and part is owned by the MOD, as is the carpark. According to solicitor Geoffrey Salvetti there are a hundred types of easements – on who can use it, on access…which will make sale and redevelopment complicated, as we discuss in our chapter on residential conversion.

Fort Cumberland

The star-shaped Fort Cumberland of at Eastney Point, a narrow spit of land on the west side of the entrance to Langstone Harbour on the south-eastern corner of Portsea Island is the oldest and largest of the forts around the harbour. The first fort on the site was constructed 1747-50 as part of the Duke of Cumberland"s scheme for extending the defences of Portsmouth, to control the

Aerial view of Fort Cumberland   Historic England

entrance to Langstone Harbour. The ordnance engineer JP Desmaretz who also designed the defences at Hilsea Lines was responsible for the design. Contemporary plans show an irregular star-shaped trace, but today all that survives above ground are the guardrooms and gunners' barracks and the storeroom. In 1782-1812 the outdated first fort was replaced by a second on the same site but extended further to the north, as part of the scheme by Charles Lennox, third Duke of Richmond and Master-General of the Ordnance (1782-95) to improve Portsmouth's defences. Seaborne invasion via Langstone Harbour was still perceived as a significant threat. This second fort was pentagonal and it was the last fully fledged bastioned fort to be built in England. It represents the culmination of two hundred years of military thinking – but it was also the first to have bombproof casemates in the bastions to house guns as well as barrack accommodation for 700 troops.

It demonstrates important changes in the design of bastioned fortification in the later eighteenth century. Examples are the separation of offensive from defensive fire, the use of bombproof casemates in the rampart for guns as well as for barracks accommodation, and active defence using ditch defences. At first no new buildings were constructed on the parade ground. The two buildings from the first fort were adapted and extended for new uses. Over time other buildings were constructed for a range of military purposes, including the cookhouse, officers' quarters, and transport-related structures - and as the original defensive principles of the fort became redundant and were replaced by others, also in the ditch. In the nineteenth and early twentieth centuries the fort's armament was changed, most notably in the 1890s-1900s when the terre-pleins (the top horizontal surface of the ramparts) and parapets of the central, south and left curtains were adapted to accommodate 6- inch breech-loading hydropneumatic guns. These were linked by lift shafts to new cartridge stores, shell stores and shifting lobbies below. Both the defences and buildings on the parade suffered bomb damage in 1940, only some of which was subsequently repaired.[22]

Infantry initially occupied the second fort and the Royal Marine Artillery companies were temporarily stationed there from 1817. In 1858 it was transferred from the War Department to the Board of Admiralty. The Royal Marines Artillery headquarters moved there in 1859 until 1867 when they were relocated to the newly completed Eastney Barracks. Fort Cumberland was kept for gunnery instruction. The Royal Marines Mobile Naval Base Defence Organisation (MNBDO) experimental unit was established at the fort in 1923-4, and continued to operate there throughout WWII. After the war Fort Cumberland became the home of the Amphibious School until 1954. It was used as the base for the Royal Marines Technical Training Wing until 1971, when the Royal Marines' long association with the fort came to an end.

The freehold of most of the fort was transferred from the MOD to the Department of the Environment in 1975, later the Department of National Heritage, then the Department of Culture, Media and Sport. Historic England now manages it on behalf of the DCMS. A limited number of its buildings were initially repaired and converted for use by the Central Excavation Unit, followed by others in the late 1990s to allow the site to become the headquarters of the Centre for Archaeology in 1999. The fort is also home to several research institutions, as explored in our chapter on research.

Events and guided tours are held at the fort on an occasional basis. Most of the fort's area is scheduled monument no. 26723 and Portsmouth 277. The areas in guardianship and scheduling do not correspond exactly. The second fort's defences are listed Grade II* and three others are Grade II.[22]

Rising sea levels are a threat to coastal areas – as the flooding of the fort's dry moat has already shown. Trees and undergrowth were killed by the salt water, which also affected Southsea's Rock Gardens when the sea and beach moved inland during winter storms.

Much remains to be done to restore the fort's many casemates and other structures, a particular challenge in a site managed by the country's national conservation agency. A Conservation Plan was prepared in 2003 in association with the move of the Ancient Monuments Laboratory into the fort. The plan's purpose was to assess the cultural significance of the fort, to provide a conservation framework for managing the fabric above and below ground and the repair, adaptation and reuse of unused areas without compromising their significance, character or visual amenity as well as providing safe access to the fort.

In an era of cuts to government spending there is a new impetus to finding revenue-generating activities to inhabit the fort. A few of the 41 casemates are rented out but the majority remain unused and in need of repair. Historic England's development of a new community of occupants with either military or naval connections or civilian ones to restore and use them is significant – not only here but as a positive and creative exemplar of how to work towards regeneration of these large defence complexes. Recent new occupants include the use of two casemates by veterans and their families: the Basha Retreat, a place for reflection and meeting. Veterans are offered access to mental health professionals and housing, benefits and healthcare services, job search and CV's. They can also learn new skills by repair work on the fort as part of veterans' stay there.[23]

Four more casemates are occupied by the Fort Distillery, which distills Fort Gin, rum and cider. Two of the three founding directors are ex-naval officers who ran the Rum Club Ltd. for the past few years. The team attended distillery courses and visited other distilleries. They joined forces with their sales and marketing director to form the Portsmouth Distillery Company Ltd.[24] Southsea Castle also has a microbrewery. David Edmundson, another occupant said: "Our relationship with Historic England has so far been very good and far from the usual commercial model. When we agreed to take over casemate 52 David Webb said that we could either take it on as a 'turn key' arrangement with it being ready to use or we could carry out some of the work in return for a rent-free period. We agreed the specification with him, waited some considerable time for approval, and when the final plans had been agreed they looked at what Tom and myself were happy to do. We couldn't carry out the electrical work (certificates being required) and decided to leave the plumbing to others as although we had both done plumbing work before, it wasn't something that we yearned to do again. We then costed the work both in materials and time, which we were prepared to and submitted it to David. There was a small amount of discussion around the sums but we quickly agreed on the total value of the work and therefore the amount of rent- free period which it equated to. We also agreed that the rent free time should start once the unit was ready for use otherwise we would use it carrying out the renovation work."

"For us this was a great solution. It removed the immediate pressure to generate income from day 1 in order to pay the rent and also means we feel an ownership in the unit. As might be expected, we carried out more work than planned (something to do with old buildings and no maintenance?) but as we felt 'ownership' we wanted to do things properly rather than just get the unit up and running. Historic England were very good about fitting out the unit for our needs, which was paramount to us and we are delighted with the result."

"They probably have a mission statement but I haven't seen it in print. They do however have a long-term plan to open up many more of the casemates for suitable businesses in order to generate income for maintenance of the Fort."[25]

Architect Deniz Beck who was also responsible for the conversion of Spitbank Fort, the Hot Walls Studios in Old Portsmouth and Fraser Battery adjacent to Fort Cumberland said: "There isn't any tenants' committee in Fort Cumberland but in the future when there are more external companies located there, I think there will be one. I have a unit there as well, but like David Edmundson said it is a long process. I am working on the conversion issues so that I can put the Scheduled

Monument Consent (SMC) application for it. I have done the SMC application for the distillery office and the store next to it as Historic England appointed me to do this. On Casemate 52 conversion, I was appointed directly by the tenants. I was also appointed by the HE to do conversion application for the veterans' three casemates and Alex's studio (Casemate 42) and the COM (Casemate 37)." [26]

As well as Fort Widley, Cumberland's large acreage has also been used for rescue and storage of architectural or engineering structures. From 2019 discussions about where to re-site the first surviving bridge by Portsmouth born Isambard Kingdom Brunel (1806-1859) somewhere in Portsmouth were on going. Its massive iron beams and brickwork had been stored in the fort since it was dismantled and brought down from its original location in the approaches to Paddington Station. Spanning the top of the canal in Gunwharf, near to Brunel's birthplace seemed an excellent location, but disappointingly its owners Land Securities declined the offer. They may be considering leaving retailing, which is declining in many centres, including Southsea. Reassembling the whole structure adjoining the King's Bastion in Old Portsmouth was proposed by Deniz Beck and endorsed by Dr. Steven Brindle, Brunel's biographer and rediscover of the bridge with industrial archaeologist Malcolm Tucker, but there are also other proposed sites including the dockyard.

Brunel's bridge from Paddington stored in Fort Cumberland

Most of the area around the fort on Eastney Point remained in military-associated usage and farmland until the later part of the twentieth century. A caravan park was established to the southwest, new housing built to the west, and a marina and a car park to the north. The area to the east continues to be used for sewage-related drainage tanks installed in the nineteenth century. Once the military role of Fraser gunnery range ceased, in 2003 QinetiQ was planning to redevelop the area to the south and west of the fort,[22] but this did not happen. The 2019 proposals for the site are discussed in our residential chapter. As we describe in our open space chapter, on the other side of the road to Hayling Ferry Southsea Marina was excavated of the dockyard's toxic waste in the former 'Glory Hole'.

The Solent Seaforts

To finish, we explore the building and subsequent uses of the four forts that sit like large cakes or plugs at Spithead in the Solent between Portsmouth and the Isle of Wight. One man, who declared his passion for them, achieved the most extraordinary transformation of two of them, from dereliction to glamorous leisure venues.

From the start it was undecided what form these outer defences of the harbour should take. There were many changes in the plans. In

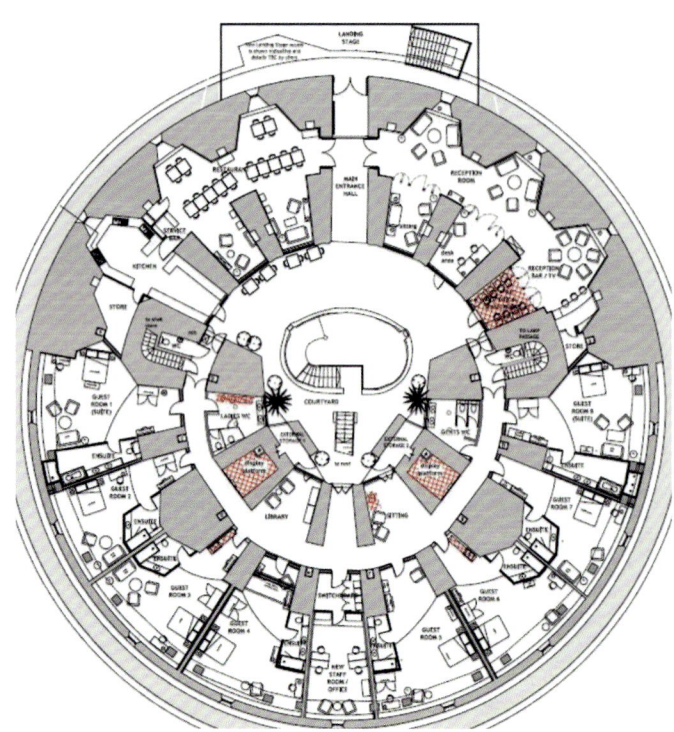

Spitbank Fort War Office drawing   Deniz Beck

1860 it was suggested that the forts should be of casemated masonry design. To meet the possibility of an enemy anchoring large ships close to any of the forts on the shoals to engage them at short range, it was recommended that they should be of large dimensions, and that the more important should have three tiers of guns in casemates so as to increase the possible amount of fire on any point, and give a command over the decks of attacking ships. The forts were to be built of masonry, faced with granite, with wrought-iron embrasures. By 1861 the plans had been reconsidered. They were to be 200 feet in diameter instead of 300 feet as it was advisable to adopt an iron construction to permit a reduction in the intervals between guns. This made the forts capable of mounting 123 guns instead of 100 in comparison with masonry construction. [27]

Four armour plated and granite structures, each sited on a pebble shoal with its own artesian well were built to defend the harbour approach to the dockyard. Their ingenious construction is a testament to the skills of Victorian military engineers and the divers who put the stone foundations in place under water. The French never came – so maybe they were a deterrent - but as they and the twelve land forts surrounding the dockyard cost over a million pounds, they were labelled 'Palmerston's Follies'. Intermittently occupied in WWI and WWII, once their military purpose was lost, the seaforts were left empty and abandoned for many years, subject to boat-bourne vandals and curious trespassers. They were scheduled as Ancient Monuments in 1967.

PLC Architects' first involvement in the sea forts was assessing how to install services in the never completed St. Helen's Fort (1867-1880) off the east of the Isle of Wight. The owner had permission to convert it to residential use for weekend stays and fishing, suitable because it's smaller than the other forts: 20m in diameter, but in the end he did not do so. It's only accessible on foot along a causeway in August low tides.

Spitbank Fort

Spitbank Fort 2015

Spitbank Fort (1861- 1878), 49.4 metres/162 feet in diameter is close to the Southsea shore. It was declared surplus to requirements in 1962 and disposed of by the Ministry of Defence twenty years later to pioneer Sean McGuire who remortgaged his house to pay the purchase price. He restored it as a museum celebrating Portsmouth's Victorian forts – on land and sea. In 1989 or 1990 Sean invited Celia, Deane and Catherine Clark to spend a night in the hut on top of the fort: a surreal experience! We painted the Solent sunset as the sun went down, but once the generator was switched off we were marooned – only a mile from home – which we could see across the strait. The graffiti and marks made by bored soldiers stationed there – maybe for weeks on end – testify to their enforced isolation, so temptingly close to the fleshpots of Portsmouth… The surprisingly loud noise of passing ships' engines woke the Clarks intermittently in the light summer night. Sean had said he would return in the morning, but in the days before mobile phones it was afternoon before his boat returned. They were glad to come home. In 1993 the Clarks rented the fort for Celia's 50th birthday party. Her fort-shaped cake, iced grey, with gunports, was on board as they approached the fort with a hundred guests. But the seas were too high for them to land, so their boat retreated to a tour of the harbour, while they drank all the drink. The party had to be transferred to their house, food, jazz band and all…

Larger and safer landing stages were the key to the new use for the forts. As architect Deniz Beck pointed out, if there is a full boat of visitors you need a nice platform for when the weather's bad. At first the entire boat was hoisted up, which limited its size as it couldn't have a lot of people on board. But now luxury seekers pay good money for a more glamorous dining experience. The view from the top cocktail lounge which replaced the hut is impressive. The voyage out by rib from the Camber and seeing our home city from this close vantage point are very special. Watching the dizzying speed of the extraordinary America's Cup yachts – built in the Camber Old Portsmouth – at close quarters as they whizzed around the fort in August 2015 and 2016, must have been an extraordinary experience.

Spitank Fort with Americas Cup spectators

Sean also worked on No Man's Land Fort off Ryde (which he didn't own). painting the huge exterior black – even though if he fell into the sea, he couldn't swim. He showed us the converted interior, with a table set for twentyfour guests in the curving mess hall. Spitbank's maintenance costs both above and below the waterline including the asphalt waterproofing made his fort increasingly difficult to maintain. In 2009 Spitbank was up for sale for £800,000 but was sold before auction reportedly for more than £1m in 2012.

The new owner, Mike Clare, founder of the firm Dreambeds sold it, and famously purchased the fort without seeing it. His firm Clarenco also bought the other two the Solent seaforts in 2009 and 2012, because of "a personal passion" – as he told Celia Clark on the way out to judge Spitbank for the Portsmouth Society's Best Restoration Award in 2013. Deniz Beck

Spitbank Fort interior

and Sam Brooks of PLC Architects were invited to design its conversion as a luxury event venue run by his firm Amazing Retreats, first based in Royal Clarence Yard, but now in Gunwharf Quays in the Gatehouse on the canal. PLC also designed the extension there.

The transformation of Spitbank into an extraordinary hospitality venue was an amazing achievement given the difficulty of transporting building materials from boats to the interior of the fort and providing modern services. The total cost of purchase and construction was £3 million. Work on the roof began in 2010 and the waterlogged building took several months to dry out. 600 sacks of rubble were removed. The company worked closely with local historical societies and English Heritage to retain and make features of importance historical details.

An 11-metre working boat belonging to Gosport based Fortress Marine Services chalked up about 1,800 trips taking supplies to and from the fort. The fort's mechanical services were renewed including a sewage plant. The massive brick, iron and teak structure was carefully repaired, while new facilities: a professional kitchen, eight glamorous bedrooms with extraordinary views out to sea from the opened up casemates, private dining room, function room, hot tub, sauna and fire pit for barbecues, games rooms, wine cellar and tasking room and bar have been ingeniously inserted. The artesian well into the chalk bed originally supplied 23.000 gallons a day. Drinking water from it is now bottled and served to guests. When it opened, the tariff for the fort – which sleeps 16 in eight rooms and can hold up to 60 people for functions and 48 for banquet dining – was £14,400 for one night. Seven nights cost visitors £61,200. But there were promotions of exclusive use stays from £8,000 and day conferencing from £150.

The interior design, reminiscent of a gentleman's club was by Carol Clare. Drayton firm ITD Consultants were responsible for the planning and design of all the engineering services in this challenging location in the middle of the Solent. Other firms involved included structural engineers Hamill Davies, service engineers Lowe & Oliver, and contractors Newman Scott. Together they have brought this neglected Ancient Monument back to life. It was the venue for the Coalition Festival in the summer of 2009 and other psytrance and hard dance parties. Police choirs sang in its beautiful oval interior, framed by iron steps, and overnighters enjoy the glitzy bedrooms with their giant mirrors and four-poster beds.

No Man's Land Fort

No Man's Land Fort   Solent Forts

The larger No Man's Land Fort (1867-1880) was built in the Solent 2.2 kilometres off the coast of the Isle of Wight. Completed in 1880 at a cost of £482,500 - a considerable sum if adjusted for inflation. 61 metres in diameter, it is almost identical in design to Horse Sand Fort. It was sold by the government in the 1960s. The 1972 *Doctor Who* serial *The Sea Devils* used the fort as a filming location for several scenes when Jon Pertwee was the Doctor. It was converted twenty years later into a luxury residence, including a curved swimming pool, two helicopter pads and golf driving- range. The central courtyard was glazed over and there are 21 bedrooms in the former gun emplacements. The fort was later used as a hospitality centre for high-paying guests – capitalising on the privacy it offers. It has an indoor swimming pool and two helipads. In July 2004, Legionella bacteria was found in the hotel's water system, forcing its closure.[28] The corporate events company that owned it went bankrupt. The fort was put up for sale in 2005 and again in 2007, but the company collapsed.[29] Banks took possession and the fort went on the market for £2-3 million. In 2006 it was acquired by Charyn part of property finance company Lexi Holdings run by a Manchester entrepreneur. The company failed in October 2006 with debts of over £100 million. In March 2008, Harmesh Pooni, claiming he was still the owner, barricaded himself inside the fort in protest against the administrators KPMichael DruMG.[29] They put it up for sale for £4 million in July 2007. In 2008 damage by vandals was being repaired in preparation for a renewed sales effort. The forts' wells silt up and need regular clearing. The property was eventually sold by KPMG for £910,000 in March 2009.[30] In March 2012 it was purchased by Clarenco with the intention of refurbishing it as a hotel.[31] The hotel opened in April 2015.[32]

Part of Clarenco's portfolio: its marketing said it all: "No Man's Fort is the perfect place for a show-stopping banquet, wedding or house party. Look up at the stars through the spectacular glass roof in the Atrium, right at the heart of our venue, or enjoy the best view in the fort from our luxurious Lighthouse Suite. With floor to ceiling windows for an unrivalled panoramic view of the fierce beauty and dramatic seascape – this really is the venue that pushed the boat out."[33] It was transformed into a four star island hotel, with a lighthouse with private dining, rooftop hot tubs, sauna, fire pit, hot pool and helipad.

In October 2016 No Mans Land Fort and Spitbank Fort were for sale for £8m.[34]

Horse Sand Fort

Horse Sand Fort – Dan Bernard

The fourth of the Palmerstonian sea forts, Horse Sand Fort has proved harder to reuse. It was designed by Captain EH Stewart overseen by Assistant Inspector General of Fortifications Col. WFD Jervois.

Construction began in 1865, but it was not completed for a cost of £424,694 until 1880, long after the threat of a seaborne invasion from France had passed. It was built on a ring of large concrete blocks with an outer skin of granite, the interior filled with clay and shingle and covered with a thick layer of concrete. The lower foundation walls of the fort are 18 metres (59 ft.) thick. The fort is split into three levels with a diameter of 62.4 metres (204 ft. 9 in.). The floors were originally used to store armoury and guns as well as supplies for the men stationed on site. According to Clarenco's 2020 report those serving on the fort were deliberately chosen for their inability to swim, to avoid any attempt to escape. By 1874 its heavy guns were changed to the latest 12.5-inch 38-ton guns on the lower gun floor and 10-inch, 18-ton guns on the floor above. The fort became a test-bed for further developments in artillery, including the use of new hydraulic systems. In the 1880s, the outer ring of magazines in the basement level was filled in to provide more substantial footings for new 12-inch 45-ton guns on special mountings.

In 1909 Horse Sand was the southernmost extension of the concrete block submarine barrier seaward of Lumps Fort designed to prevent to Portsmouth Harbour. Some of the heavier guns were removed or replaced in following years and new quick-firing, breech-loading guns were added to counter the threat of attack by smaller craft. The top has a lighthouse, chimneys and ventilators. The seaward side was covered with heavy iron-armoured plating to protect it from seaborne attack. Access to the fort was by a wooden-decked landing stage supported on cast-iron piles. In the late 19th century the Solent forts were painted in a black and white checkered paint scheme as an early form of dazzle camouflage. In its unrestored state part of this pattern is still visible. The fort was equipped with anti-aircraft guns and armed during both world wars. During WWII a boom was laid across the open water between Horse Sand and No Mans Land forts to improve the security of the harbour approaches. Early electronic equipment was also installed to detect the approach of enemy craft. The last gun was removed in 1951 and the MOD declared it surplus in 1963. It was listed as a scheduled monument in 1967.

Sailor P Madge née Nicholson cooked for members of the Royal Ocean Racing Club on top of Horse Sand Fort, when cross-channel races ended there. Deane and Celia Clark remember visiting the fort in 2014 after the Portsmouth Naval Base Property Trust in 2014 purchased it at auction for £50,000. The boatman carried a rifle to shoot the hundreds of pigeons nesting there: the smell of guano was overwhelming. It had not been occupied for decades. The financial overrun of the Portsmouth Harbour Renaissance scheme obliged the trust to sell it – at a considerable profit: £350,000. A subsequent owner applied for planning permission to turn it into flats, but did not do so, and in March 2012 AmaZing Venues, Mike Clare's firm, who intended to convert it into a museum, purchased the fort.[35] Many more original features remain since it is more remote than the other forts: 100 chambers, living quarters, original gun carriages, signage and wall hooks. In January 2015 the BBC TV programme *Antiques Road Trip* included a clip of a visit to the fort.[36] AmaZing Ventures' planned exhibition would have offered visitors the opportunity to learn more about the monumental transformation undertaken on both Spitbank and No Man's Land. They planned to open it to the public in 2016,[3] but this did not happen.

In July 2020 *Country Life* reported that all three of Mike Clare's forts were for sale, either individually or as a group of two or three through estate agents Strutt & Parker. Offers of around £4m for Spitbank: 'a 33,000 sq ft boutique retreat on three floors', £4.25 for NoMan's Fort: a 99,000 sg ft hotel, restaurant and leisure complex on four floors, including a helipad; and £750,000 for Horse Sand Fort 'a blank canvas, with 1000 chambers and living quarters, plus the original gun carriages".[38]

The sheer number of the harbour's forts has been a stimulus to creative reuse of most of those that survive, while others are in a poor state or disappearing. Current uses are industrial estate

(Forts Fareham and Wallington): Royal Armouries artillery museum (Fort Nelson); paint-ball venue and car sales (Fort Southwick): Activities Centres and stables (Forts Purbook and Widley): Historic England Centre for Archaeology, Maritime Archaeology, Veterans' Centre, distillery, boat builders and restorers, architect's office (Fort Cumberland); derelict (Eastney Fort East, Horse Sands Fort): private garden (Eastney Fort West); luxury event venue (Spitbank Fort): hotel (No Mans Land Fort).

Sources

Norman Vidler 1969 *The story of ASWE*

Portsmouth Society News Autumn 2009 Page 4 www.portsmouthsociety.org.uk http:// www.portsmouth.co.uk/news/defence/portsmouth-picked-for-new-4m-defence-research- base-1-6176234

http://www.portsdown-tunnels.org.uk/index.html

Ordnance Survey 1:25,000 scale Explorer mapping accessed 02 November 2014

John Webb 1977 *The Siege of Portsmouth in the Civil War*. Portsmouth City Council. p. 14. ISBN 0-901559-33-4.

Penny Legg 2010) *Folklore of Hampshire*. The History Press. p. 18. ISBN 978-0- 7524-5179-4. http://www.subterraneanhistory.co.uk/2009/03/fort-fareham-hampshire.html

Historic England. "Details from listed building database (1094240)". *National Heritage List for England*. Retrieved 26 June 2015.

http://fwie.co.uk/

Historic England "Details from listed building database (1094233)". *National Heritage List for England*. Retrieved 26 June 2015.

Historic England "Details from listed building database (1350616)". *National Heritage List for England*. Retrieved 26 June 2015. http://www.subbrit.org.uk/rsg/sites/f/fort_southwick_comcen/ index.html

Portsmouth Society News Winter 2012 Page 11

Dockyards The Naval Dockyards Society July 2009 Volume 14, Number 1 pages 4-5 www. navaldockyards.org

Portsmouth Naval and Defence Heritage. Page retrieved 29 July 2005.

Fareham Borough Council Page on Fort Nelson. Page retrieved at 12.20pm 29 July 2005. Portsdown Hill article.

David Moore 1994 *Arming the Forts The Artillery of the Victorian Land Forts*. The Palmerston Forts Society p. 4. ISBN 0-9523634-0-2

David Moore, Geoffrey Salter 1995 *Mallet's great mortars (Great Victorian guns-1)*. Palmerston Forts Society. pp. 8–9. ISBN 0-9523634-3-7

https://royalarmouries.org/projects-and-associations/palmerston-forts-society/

"Subterranea Britannica: Research Study Group: Sites: Fort Southwick NATO Communications Centre". Retrieved 30 December 2014.

Historic England. "Details from listed building database (1167213)". *National Heritage List for England*. Retrieved 29 September 2015.

Historic England. "Details from listed building database (1104368)". *National Heritage List for England*. Retrieved 29 September 2015. http://www.subbrit.org.uk/rsg/sites/f/fort_southwick_ comcen/index.html http://ucap.co.uk/ucap-airsoft-venues/the-bunker/

https://historicengland.org.uk/listing/the-list/list-entry/1092134).

Arthur Corney 1984 *Fort Widley and the Great Forts on Portsdown*. Portsmouth City Museums. pp. 5–6, 14, 21–25.

Garry Mitchell, Peter Cobb 1986 *Fort Nelson: History & Description*. p. 8. ISBN 0-947605-04-5.

Army List, HMSO, 1875

Dominic Blake 8 September 2010 "Hampshire's rescue service skills put to the test". *BBC news*. Retrieved 24 February 2012.

"Mock earthquake test for rescue teams in Portsmouth". *BBC News*. 8 September 2010. Retrieved 24 February 2012.

Historic England. "Details from listed building database (1387128)". *National Heritage List for England*. Retrieved 29 September 2015.

Historic England. "Details from listed building database (1387129)". *National Heritage List for England*. Retrieved 29 September 2015.

http://www.peterashleyactivitycentres.co.uk/horse_riding_portsmouth.asp

http://www.peterashleyactivitycentres.co.uk/index-ww1_rememberance_centre_widley_portsmouth.asp

http://www.subterraneanhistory.co.uk/2008/10/fort-purbrook-portsmouth.html

Arthur Corney 1965 *Fortifications in Old Portsmouth - a guide*. Portsmouth City Museums. pp. 10– 12.

English Heritage – Lumps Fort

Castles and Fortifications of England and Wales

Army List, HMSO, July 1885

The Graphic, 27 April 1889, p.445

John Webb, Sarah Quail, P Haskell, Ray Riley 1997 *The Spirit of Portsmouth: A history*. Phillimore & Co. p. 66. ISBN 0-85033-617-1

Ken Ford 2010 *The Cockleshell Raid – Bordeaux 1942*, Osprey Publishing ISBN 9781846036934 p.11

I V Hogg and LF Thurston 1972 *British Artillery Weapons & Ammunition 1914-1918*, pages 188-189. Ian Allan, London. ISBN 978-0-7110-0381-1

The Coalition Festival, 25 July 2009: https://www.efestivals.co.uk/festivals/others2009/coalitionpeace Overload, HarderFaster: http://www.harderfaster.net/?section=whatson&action=showevent&eventid=57a14424308a61a2116c6e95be87ca1c

The Sun, 12 October 2009, "£800k fort for sale ... lovely sea views", accessed 12 October 2009.

Metro, 12 October 2009 "Sea fort put on the market for £800,000", , accessed 12 October 2009.

Daily Telegraph, 4 November 2009 Historic Spitbank Fort sells for £1m,

English Heritage. "Spitbank Fort". Retrieved 17 January 2015.

Tom Robbins (April 28, 2012). "Naval gazing: A Victorian fort on the high seas has been reopened as Britain's most unusual luxury hotel". Ft.com.

https://www.historicengland.org.uk/listing/the-list/list-entry/1018587

www.portsmouthsociety.org.uk/docs2013/Design%20Awards%202013.htm) Amazingretreats.com 23 November 2011 09.58

https://historicengland.org.uk/listing/the-list/list-entry/1018588 accessed 15 August 2020

https://wetwheelsmaritimeadventures.co.uk/discover/solent-forts/ accessed 15 August 2020

Portsmouth Naval Base Property Trust *20 Year Review 1986-2006*

References

1. National Archive Plans of Fort Fareham held in WO78/ 3670 and WORK 43/240 to 244
2. Solent papers No.7: Fort Fareham by David Moore page 24 ISBN 0-9513234-8-2
3. Historic England. "Details from listed building database (1094240)". National Heritage List for England. Retrieved 1 October 2015.
4. Listing NGR: SU5862006786 https://historicengland.org.uk/listing/the- list/list-entry/1094233. https://en.wikipedia.org/wiki/Wallington,_Hampshire
5. DSTL "About us" Dstl Archived on 21 October 2013 Retrieved 16 February 2014
6. Geoff Hallett 2016 'Fareham Reds' *The Redan* Palmerston Forts Society Number 77 2016 p.68
7. David Moore 1992 'Guthrie's Rolling Bridge' *The Redan* No. 26 October 1992: Journal of The Palmerston Forts Society
8. Geoffrey Salvetti 2009 *Palmerston Forts Society Forts, Guns and Holes A Concise History of the Palmerston Forts Society 1984- 2009* The Palmerston Forts Society
9. https://historicengland.org.uk/listing/the-list/list-entry/1167213
10. https://www.portsmouth.co.uk/news/historic-fort-among-buildings-high-risk-named-historic-england-1188085
11. https://historicengland.org.uk/listing/the-list/list-entry/1092134)
12. (http://pompeypals.org.uk/latest- news/).
13. James Thomson Email to Celia Clark 5 May 2017

14 Email from David Horne to Celia Clark 12 January 2017

15 C. E. Lucas Phillips. *Cockleshell Heroes*. William Heinemann, 1956. Pan reprint 2000 ISBN 0 330 48069 3; Quentin Rees *2011 Cockleshell Heroes: The Final Witness. Amberley ISBN 978-1445605951;* Quentin Rees 2008 *The Cockleshell Canoes: British Military Canoes of World War Two* Amberley ISBN 978-1848680654; "Operation Frankton" Royal Marines. Archived from the original on 8 September 2008 Retrieved 13 May 2010

16 http:// historyinportsmouth.co.uk/places/eastney-fort-east.htm © Tim Backhouse 2007-2019

17 https:// en.wikipedia.org › wiki › Disappearing gun

18 http:// historyinportsmouth.co.uk/places/eastney-fort-east.htm© Tim Backhouse 2007-2019

19 https://www.portsmouth-guide.co.uk/local/ eastfort.htm

20 Geoffrey Salvetti *Palmerston Forts Society Forts, Guns and Holes A Concise History of the Palmerston Forts Society 1984-2009* The Palmerston Forts Society 2009 pp.28, 31

21 Geoffrey Salvetti Interview with Celia Clark 18 March 2019

22 Liv Gibbs and English Heritage 2003 *Fort Cumberland, Hampshire Conservation Plan* Adopted 2003 Volume 1 of 11

23 https:// www.forgottenveteransuk.com/basha-retreat/

24 https://www.teamlocals.co.uk/news/portsmouth-distillery-open-for-business).

25 David Edmonson Email to Celia Clark 2 April 2020

26 Deniz Beck Email to Celia Clark 3 Aoril 2020

27 http://www.palmerstonforts. org.uk/redan/spit.htm); The Redan Palmerston Forts Society Number 77 2016 p.15

28 *BBC News* 9 July, 2004 "Legionella bacteria found at fort now off limits until the health risk is controlled" External link in |publisher= (help)

29 *BBC News* 10th July 2007; "Who will buy fort used in Doctor's adventure?",*The News*, Portsmouth, July 9, 2007; David Ward "Bolthole in the Solent goes on sale for £4m" *The Guardian*, July 16, 2007 Retrieved 2010-05-; "Besieged man barricaded in fort". *BBC News*. 7 March 2008. "A sea fort [...] is under siege from creditors, prompting a businessman to barricade himself inside." External link in|publisher= (help)

30 "'Bargain' sale of £14m Victorian seafort". *BBC News*. 2009-07-03. Retrieved 2010- 05-20

31 "Millionaire snaps up three forts off Portsmouth" *BBC News*. 26 March 2012. Retrieved 27 March 2012

32 "Historic No Man's Fort in the Solent re-opens as hotel" *BBC News*. 23 April 2015. Retrieved 23 April 2015

33 https://solentforts.com/about/solent-fort-history/

34 Matthew Mohan-Hickson *The News* 11 October 2019 'inside the £8m Solent Forts which come with a pub, lighthouse, spa and more'; Michael Drummond 12 October 2019 'Do these float your boat? Forts up for sale' i Newspaper

35 "Millionaire snaps up three forts off Portsmouth". *BBC News*. 26 March 2012. Retrieved 27 March 2012

36 "Antinques Road Trip". BBC. Retrieved 17 January 2015

37 "Horse Sand Fort". AmaZing Venues. Retrieved 17 January 2015 https://solentforts.com/horse-sand-fort/

38 Penny Churchill 11 July 2020 'Live in your choice of Victorian sea forts, from a boutique delight with helipad to a crubling wtect that's a blank canvas" https://www.countrylife.co.uk/property/live-in-your-choice-of-victorian-sea-forts-from-a-boutique-delight-with-helipad-to-a-crumbling-wreck-thats-a-blank-canvas-216940 accessed 14 August 2020

Chapter 14
Education: University and Schools in former defence buildings

This chapter explores how education is a key activity on former defence sites on both sides of the harbour. Instead of soldiers' shouts and drills, school and college pupils and university students now throng former barracks in both Gosport and Portsmouth. The research chapter explores sites used for research: into defence development, marine and land-based archaeology and heritage conservation. The university's many-stranded research is not examined here, except for work on the economic impact of the dockyard and associated industries on the local economy.

Barrack building - and demolition

"Before the nineteenth century solders were generally billeted in private homes or in local inns and hostelries; barracks buildings only existed within large fortifications. However, during the Napoleonic Wars the sheer numbers of soldiers on active service, particularly on the south coast where troops were stationed to repel the threat of a French invasion, forced the government to provide alternative accommodation. A Barrack Department was established in 1792, and standard plans and types were produced. It was then general practice to convert existing buildings into soldiers' accommodation and, according to English Heritage, the small former warehouses now occupied by the [Portsmouth] Lower Junior School are 'rare and important examples of industrial buildings for soldiers' living quarters.' "The warehouses had been converted to barracks in 1875… Soldiers marched from here to all corners of the British Empire, to the Boer Wars, to India, to the Sudan, to the Far East, and to the battlefields of the First World War."[1]

It is not generally appreciated just how significant Portsmouth and Southsea were to the Army, even into the immediate post World War II period. Right up to after 1945 the scale of the military presence in Portsmouth and the surrounding areas equalled that of the navy. Soldiers and their barracks were highly visible on both sides of the harbour. Portsmouth itself was a garrison town. It served as the HQ of Southern Command in the late 1930s. Just before WWII the future Field Marshall Montgomery resided at Ravelin House during his brief tenure as its Commanding Officer. In addition, not only was there a strong Royal Marine presence in Eastney: their Artillery Barracks, Officers Mess were there; they also manned the Palmerston forts. On the Gosport shore, St Vincent Barracks in Forton Road served as the Depot for their Light Infantry wing.

There were three major military barrack complexes in Old Portsmouth: Colewort Barracks in St. George's Road and Clarence and Victoria Barracks to the east in Alexandra Road, renamed Museum Road when Clarence Barracks was converted into the City Museum and Art Gallery in the 1970s. Milldam Barracks in Portsea north of the railway line was built for the Royal Engineers. It is now used as polytechnic/university lecture rooms – as explored later in this chapter. The Engineers' Drawing Office is now the City Register Office.

A significant proportion of historic defence sites on the Gosport peninsula became redundant as a result of the dramatic reduction in the military rather than in the naval role of the area. Once the soldiers left, not all the barrack blocks on either side of the harbour survived. Point Barracks which closed off the southern side of Broad Street in Old Portsmouth until the 1960s has completely disappeared, as have most of the buildings in Gunwharf. Only the outer wall of Colewort Barracks still stands in St. George's Road Old Portsmouth. Hilsea Barracks' redevelopment for housing is described in the chapter on residential conversions.

When Gosport's St. Vincent Barracks was converted into a comprehensive school, now St. Vincent College, only the frontage buildings and entrance gateway survived – due to the determined efforts of conservation architects and planners at Hampshire County Council – as explored later in this chapter.

Cambridge Barracks: Portsmouth Grammar School

Portsmouth Grammar School was the first institution to seize the opportunity offered by empty barrack blocks and land cleared of the town's defences. When Portsmouth's fortifications were demolished in the 1870s the Grammar School built a new building over the old foundations (on the opposite side of the roundabout) for the Junior School in 1879, designed by architects Davis & Emmanuel. Barrow Emmanuel was the son of Portsmouth's first Jewish mayor, Emmanuel Emmanuel. They also built synagogues and offices, banks flats and warehouses in London.[2]

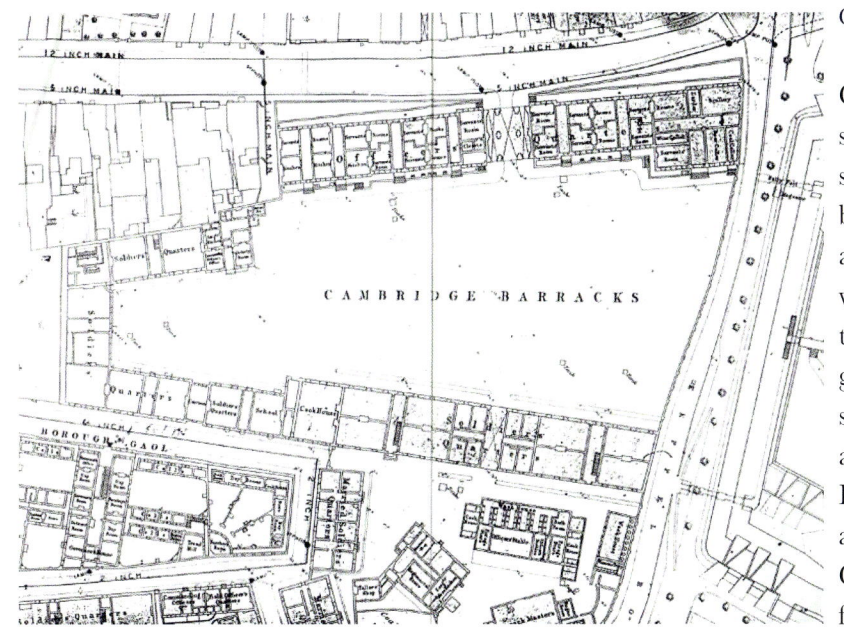

Cambridge Barracks 1861 OS Map

Cambridge Barracks was built in several phases. The plain three-storey back-to-back barrack block of the mid 1850s named after the first Duke of Cambridge was built to house soldiers forms the eastern side of the parade ground. It had a central office section with a through arch and a cookhouse at the southern end. Beyond it was the laundry of the adjoining Clarence Barracks. Concern for the welfare of soldiers following the Crimean War led to the addition of married quarters. The main entrance was in Penny Street with a second entrance in the High Street with a guardhouse.

Portsmouth Grammar School Main Entrance By kind permission of John Sadden PGS Archivist

The imposing three-storey front range for officers facing the High Street was designed and constructed by the Royal Engineers rather than by a local contractor, reflecting the importance attached to it. In yellow brick c.1855-60, it is forty bays long with a stone cornice and frieze. The entrance is through round stone-framed arches with elegant red brick cross-vaults above. This finely detailed block, which steps forward in the centre and the ends, was an addition to the existing barracks built

339

around three sides of the wedge-shaped barrack yard to the rear described above.

The 'Gatehouse' of 1891-93 facing Museum Road was originally an adults' school for the adjoining Clarence Barracks. Literacy and numeracy skills were an important social provision when a large proportion of the adult population, especially the military, could not read. In 1949 the building was used as a city council infants and junior school, which closed in 1974 when it was converted to the Navy Records Office. Next door is the 'Ranch House', designed by Col. Edward Hunt RE, a plain red brick block for orderlies of 1904. There are similar designs at St. George Barracks in Gosport and at Tidworth.

By 1926 the officers' building was derelict. The main frontage of Cambridge Barracks on Portsmouth High Street (c. 1855-60) was purchased from the War Office in 1926 for £10,000 and converted for the Grammar School's senior school. In 1927 the school moved in. The Mayor, Frank Privett, who was also the contractor and paid for the conversion work, donated the cast-iron gates. The Home Secretary Sir William Joynson-Hicks ceremoniously opened them, accompanied by the mayor. The school memorial library was originally the officers' mess; the staff common room was the barracks library.

Portsmouth Grammar School Courtyard/ex Parade Ground and rear block

The Grammar School purchased the rear block, but it was separated from the front one by a chain link fence and barbed wire until 2000, when the two halves were reunited. The northern block facing Alexandra Road was also bought and converted into classrooms.

In WWII the school moved to Bournemouth to avoid enemy bombers. The Army Signal Corps occupied the main building. Their duties included operating the defensive electric light emplacements "intended to deceive German bombers into dropping their bombs harmlessly into the Solent." HMS Vernon's research scientists moved into the barracks before transferring to Leigh Park, ASWE on Portsdown Hill and Roedean School in Brighton.

In 1945 when Deane Clark was a pupil, the barrack rooms were as austere as they had been when they echoed to soldiers' voices: the splintery floors and rudimentary heating unchanged from military days, and the barrack square just bare tarmac. "There were no younger teachers because they hadn't come back from the war. There was a static water supply tank in the playground. There was no library and it was a struggle to get school uniform."

David Sherren said that when he spent ten years of his life at PGS from 1970 many parts of the school were just as they were in Deane's time. Only part of the barrack block was in use during his time there, and he remembers finding a hole in the wall of a storeroom which led him into the empty and dusty rooms; "All very exciting to explore without being caught!"

Portsmouth Grammar School Aerial View in 1970

The MOD still had a presence there. In 1983 Cambridge House was taken over by the Naval Surface Weapon Engineering Purchasing Office which was responsible for ordering and supplying Trident, Seawolf and Seadart missiles. The end of the Cold War resulted in cuts throughout the armed forces and military premises no longer required were sold off. The navy vacated Cambridge House, and after prolonged negotiations the school purchased the site in 2000.

From the early 21st century new buildings have been added into the square: a sixth form centre, soon demolished was replaced in 2010 by a larger building. The 1980s Michael Nott Music School incorporated the laundry – which had roof vents to let out the steam. A new sports hall stands on the foundations of the NCO's married quarters, music resounds in the laundry and small children in the nursery school laugh and play in the Napoleonic warehouse.

Bristow-Clavell Science Centre Architect Harrington Design Bosham
Celestial Microscope in front 2011

The Bristow-Clavell science block with a dramatic full height entrance atrium, classrooms and labs facing Museum Road was added at the northern end of the site in 2013. The playground now has green edges and gardens for the junior school, but it is still used for military drill by the school's Combined Cadet Force. In 2020 a new Performing Arts Centre was finished. The school's numbers have grown by a third since the 1980s music school was opened, so new music facilities are needed. The new centre has a 260-seat concert hall with a choir gallery, three spacious classrooms, 18 practice rooms, a dedicated percussion studio, a fully equipped rock music suite and a professional recording studio.

Portsmouth owes an enormous debt to the Grammar School for organising the city's Festivities including music, drama and lectures in June every year.

Municipal College – Technical Institute – Polytechnic – University

Ordnance Survey air photo mosaic of Ravelin Park, sheet SZ6399NE (1945)
The Map Library, University Library, University of Portsmouth

Portsmouth Polytechnic which developed from the Municipal College, later the Technical Institute founded in 1894 became a university in 1992. In the nineteenth and twentieth centuries it has gradually built over large stretches of former military space north and south of the railway line with new facilities. It also constitutes an important strand in the modern city's economy. In 2015/6 the university generated £264m in the Solent region, including £476m in Portsmouth. 7,900 people work for the university in the city.[3]

The land north of Museum Road to the city centre now called Ravelin Park was cleared of fortifications in the 1880s, although the zigzag trace of their shape is still echoed in the terraces across the road, and as described in Chapter 1 the fortifications' base still existed deep below ground level in the northwest corner of the park.

'Near Ravelin House was an octagonal structure opposite the former City Records Office known as the Rotunda. This opened in January 1941 as a YWCA United Services Club, a £10,000 gift from the South African Mayors' Fund named Cape Town House. The Portsmouth *Evening News* of 21 January 1941 describes it as 'a new building, octagonal in shape with canteen, reading and writing rooms, and stage for dance band and concert parties.' In 1953 *The News* reported that it 'was lying in a dilapidated condition with the rooms inches deep in water. The building... had been unoccupied for a considerable time. Three weeks ago restoration work began. The building is suitable for a small unit. There is a large central hall with a platform suitable for lectures, and several smaller rooms, which can be used as offices, an officers' room, a bar and a canteen.' It opened again as the headquarters of the 21st Battalion (Portsmouth West) Home Guard, on 21st April 1953. The polytechnic, later the university –

Rotunda   University of Portsmouth Archive

used it as a television studio. Empty for several years, it was demolished in July 2019 in the university's sports centre redevelopment.

Rotunda interior   William Hutchin

University Library interior 2009

Rotunda in use as a television studio   William Hutchin

The largest and most imposing new structure is the ziggurat form of the University Library (formerly known as the Frewen Library) by Ahrends, Burton & Koralek of 1975-7. It has been extended twice, once by ABK and then with an exhilarating sawtoothed south wing by Penoyre and Prasad in 2008, its sloping black brick base echoing the revetments of the old defences.

The William Beatty Building dates from 2005: the dental academy with its top floor sawtoothed laboratory where trainee dental technicians drill the teeth of model heads is by Miller Hughes.

William Beatty Wing extension link

A new wing linked by a Germolene pink glazed first floor corridor to the rear in 2010 extended it westwards.

In St. Paul's Road is the university's St. Paul's Gym, a former military drill hall, offering badminton, basketball, table tennis, netball, squash, volleyball and five-a side football.[4]

Nuffield United Service Officers' Club University Archives

In St. Michael's Road is the Nuffield United Services Officers Club which was designed in 1940 by E Berry Webber, architect of the burnt out Guildhall. It has long windows with thin stone frames and geometric glazing patterns and a curved glazed first floor extension facing the sports field. It was last used for student services and the Medical Centre, but the university plans to demolish it as part of a new sports development scheme. Behind it is the cricket ground, bounded to the north by the railway line, and next to it in Cambridge Road is a new sports centre and an inflated tennis dome.

The relocated King James's Gate, shorn of its scrolled pediment faces Burnaby Road, the cricket ground's western edge. The land on the other side of that road is still a naval sports ground.

Pitt Street Baths 2005

The relocated walls and art nouveau railings from the military Pitt Street Baths enclose HMS Temeraire's swimming pool and changing rooms. Next door is the yellow brick Grade II military Gymnasium divided by rough concrete blocks into squash courts.

North of the railway line on the west side of Anglesea Road once occupied by the Portsea defences the technical college constructed the Anglesea Building in 1952 by F Mellor, the city architect, in Festival of Britain style with 'comically idealised' sculptures of a bricklayer and an engineer (O'Brien 2018). Behind it is the Burnaby Building (civil engineering) of 1967 by RW Leggatt & Partners with single storey curved concrete vaulted workshops.

Lion Terrace Portsea prior to demolition   Deane Clark

Anglesea Building 1952   University Archives

344

Grand scale Georgian houses in Lion Gate, once looking over the Portsea ramparts, were cleared in the late C20 for the red brick polytechnic Buckingham Building designed by the city Department of Architecture and Civic Design in 1973-4, entered under the cantilevered lecture theatre. This was the Geography department where Ray Riley and Michael Bateman worked.

Dennis Sciama Building 2009

To its south the small square at the end of Burnaby Road has the white Dennis Sciama Building of 2008-9 by Van Heyningen & Haward for the Institute of Cosmology on one side and the seven-storey brightly panelled Richmond Building of 2002 by Hawkins\ Brown for Business Studies on the other. The Sciama building was the first in the university to incorporate energy generation on its roof as well as using the stack effect for ventilation. In 2009 the Portsmouth Society gave it the Best New Building Award.

University Business School 2009

Celia Clark remembers the extraordinary shaped tapering central atrium of the Business School when we judged it for the Portsmouth Society's design competition. We enjoyed the bright colours on the outside and the south facing external balcony.

Portland Building

To the west of the Buckingham Building is the white-faced Portland Building which occupies an entire block, enclosing a courtyard and pond between it and the Buckingham Building. The Portland building was designed by Sir Colin Stansfield Smith in 1993-6 as an experiment in low energy architecture using the stack effect for ventilation and cooling, though this did not always work well in high summer. Sir Colin's office as professor of architecture was in the south wing; he was also Hampshire County Architect. The central glazed atrium with its high level bridges is much enjoyed – for exhibitions as well as parties!

Milldam Barracks from Dennis Sciama Building 2009

To the south of the square in Richmond Place are Milldam Barracks. These were built c.1800-10 for the Royal Engineers on the northwest edge of Milldam, the tidal pond between the towns of Portsea and Portsmouth. The red brick barrack block is 15 bays long and two storeys high with a central pediment over three bays. The southern end was damaged by WWII bombing and replaced in 1983 when the barracks were converted for the polytechnic. Light metal galleries and a glassed-in staircase were added to LE Block facing the parade ground to provided disabled access after the Disability Discrimination Act of 2005. There is a lift in LC block, and when the link was made it became possible for a disabled student to gain access to the lecture rooms on the upper floor of LE. "After that we saw quite a few students in wheel chairs, sometimes with carers."[5]

Mark Jones said: "We who work at Milldam appreciate the tranquil space it provides at its centre, what we now term the quad but I believe was once a parade ground. Speaking personally I wouldn't want to work in any other University Building". The centrepiece of the landscaped garden that replaced the parade ground was a massive granite gun emplacement plinth – which has since disappeared. "I think the gun emplacement was removed as many people felt that Milldam should not have such a symbol of militarism. I don't know who exactly made the decision, but I agree with it. It was done a few years ago at the same time as the area was landscaped with new benches and a pagoda which has a magnificent wisteria - looks fabulous when in flower."

A tall block of married quarters was added in the later nineteenth century, forming the south side of the barrack square. "As you would expect the interiors are very different to how they would have been, but they suit our purposes quite well really. The heft of the walls insulates quite well in winter, and, in my particular case, being in one of the South facing blocks, heats up quickly in the summer."[6]

Sue Bruley said "I came to Milldam in 1988. Originally the LB and LC blocks were flats for engineers' married quarters, with four flats per floor, leading off from the two staircases. At this time some of the rooms were used for teaching and the others were staff offices. Teaching was difficult as the rooms were so cramped. A maximum of 12 students in each room plus the teacher. In hot weather it was really stifling as we were practically on top of each other. There was no central heating, we only had gas fires so the last student in the winter the last student in the room had to sit right in front of the fire and roast! After the Sciama Building went up classes were moved to the rooms there which are much more spacious (hence we now have over 20 in a class)."

In the northwest corner of the square in the Portsmouth & Portsea Free Ragged School building is the university nursery. It was built in 1863 and was used as a school until about 1935.[7] Its predecessor, the Polytechnic nursery occupied a hall in the lower ground floor of Trafalgar House in Edinburgh Road, the former YMCA.

Eric Rimmington Mural Trafalgar House 1949

That building is listed because Eric Rimmington painted a dramatic mural occupying an entire wall in 1949 depicting Portsmouth history and the Solent. The central figures including sailors and marines in uniform look wistfully seaward from Portsmouth and Southsea Station. The mural is of considerable interest as an example of postwar public mural painting.[8] Yet when she was at the nursery Celia's daughter does not remember it, but her mother loved seeing it every time she collected her at the end of the day. It is now glassed in inside a bar, which makes it more difficult to see.

Royal Engineers Barracks Portsea converted to Portsmouth Register Office
Architect: Hampshire County Council

Nearer the railway line is Mill Dam House, built before 1825 as the Chief Engineer's Office. The Royal Engineers designed many of the functional buildings in the dockyard and the dry docks. Their records are now kept at Brompton Barracks, their Chatham headquarters. Copies proved useful to the dockyard managers when a filled in lock kept flooding. The drawing office is square with four identical sides, chequer patterned in red brick and grey glazed headers. The porch has the Royal Engineers' monogram which may date from about 1900. The building's proximity to Milldam and perhaps to the railway line to Portsmouth Harbour station severely damaged its foundations. Paul Davies, Hampshire historic buildings architect reinforced its underpinnings and restored the building as Portsmouth's Register Office, where we go to register key events in our lives: births, deaths and marriages - and new citizenship ceremonies. The dramatic cast iron central staircase in the entrance hall takes you to the upper galleries. Behind the building is a lovely garden designed by the county landscape architects – just right for wedding photographs. Opposite is Burnaby Terrace, eight yellow brick quarters now used as university offices.

Forton Hospital - Barracks - HMS St Vincent – St. Vincent College Forton Road Gosport

The main gate of St. Vincent Barracks Gosport. George Millener Collection

Forton Barracks   Martin Marks

We now explore the development over more than two centuries of another educational establishment: the sixth form college in Gosport. Its origins go back to the late 18th century. In 1795 land was purchased in the Forton area to build a military hospital. John Marius Wilson in the Imperial Gazetteer of England and Wales (1870-72) notes that: "FORTON, a village and a chapelry in Alverstoke parish, Hants. The name is a corruption of Fort-town." The hospital was in use from 1796 to 1806. It consisted of four three-storey blocks and several outbuildings linked by a colonnade. It was the first purpose-built army hospital in the country and was the work of John Sanders and James Johnson, who between them were responsible for the Duke of York's Headquarters (now the Saatchi Gallery) and the Royal Mint.

In 1807 the hospital buildings were altered during construction to house a Royal Marine Barracks. They were next used by the army. In March 1848 the Royal Marines Light Infantry moved to Forton from Eastney in Portsmouth. The impressive frontage dates from this time. Its tall rusticated central entrance archway with big scrolled keystones is topped by a wooden cupola with a square dome and gilded ball finial. Captain H James RE designed it. On either side there are curved wings of gates with pediments linked to symmetrical low, singe-storey blocks, originally of eleven bays and later extended. A few other buildings are preserved within the college campus, including the former Commandant's House also of 1848 to the east with a three-bay elevation to the road designed by Captain James. The house is dominated by the enclosed stone porch and above the eaves it has a strange pyramid-roofed square tower at one corner and stone chimneys with arches at the others.

In 1857 when the Admiralty took over the site, sheds were added to house a variety of small and large calibre guns for drills - including a rolling motion platform to simulate sea conditions. Maxim machine gun practice was provided, using blanks. Further into Gosport a new barracks, later called St George Barracks was built in 1859, north and south of Mumby Road. Its conversion is described in the residential chapter. In 1893 a 400-seater theatre was added within Forton barracks. The Royal Marine Light Infantry and the Royal Marine Artillery who had given a good account of themselves during the Great War especially at the Zeebrugge landings, amalgamated in 1923, and completed a move back to Clarence Barracks, Southsea by September. Forton Barracks was then left to the caretaking of naval pensioners for four years.

A decision by an Admiralty Committee in 1926 that Forton be used for Boys' Training set in motion a spate of refurbishment and modernisation, which included the fitting of electric lights and building of a new boiler complex. Moored off Gosport for a good few years had been *HMS St Vincent*, the Boy's Training Ship, a place for homeless lads. This was thought a good reason for re-using the name by conferring it on to the new establishment. In 1927 the first 434 boys arrived by special train at Gosport Station (1841-1969). This closed and was derelict for many years. Grade II* listed, it was redeveloped into residential properties and offices in about 2010. Along with their officers and instructors the boys marched to their new barracks. They had come from HMS Ganges to form the backbone of the 684 trainees and were soon to be enrolled on HMS St Vincent's first main training course. The boys were expected to climb the tall four-stage mast which was rigged like a real ship's mast with full yardarms. The worst part was going up and over the devil's elbow under the first platform. They spread out along each yardarm. The button boy at the top, standing to attention, gripping the lightening conductor with his knees was paid 6 pence. They performed the same feat in Portsmouth Guildhall Square… Goodness knows what Health and Safety would think of it now! At the beginning of WWII, the site became a training centre for officers of the Fleet Air Arm and the boys were evacuated to HMS George on the Isle of Man. A torpedo training section was opened in 1940 but 1945 it reverted to training boy seamen, remaining as such until the establishment was officially closed in 1968.

The Admiralty transferred new entry training to HMS Raleigh at Torpoint and pulled out in 1969. The site was purchased jointly by Hampshire County Council and Gosport Borough Council for a new comprehensive school, which opened in 1974. Sadly, unlike Portsmouth Grammar School which reused its barrack blocks, most of the 18th century buildings were demolished, but Deane Clark, head of Hampshire's Historic Buildings Bureau intervened to preserve the impressive frontage. A cast iron canopy was reused at the entrance to the new block in the right hand side. As part of the reorganisation of secondary education it became Gosport Sixth Form College, initially sharing the site with St. Vincent Secondary School, but when the school's final year left in 1987 it was renamed St Vincent College, a co-educational sixth form college for 14 to 19 year olds.

St. Vincent College   Martin Marks

In 2007 a proposal to build a 'super college' on the HMS Daedalus site was strongly resisted – and rejected. Today it has some 1,200 full-time students and offers part time adult education, NVQ courses, a Foundation Degree in Early Years Care and Education, two levels of Apprenticeships, a range of Teacher Training courses and access to Higher Education Diplomas. There is a small museum on the site that has artefacts and pictures of the site's time as a naval establishment, although this has limited opening hours only at the weekend.

Bay House, Gomer Lane and Alverbank House, Alverstoke

We now explore two properties associated with military use, one of which is now the centrepiece of Gosport's other sixth form college. Bay House in Alverstoke has a rich history which can be traced back to 1838. It was designed by Decimus Burton, a fashionable architect of the day as a seaside getaway for Alexander Baring, the Ist Lord Ashburton on land that became redundant post –Waterloo when hostilities with the French ceased. Ashburton already had three houses but this new one was intended as a retreat. An asymmetrical Tudor composition in rough white stone, it has three storeys with a central gable to the entrance façade and a taller side turret. The service wing to the left once had a stable courtyard. Three rooms face south; the middle one has a large square bay window with a tented roofed veranda above it. Behind a screen wall at the rear of the main approach is a long conservatory, restored in 2015. Burton's entrance gates in scrolled ironwork are set into a wall recessed from the main road frontage with a small lodge set in the angle. The house had a garden that reached down to the sea. Ashburton's friend Rt. Hon J Wilson Croker, former Secretary to the Admiralty, was persuaded to buy land adjacent to it and built Alverbank House next door. This also designed by Decimus Burton in 1842 in picturesque Tudor Gothic style with a timber veranda facing the sea.

In 1870 the Admiralty took out a lease on the Bay House land after the War office purchased it, and a Naval College was set up there, later known as Ashburton House, under the control of the Rev. Edward Burney. Prince Alfred Ernest, Queen Victoria's second son went to the academy and stayed at Alverbank House whilst he was studying there. In 1892 Bay House passed to Col. Francis Sloane-Stanley, a keen yachtsman and a very well-connected man, being a friend of the Prince of Wales.

In 1943 the house was sold to Gosport Borough Council, and the 17 acres of parkland were donated to the people of the borough. The house became a base for the Royal Engineers during the war. In 1949 it was described as 'shabby and in some places almost decrepit'. Gosport County Grammar School moved here from 1949 and a major expansion began.

According to O'Brien, editor of the *South Hampshire Buildings of England*, the school buildings added in 1956-8 by Louis de Soissons, Peacock, Hodges and Robinson are sympathetic to Bay House, built in stock brick with low copper roofs, but the spacious layout has gradually been filled up with later additions. In 1984 a fire caused extensive damage, but the house was well restored and the work received an award from the Gosport Society. Today it is a busy sixth form college with a good reputation.

Alverbank House was sold in 1912 to Winifred Platt. During World War II. It was used as her home and an officers' mess for those working at Grange Airfield described in our aviation chapter which was located within the grounds of today's HMS Sultan. In 1947 the house became the property of Gosport Borough Council via a compulsory purchase order. Some of the land was added to Stanley Park. In 1983 Alverbank House was Grade ll listed. It has been used as a guesthouse and a restaurant and in the late 20th century it became known as the Alverbank Hotel and more recently the Alverbank Country House Hotel. In January 2019 the company running the hotel was wound up and the chief executive of Gosport (and Portsmouth – a combined post) used controversial delegated powers to remove that company and contract Sisters Inn Arms Venues to take over for a 70-year lease without a tendering process.[8]

Barracks and also historic houses with defence associations have proved adaptable for the completely different activity of education, and the university has filled up the space left by the demolition of Portsmouth and Portsea's defences with specialised new facilities, making major contributions to the local economy.

Sources

https://historicengland.org.uk/listing/the-list/list-entry/1104363 https://www.pgs.org.uk/pac/ accessed 20 January 2019

http://www.hmsstvincentassociation.com/ http://www.stvincent.ac.uk/ https://en.wikipedia.org/wiki/Forton_Barracks https://www.gosportsociety.co.uk/fortonbarracks.htm

https://www.portsmouth.co.uk/lifestyle/heritage/boys-on-top-of-the-world-1- 2763673

Sion Donovan 2008 "College merger plan crashes into buffers". Johnston Press Digital Publishing. Portsmouth Today November 19, Retrieved May 24, 2010

Charles O'Brien et al. 2018 *Hampshire South. Buildings of England* Yale University Press p. 295, pp.480-481 http://www.mygosport.org/info_bay_house.htm

"Ashburton House, Gosport" from 'The Illustrated London News', April 12th, 1863 http://research.hgt.org.uk/item/alverbank-hotel/

Millie Salkeld 2020 'Plans revealed to demolish Portsmouth university building as part of sport scheme and move GP surgery to Commercial Road' *The News* March 20

References

1. *A History of Cambridge Barracks* Catherine Smith Portsmouth Grammar School Monographs 7 2001 [2] *The Palgrave Dictionary of Anglo-Jewish History* p. 203

2. https://www.port.ac.uk/about-us/our-impact/local-impact accessed 20 February 2019

3. https://www.sportportsmouth.co.uk/facilities/st-pauls-gym/ accessed 20 February 2019 [5]

4. Sue Bruley Email to Celia Clark 20 February 2019

5. Mark Jones email to Celia Clark 26 February 2019

6. http://discovery.nationalarchives.gov.uk/details/r/ea37227f-aaed-4bd6-8e3b-b79a3d4cbffb

7. https://historicengland.org.uk/listing/the-list/list-entry/1360779

8. Minutes of a GBC Development Board in 2018:- Page 5
"LEASE OF THE ALVERBANK HOUSE HOTEL Consideration was given to an exempt report of the Chief Executive advising the Economic Development Board of his decision (under paragraph 3.7 of the officer scheme of delegation in the Gosport Borough Council Constitution) to approve the grant a new lease of the Alverbank Hotel to Sisters Inn Arms Venues Limited (SIAV). The report of the Chief Executive was reported to the Board."

Chapter 15
Research – cutting edge developments in two centuries – and into a third

"Given the enormous scale of naval dockyards in the eighteenth century, making them unequivocally the largest industrial enterprises in the country, it is to be expected that they would have had an important impact upon contemporary innovation. The part played by Dummer and later Bentham in dock design and caissons, and by Marc Brunel and Maudslay in machine tool manufacture and the establishment of the principles of production engineering in the Portsmouth block mill, is well known. Nonetheless, the way the dockyards, drawing on public funds, were able to influence private sector enterprise is a somewhat under-researched area." Ray Riley 'Henry Cort and the Development of Wrought Iron Manufacture in the 1780s: the Naval Collection' in *Transactions of the Naval Dockyards Society Volume 3 November 2007.*

We explore defence sites where research takes place now and where it used to do so. As Ray Riley said, over the centuries the presence of the dockyard and its associated establishments stimulated research and technological development in many different fields. This history of innovation was the basis for the bid in 2006 - 2012 to inscribe Portsmouth Harbour on the World Heritage list - as the world's first cultural seascape, focused on the theme of defence of the realm.[1] An extraordinary range of inventions have been developed around the harbour. James Lind (1716-1794) Chief Physician at Haslar Royal Naval Hospital from 1762-1772 published 'A treatise of the scurvy' in 1753 based on comparative clinical trials; he also proposed distilling fresh water from seawater. As well as mass production of pulley blocks in Block Mills, also in the dockyard Samuel Bentham devised the first working caissons to close dry docks; the first use of circular saws also occurred there. There were also architectural and engineering innovations in building such as the early use of large span prefabricated cast iron. The Whitehead torpedo was developed in the diving lake at Horsea Island. During WWII degaussing of ships was developed at HMS Vernon, and shipborne radar to detect aircraft in Eastney Fort East. The ship-testing tanks in Haslar, Gosport built by William Froude and his son in the 1880s are still in operation. In the twentieth century new techniques including freeze-drying, refrigeration and experimental methods of food preservation were pioneered in Royal Clarence Victualling Yard.

New discoveries and innovation – even without including research by the university - are still major local activities. Today specialist conservation experts and land and marine archaeologists in Historic England's Fort Cumberland conserve and repair objects, research archaeological digs, write policies; and the Centre for Marine Archaeology there researches marine archaeology and trains marine archaeologists; while specialists at Haslar model ship hull design using QinetiQ's ship testing tanks, test new weapons systems and play war games on the crest of Portsdown Hill. This chapter explores some of these sites' history, and what happens there now.

Research into food preservation technology at Royal Clarence Victualling Yard

The Royal Clarence Victualling Yard was one of the first sites to mass produce food on an industrial scale. It supplied the navy with fresh water, salt meat, biscuits and rum. Pioneering inventions in food production and preservation took place here in the nineteenth and twentieth centuries. Following his appointment as storekeeper at the Royal Clarence Victualling Yard in 1828 Sir Thomas Grant (1795-1859) began his career of inventions related to food processing and many other beneficial improvements to naval life. In 1829 he devised machinery for making ships' biscuits which was implemented by George Rennie and installed in the yard in about 1831 under the direction of Sir John Rennie. This mechanisation speeded up the production process and substantially reduced costs. In recognition of his achievement and of the savings made, the government awarded Grant £2000. He also received a medal from King Louise Philippe and a gold medal from the Society of Arts in London. In 1834 he invented a desalination plant which distilled fresh water at sea. This was adopted by the navy 14 years later. According to *The Times* of 19 October 1859 it was "the greatest benefit ever conferred on the sailor, materially advancing the sanitary and moral condition of the navy." In about 1839 Grant devised a patent naval fuel. A year later he was elected a fellow of the Royal Society. In the 1840s he developed a steam kitchen, first tried aboard *HMS Illustrious*. He also constructed a new type of lifebuoy and a feathering paddle wheel. In 1850 he was promoted to the comptrollership of the Admiralty's Victualling and Transport service. In 1854 the Crimean War provided many opportunities that made use of Grant's abilities. Victualling yards at home worked day and night turning out machine-made biscuits for the army and navy, while Grant's distilling apparatus provided fresh and clean water on the spot. His public services were recognized when he was knighted: Queen Victoria presented him with a gold and silver vase.[2] The redevelopment of Royal Clarence Yard is described in the next chapter.

The Admiralty Experimental Works/QinetiQ Marine Technology Park Haslar
Modelling the design and performance of ship hulls on a miniature scale to test a ship's behaviour and other physical characteristics before building it is a key stage in the design process. The pioneer inventor in this was William Froude (1810-1879) a railway engineer who turned to the study of what is now called fluid mechanics at the invitation of Isambard Kingdom Brunel. He devised the Froude Number by which the results of small-scale ship model tests could be used to predict the behaviour of full-sized hulls, based on a sequence of models as different scales to predict their behaviour. He was given a grant by the Admiralty to build the world's first experimental ship testing tank at his home in Torquay. He combined mathematical expertise with practical experimentation to such good effect that his methods are still followed today.

William died suddenly in 1879, and his son Robert carried on with his work. But by 1886 the site proved to be too small and the work was moved to Haslar, next door to the Haslar Gunboat Yard. Behind a high red brick wall parallel to the Royal Naval Hospital Robert supervised the building of the testing tank. Work expanded to all types of ships and submarines under the title of the Admiralty Experiment Works. Wax models were towed through the water by an overhead carriage running on rails. Further work on fuels and oils was undertaken from 1902 as the Admiralty Liquid Fuel Experimentation Station (ALFES), soon to become the Admiralty Fuel Experimentation Station (AFES).

Research shifted to focus on predictions of ship power, hydrodynamics, submarine design and later propeller design in relation to ship manoeuvrability and sea-keeping. As its work began to expand, new research testing facilities were established between 1930-1972 that enabled the works to extend their tests on models in a ship tank covering all classes of battleships, cruisers, destroyers, submarines and miscellaneous vessels. The results of these experiments were then used by the Naval Construction Department to enable it to improve ship design and performance.

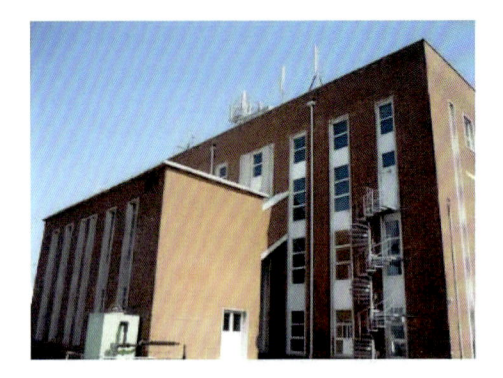

Cavitation Tunnel and the building housing it

Figure 6 The second HSVA-Tunnel (1941)

As part of war reparations in 1947 No. 2 Cavitation Tunnel was brought by the Admiralty from Germany to Haslar. A cavitation tunnel is similar in function to a wind tunnel but water filled. It is housed in Buildings No. 46 and 47 and was designed by Dr Herman Lerbs, a leading German cavitation and propeller research and experimentation scientist and built by Blohm and Voss of Hamburg. This company built the battleship *Bismark* among others, U-Boats and also aircraft. Cavitation is the sudden formation and collapse of bubbles in a fluid through mechanical forces, such as those caused by a propeller in water. A tunnel is used for testing scale model propellers and hulls to consider the impacts on propulsion, wake, vibration and noise. Cavitation causes underwater noise. Minimising or eliminating this is an important component of submarine design. Cavitation also has an erosive effect on propellers and thus on efficiency. The No. 2 Cavitation Tunnel was operational until circa 2008 primarily testing propellers for submarines and other vessels including warships and minor craft. No. 1 Cavitation Tunnel, smaller in scale than No. 2, was completed in 1941 and operational during the war. It is understood to have been housed in a building immediately southeast of No. 2 Cavitation Building but it was demolished in 1993. In May 2017 there was some commercial interest in reactivating No2.

The 270m long Towing Tank is by far the largest in the UK.

Marine Towing Tank   Copyright QinetiQ

In 1958 the functions of the Director of Naval Construction became a division of the new Ship Department. In 1966 AFES Haslar was renamed the Admiralty Marine Engineering Establishment (AMEE), before being taken under the wing of the National Gas Turbine Establishment (NGTE) in 1965. A 1970 report noted that "The Admiralty Experimental Works is primarily an arm of the Royal Navy Ship Department at Bath and does extensive propeller, hull, sonar dome and rudder design work as well as related model testing. Their research program, constituting some 25% of their total effort, is also entirely in support of their primary function. About 10% of AEW's work is undertaken as contract work for private industry or other government agencies." This was an early indication of the way things were to develop in the future.

It was renamed the Admiralty Marine Technology Establishment (AMTE) in 1979 and then absorbed by the Admiralty Research Establishment (ARE) in April 1984 with various other establishments. It became part of the Defence Research Agency (DRA) in April 1991, and in 1995 DRA became a division of Defence Evaluation and Research Establishment (DERA). DERA was part part-privatised in 2001, when the Haslar establishment became part of a commercial company QinetiQ plc. The Ministry of Defence kept a 'special share' in the company, and safeguards were put in place to prevent conflicts of interest. The unusual name "QinetiQ" is invented. "Qi" reflects the company's energy, "net" its networking ability, and "iQ" its intellectual resources.

In February 2003, the US private equity firm the Carlyle Group acquired a 33.8% share for £42m. Prior to stock market flotation, ownership was split between the MoD (56%), Carlyle Group (31%) and staff (13%). In 2007, the National Audit Office (NAO) conducted an inquiry into the privatisation to determine whether UK taxpayers got good value for money. The NAO reported that taxpayers could have gained "tens of millions" more and was critical of the incentive scheme given to QinetiQ managers, the 10 most senior of whom gained £107.5m on an investment of £540,000 in the company's shares. The return of 19,990% was described as "excessive" by the NAO. The role of QinetiQ's management in negotiating terms with the Carlyle Group while the private equity company was bidding for the business was also criticised by the NAO.

Ocean Basin and Rotating Arm   Copyright QinetiQ

Now called Haslar Marine Technology Park, the site offers the UK's largest ship tank, the Ocean Basin where buoyancy, dynamics and manoeuvrability tests are run. It measures 122m by 61m, and is 5.5m deep. The tank holds 40,000 tonnes of water and is available for hire. The latest development in

Marine Autonomous Robotic Systems is a Hyperbaric Trials Unit (HTU) used for testing submarine equipment. Its chamber recreates pressures equivalent to a depth of 1,500 m. Overall, QinetiQ is one of the top ten largest UK employers of science and engineering graduates, recruiting around 150 a year. This premier research firm also has extensive facilities along the top of Portsdown Hill.

Defence research on Portsdown Hill

Next we consider the active defence research establishments. The Admiralty Surface Weapons Establishment. (ASWE), a prominent building between these active facilities - where 2,000 scientists, technologists and support staff once worked, disappeared in 2012. On the western flank are the modern buildings of the Defence Science and Technology Laboratory (DSTL), a Centre of Marine Intelligent Systems developing autonomous boats and submarines. DSTL is an agency of the Ministry of Defence formed in July 2001 from the splitting up of the Defence Evaluation and Research Agency. Its purpose is "to maximise the impact of science and technology for the defence and security of the UK by carrying out work best done within government.[5] DSTL is housed in a series of two-storey red brick buildings within a secure perimeter along the top of the hill. The majority of DERA's work, suitable for industry, was transferred to QinetiQ, at the branch based in Haslar already discussed.

ASWE

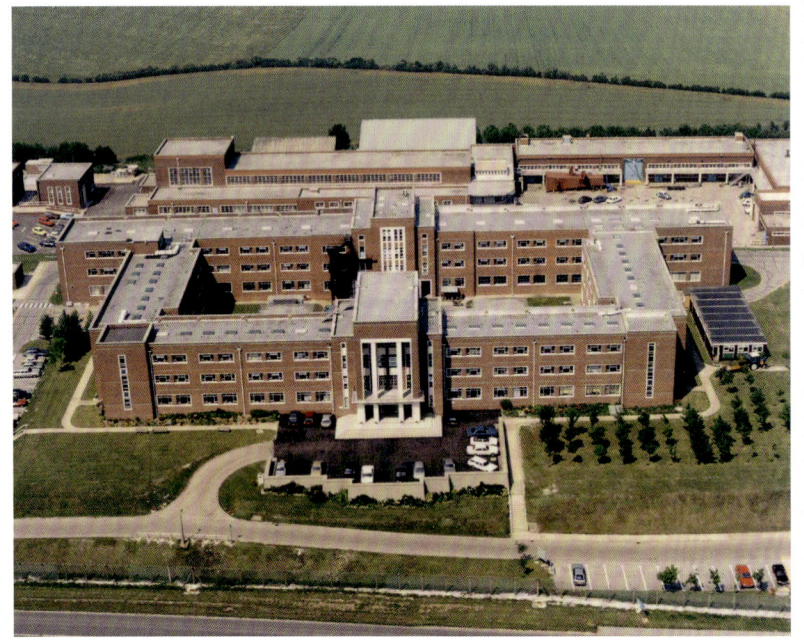

Portsdown Main

Portsdown-tunnels.org.uk

From 1951-2012 a prominent landmark on the Portsmouth skyline was the fine Art Deco building of ASWE, the Admiralty Surface Weapons Establishment, also known as Portsdown Main. Its history was poorly documented because of the nature of the work, which went on there. ASWE's origin in the mid-1930s was the Experimental Department of HM Signal School in the Royal Naval Barracks Portsmouth. In the year or so prior to 1939 the department moved some units to other places in Portsmouth: Eastney Fort East, the old school at Onslow Road where RDF (later known by the American term 'Radar') was developed and Nutbourne. Rapid expansion of the department to 26 other sites including universities and country houses including Lythe Hill House at Haslemere and King Edward's School Witley, as well as trial sites such as Tantallon - and even, for a short time, the Summit Hotel, Mount Snowdon.

The establishment eventually built on Portsdown Hill was designed in the 1930s - to be located somewhere in the UK. WWII prevented further progress. In 1944 a committee of representatives of the Communications and Radar Laboratories, the Production Department and Test Rooms, Workshops, Naval Stores and the Naval and Secretarial Groups was set up to plan the future Admiralty Signals

Establishment. This Committee produced and published for limited circulation an overall plan reconciling the needs of the many divisions and their combination on one site in October 1944 – an act of faith at such a critical stage of the war. The east and south coasts were considered too near to enemy territory as far as our radar was concerned. Proximity to the sea and to a major naval port was the overriding consideration. In 1945 one important site dominated the naval war scene: Fort Southwick, which had developed as the Combined Communications Headquarters, from which the invasion of Europe had been controlled. The grand plan envisaged that Portsdown Hill would house, from east to west, ASE, Tactical School in Fort Southwick, Signal School, AGE, and to the west ASE with its 'Quiet' work and a sports area. In the event, only ASE was built, in about six years.

On 3 September 1948 Vice Admiral C B Daniel, Controller of the Navy visited the construction site of what was to become the Admiralty Surface Weapons Establishment on the summit of Portsdown Hill. The overall cost of the monumental quadrangle building and canteen in Art Deco style was about £2.5m for the buildings and site, and about £1m for the equipment installed inside it. A total of 32 buildings were built on the 46-acre site between 1946 and 1955, the main block, Portsdown Main being the most prominent. The complex housed the Admiralty Signals Establishment (ASE). Successor bodies were the Admiralty Signals and Radar Establishment (ASRE) 1948 – 1959; the Admiralty Surface Weapons Establishment (ASWE) 1959 – 1984; the Admiralty Research Establishment (ARE) 1984 – 1991; the Defence Research Agency (DRA) 1991-1995; the Defence Evaluation and Research Agency (DERA) - 1995 – 2001.

An objector to the spoiling of the skyline of Portsdown led to the architect's drawing of the south elevation being shown to the Fine Arts Commission, who raised no objection. In *The story of ASWE* (1969) Norman Vidler gives details of the construction and the provision of essential services to this isolated and exposed site. The modernist main building, canteen and entrance in red brick had fine details such as the curving steel and brass main banisters to the main staircase which used to be polished bright. The building had three full floors, two levels of penthouses and a basement occupying a quarter of the building area at the western end. The layout of the floors was uniform throughout the building with one corridor looking much like another. The floors are laid with parquet tiles. The building was rectangular with a central quadrangle or light well. Its 1930s design made it an ideal location for a film crew in the making of an Agatha Christie film.

By summer 1949 Stage II of the project, the three laboratory blocks and the canteen were under way. Its exposed position meant it was always a cold and windy place to get to. The entrance to the site was between the two guardhouses. The bus shelters were for the employees. Traffic to the site was so heavy that it had its own set of traffic lights.

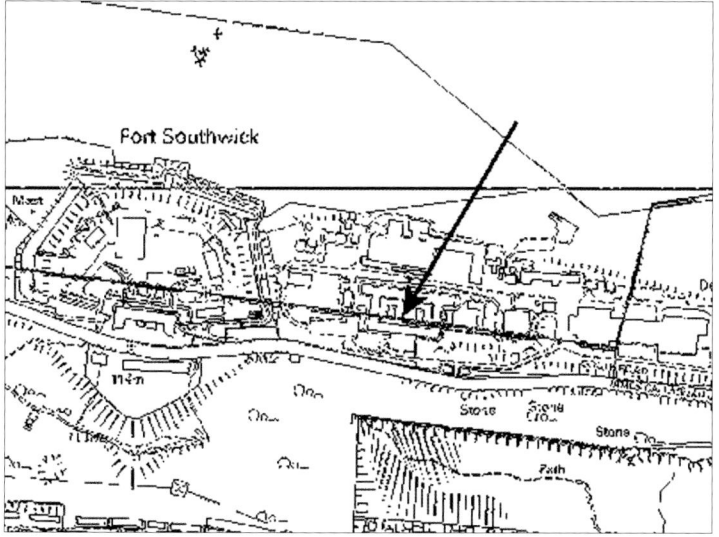

Site of Portsdown Main
Map Grid Ref SU631068   Source: Bob Hunt

This map shows the position of Portsdown Main (arrowed) in relation to Fort Southwick to the west and Paulsgrove Chalk pit to the south. To the west is Fort Southwick, the former Royal Navy Command Centre COMMCEN.

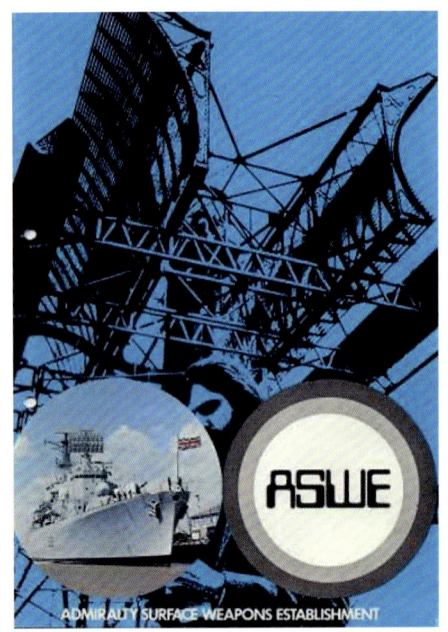

Photo Interpretation: Dave Spencer.
Source: Andrew Taylor

This is the front cover illustration from an ASWE recruitment information pack of the early 1970s. The ship is *HMS Bristol* (D23) the only Type-82 Destroyer to be built. The featured radar is either a Type-901 fire control radar or the Type-965 air search radar.

Portsmouth Main from Cosham

ASWE Main staircase 2012

Portsdown Main was occupied by many different departments including the Admiralty Signals Establishment (ASE) in 1952, the Admiralty Surface Weapons Establishment (ASWE), and the Defence Evaluation and Research Agency (DERA) in 1997. Deane Clark's father worked there as a senior scientific officer, a photographer taking pictures of mines exploding, working with navy divers including Commander Crabb.

By April 1959 there were 1,700 staff. Space in all buildings was at a premium. Ten years later there were 2,200 people working there, as well as at Funtingdon, Eastney Fort East and at 511 London Road Hilsea. Care had to be taken to preserve the 180-degree working arc to the north of Block 3 for trials, and also to avoid radiation hazards. A temporary building housed the Polaris submarine group at Portsdown.

Portsdown Main finally closed its gates on 1 January 1997. There were a number of government departments interested in the site. On 3 October 2002 it was announced that the Defence Science and Technology Laboratory (DSTL) formed by splitting DERA into DSTL and QinetiQ on 2 July 2001 - was going to move back, but this was cancelled in April 2004. The grand office building was last occupied in 2007, when its last owner was DSTL. Some years before Defence Estates had put the building on the market, a Bible College expressed interest in buying it, but it needed maintenance if it was to survive on its exposed site. Clearly the best new occupant would have been a large institution

– such as a commercial office, residential development, college or university – though its exposed position made for a chilly microclimate.

In July 2012 as part of the campaign to inscribe Portsmouth Harbour, Spithead and Ryde on the northern coast of the Isle of Wight onto the World Heritage List, Celia Clark and Portsmouth Grammar School head of history Simon Lemieux persuaded the managers of DSTL and QinetiQ, where so much technological innovation continues to take place, to let a group of teenagers visit the ten key sites involved in the bid including ASWE and the ship-testing tanks at Haslar. Guided by scientist Dr. Tim Crowfoot, one of the two thousand scientists who had worked at ASWE, fifty young people aged 13 and 14 noted, photographed and videoed their impressions of what each site contributed to the overall theme of the bid: defence of the realm. The main building and the canteen were empty and deteriorating.

Dug into the northern flank of the hill below is the modern Dr. Strangelove-like recent control room, which enables QinetiQ to conduct simultaneous war games or real scenarios with a ship in the harbour, the Gulf, the MOD and Seattle. It's an interservice facility. Celia Clark's nephew, Colonel of the 4th Rifles took part in a war games scenario. The school's visit to this complex was by far the most surreal experience of the trip.

Portsdown Main being demolished

DSTL declared the main structure of Portsdown Main unsafe and demolished it in 2012. King Sturge, international property consultants, put up the site for sale on the open market. By 2010 the northern profile of Portsdown Hill had been recreated to its original shape using rubble from lesser structures on the site. According to the Defence Infrastructure website, with a guide price of £0.5m, of the mainly cleared site with 'development potential (subject to planning)", 15 different bidders offered between £2m and £2.517 million on 12 June 2019 when the auction was held. We have not been able to establish who owns the site now.

The Centre of Marine Intelligent Systems

Research on Portsdown Hill is still of national strategic importance. In 2014 the government opened its defence research station, the $3m Centre of Marine Intelligent Systems (CMIS). The facility brought together specialist teams including academics, scientists, engineers and naval experts to research into unmanned autonomous boats, underwater vehicles and aircraft in a maritime environment. It was developed by the Defence Growth Partnership to build on the strong maritime experience and skilled workforce in the area, supported by government funding, the Solent Local Enterprise Partnership and the University of Southampton. Using equipment loaned from Thales UK it focused on the Unmanned Warrior 16, assessing the capability of autonomous maritime mine countermeasures in partnership with the Royal Navy. It was hoped that it would bring in £72bn worth of exports to the global market by 2021. However the CMIS ceased operations in 2017, and instead a Maritime Community of Interest was established which continued to engage with the navy to simulate maritime autonomous systems contracts for UK industry. The CMIS labs at Portsdown Technology Park were returned to QinetiQ, which intended to develop and modernise the site.

Land Based Test Site

Between Forts Southwick and Widley is a full size mast of a Type 45 mast bristling with electronics sailing along the crest of the hill, where radar systems are tested before they go to sea. Further east on the crest of the hill is the Land Based Test Site (LBTS). "The LBTS facilities at Portsdown have been a landmark on the Portsmouth skyline for some twenty years and will continue to evolve" as the study by Danny Atkins and Ken Tout said in 2002. The superstructure looks as if a ship is sailing along the top of Portsdown Hill. The LBTS was owned by the MOD until management was passed to DERA in 2000. In 2004 a multi-million pound MOD contract was awarded to QinetiQ to operate the LBTS, since employing civilians is significantly cheaper than sailors – as calculated by the National Audit Office: the cost of using a frigate per day is about 20 times the cost of using the T23 Shore Integration Facilities such as LBTS.

The structure is meant to look like a ship as it replicates the fit on the carriers and the Type 45s exactly. A ship's superstructure can impact on radar performance. The LBTS builds working simulations of large bits of kit – mainly radar in this case – fitted to new classes of ships and operates them to build experience ahead of the ships, hoping to predict problems. Ships' operation rooms are replicated exactly. Later on it can be used to develop modifications if needed or to examine ship-reported defects under more flexible and controlled conditions.

The LBTS is evolving and expanding as the Royal Navy's prime facility for integrating, proving and accepting new Combat System equipment into the surface fleet. The LBTS supports the T23 Frigates, T42 Destroyers and Aircraft Carriers. It uses simulator equipment to integrate new equipment. It is also the leader on the Combat System Integration programme. It has associated satellite sites around the UK and is the nucleus of a National Maritime Reference Centre (NMSC), which is intended to ensure that the Solent remains the focus for design, integration, testing, evaluation and frontline support of the complex software intensive systems that deliver the Royal Navy's fighting capability. The Virtual range concept enables ships to carry out trials to include virtual threats in a more realistic environment. New software for radar tracking system can spot otherwise difficult to detect missiles. Testing ensures that when installed at sea, the software will work first time and protect onboard personnel.

In 2018 the defence contractor BAE Systems which is also heavily involved in the dockyard announced a £10m investment to upgrade its Maritime Integration and Support Centre (MISC). New technologies such as artificial intelligence, information and electronic warfare, unmanned vehicles and new weapons was to be researched using new facilities including a state of the art visualisation suite able to display tactical data from any Royal Navy ship anywhere in the world.

LBTS has links with the Land Reference Centre at Blandford to provide a test environment for joint naval/army trials as well as the established links to US facilities.

In 2019 the Innovation and Collaboration Hub, a large black glass office building was constructed on the north side of the hill, following a business case made to the Local Growth fund to leverage QinetiQ plans to redevelop Portsdown Technology Park. Its aim was to foster growth and offer opportunities for small and medium sized enterprises by encouraging greater collaboration, innovation and enterprise across UK maritime mission systems. Global and local occupiers/tenants keen to collaborate with QinetiQ were identified. The development was intended to enhance the benefit of the NMSC for the Solent region significantly and to stimulate the UK maritime defence export market.

The new building is brightly lit up at night, a glaring exception to the generally careful control exercised by local planning authorities to protect the Hampshire countryside from intrusive development. Planning consent for QinetiQ's earlier plan was granted by Portsmouth City Council in 2016. Discussion took place in 2017 about how the skyline of Portsdown Hill would be affected by the new building. The design was by WYG and architects Scott Brownrigg. It was said to maximise the natural topography of the land, a claim that is open to question!

Fort Cumberland

Fort Cumberland's history is set out in the second of our forts chapters. It is home to two areas of research and their related institutions: land based and maritime archaeology. The Nautical Archaeology Society, "dedicated to innovative, fit for purpose research into all aspects of maritime archaeology with strong links to like-minded organisations around the world" is a long-term tenant there. The Society also offers training to maritime archaeologists and publications including the *International Journal of Nautical Archaeology,* as well as investigations of wreck sites. The Institute of Nautical Archaeology was jointly responsible jointly with the Hampshire and Wight Trust for Maritime Archaeology based in the Southampton Oceanography Centre for the Forton Lake project from 2006 to 2008. This was a community-based recording of the hulks of vessels abandoned at Forton Lake Gosport where the Maritime Workshop mentioned earlier was located. The fieldwork and community training for the project was supported by the Heritage Lottery Fund, and following the final year of fieldwork the Crown Estate supported publication of the findings.

From 1975 when Fort Cumberland was taken into the guardianship of English Heritage the Central Archaeological Service (formerly the Central Excavation Unit) has been based there. In 1998 the Ancient Monuments Laboratory of English Heritage was moved out of London to Fort Cumberland; its facilities were combined with those of the Central Archaeological Service as the Centre for Archaeology (CA) in the highly serviced Transit Shed inside the fort. Historic England supports English Heritage properties with research and services to support their conservation, improvement and presentation. It offers a better understanding of their significance as they may be subject to change through repair. It assists in the development of new visitor facilities and provides new information and illustrative material to contribute to their understanding and presentation.

The wide scope of research includes heritage science: technological and scientific work of benefit to the heritage sector through improved management decisions, enhanced understanding of significance and cultural value or increased public engagement. The centre covers both conservation research and archaeological science, including remote sensing techniques, scientific dating, environmental archaeology, investigative conservation and materials science. Understanding materials and environments, raising awareness, improving methods, access to information and advice are also covered, in the process addressing issues of capacity, capability and public benefit.

In 2001 in its Annual Design Awards the Portsmouth Society commended the creation of the high tech laboratory created inside the former Transit Shed where specialist conservation experts are at work. An extraordinary range of objects is conserved there, whether they are Anglo Saxon jewellery or a Roman sword. In 2010 chemical analysis of fifteen fragments of window glass from the fort used in Casemate 54 and the repair of the Guardhouse of c.1940 took place there. Exemplary restoration practice and reuse of the fort's complex structures have been on-going for many years as we explored in our earlier chapter about the fort.

The impressively wide range of research currently undertaken around the harbour, whether defence-related, marine or land-based archaeology or heritage conservation, is evidence of continuity in the area's creativity and of continuing innovation.

Sources

The News 2018 "Dockyard was the Navy's Larder". *The News*. Retrieved 28 May 2018.

MS Architecture http://msharchitecture.co.uk/prjHGY.html accessed 20January 2019

https://www.flickr.com/photos/94044317@N00/9965366264 Cavitation tunnel https://en.wikipedia.org/wiki/ Qinetiq - cite_note-44

Berry, William John. The Influence of Mathematics on the Development of Naval Architecture. In: Proceedings of the International Congress of Mathematicians in Toronto, August 11–16. 1924. vol. 2. pp. 719–736. (discussion of Froude's research on rolling motion, pp. 724–726)

Hunt, P. W. (1 December 1972). "Paper 13. Manoeuvring Research at Admiralty Experiment Works Using Free- Sailing Models". *Journal of Mechanical Engineering Science*. Sage Publishing. pp. 85–90. doi:10.1243/JMES_JOUR_1972_014_068_02. Retrieved 24 December 2017.

The National Archives, "Admiralty: Admiralty Experiment Works: Reports". discovery.nationalarchives.gov.uk.

The National Archives, UK, 1872-1977, ADM 226. Retrieved 24 December 2017.

Archives, The National. "Admiralty and Ministry of Defence: Admiralty Marine Technology Establishment: Reports and Files". discovery.nationalarchives.gov.uk. The National Archives, UK, 1975- 1984, ADM 341. Retrieved 24 December 2017.

The National Archives "Admiralty: Admiralty Experiment Works: Reports". discovery.nationalarchives.gov.uk.

The National Archives, UK, 1872-1977, ADM 226. Retrieved 24 December 2017

Gosport Borough Council Modified Sustainability Appraisal. Supplement to Annex C: Assessment of Options: Spatial Strategy, Regeneration Areas and Allocation of the Sustainability Appraisal which supports the Gosport Borough Local Plan (2011-2029): Publication Version. Haslar Gunboat Yard

http://www.portsmouth.co.uk/news/defence/portsmouth-picked-for-new-4m- defence-research-base-1-6176234

The News 2018 'Revealed: Ambitious plans for Haslar Gunboat Yard development. A development could help to 'breathe new life' into a historic military site. 13 February

Royal Naval Sailing Association: https://www.rnsa.org.uk/RCs.aspx?SectionID=60

UK defence solutions centrehttps://www.ukdsc.org/press-releases/2341/: Update: Centre for Maritime

Intelligence Systems (CMIS) transition

Royal Navy 2015 https://www.royalnavy.mod.uk/news-and-latest- activity/news/2015/February/10/150210-first-sea-lord-visits-centre-for- maritime-intelligent-systems 10 February

https://www.hornetservicessailing.org.uk/history.aspx accessed 5 May 2020

Danny Atkins and Ken Tout 2002 'QinetiQ Land Based Test Site – The Direction of Integration' QinetiQ Portsdown Technology Park Southwick Road, Cosham Portsmouth, PO6 3RU, England accessed at MP-097(I)-11.pdf

https://historicengland.org.uk/listing/the-list/list-entry/1413978

https://imcs.qinetiq.com/downloads/facilities/blue-top-cavitation-tunnel.pdf

https://britishlistedbuildings.co.uk/101413978-no-2-cavitation-tunnel-buildings-46-and-47-haslar-road-gosport-gosport-anglesey-ward#.X5G-alDTV3g

http://wikimapia.org/11019362/QinetiQ-Portsdown-Technology-Park https://www.theyworkforyou.com/wrans/?id=2007-07-18b.149940.h Defence disposals https://www.baesystems.com/en/article/bae-systems-invests--10-million-to-develop-new-technologies-for-the-maritime-integration---support-centre

http://www.portsdown-tunnels.org.uk/surface_sites/portsdown_main_p1.html https://onlineauction.574.co.uk/ lot/details/10746 accessed 10 May 2020

https://www.portsmouthsociety.org.uk/former-are-building-portsdown-hill/ accessed 10 May 2020

https://www.portsmouth.gov.uk/ext/leisure/parks/portsdown-hill accessed 10 May 2020

https://www.nauticalarchaeologysociety.org/Pages/Category/research http://www.solentforum.org/membership/members/Members_directory/hwtma. php https://historicengland.org.uk/about/contact-us/

national-offices/fort- cumberland/ accessed 10 May 2020 https://historicengland.org.uk/research/support-and-collaboration/research- and-english-heritage-trust/ Jim Williams, Edmund Lee, Gill Campbell 2013 *English Heritage Science Strategy* JEH Science Network

Jim Williams, Edmund Lee, Gill Campbell 2015 English Heritage Science Strategy - tables and annexes

David Gunworth RESEARCH DEPARTMENT REPORT SERIES no. 20-2010 ISSN 1749-8775 AN INVESTIGATION OF SOME WINDOW GLASS TECHNOLOGY REPORT

Historic England 2009 *The role of science in the management of the UK's heritage* Historic England 2009 *The use of science to enhance our understanding of the past* Historic England 2009 *Understanding capacity in the heritage science sector*

Historic England/Museum of London Archaeology 2016 *Portsmouth Harbour Hinterland Project. Project Report*

https://www.facebook.com/nauticalarch

https://www.gracesguide.co.uk/Thomas_Grant accessed 21 September 2020

https://www.pastscape.org.uk/hob.aspx?hob_id=238800&sort=4&search=all&criteria=royal%20william&rational=q&recordsperpage=10&p=12&move=p&nor=227&recfc=

Chapter 16
Defence sites repurposed as housing

Royal Crest Eastney Barracks

Both local and national developers have taken on large military complexes around the harbour to convert them to housing, raising substantial funds to do so, and thereby considerably increasing their value. Residential conversion might seem the obvious way of reusing the harbour's legacy of army, navy and marine barracks, but in the earlier redevelopments of defence complexes around the harbour the standing buildings did not always survive and they were demolished. Here we explore how surviving barracks and industrial buildings have been adapted for civilians to live in. Key factors are the degrees of intervention: how much of the original fabric and layout remains after conversion, given the need to insert complex new services, internal communications: lifts, stairs and subdivisions for kitchens and bathrooms.

Perhaps surprisingly, barracks' austere utilitarian design and surroundings do not deter new owner-occupiers and tenants. Where once twenty soldiers or Royal Marines were billeted in an Eastney barrack room, perhaps a couple or small families now live, revelling in the space, period features and high ceilings of these still gated communities. Given their military associations, not surprisingly, ex-service people enjoy living in former military buildings. Civilian or military, the new residents share the cost of maintaining the surviving perimeter walls and of the new landscaping which may be very different from the large expanses of tarmac or gravel of the parade grounds that once surrounded the barracks.

Gunner's Row Eastney Barracks   Martin Marks

We now explore the wealth of local examples of both demolition and newbuilds and conversions around the harbour - in the order in which they were either cleared or converted.

Point Barracks, constructed in 1847 which closed off the seaward side of Broad Street in Old Portsmouth until the 1960s has almost completely disappeared, leaving only the gun casemates and the rear halves of the 'bomb proof' barrack rooms. Also in Old Portsmouth only the outer wall of Colewort Barracks in St. George's Road still stands. A new housing estate: Gunwharf Gate, was built inside the wall, opened by the Lord Mayor in 1985. Victoria and Clarence Barracks were demolished in the 1970s, except for the northern block converted into Portsmouth City Museum and the NAAFI into the City Records Office.

Hilsea Barracks has also disappeared. Originally constructed in or about 1757, from 1780 to 1794 rows of long wooden huts were built arranged around three sides of a parade ground to house soldiers to man Hilsea Lines.[1] The barracks had been rebuilt in permanent form in 1854 for occupation by the Royal Field Artillery.[2] The unit left the site in 1921 when it became the main headquarters for the Royal Army Ordnance Corps.[3] During WWII the United States army used the site[4] The RAOC left in 1962.[5] The barracks were demolished from 1965 to 1970 to make way for the Gatcombe Park housing development,[6] still enclosed by the high brick wall on London Road and railings in Copnor Road.

A survivor of the barracks, Gatcome House is at the end of an avenue of horse chestnut trees from London Road. Gatcombe Manor, a medieval house acquired through marriage by Admiral Sir Roger Curtis in the 18th century was the first on the site.[7] In the 1770s the War Office requisitioned it from Curtis for military purposes. The classical Commanding Officer's house in red and yellow brick was rebuilt from 1780 to 1800; additional wings were added in 1887. Gatcombe House evolved to become the officers' mess. It is now listed Grade II.[8]

The Army left Hilsea barracks in 1973 and the city of Portsmouth acquired it as accommodation for students. The city architects re-erected the temple and balustrade from Crichel Down in the western gardens. But the house was later left vacant, falling into an almost derelict state, until about 1985 when a construction company, Warings Ltd. bought and restored it for use as their headquarters. In 2020 serviced offices, conference facilities and meeting rooms were available for hire there.[9] Nearby the coach house/stables and former officer's quarters building as well as the large Riding School survive, the latter used as a city museum store.

Fort Widley

Forts Widley and Purbrook, explored in the second forts chapter, have extensive barracks whose red silhouettes are prominent in the city's skyline. Purbrook's are used to accommodate residential youth groups and provide meeting room facilities, while Widley's are used for a wide range of community groups and to provide space to support the charity's equestrian provision. Both forts are currently managed by the Peter Ashley Activity Centres Trust whose mission is to "contribute to the development of young people to be the best they can be whilst using the Victorian Forts for the benefit of all."[10]

Once the conservation tide took hold from the early 1970s, most barrack complexes around the harbour still standing had to remain, making them available for other uses. Residential conversions examined in detail here are: Eastney Barracks converted from 1995, St. Andrews Garrison church in 1997, Royal Clarence Yard in 1998, St. George Barracks East and West in 2002, as well as Fraser Range, Tipner West and Haslar Barracks – still to come.

The financial challenge of acquiring and converting these large buildings is considerable. After the MOD released Eastney barracks for sale in about 1994, Gudgeons Ltd., a local firm first bought it. In a recession the work was paused, and a national firm Redrow PLC completed the work.

Another issue is to what extent local people can now explore these spaces. When the Army, Royal Navy or Royal Marines occupied them, service guards were stationed in entrance guardhouses. In some cases these controllers have been replaced with civilian security guards. Other sites, while reminding entrants that they are private property, such as Royal Clarence Yard do not bar visits from the public.

Eastney Barracks

Eastney Barracks was built largely from 1863 to 1870 along with the two small forts, Eastney East and West (1861-1863) to provide the forts with purpose-built barracks for the Royal Marine Artillery. It was designed by William Scamp, assistant director of the Admiralty Works and Engineering Department and Col. G T Greene of the Royal Engineers as headquarters for the Royal Marine Artillery who moved into the barracks from Fort Cumberland in 1867.[11]

The soldiers' accommodation in the long range had a fine view over the parade ground, esplanade and shore to the Solent - the departure setting for imperial points East.[12] The three-storey six-bay block was built by civilian contractors in brown brick from Fareham Brick Company with decorative red rubbed brick arches over round-headed windows on the ground floor, string courses on the intermediate floor levels and hipped roofs with a fine row of chimney pots.[13]

Attached at the back was the long glass-covered drill shed, which was demolished in favour of landscaped parking space for the new residents of Gunners Row and the town houses.

Eastney Barracks and Brent geese.   Martin Marks

New pedimented entrances to the apartments were constructed from iron salvaged from the drill shed, cast by Henry Chissell of London and dated 1863 and 1864. The wooden floor planks were individually lettered and numbered. A sample were collected and given to the Royal Marines Museum, but these have now been misplaced.

Central to the eastern flank of the square is the imposing stone-faced Officers' Mess, with a double flight of steps and loggia over the main floor. It has a particularly opulent interior: a fine wide staircase and a large and splendid rectangular dining room with a massive coved ceiling, big rounded-headed plate-glass windows and a central semi-circular bay. This would have looked over the eastern redoubt constructed on the Exercise Field or training ground. The dining hall was reputedly so cold in winter that the officers would wear their greatcoats to dinner. The Kaiser, later a leading player in the Great War, watched a mock attack from the dining hall staged on the sports field. His comments on British military prowess are not recorded.[14]

At the western end of the square are the four five-storey service officers' houses and headquarters building known as Fortview House facing east, south and west in red brick with rusticated yellow brick pilasters and stone-framed windows.[13] The Colonel, Second Commandant of the Royal Marines Artillery lived with his family and servants in the corner house from 1866, which was later occupied by many other senior officers responsible for the barracks, its personnel, training and deployment around the world.[15] The War Office later divided the houses vertically. Teapot Row to the west, the former Field Officers' Quarters was subdivided horizontally into apartments.[16] A new block of flats built by Warings Ltd. between Fortview House and the Colonnades (former Quartermaster's building) and the Drill House infills the square.

The colonnaded guardroom to the south of the main entrance inside the tall flint wall enclosing the barracks from Cromwell Road used to face the school for the Royal Marines' children. It was demolished and replaced with two-storey town houses. Across the road were three strategically placed pubs. The six-storey water tower of 1870-1 supplied the barracks complex. It has circular windows and white stone machiolation. It still houses the three-faced clock, no longer powered by three rods across the space (now part of an apartment) but by a much smaller wall-mounted mechanism. Paired chimneystacks rise through the roofline like pinnacles at three of the angles of the flat pyramid roof.[13]

Forton Barracks in Gosport originally housed the Royal Marine Artillery and later the 17 Ports Maritime Regiment and its predecessors. After the amalgamation of the Royal Marine Light Infantry and the Royal Marine Artillery in 1923, Forton Barracks was closed, and Eastney Barracks served as headquarters for the Portsmouth Division of the Corps for training, reserve and special forces.

Eastney Barracks was a classic military enclave within a civilian world, with its own church, water tower, library, gym, theatre (later cinema), school, drill hall, drill field (later a sports field) and officers' mess. Ironically the barracks like the new universities of the 1960s were strong examples of coherent urban design rarely found in cities themselves.[17]

The public visibility of the Royal Marines was enhanced by their marching band performances at sporting and public events. The origins of this lie in King Edward VII's appointment of the Royal Marine Artillery Divisional Band as the Royal Yacht band in 1904.[17]

Eastney Barracks become the HQ for all Royal Marines Hampshire establishments in 1947 and finally Headquarters for training, Reserve and Special Forces. After WWII as the role of the Royal Marines changed the number of personnel using the barracks fell, until by 1973 there were only 200 marines on site. In 1991 the last marines left and the barracks were closed.[18] The site was sold to Gudgeons of Fareham in 1995.

Officers' Mess Eastney Barracks
Royal Marines Museum 1972-2-17
Martin Marks

The Royal Marines Museum, established in 1958 was in the former Officers' Mess from 1972. It covered the history of the Royal Marines from their beginnings in 1664 through to the present day. One display highlighted the demands of the 32-week training course undertaken by all Royal Marine recruits. Personal stories of wars, battles and significant events brought the history of the corps to life. An impressive medal room contained rare and valuable medals. Paintings on military themes decorated the walls, many of them donated by the public or serving and former Royal Marines.[19]

Landing craft 2019

Outside was a landing craft, various field guns and light anti-aircraft guns as captured trophies from the assault ship HMS *Fearless* of the Falklands War of 1982. The Memorial Garden along the front of Gunners Row contained many moving tributes to Royal Marines from their campaigns in different parts of the world, including stone panels from the Royal Marines Barracks at Chatham demolished in 1956. There was a stand of 13-inch mortar projectiles with an explanatory plaque. The mortar 'bombs' were of a type used in the 13 inch Sea Service Mortar displayed by the saluting platform on the edge of the former parade ground.

Yomper 2019   Martin Marks

The giant statue of a Falklands conflict 'Yomper' by Philip Jackson signalled the public entrance to the museum from the Esplanade. The museum closed in 2017 in preparation for a transfer into Boathouse 6 in the Historic Dockyard as part of the National Museum of the Royal Navy in order to benefit from the large tourist footfall there. Failure of the lottery bid delayed this move. A strong public outcry greeted the news that the statue too might be moved to the dockyard. The consensus was that it should stay where it is. This was confirmed by the National Museum of the Royal Navy in 2018. The 13 inch Sea Service mortar, used for practice firings around the firing ranges towards Fort Cumberland was also the subject of possible relocation and concern.

The Barracks were a good example of what Erving Goffman called total institutions where the daily round was totally prescribed. This contained world operated according to different norms and rules - mental institutions, prisons, military establishments.[20] The Royal Marine soldiers in their spartan mess rooms in 1903 might have been amazed to see their billet turned into desirable apartments a century later. The authors can vouch for the effectiveness of the acoustics of the barrack square, which once echoed to shouted parade commands. Gunners Row acted as a sounding board as they called to each other across the long expanse.

"This barracks has gone the way of so many military townships and become part upmarket housing, part museum." [21] When demolition and clearance of Eastney Barracks was first proposed, Rosemary Dunne whose husband was in the Navy remembers writing to protest – to the C in C or to the person responsible for Naval property - that it would be an ultimate act of vandalism if the barracks were flattened. She said that they were very well built, and that they were terribly important to anyone currently serving in the Royal Marines or who had done so. Prophetically, she suggested that the whole complex with its space open to the sea would make a fabulous development.[22]

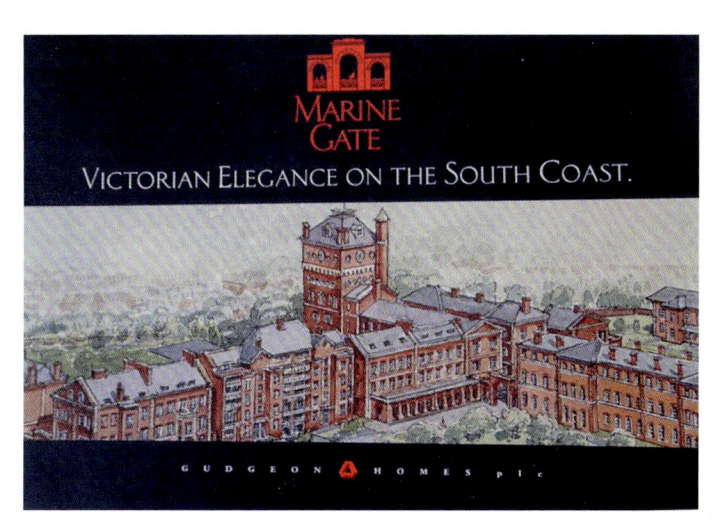

Conversion

Gudgeon Homes plc sales brochure
Geoffrey Salvetti

Gudgeon Homes Ltd., a Fareham company, "saw there was an opportunity here. They couldn't pull it down. They were one of the first local firms to realise residential could work."[23] They marketed the site which they renamed 'Marine Gate' as 'Victorian Elegance on the South Coast'. Their brochure said that they enjoyed 'the additional challenge presented by working to enhance conservation sectors and areas of natural beauty or special historic interest.' Gudgeons worked on different sections of the development all at the same time, starting on the long block renamed Gunners' Row and also on Teapot row and Fairview House. Perhaps this led to financial difficulties, and the eventual sale to Redrow, who completed the job. In the redevelopment the six houses in Gunners' Row were named Bamford House, Dowell House, Finch House, Halliday House, Prettyjohn House, and Wilkinson House - after winners of the Victoria Cross.

John Bell and his wife were the first private owners to move into the converted Gunners' Row."We were the first ones in! We were sad to see the Royal Marines going. What we were pleased about was that the building was staying. I'm an Eastney boy and it was nice to come back. My son was a good footballer until the age of 14 or 15 and he used to play football in the Royal Marine Barracks when it was still operative. We were just strolling along the front and we saw the development. We looked at the show flat on the ground floor of Bamford House. We couldn't believe how nice that was. We managed to get into this one [No.3 on the first floor]. I can remember walking over the floor joists. We just fell in love with it, and the lovely view. When we first moved in there were no roads. They gave us a torch, because there were no lights to come in!"

Every apartment was originally a mess for up to 20 soldiers in one big space. Photographs show what it looked like, with shelves at the end, and "all these big windows", with views both ways through the narrow block. The developers raised the floor and dropped the ceilings. The original windows had rope sash cords, but they were replaced after the first two blocks with a different mechanism. John Bell said that the original specification was for secondary glazing. "When the wind blows off the sea it's really cold, despite good heating. It's really draughty". Listed status put paid to that, until 2020, when Historic England agreed that double glazing was permitted if it was to the exact dimensions of the previous windows.

"The surgeon who first lived in Bamford House was Surgeon Vice-Admiral Ian Jenkins. After retirement from the Royal Navy HM The Queen appointed him Constable Governor of Windsor Castle from 2008-9. He was the first Chair of the Marine Gate Residents Association in the 1990s. There's a memorial stone to him in front of Halliday House. If you walk along Gunners Row it's on the edge of the parade ground. The Brent Geese are always there every year."

John Bell who worked as a property surveyor for the city council paid £90,000 for his three-bedroom apartment on the first floor of Bamford House in 1995. Cheryl Jewitt, who bought No.3 Wilkinson House for £130,000 within a year of the Bells, said "they didn't know what they had… They hadn't priced it properly, and because they realised the demand, they put up the prices. Because it is listed, some people regretted that there were no balconies and no private outside space. You had to dress up to go and take your deckchair outside. At first it was like a retirement development, very quiet. But when they built the houses such as Churchill Square we had children on bikes… Some older people didn't like that."

The tennis courts and the parade ground belong to the residents. They had to pay £6000 to have the parade ground resurfaced. The service charges are £3000 a year for looking after the lawns and trees and everything else. As John Bell says, it's not a gated community. "People come in and their dogs leave a mess we're got to clear up. I say to them: 'If you like coming in here come and purchase an apartment and pay the serve charge we have to pay!"[23] He has a collection of documents including plans by the first developer Gudgeons and their successor Redrows' and sales brochures. He was the secretary of the Marine Gate Residents' Association for five years. They gave him a handsome engraved cut glass decanter in gratitude for his service. The Portsmouth Society awarded Marine Gate their Best Restoration award in 1995 and Best Landscaping in 2008. Their two blue ceramic plaques are on the west wall of Gunners Row.

The corner buildings in Royal Gate at the southwestern corner of the square remained as houses rather than being converted into apartments. Gudgeons used modern building materials and methods and failed to check that the Victorian damp course in the corner houses was clear. Designed by Royal Engineers the walls are 2 ft 6 inches thick. The damp course consists of horizontal courses of huge triangular stone blocks drilled with holes. To ensure flat surfaces on the walls inside the builders put large sheets of plasterboard straight onto the wall with blobs of 'pug'. Blocked by debris, water leached through the inner surface and large damp patches appeared. The occupants of the two south facing properties had to move out for 6 or 7 months, for repairs undertaken by the National House Building Council supervised by the architect.[16]

By 2019 the leasehold owners of No. 1 Royal Gate, Mr. and Mrs. Salvetti, had lived in Fortview House for nearly 19 years. As a pupil at Portsmouth Grammar School Geoffrey Salvetti had a long had a connection with the Royal Marines – first as a member of the Portsmouth Grammar School Combined Cadet Force. The CCF was supported by Sgt. O'Toole of the Signals Troop Force from 1961-2. Geoffrey knew the barracks because he was a member of the Commando Logistic Regiment Royal Marines for some six years. As a member of the legal profession since 1968 and a solicitor since 1976 Geoffrey specialised in Armed Forces legal work at the Royal Marine Barracks Eastney.

He was involved in the last two court-martials in the barracks: a Band Corporal who sold band instruments and a Captain who falsified his expenses. Both courts-martials took place on the top floor of Bamford House in a barrack room that ran the length of what are now two apartments. The Captain was found guilty but fined and allowed to continue in service, but the Corporal pleaded guilty and was sentenced to a period of detention and dismissed the service. He was taken to the cells, which were by the guardroom.

The Salvetti family chose to live at Eastney Barracks because it was the quietest place in Southsea, with the best views, access to the beach, beautiful lawns and on-site carparking. As already said residents pay for maintenance of the site, street and window cleaning and street lighting through the service charge – in addition to the council tax. There are a few ex-service residents – former members or some still serving. It's a semi-gated area. There is no public access to the residential area from the seafront: the entrance to Marine Gate is via Cromwell Road. The Teapot Row entrance is from the seafront via electronic gates. "We like the peace and quiet. No hordes of tourists, unlike the noise and hubbub of Gunwharf Quays – or Old Portsmouth. Gunwharf Quays is too noisy."

According to Geoffrey, who was a trustee of the Royal Marines Museum, the visitor footfall to the Royal Marines museum was not large: 12,500 a year, but only 4000 paid; the rest were service personnel including Royal Marines bussed from the Commando Training Centre Royal Marines at Lympstone in Devon on day visits.

Until 2017 there used to be a band concert and Beating the Retreat on the lawn, an annual event when the Royal Marines recruiting team spent the day there, bringing massive numbers of people to the museum. The Royal Marines Association Concert Band of former NCOs and invited supplementary musicians still played popular concerts there. John Bell said: "Every Remembrance Day there was a ceremony outside the Museum. The Royal Marines Juniors paraded, and usually a Royal Marines bugler would play the Last Post. They've been holding weddings in there ever since the museum opened. I can remember a wedding when a helicopter landed on the green and the bride got out of it! There were open air concerts. The first one was celebrating 50 years after D-Day. I can remember people came up from London and a BBC commentator ran the show. The green was covered in cars. The Marine Gate Residents' Association had to agree that they could use it. It was a lovely day – and the next day you would not think it had happened. There were hundreds of people there. Every August up to two or so years ago we had a Royal Marines concert on the green when the junior band played outside the museum. There were loads of people taking picnics and drinks."

"The Antiques Roadshow was there about ten years ago. I was standing watching. They had some furniture from Fratton Park which they were going to throw away because Pompey wasn't doing very well and someone took over the club and said 'We don't want all this furniture'. But someone picked it up and took it down to the show. The Antiques Roadshow people were very interested in it. I was watching, and a number of people said 'I saw you on TV!' 'When I was in Australia I saw you on television…'"[23]

But once the NMRN closed the museum, the outside events also ceased.

Sale of the Royal Marines Museum and Eastney Fort East

For Sale Notice March 2019

Geoffrey Salvetti said that part of the north wing of the officers' mess was owned by the museum, later subsumed into the National Museum of the Royal Navy, which also owns the adjoining Eastney Fort East – whose history, restoration by the Palmerston Forts Society and subsequent deterioration is explored in our second forts chapter. He was the founding chair and member of the Society for 26 years. The MOD also owned part of the Mess, together with the carpark, but delegated dealing with it to the NMRN. There are many easements on who can use and access the site. There is a banquette (rampart with firing step), berm and ditch (dry moat) approximately 10 ft. deep and 30 ft. wide under the carpark. In 1773 a Major Archer designed an early Eastney Fort on the site of an earlier Tudor gabioned fortification of 1545 and replaced in 1587/1588.

The Officers' Mess is in need of serious investment. Since the museum left in 2017 there has been emergency maintenance but no significant upgrades. In 2019 the central core that housed the museum and the two adjoining wings, a total area 3.35 hectares and an internal floor area of 5.500 sq.m. was marketed for sale for commercial development by Lambert Smith Hampton estate agents of Barnes Wallis Road in Fareham. They suggested that the listed building could be a 'conference centre, hotel, education centre, healthcare hub or exclusive residential resort.'[25]

In April 2019 residents of Southsea and Marine Gate objected to the most likely use: expensive private housing. They preferred to see it transformed into a community facility. Covenants protect the memorials, gardens and parade grounds and any developer would have to follow strict rules to preserve its history and context, as a prominent part of the city's landscape.[25] When it was put up for sale in summer 2019 it had been closed as a museum for two years, while NMRN staff packed up and removed the collection and artefacts. The failure of the lottery bid for the conversion of Boathouse 6 for the museum and work on the NMRN's library meant that the Royal Marines archives was not available for researchers for at least two years. In summer 2020 the For Sale notice was still there.

We now consider the conversion of the Marines' church nearby into unusual houses.

St. Andrew's Garrison Church Eastney Barracks

Royal Marine Artillery Church Eastney Postcard January 18 1910

The Crinoline Church, a polygonal temporary timber structure was originally built as a 'temporary' hospital for the Crimean campaign of 1854 to 1856 and then utilised at Eastney as a church for the Royal Marines barracks. It was twice relocated, and then replaced by the permanent brick church, St. Andrew's in Henderson Road. Designed by the Works Dept. of the Admiralty it was dedicated in 1905, one of a series of churches sometimes referred to as 'Admiralty Pattern Churches'. There are near identical ones in Keyham Devonport and Chatham. St. Andrews was built in brick with stone dressings and lancet windows. It was closed in 1973 and used for band practice until its conversion in 1997.

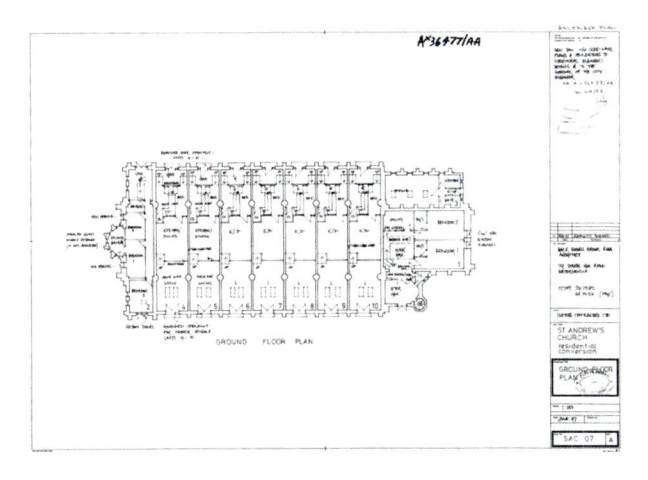

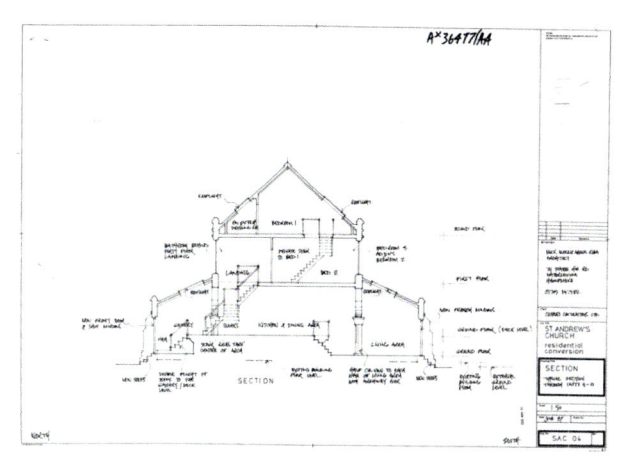

Conversion of St. Andrew's Church to houses Plan and Section   Mick Morris RIBA

Architect Mick Morris subdivided it vertically into ten dramatic three-storey houses in the renamed Grand Division Row. The interiors have exposed brick walls, stone columns, high ceilings, oak floors and religious mosaics. The property is the smallest of Portsmouth's 22 conservation areas. The conversion won the Portsmouth Society Best Restoration Award in 1998.

Royal Clarence Victualling Yard

Royal Clarence Yard Granary, Bakery and slaughterhouse and tug 2019

We now move across the water to explore the history and transformation of the navy's specialised food factory. To control quality and costs and minimise fraud, the navy's Victualling Board preferred to provision warships directly from its own yards or depots. By 1700 the variety of provisions had become largely standardised, reflecting the fairly rudimentary food preservation technology of the time. Beef, pork, fish, biscuits, butter, cheese, peas and beer formed a staple fare. In the course of the 18th century oatmeal, sugar, sauerkraut and cocoa were added. Between 1774 and 1783 the number of naval men at sea rose from just over 17,000 to 105,000 and grew still further in the Napoleonic Wars, making hugely increased demands on the Victualling Board. After the Square Tower ceased use as a powder magazine in 1779, it became a meat store for the Victualling Board; hooks from this use still survive.[26]

The Gosport yard was originally known as Weevil Brewery, a naval brewery established by Captain Henry Player in 1690. The source of the name Weevil is not clear, though it's those insects that infect ships' biscuits. An early 17th century chart, long before the victualling yard existed, shows a reference to "Weevel Wel Spring", while an 1844 account of Gosport refers to a landowner named "Weovill". By 1716 a good supply of fresh water used for supplying the fleet encouraged the Victualling Board to establish a brewery and cooperage on the Gosport waterfront. During the early 18th century the site was sold to the Board of Ordnance and by 1761 it was the property of the Admiralty. It became Weevil Yard, responsible for provisioning the Royal Navy's major dockyard in Portsmouth.

By the mid-1760s the Board decided to concentrate all its local cooperage for the construction of barrels to store food and water there next to the Weevil brewery. Single-storey workshops in brick, each centred on a hearth, with timber-framed and timber-clad upper parts and tiled roofs were built in 1766 around a quadrangle, a large open space used to store barrels, now called Cooperage Green. The workshops are the oldest buildings to survive in any victualling yard. Remarkably, a few remained in use for constructing barrels until the abolition of the rum ration in 1970.

Inside the square, the original octagonal pumphouse of 1778 was powered by a horse gin pumping water from the well below to the Weevil brewery. According to historian Jonathan Coad these buildings are the oldest to survive in any victualling yard.[26]

Cooperage Pump House 2017

Slaughterhouse 2014

In 1820 other victualling functions were also transferred from Portsmouth, including the King's Mill at the entrance to Gunwharf, a slaughterhouse near the Square Tower and a bakery and storehouse of 1513 in King Street. The whole complex was much expanded in 1828-32 to the designs of George Ledwell Taylor, Civil Architect to the Navy under the direction of Jeremy Bentham.

In 1831 it was renamed the Royal Clarence Victualling Yard in honour of the Duke of Clarence, the Lord High Admiral (later King William IV), without whose support the scheme would probably have foundered.

Officers' houses and colonnade

The Yard has an impressive classical gateway from Weevil Lane of 1828 designed by George Taylor. Surmounted by the duke's coat of arms in Coade stone, it is flanked on either side by curved guardhouses with Greek Doric colonnades on the inner side. Two large white stucco houses for the Superintendent and Deputy were built just inside to the north. Behind them is the walled triangular reservoir, flanked by early artificers' workshops whose function was rediscovered in research anticipating the redevelopment.

Granary

The most impressive building is the red brick four-storey Granary built in 1828-30 facing the harbour. The ground floor is open, supporting the floors above on iron Doric columns around the outer edge with a stone frieze similar to the one at St. Katherine's Dock on the Thames. Slender iron columns support the storage floors above. Joined to it to the north is the long three-story Bakery for producing ships' biscuits powered by a steam engine designed by Sir John Rennie.

It still contains the original ovens for baking ships' biscuits. The bakery is now called The Mill and Grade II* listed. To the rear of the bakery, an area once known as North Meadow was used as cattle lairage (holding area), and former stables still exist here. On the waterfront is the long Italianate Slaughterhouse of 1854-5, now called The Old Storehouse, Grade II. The Bakery won the Gosport Society's Restoration Award in 2011

The Tank Store was built in 1862-3, the only known surviving example of a purpose-built tank storage unit. The Steam Fire Engine House was added in 1862-3 designed by Andrew Murray, chief engineer in the dockyard under the direction of Col G T Greene.[27] Parallel to the Tank Store is the New South Store, parts of which date from 1758, altered in 1830, remodelled in 1897-8, damaged in 1940. It was originally used to store dry goods for the brewery and later for storing rum and sugar.

Oven for baking ships' biscuit   Jim Humberstone

A railway to Gosport was first promoted in 1836 as part of a plan by the London and Southampton Railway to connect Portsmouth to London but a branch line to Portsmouth was rejected. Instead a terminus was placed just short of the town centre as Gosport was a fortified town and, as later in Portsmouth, the Commanding Officer refused to allow the walls to be breached to bring the railway closer to the town centre. The line opened on 29th November 1841.

Rail link to Royal Clarence Yard 2014

The royal family first used the station on 8th October 1843 when Prince Albert arrived at Gosport by train to greet King Louis - Phillipe of France on a state visit. The French king landed at Royal

Clarence Yard on his large hybrid sail-steamship, Gomer. He and his entourage were driven by horse and carriage along Weevil Lane to Gosport station where they were taken by train to London for a state visit. The painter JMW Turner recorded his arrival in atmospheric sketches and oil paintings, as did an engraver who showed the Royal Clarence Yard in detail with the waterfront thronged with boats and flags. Queen Victoria came to the station six days later when she accompanied the King on his return to France.

The following year the Queen purchased the Osborne Estate on the Isle of Wight and the Clarence Victualling Yard at Gosport was her favoured point of departure in the Royal Yacht rather than Portsmouth or Southampton. Shortly afterwards Prince Albert and the London and South Western Railway agreed to build a 605-yard extension from Gosport through the ramparts of the Gosport Lines to a new station closer to the pier. Gosport Clarence Yard Station, also known as Royal Victoria Station, opened in 1845. The station consisted of a single curved platform flanked on one side by a long wall. There was a waiting room for royal travellers, but it is said that Queen Victoria never used it. When the Queen was passing through the station a carpet was laid between the train and the Royal Yacht. The line of the extension from Gosport Station through the town ramparts to Weevil Lane was restored in 2006 and new track was laid along this short section. "Completion of the extended yard in 1832 presented the same problems of over-capacity that faced the Royal William Yard [in Plymouth], and by the 1850s the brewery had ceased to function and had become a slop store for clothing. The bakery appears to have remained in use, with extra ovens being recommended in 1861 to keep pace with demand but was closed after the First World War and used as a storehouse. By the 1960s the only buildings still used for their original purpose were the cooperages."[27]

> There were three broad ranges of victualling stores:
>
> *food: tinned, cased and bagged dry provisions, fresh and later frozen food, rum, tobacco and duty-free cigarettes;
>
> *clothing: uniforms for RN Officers and ratings, Royal Marines (from 1951, WRNS, QARNNS, Fleet Air Arm flying clothing;
>
> *Mess Gear: utensils, implements and galley gear – necessary for preparing, cooking, serving and eating food.[28]

An Army Mobile Field Bakery worked under great pressure to provide bread for the soldiers preparing for D-Day. Until after WW1 the Victualling Department was also responsible for the supply of water at the dockyard ports, including water from the reservoir in Royal Clarence Yard which was transferred to ships by water boats and lighters crewed by dockyard personnel.[29]

As a military target the yard was severely damaged during World War II. From late 1940 onwards the Yard suffered increasing air raids, climaxing in a very severe raid on 10 March 1941 when hundreds of incendiaries and high explosives caused enormous destruction, loss of stores and half of the total storage space. As well as the southern wing of the Mill/Granary building, the Salt Meat Store, Implement Store, Miscellaneous Store, Clothing Store, Cooperage Shed and other smaller buildings were completely destroyed. 'The imposing office building with all its records and stock accounts was also burnt out'.[29] Stocks were dispersed to locations away from Gosport.

The removal of modern storage buildings left gaps in the historic layout, "resulting in under utilised and undefined spaces."[30] The function of the Yard was subsequently modernised; after the war its main function was storage. From 1961 the Navy's food laboratories were based there; innovations in food production and preservation were developed. Royal Clarence Yard was one of the first sites to mass-produce food on an industrial scale. In the 20th century, refrigeration techniques and experimental methods of food preservation were pioneered here. Many specialist buildings survive.

The creation of the unified Ministry of Defence on 1 April 1964 heralded studies to rationalize supply to the three services. In January 1965 the former Admiralty Victualling Department became part of the RN Supply and Transport Service. In 1966 the Navy Department was made responsible for food supply policy and provisioning, storage and distribution of food for all three services.

Although the yard officially closed in 1970 it was reopened for the Falklands War (1982), the first Gulf War (1990-91) and the former Yugoslavia War, supplying the navy with a variety of essential equipment and goods. The yard closed for good in 1991. 'Market testing' the supply of food to the Armed Services resulted in the transfer of food supply to the NAAFI from 1994, although they subsequently lost the contact and the MOD has used commercial suppliers ever since.[31]

In 1995 the MOD declared 16.26 hectares of Royal Clarence Yard surplus to requirements. Berkeley Homes bought the northern part of the site in 1998. As *"one of the first large industrial food processing plants in the country"* it has an important place in British industrial history.[32] The Yard was designated a conservation area by Gosport Borough Council in 1990. It is adjacent to a Site of Special Scientific Interest. It contains 18 Grade II and II* listed buildings and Scheduled Monuments.[33]

Bakery converted into pub 2017

From the conservation point of view two factors improved the course of redevelopment compared with Gunwharf (also developed by Berkeley Homes) across the harbour. It was agreed that a proper research study of historic sites by the MOD before sale would save thousands of pounds and several years of negotiations. Gosport Borough Council and English Heritage commissioned a historian Dr. David Evans to research the history and evolution of the structures – brewery, bakery, cooperage, artificers' workshops, reservoir, landing stage, railway station – an independent analysis of the sites and its heritage significance - so that development proposals by the purchaser Berkeley Homes had to respect them. In 2002 Cotswold Archaeology conducted an excavation for Gifford and Partners on behalf of Berkeley Homes (Hampshire) Ltd. including Flagstaff Green, the Salt Meat Store and Brewhouse Square.[34]

Secondly, Berkeley Homes may have learnt from the opposition in Portsmouth to the clearances in Gunwharf – or perhaps at the insistence of Gosport's planning department - they commissioned John Thompson and Partners to hold two Community Planning Weekends in the Slaughterhouse in 1998: hands-on planning of the site's future by 500 people, most of whom had never been on the site before.[35] The leader of Gosport council at the time remarked to Celia Clark that he had lived in Gosport all his life but had never seen it. Those who attended put up many Post-It notes on the site plan with their hopes for new uses for the buildings and spaces on the site.

JTP architects were commissioned to develop a masterplan "through a Collaborative Placemaking process" which they then took through to detailed planning and listed building applications, working drawings and the implementation of the plans on site. Mixed uses including leisure facilities and employment, with 698 apartments and houses, both converted and newbuild were planned in a mixture of tenures including rented and shared ownership. Contemporary architecture was to be introduced. Public spaces and pedestrian routes and the recreation of lost spaces would knit the yard back together. Full public access to the previously inaccessible waterfront and a new marina would be a new destination which would regularly host major yachting events and festivals.[36]

There were regular meetings with Gosport, English Heritage and local people. Marcus Adams of jtp architects described how the masterplan established principles of scale and massing; conversion, new build and new uses; allowing a phased approach, establishing certainty for the developers, who were able to submit a single planning application. The scheme was due for completion in 2006, allowing access to the waterfront for the first time, employment in the cooperage, social housing by Portsmouth Housing Association, and opportunities for leisure and recreation which Gosport previously lacked.[37] Conversion of the brewery involved deep plan flats with only one or sometimes two windows to light and ventilate the whole space. A replica of the bombed south wing of the Granary also containing flats was built facing the harbour.

Royal Clarence Yard Location Map 2016

In 2003, the 145-berth Royal Clarence Marina opened[38] and in 2005 Berkeley Homes submitted a planning application to turn Royal Clarence Yard into a hub of bars, restaurants, housing and retail outlets. In the original planning application, Berkeley Homes wanted 380 houses with 17,000sqm of commercial floor space. Also included were a cinema, restaurants, bars and retail outlets. The firm then put in a new application, increasing the houses to 698 and reducing the commercial space to 10,203sqm, later revised again to around 500 homes.[39] The new application was refused by Gosport Borough Council's planning board, but, after appeal, the Secretary of State agreed to the application.[40] This permission involved the refurbishment and conversion of many of the existing buildings and construction of new ones.

Between 2004 and 2008 the development won several awards for planning and restoration. Much of the scheme was completed by 2012. In 2015 car parking issues were raised and a revised plan approved. But the waterfront area and the slaughterhouse have remained empty, apart from the marina. A bar in the ground floor of the bakery failed. Residents said the yard was still not living up to its potential and councillors are angry at the lack of businesses. Marine Artist Colin Baxter has his studio in the Yard. He commented: "Royal Clarence Yard, it's more than a Marina. Yes, Berkeley's have failed to fill the units they built facing the Marina, but the units in North Meadow and Cooperage Green are mostly owned by a private landlord and are FULL. Up in North Meadow we are a bit more Arty/Artisans also with offices for sail training. Cooperage Green has a mix of all sorts. The biggest battle we have is letting people know we are all there. Last week a few signs we had on railings at the town end of Weevil Lane were taken down en mass by someone; not the council. We have asked for proper signage but no luck so far."[41]

In September 2020 on a visit during the Heritage Open Days the Slaughterhouse was full of fitness equipment, Arty's café which had been there two and a half years had also taken over outside space to eat outside, as had, more recently, Baker's Bar and mezzanine café. Customers sitting inside and out were a mix of local residents, people holidaying in the Air BNBs on site and yachtspeople from the marina. The public are free to explore and enjoy the yard – a contrast to the closed mainly residential Haslar Hospital.

Royal Clarence Yard southeast area 2019

One area, the southeast part of the yard, which has the navy's oil fuel jetty linked to the storage depot west of the Gosport lines and St. George Barracks as its northern boundary, remained unrestored in 2020. Historic buildings there include the former South Stores. The single-storey west range was built by Samuel Wyatt in 1758 for the storage of rum. Extended to three storeys in the nineteenth century, it was reduced to two storeys by bombing. Parallel with this is the former Tank Store converted from the boom defence depot of 1832 for the storage of water for ships, fed by the heavily buttressed reservoir. Of cast-iron construction with gantries it has a brick and timber-clad exterior and a double-pitch roof.

East of this is 'the heavily truncated remains of the Royal Station, terminus of the special branch line from Gosport Station used by Queen Victoria from 1845' to embark on the Royal Yacht to sail to and from Osborne House on the Isle of Wight. The tank store was expanded to include 'Her Majesty's Landing Furniture Store' where the red carpet was stored. The building was originally 520 feet (158m) long, with the canopy over the track supported on cast-iron columns.

'The last piece of the jigsaw' of MOD disposals on the Gosport shore according to the leader of Gosport Borough Council, the Ministry of Defence announced its intention to dispose of it in 2014 but this did not happen immediately. In June 2017 Gosport Borough Council included Royal Clarence Yard as a "Character Area" and proposals for its eventual development in a draft "*Waterfront and Town Centre Supplementary Planning Document*".[43] In November 2018 the MOD put the 5.2-acre (2.1 ha) site up for sale, emphasizing its 'full deep-water access to Portsmouth Harbour, making it ideally suitable for commercial marine activity'. Local representatives expressed the hope that this would revitalise the area, and local businesses were even seeing it as a chance to rival Gunwharf over in Portsmouth.[42] Gosport Borough Council was offered the site for £1 but rejected it because of the estimated millions of pounds it would cost for repairs to the site, renewal of its utility serves and conservation of its listed buildings before it could be used. When it was put up for sale in 2019 the Naval Base Property Trust expressed interest in buying it for £1 plus 50% of development value after deduction of all costs associated with the development.[44] They proposed that the former cooperage might well be used for the purpose for which it was originally built, given the increased demand from the craft spirit industry for English oak barrels which were unavailable; the only operational UK cooperage was in Scotland. They proposed to demolish the remains of Queen Victoria's Railway Station and relocate it. There was a precedent for this when her Railway Shelter in the dockyard was shifted on rollers to a new position on South Railway Jetty. There were ownership matters to be resolved between the MOD and the Crown Estate connected with the jetty and slipway and contamination from a previous fuel pipe. They intended to work in partnership with Elite Homes with whom they were already working at Priddy's.

Instead, the MOD sold the 5.7-acre site in November 2018 to UK Docks based in Tyneside in September 2019. They provide a comprehensive range of dry docks and marine services in the Tamar estuary and Endeavour Quay in Southampton. They planned to expand their maritime servicing and repair business in Royal Clarence Yard. The selling agent acting on behalf of the MOD said the five and a half acre site was one of the largest sold in the last decade.[45] In September 2020 UK Docks applied for planning permission to pull down five buildings including the railway station.[46] The heritage statement they commissioned from Giles Pritchard suggested that the station be fully recorded by a conservation architect and preparation of a method statement for its careful dismantling, storage on pallets and reassembly once proposed uses are found for the remaining heritage. The application to pull down the railway shelter was withdrawn.

Gosport Oil Fuel Depot

The south-eastern part of the Yard (approx. 3,74 hectares), which includes the Oil and Pipelines Agency access to the Gosport Oil Fuel Depot, was retained by the MOD for operational reasons.[47] A multi-million pound redevelopment project to replace the tanks and upgrade the buildings on Forton Road was due to begin in 2018 and finish in 2021. The fuel tanks are over 100 years old and are approaching their end of service life. The project is said to ensure that the depot will be fully equipped to meet the requirements of the Royal Navy's new Queen Elizabeth Class carriers.[48]

Royal Engineers Mews

On the seaward side of the beginning of Weevil Lane is a row of single storey red brick workshops with one at right angles at the end and a two-storey office facing the entrance. These were converted into houses, and won the Gosport Society's Restoration Award in 2012.

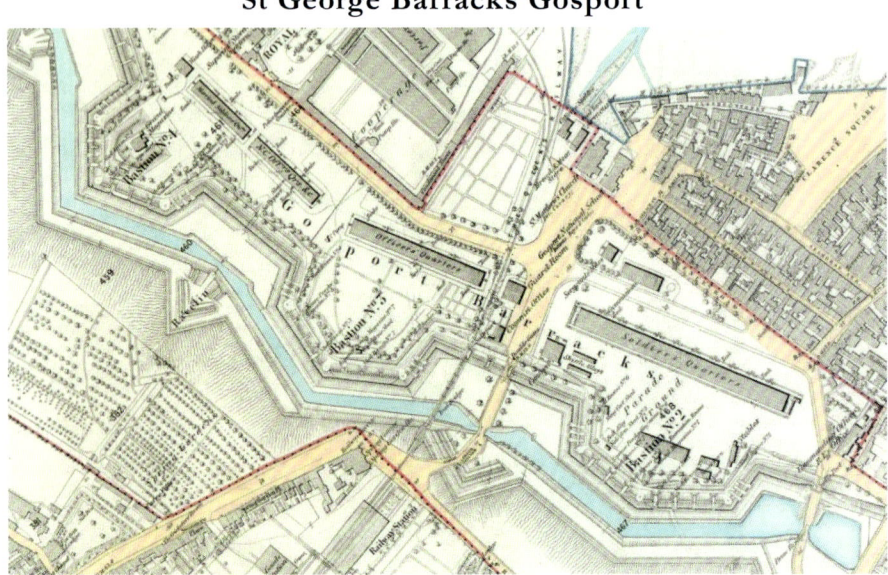

St George Barracks Gosport

New Barracks 1891 https://historicengland.org.uk/listing/the-list/list-entry/1233824

St. George or New Barracks   Postcard dated May 28 1918

This pair of barracks was built in 1856-59 as a transit barracks for troops moving to or returning from overseas stations. They were built by Messrs. Lucas Brothers, a Suffolk firm, at a cost of about £100,000 on land purchased in the mid-17th Century for the construction of the Gosport Lines fortifications. Designed to accommodate 2000 men, 44 officers and separate apartments for about 60 married soldiers and their families, this is the only example in the country of a highly developed design, combining cross-lit rooms with an external veranda, thereby answering some of the criticisms levelled at older barracks by contemporary barracks reformers. More than 10 million bricks were used in their construction, and the main accommodation block was said to be the longest single accommodation block in the UK. The veranda was used for wet weather exercising. The barracks were a transit station for infantry built after the Crimean War. They were designed to be bomb-proof, against mortar attack, because of their location just inside the Gosport Lines. This is the only example of this type of barracks in the country, forming a complete group with the Lines earthworks, and marking Gosport's importance in the defences around Portsmouth dockyard.[49]

St. George Barracks North   2014

A myth exists about the barracks that they were apparently built to a design that was meant for barracks in India. It says that the designs for two sets of barracks - one for India, the other for Gosport - were muddled up. St George's Barracks seems more suited for troops based in a hotter climate than Gosport. However, the real reason for its design is that at the time the barracks were built they were part of the defences around Portsmouth Harbour. One threat was of mortar attack from French ships in the Solent, so the main barrack blocks were set below ground level to keep the buildings below the skyline, so that the ground floor was semi-basement. The purpose of the flat roof was to provide further protection against mortar attack, with a mound of earth grassed over, much as some of the forts up on Portsdown Hill. However, the buildings could not take the weight and this part of the design was abandoned when cracks began to appear. An anonymous letter to *The Builder* magazine in May 1859 alludes to problems encountered when the earth was placed on the flat roof of one building, and this letter criticised the design.[49] The first-floor room nearest the steps was the CO's office and in the wall outside there was a large crack that was believed to go back to the attempt to put earth and grass on the roof with consequences. Berkeley Homes skilfully repaired the crack when they converted the block into flats.

The barracks were first occupied in August 1859 to April 1860 by the 86th (Royal County Down) Regiment of Foot on their return from India and later by the Hampshire Regiment. The gymnasium was added in 1868.

St George Barracks
post conversion to flats
Martin Marks

They were called the New Barracks until 1941, when they were transferred to the Royal Navy, renamed HMS St George and used as "New Entry" barracks. They were returned to the Army in 1947, keeping the name St George. They stayed in Army use until the early 1990s.

The North site was empty for several years before being sold for housing to Sunley Estates in 2002. A brochure describes 160 new homes. Sunley's Pavilions are a mixed-use development with studio flats, three-bed townhouses and four-beds houses set in the barracks block described above. Admiral's Keep is a development of 33 three- and four-bed detached houses and townhouses.

The smaller South site, which is Grade II listed, remained in MOD use until 1998 when the site was closed, handed back to the Crown Estate and sold for development.[50]

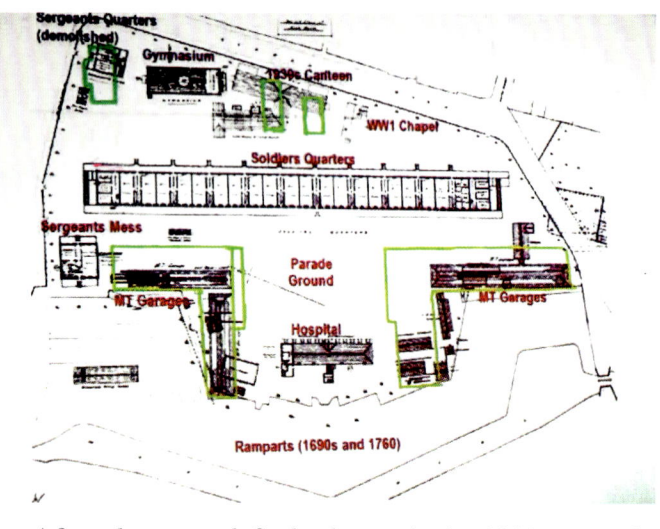

St. George Barracks South – plan

Chris Haslam, who served there in the early 1970's notes that the building called the Gatehouse/Captain's House was the Sergeants' Mess, while the North Building was the guardhouse and armoury. During his time there, the barracks were very lightly used and most of the space was empty. This was because most of the soldiers were away much of the time on-board ship and very rarely stayed in barracks.[51]

After the army left the barracks in 1991, several charities used the Old Chapel on the site for voluntary and back to work programmes including Gosport Voluntary Action (circa 2004) and the Gosport Development Trust. The Disability Information Centre moved in 2005 and out early in 2008. Then in 2010 the Gosport Employment Access Centre took over in a joint initiative between the Wheatsheaf Trust and Gosport Borough Council. The chapel was renovated to provide a centre offering free facilities to enable unemployed and low-paid people search for work and training opportunities. Visitors to the centre were offered an extensive range of free facilities and advisors, fully trained and happy to offer tailored, one-to-one assistance.[52]

St. George Barracks North Guardhouse. Converted to Hopscotch Day Nurseries 2014

The development here was led by Berkeley Homes (8 houses, 39 flats and 26 "residential units") and Portsmouth Housing Association (81 affordable housing units).

Fraser Range

Aerial view showing Fort Cumberland behind Laboratory/Admin Building 2018
Shaun Roster http://www.shaunroster.com

The strangest coastal walk in Portsmouth is to continue eastwards along the Eastney coastline to the point where it meets the Langstone Harbour channel. Rows of concrete blocks are buried into the bank protecting the caravan park, but seaward of derelict Fraser Battery the hastily put together WWII anti-tank traps have eroded to their pebble interiors. No longer lined up along the perimeter, they have been sliding down the beach for many years, pulled by the tides, until some are lapped and worn away by the ceaseless waves. If you keep to the concrete wall you can see through the fence into the heavily damaged battery, with twisted metal shapes and two ranges of enigmatic broken-windowed brick buildings, guarded by teams of dogs. Behind the battery is the Grade II* listed Fort Cumberland, the oldest fort in Portsmouth, constructed to defend the harbour entrance, described in our second forts chapter.

Fraser Battery was a research station, a firing range and a training base.[53] The close relationship of Eastney and nautical gunnery dates back to 1859 when Fort Cumberland became the headquarters of the Royal Marine Artillery and a Sea Service training battery was established there. Between 1924 and 1937 the foreshore in front of the fort was used for searchlight and sound locator training, pier building and landing craft development. Fraser Range was established either around this time or shortly afterwards. It is clearly shown on aerial photographs of 1946. After 1945 it was named after Admiral of the Fleet, Baron Bruce Fraser of North Cape.

The range, initially called Fleet Assessment Unit Fraser, specialised in training naval gunnery personnel.[54] It has two main ranges of buildings; the inland one used as a laboratory.

Type 993 'quarter cheese' radar left behind from range days

The site had 4.5" and 4.7" Quick Firing as well as 40mm Bofors guns configured to simulate a shipboard-firing scenario. By the 1960s, there was also a Seacat Missile launcher for training purposes, replaced by a Sea Wolf c.1980. The guns were mounted on the foreshore pointing seawards, requiring warning markers to be posted in the Channel to keep shipping clear of the nine-mile range. If they were not aware of its function, locals were startled by regular firings three times a week nine miles out to sea at a plane pulling a large windsock behind it. Missiles were fired out to sea, startling Celia Clark with a tremendous Whoosh when she was delivering leaflets to the MOD housing to the north.

During the 1960s, Fraser was also home to HMS St George, the Royal Navy's Special Duties Officers' School where Senior Ratings who had been selected for promotion were given nine months of specialist officer training. The school moved to Britannia Royal Naval College, Dartmouth in 1974. In 1972, the base became briefly famous when it stood in for 'HMS Seaspite' in the Doctor Who episode *The Sea Devils*, starring Jon Pertwee and Katy Manning.[54] Fraser Gunnery Range closed in 1986, but the base was put to use again as the Civil Marine Division of the Admiralty Research Establishment, moved there from Eastney Fort East in 1989. The ARE's work at Fraser consisted mostly of testing RADAR equipment using large steel-lattice towers to improve range. In 1995 the ARE became part of the Defence Evaluation and Research Agency and in 2001, not being of particular strategic importance, Fraser was transferred to the privatised defence company QinetiQ, who slowly wound down operations, closing the site in 2006. Because the site was not secure, considerable vandalism and damage ensued.

Laboratory/Admin Building

North Cape Building: range control tower

QinetiQ proposed to build three oval blocks of flats – around which the winds in this very exposed location would have whistled; but a dispute with the council over the access road prevented this development. By 2011 no progress had been made and the plans were withdrawn. Since then the site became an unofficial tourist attraction for bored young people and curious visitors – and graffiti artists, some of whose work is worth preserving - according to architect Deniz Beck, who was also responsible for the transformation of Old Portsmouth's Hot Walls and Spitbank Fort and is the architect for the latest scheme for Fraser Battery. In 2013, aware of the dangers of derelict buildings QinetiQ had the base's two iconic RADAR towers demolished to stop people from climbing them.

Radar tower at Fraser Battery from Fort Cumberland

In 2017 the site was bought by the National Regional Property Group, which is based in the old Recruiting Office (now Victory Gate Lodge) outside the dockyard gate on the Hard, Portsea. This firm is also involved in redeveloping Fort Gilkicker and Debenhams department store in Southsea They propose to convert Fraser's two 1950s main buildings into flats and to add new houses. A new coastal path along the seaward edge would incorporate new flood defences and protect the special maritime ecology[5]

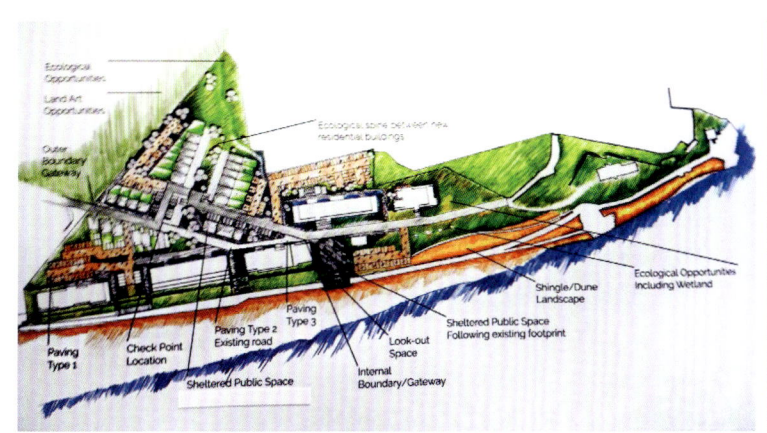

National Regional Property Group
Site Plan Public exhibition 2019

High ceilings in rear block would facilitate residential conversion

Shaun Adams, chief executive and owner of the National Regional Property Group, said: "We fully understand the responsibilities we have to this site as developers. We have taken care in how we manage the flood defences, the coastal path, the ecology and most importantly the heritage asset of Fort Cumberland which sits behind our site. It's taken us two-and-a-half years to get where we are today so this has been a long journey with a lot of effort and thought gone in from the team. I think these proposals will be the catalyst towards the regeneration of the wider area and I hope the investment here will increase the value of surrounding homes."

The new coastal path along the seaward edge was to incorporate new flood defences and protect the special maritime ecology. The application to convert three existing structures and construct new ones to create 108 new apartments, 26 new houses, a new sea wall for flood defence and a walkway, an access road, parking and landscaping was submitted to Portsmouth City Council in March 2019.[60] However, this and other coastal developments, including 12,000 potential homes in south Hampshire were held up by a European Court of Justice decision in November 2018 and other judgements that led the government agency Natural England to recommend in May 2019 that all new-build homes would have to meet strict environmental regulations over water-borne nitrate levels, because high levels of nitrogen pollution were affecting ecologically significant protected sites in the Solent Area. According to the Environment Agency new housing as well as the major source: agriculture, contributes additional nitrogen to the water draining from the catchment area, causing excessive growth of green algae. There is uncertainty as to whether new growth will further deteriorate designated coastal waters. The potential for future housing developments across the Solent region to exacerbate these impacts created a risk to their potential future conservation status. "The Solent water environment is internationally important for its wildlife and is protected under the Water Environment Regulations and the Conservation of Habitats and Species Regulations as well as national protection for many parts of the coastline and their sea. There are high levels of nitrogen and phosphorus input to this water environment with sound evidence that these nutrients are causing eutrophication at these designated sites."[61]

Despite the many thousands in need of new homes and the government's pressure to meet housing targets, local authorities including Portsmouth, Gosport, Havant and Fareham temporarily ceased giving planning permission while they looked for ways to enable development to take place while ensuring that the water quality in the Solent's internationally protected sites was preserved. Delays in getting planning permission lead to an absence of new housing for the thousands of people on waiting lists, developers losing money, contributing to high house prices and contractors (and potentially planners) losing work.[62] In August 2019 Portsmouth Director of Regeneration proposed a 'nitrate neutrality' mitigation strategy to enable both city and private development proposals. This was intended to meet the tests of Habitat Regulations and avert the potential risk of legal challenge. The Partnership of South Hampshire (PUSH) worked towards a sub-regional long-term strategy to address the problem with central government agencies.

In December 2019 having agreed a new strategy to reduce nitrate output from new development by offsetting water-saving measures in the council's existing housing stock, Portsmouth city council again began granting planning permission for new housing developments. They were also considering introducing oyster beds because they feed on harmful algae and store nitrogen in their shells and tissue. Havant Borough Council proposed to install reed beds to filter runoff, but both these authorities' measures can of course only be temporary. Fareham's housing stock was already said to be water efficient. In early 2020 PUSH, working with the Hampshire and Isle of Wight Wildlife Trust, Natural England and the Environment Agency hoped to buy areas of farmland that contribute large amounts of nitrogen through the use of fertilisers, to green it by planting trees, plants and introducing animals. Their problem was how to find the money to buy the land.[63]

Transit shed and ravelins of Fort Cumberland from roof of Fraser Battery

In July 2019 Historic England said the development would cause harm to the ravelin and western ramparts and to the fort's southern defences because of the proposed coastal defence scheme – where there may be as yet unidentified archaeology. This harm could be mitigated through careful design of the block plans and landscaping of residential properties. They said that although the Fraser buildings interrupt the fort's sightlines, fields of fire – originally 360 degrees via its star-shape, and its connectivity with the sea, "the stark industrial character" of the Fraser Battery buildings "is not entirely out of keeping with the earlier military defences represented by the scheduled monument, and their current dereliction adds a circumstantial but evocative backdrop when experiencing the isolation of the Fort and its southern-most defences." They asked for a conservation management plan to identify archaeological recording and conservation works and provision for future maintenance of the heritage resource as well as opportunities for better interpretation and public access.[64]

Eroded concrete blocks sliding down Eastney foreshore

A particular focus of conflict was the proposal to lift, store and reinstate 30 of the WWII 280 anti-tank concrete blocks extending 400m along the beach in front of the battery erected in early 1940. Their removal was considered necessary because the Environment Agency does not fund coastal defences for private development, so the developers must construct their own. The blocks were listed Grade II in response to members of the Council for British Archaeology (CBA) highlighting them as "Among the best preserved anti-invasion remains on the south coast of England."[65] The CBA believed that the significance and context of such superficially unattractive utilitarian structures was not often fully recognised. They recommended that a detailed heritage statement prepared by a specialist in WWII defence heritage should accompany the planning and listed building applications.

"As a key part of the visible evidence for a particularly challenging period of recent history, the Listed Grade II Anti-Tank Defences retain very high communal and historic value for residents as well as casual visitors and heritage professionals and others with an interest in modern history. They therefore retain very high significance, much of which is related to their context."[65]

SAVE Britain's Heritage objected to removal of the blocks.[66] But Historic England (whose Centre for Archaeology is in the adjacent Fort Cumberland) said moving and reinstating them would be "less than substantial harm". In February 2020 the Naval Dockyards Society said that the proposal would cause substantial harm to the significance of Fort Cumberland, and that a topographical survey and an archaeological management plan to protect and manage below- ground archaeology should be carried out. The battery's history and archaeological features should be integrated into the development's design and interpretation. As other objectors said, it should also address the community need for affordable housing. Another conflict is between conservation of a historic site and wildlife: it is adjacent to two Sites of Importance for Nature Conservation and migratory bird flightpaths.

In May 2020 a scheme to offset the ecological damage to the Solent caused by nitrates in waste water from new homes, via the purchase of nitrate credits available for developers was put together. The Hampshire and Isle of Wight Wildlife Trust bought Little Duxmore Farm in Wootton Isle of Wight for £950,000 with a plan to rewild it, stopping its fertiliser-rich soil pouring more nitrates into the Solent. The chief executive of the trust said that there was a real risk of legal challenges or environmental regulations being ignored that would set a really bad precedent. A house was predicted to cause 3kg of nitrates a year; the certificates will cost about £6,000. The Solent is badly polluted because of fertiliser and run-off, and so taking farmland back to nature and restoring habitats will achieve an offsetting effect that would be more than neutral. The trust was looking for more farmland in the north of the Isle of Wight and south Hampshire. Cllr. Sean Woodward, leader of Fareham Borough Council and chair of the Partnership for South Hampshire drew up a legal agreement between Fareham and the trust which would prevent developments on the land for 82 years and ensures that rewilding takes place. He urged the housing minister to ensure that any fines levied in Southern Water's prosecution were spent on solving the nitrates issue, buying more farmland. He hoped that this initiative would be a boost to the recovery from the Covid-19 outbreak.

This paradoxical solution – mitigating one source of pollution by enabling another may allow housebuilding in the area to begin again, although building on the diminishing open space left along this densely developed strip of the south coast rather than in existing built up areas is likely to result in other problems such as increases in road traffic, air pollution and diminished quality of life.

Future Conversions

We include Haslar Barracks in Alverstoke which adjoins Haslar Hospital, because once sold by the Ministry of Justice, its future use might also be residential. It was built in 1802-3 on an area known as Camp Field, not far from the major anchorage at Spithead, since Gosport was a major focus for troop movements, coastal defence and garrison duty during the French Revolution and Napoleonic eras.

According to Gosport Borough Council's Conservation Area Appraisal of March 2018, "it appears to be the only significant regimental infantry barracks complex that survives in England relating to the threat of invasion in the years leading up to the Battle of Trafalgar (1805): a crucial and nationally significant period, making the site of particular historic value. Many of the original buildings appear to survive." It was converted to a military hospital in 1864 "following major reforms to the design of military hospitals after the Crimean War, and around 1892 became the home of the Royal Engineers who have a long and close association with the extensive national important military defences in the area." In the early 1950s the Royal Army Ordnance Corps occupied it. At some point after that it became a Youth Offenders' Centre, "before its final use as an Immigration Holding Centre, which closed in 2017."[67]

As they usually do, Gosport council anticipated its release by designating the site a conservation area, stipulating that the 'regimented and balanced geometric form should be 'strictly maintained and enhanced" where any new buildings are considered. Demolitions will be considered on the basis of their impact on the special interest of the area. The policy statement's attention to detail is characteristic – and exemplary!

Tipner

Pounds scrapyard, Greyhound Stadium in 1973 before M275. JPIMedia Limited

"The redevelopment of this under-used and heavily contaminated site is an important element in the Portsmouth City Local Plan's coastal strategy of promoting both the physical and economic regeneration of Portsmouth. The location of the site in a prominent position adjacent to the M275 Bridge at the gateway to

Portsea Island creates a poor visual impression on visitors entering Portsmouth. Present uses of the site include part of Pounds Shipowners and Ship Breakers Ltd., the Greyhound stadium, and a number of largely industrial employers."

"Proposals for the redevelopment of Tipner have been put forward since the early 1970s, when the construction of the M275 motorway made the site (which had previously been largely hidden from view) far more visually prominent. Despite a significant level of interest and effort both on the part of the City Council and prospective developers in discussion with the various land owners, the site has proved difficult to redevelop..."[68]

Tipner, Horsea Island, Paulsgove tip and Port Solent   JPIMedia Limited

In 2014 6.5 hectares of the Tipner peninsular on the eastern side of the harbour bisected by the M275 in 1976 was proposed as a site to provide 2,370 new homes and 58,000 sq m of employment space. Tipners firing range was part of the City Deal Portsmouth and Southampton secured with the government in 2013. They gave the cities £48.75m and part of the deal was that Portsmouth City Council would then buy the firing range for a nominal fee and pay £3.75m to the MOD to relocate its facilities to Longmoor Camp.[69]

As we've said, free transfer of government assets is a rare occurrence: in the harbour area only Priddy's Hard, also with severe contamination and Gosport Railway Station have arrived into local authority ownership via free transfer. The reason for its use for housing was that Portsmouth as an island city with little undeveloped land has difficulty in meeting government house building targets. The city council invited local input to the Tipner Strategic Development Area or 'Super Peninsula' in 2019.[69] This document identified local housing need as 17,260 dwellings for the period 2016-2036. "30,000 sq. metres of employment floorspace could be provided creating a significant number of new jobs, which due to its unique location, including potential access to deep water, would be positioned to support the economically important marine industries." However, "It is also understood that in the longer term some parts of these areas will be significantly affected by predicted rise in sea levels." The proposed housing numbers on the whole 140-acre site were increased to 4,000 homes and a school, with cars parked underground to leave streets pedestrian-friendly.[70] "The huge scheme will involve reclaiming 22ha of land from Portsmouth Harbour to create a new 2.2km stretch of waterfront, which will be connected to the rest of the city by a new bridge... The aspiration for Tipner West was to create an entirely new district for the city that balances the employment needs of the rapidly expanding maritime technology cluster while also ensuring a complete living environment with new housing and the necessary social infrastructure. The largest development scheme the city has ever seen, Portsmouth City Council claimed that it could be a "beacon for the whole of Portsmouth and the rest of the country".

If the plans were approved work could start in 2023. A bridge estimated to cost £31.2m from Tipner West would link it to nearby Horsea Island, only to be used by buses and bicycles. Access to the site has always been a problem. There is no direct road link to the motorway, so traffic had to make its way through Stamshaw and Tipner East.

When it opened in 1976 the M275, a southward spur off the M27 cut the peninsula off from nearby communities. It also revealed a site that had been largely hidden from view. In 1991 the Mowatt Group proposed to develop it, which prompted the city to prepare development guidelines.[68]

The east of Tipner, occupied in 2002 by Cox's Plant Hire and the former Ready Mix Concrete premises, was used as a chemical works, brickworks and timber yard in the 19th and early 20th centuries, and more recently by Corral's Coal Depot. The northern part was gradually reclaimed from Tipner Lake, and closer to the M275, north of Tipner Lane was a former Admiralty tip, which took MOD spoil from the dockyard.

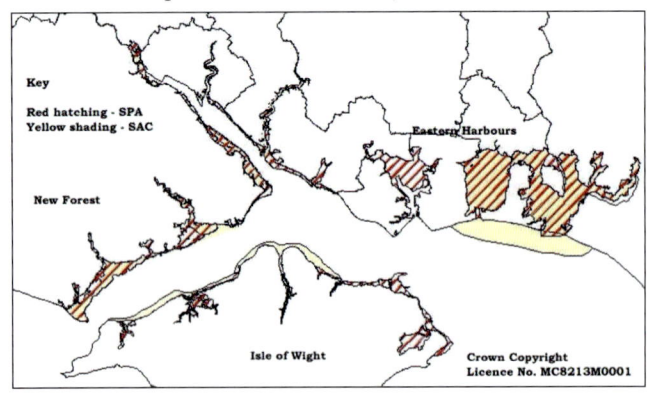

Special Protection Areas and SAC map
Crown Copyright

Tipner Lake is part of the SSSI, SPA and Wetland of International importance (RAMSAR site). In 1999 there was a proposal to build a 9000 seat arena for major leisure, sporting and commercial events with commercial premises at Horsea Island. This was subject to a bid under the government's Private Finance Initiative, which needed permission from the Treasury. In 2002 a retail scheme including an Ikea and Decathlon stores and park and ride was proposed. It was hoped to coincide a 'landmark' development with the Portsmouth Gateway Project to improve the image and appearance of the M275 corridor. Neither of these proposals materialised.

Given its history, the Tipner peninsula would not seem to be a suitable place for people to live. It was developed for the safe processing and storage of ammunition, far away from where people lived. War with France and the invasion scare of 1779 led to concerns about the state of the nation's gunpowder, resulting in the Master of Ordnance's strategy to distribute gunpowder stored near the dockyards to a few scattered locations. Isolated from residential development, land at Tipner Point was acquired between 1789 and 1793 for the construction of a gunpowder magazine with groined arches and a copper-clad wooden roof. From 1805 until the mid 1820s Tipner acted as deposit magazine for damp or damaged gunpowder offloaded from ships to be restored and then reused.[71]

By 1804 facilities had been built nearby on Stamshaw Point and Horsea Island. From Tipner, the powder was taken first to Stamshaw, where it was unpacked from its barrels, assessed and sieved ("redusted"); from there it was taken to Horsea for the dangerous process of "restoving" (drying the damp powder in specialized ovens), after which it would either be returned directly to Tipner magazine ready for re-use, or else would go via the "mixing house" at Stamshaw to be blended with fresh powder.[72] There were three later phases of development: from 1788-1827, 1856-7 and 1891-1910. When the ordnance depots were divided between the two services in 1890 Tipner passed to the army. The magazines were converted into general ordnance storage and the present iron doors were then inserted. The stores were internally remodelled after WWII. The Shifting House was converted into a magazine in 1827 and it was still used to house explosives in WWII.[73]

At one time there was also a barracks to house the troops tasked with protecting the facility. While the Watch House and Guard House have been demolished, the magazine of 1796 still stands at Tipner, together with the southern magazine with parabolic arches of 1856-7, the Shifting House of 1800, and some ancillary buildings. The cooperage, shifting room and perimeter walls of 1800 are all Listed Grade II.[73]

Pounds Scrapyard has been breaking up ships, linkspans, submarines, concrete pontoons and tanks on the mud flats at Tipner from 1953, before actually buying the site from the MOD at auction on 28.10.1964 When Celia Clark was training in the first group of Portsmouth tourist guides she was told not to mention that the city only broke up ships, no longer building them. The site's accumulation of discarded military equipment, submarines, floating cranes and broken metal visible from the motorway exerts a powerful fascination over photographers, filmmakers and urban explorers.[74]

Ken Russell's film Tommy Filmed in Pounds Yard
JPIMedia Limited

At the climax to Ken Russell's 1975 film *Tommy* about a pinball wizard, "scrap maritime marker buoys were spray-painted silver, as Tommy navigates his way through a surreal scene of giant pinballs and collections of burning pinball machines, after his followers riot and destroy the holiday camp." filmed 'at Pounds Yard[74] Deane Clark won a free photographic trip to Venice with his photo of the film extras in 1940s gear eating their lunch during a break in filming at Fort Widley.

The magazines are used for storage by Pounds. They were damaged in the 1970s; one was roofless for a time. Rumour has it that a helicopter was hired to assess their condition, since access was debarred. In any new development their setting would have to be enhanced.

Until recently there was also an active firing range, extending to 2500m from the range described in Danger Area notices from the Queen's Harbour Master. "In my day as a range qualified officer in the cadet forces I have conducted firing at Tipner. After firing all the brass was collected and bagged. I always assumed it went back to be recycled rather than buried on site. Bullets after hitting their target would go into the butts which is basically a lot of sand. What happened to the sand? It would contain lead without a doubt."[76]

South Parade Pier burning after an overheated cable caught fire: the climax of the film

As the Planning Service information sheet (2002) says, "Tipner has been home to a variety of industries, many predating existing occupiers and modern pollution control measures and regulations."[68] Details about where spent ammunition is now and any associated contamination should be requested from the MOD before any development.

According to *The Torpedomen: HMS Vernon's Story 1872-1968* by Admiral Poland the establishment of a Chemical Warfare Department in 1927 'Valuable information was obtained concerning the travel of gas clouds over the sea and training in mustard gas decontamination was introduced at *Vernon*.' The shore of the Tipner peninsula was a test site. A document in the Public Record Office says 'Contamination of beaches trials at Tipner Point, Portsmouth' mustard gas was tested between January 1938 and December 1939 there to ascertain whether it might hold up an invasion force or be deployed by one. Unfortunate animals were subjected to it.'[77] A further report 'Persistency (including reports on other gases) includes 13 photographs depicting Chemical Defence Research Department: chemical weapons - mustard gas: contaminated beaches; trials at Tipner Point, Portsmouth; blisters on skin; plate glass and raffia mat used in evaporation experiments; mustard gas under tropical conditions' is dated 1932-1938. This evidence shows that testing continued at Tipner for a number of years. In 1999 more than 60 sites across Britain, officially declared "safe" were still contaminated by mustard gas, which remains effective for decades. Although chemical weapons were very rarely used in action, in WWII Britain and the US stockpiled huge quantities in case they were needed.[78]

Before Tipner is developed for people to live there, it is important that soil tests are carried out to establish whether there are residues left in clay soil there. As David Jenkins says: "One component of soil is clay and it is composed of thin layers of minerals called micelles. A molecule that gets into a micelle could be held almost indefinitely. How serious the risk is would depend in part I think on how much contamination occurred, how long ago it took place and what proportion of the soil is composed of micellar clay. I would want to see the results of mass spectrometry or a GLC before deciding to site houses at Tipner."[76]

How will pollution from all these sources be tackled? The first stage of decontamination was carried out by the Homes and Communities Agency, now Homes England, which spent £10m on remediating the eastern part of the site so it could be reused for development, because there had been a chemical factory next to the greyhound stadium. Engineers Campbell Reith bio-washed the soil; the toxic residue was collected and disposed of and crushed concrete and clay to act as a barrier was spread over the surface. A similar process is needed on the western side.

The giant Goliath floating crane visible from all over the area arrived from Southampton arrived a few years ago. A field near Waterside School next to the sailing club was levelled in 2019 as a temporary lorry park in case Brexit led to delays at the ferry port. In 2020 until late September Covid-19 testing was available there. While the Park and Ride site on the motorway's east side did not involve any excavation, foundations for the proposed seven-storey carpark to replace the existing bus shelter will of course involve piling.[79] As the Naval Dockyards Society pointed out, this would affect buried heritage features which were largely unknown because no archaeological investigation had been carried out.[80]

English Nature warned that plans to reclaim land at Tipner could 'damage or destroy' animal habitats and disturb Brent Geese, because the head of Portsmouth Harbour is an internationally designated area under the Convention on Wetlands (Ramsar 1971) and an EU Special Protection Area. There are vulnerable bird species that feed in the harbour: Black Tailed Godwits and Eurasian Curlews, which are on the Red List of the International Union for Conservation of Nature, along with others that could be affected. It is also recognised nationally as a Site of Special Scientific Interest for its rare salt marsh species.[81]

Underground parking, which would have to penetrate the capping would be vulnerable to flooding as the sea level is expected to rise between 0.7 and 1.9 metres by 2100. The council's report 'Planning to adapt to climate change' acknowledges that by 2080 heavy winter rain, sleet and snow is likely to increase by 50% which will also affect water levels.

'Critical questions need to be asked about the estimated £1.3bn cost of the project: whether costs will escalate – they often do – and if the homes will be affordable, which sounds unlikely. In the face of today's best scientific knowledge and criticism from Natural England and other conservation organisations, Portsmouth City Council's current Tipner housing project looks highly unsustainable and environmentally damaging.'[82]

The third constraint is the emerging guideline on air quality - already poor in Portsmouth, particularly near main roads. In 2002 the city council identified issues of noise and vibration from motorway traffic as one of the constraints on development, even before residential use was proposed.[83] In 2020 London, Slough, Chatham and Portsmouth had the highest proportion of deaths attributable to air pollution according to the Centre for Cities thinktank.[84] In London estate agents now have to declare the levels of air pollution in properties they are selling. Apart from traffic fumes from the motorway, the fourth reason is noise. When the Portsmouth Society has judging the outside of the Mountbatten Centre for our annual Design Awards, we could not hear ourselves speak - and that's some distance away.[85] Any impact on the historic magazines or their settings will also have to be carefully considered.

On 30 January 2020 a six-week consultation to allow the public to find out more about the plans and give their feedback began at Port Solent, to be followed by other venues. "Portsmouth City Council is currently engaged with a wide range of stakeholders, including environmental agencies to progress with plans for Tipner West. It is thought the scheme could cost upwards of £1bn."[86] In May 2020 an outline application to construct a multi-storey transport interchange of up to 34.8m incorporating a park and ride facility for up to 2,650 cars and 50 bicycles, taxi rank, public conveniences, landscaping, ancillary offices and units within use classes A1, A2, A3, D1 and D2 with access from the M275 was submitted.[87] On 18 June 2020 the Naval Dockyards Society and Hampshire Buildings Preservation Trust objected to the application, on the grounds that it was likely to cause historical and environmental harm to Tipner Peninsula; that insufficient attention had been paid to the richness of local historic assets such as Portchester Castle, the forts on Portsdown Hill and the Nelson Monument or to the impact of the 34.8m high 'Citadel' on the views across Portsmouth harbour; that the proposed scheme failed to incorporate archaeological investigation of the area, currently lacking; and failed to show how any archaeological asset that might be discovered would be conserved or interpreted.[88]

In October 2020 Portsmouth City Council cabinet gave the Tipner West planning submission the green light to go ahead on the work required to prepare a concept plan and engage formally with the statutory environmental bodies on the environmental surveys already carried out. The scheme required £8m from the £48m City Deal, a central government grant awarded to the council in 2015 to prepare for the necessary planning applications. They were also considering marketing and branding for Tipner West for a public consultation in January 2021.[89]

As these case studies have demonstrated, residential conversion is not the easy option some might expect. Respect for the barracks' and other military sites' history coupled with exemplary conservation practice may produce the best results, judged both from the environmental and economic points of view.

Sources

https://marinafitness.co.uk

References

1 *James Douet 1998 British Barracks 1600-1914: their architecture and role in society The Stationery Office*

2 *History in Portsmouth.* Retrieved 4 December 2016

3 Hilsea Barracks *History in Portsmouth.* Retrieved 4 December 2016; The Book of Hilsea: Gateway to Portsmouth Jane Smith 2002

4 The News 17 March 2013 Retrieved 4 December 2016

5 Brigadier Frank Steer 2005 To The Warrior His Arms: the story of the Royal Army Ordnance Corps 1918-1993 Pen & Sword

6 Garry *Mitchell 1988 Hilsea Lines and Portsbridge* ISBN 0-947605-06-1

7 William *Page 1908 London p. 165-170* Retrieved 5 December 2016

8 Gatcombe House *British Listed Buildings.* Retrieved 5 December 2016; David Lloyd 1967 Buildings of Portsmouth and its Environs: Portsmouth: 1974-: 26, 28, 31, 153; David Lloyd The Buildings of England Hampshire and the Isle of Wight: Harmondsworth p. 430

9 https://www.gatcombehouse.co.uk

10 Darren Bridgeman email to Celia Clark 24 July 2019

11 "Royal Marines Museum - Commandants of the Royal Marines Portsmouth Division". Memorials in Portsmouth. Archived from the original on October 6, 2014. Retrieved 28 May 2016;
"Royal Marines Museum - Commandants of the Royal Marines Portsmouth Division". Memorials in Portsmouth. Archived from the original on October 6, 2014. Retrieved 28 May 2016;
"Bamford House, Dowell House, Finch House, Halliday House Former Long Barracks and Screen Walls to E, Portsmouth". British Listed Buildings. Retrieved 28 May 2016;
Pevsner, Nikolaus & Lloyd David 1967 The Buildings of England: Hampshire and the Isle of Wight. Penguin Press;
"Inland Planning" (PDF) Portsmouth Society News. August 1995 Retrieved 22 May 2016;
"Royal Marines Museum to be moved to new home in Portsmouth Historic Dockyard". Portsmouth News 17 October 2013. Retrieved 22 May 2016;
Ambler, John and Little, Matthew 2008 Sea Soldiers of Portsmouth. A pictorial History of the Royal Marines at Eastney and Fort Cumberland, Halsgrove, Somerset, ISBN 978-1841147437;
Lane, Andrew, 1998. The Royal Marines Barracks Eastney. A pictorial history, Halsgrove Publishing, ISBN 1874448922

12 http://historyinportsmouth.co.uk/places/eastney-barracks.htm accessed 17 July 2019

13 Charles O'Brien et al *Hampshire South The Buildings of England Yale* University Press p. 536

14 http://historyinportsmouth.co.uk/places/eastney-barracks.htm accessed 17 July 2019

15 Marine Gate Victorian Elegance on the South Coast sales brochure Fortview House Gudgeon Homes plc

16 Geoffrey Salvetti interview with Celia Clark 19 March 2019

17 http://historyinportsmouth.co.uk/places/eastney-barracks.htm accessed 17 July 2019

18 http://portsmouthmuseums.co.uk/collections/collection-a-district-of-eastney.html

19 https://www.tripadvisor.co.uk/LocationPhotoDirectLink-g186298-d216600-i115791753-Royal_Marines_Museum-Portsmouth_Hampshire_England.html

20 Gambino M. 2013 'Erving Goffman's asylums and institutional culture in the mid-twentieth-century United States' Harvard Review of Psychiatry. 2013Jan-Feb;21(1):52-7.doi: 10.1097/HRP.0b013e31827d7df4. https://www.ncbi.nlm.nih.gov/pubmed/23656762 accessed 29 July 2019

21 http://historyinportsmouth.co.uk/places/eastney-barracks.htm accessed 17 July 2019

22 Rosemary Dunne discussion with Celia Clark 20 February 2019

23 John Bell and Cheryl Jewitt interview with Celia Clark 13 January 2019

24 https://www.rightmove.co.uk/commercial-property-for-sale/property-79831904.html

25 Jonathan Coad 2013 Support for the Fleet. Architecture and engineering of the Royal Navy's Bases 1700 – 1914 English Heritage pp. 299-309; Royal Clarence Yard https://www.rcyard.org.uk

26 https://historicengland.org.uk/listing/the-list/list-entry/1246651; Historic England http://www.pastscape.org.uk/hob.aspx?hob_id=238800&sort=4&search=all&criteria=royal%20clarence%20yard&rational=q&recordsperpage=10

27 Coad op cit p. 31228 Bernard Mennell 2019 'Royal Clarence Yard Gosport, the 'dispersed depots' and Victualling Stores supply from WW2 to closure' *Dockyards* The Naval Dockyards Society November 2019 Volume 24 Number 2 navaldockyards.org pp.19-24

29 Mennell op cit p.20

30 https://www.jtp.co.uk/projects/royal-clarence-yard

31 Mennell p.24

32 Main Gate and Two lodges: List Entry Number 1272344". Historic England. Retrieved 29 December 2017

33 https://www.jtp.co.uk/projects/royal-clarence-yard accessed 15 July 2019

34 https://legacyreports.cotswoldarchaeology.co.uk/content/uploads/2014/01/1423-royal-clarence-excavation-report-02109-complete.pdf

35 https://www.jtp.co.uk/projects/royal-clarence-yard accessed 23 June 2020

36 https://www.jtp.co.uk/projects/royal-clarence-yard

37 Celia Clark 'Ministry of Defence land sales' - Report to Hampshire Buildings Trust August 2003

38 Royal Clarence Marina http://www.yachtingmonthly.com/news/new-portsmouth-marina-opens-26942

39 Berkeley Commercial brochure http://caroline4gosport.co.uk/downloads/rcm-commercial-brochure.pdf

40 "Report to the Secretary of State: APP/J1725/A/05/1185799". The Planning Inspectorate. 2 December 2005

41 Colin Baxter interview with Martin Marks 14 February 2019

42 The News 10 October 2015; https://www.portsmouth.co.uk/news/people/ten-years-on-will-royal-yard-in-gosport-ever-realise-its-potential-as-a-leisure-hub-1-7029710;
The News 21 Nov 18 https://www.portsmouth.co.uk/our-region/gosport/excitement-as-royal-clarence-yard-in-gosport-is-put-up-for-sale-but-council-says-there-s-a-catch-1-8712779

43 "Gosport Waterfront and Town Centre Supplementary Document (SPD)". *Gosport Borough Council*. Retrieved 30 December 2017.

43 "MoD to sell off historic former naval site in Portsmouth Harbour". *The Institute of Marine Engineering, Science and Technology*. Retrieved 14 July 2020

44 PNBPT RCY offer and supporting docs 03 04 19.pdf

45 "Regional Ship Repair Firm Purchases Historic Former Royal Clarence Yard from MoD". *Minicoffs Solicitors*. Retrieved 14 July 2020.

46 Disused Stations http://www.disused-stations.org.uk/g/gosport_clarence_yard/

47 Martin Marks *Dockyards* The Naval Dockyards Society November 2019 Volume 24 Number 2 navaldockyards.org p.25; https//www.ukdocks.com/our-yards-facilities

48 The Oil and Pipelines Agency: About Us". Retrieved 27 December 2017; https://www.gov.uk/guidance/gosport-oil-fuel-depot-redevelopment accessed 17 July 2019

49 https://historicgosport.uk/new-barracks/ 1891 webmaster@historicgosport.uk; https://historicgosport.uk/new-barracks/;
Sense of Place, South East 50http://www.sopse.org.uk/ixbin/hixclient.exe?a=query&p=gateway&f=generic_objectrecord_postsearch.htm&_IXFIRST_=32287&_IXMAXHITS_=1&m=quick_sform&tc1=i&tc2=e&s=hVKBlP3ejIk
51http://s599972214.websitehome.co.uk/gosportinfo/History/Stories__Memories/St_Georges_Barracks_Memories/st_georges_barracks_memories.html; http://www.push.gov.uk/gosport_employment_access_center.pdf

52 Gosport Voluntary Action

53 http://derelictmisc.org.uk/fraser.html;

54 Briggs, J.N., 2004, 'Target Detection by Marine Radar' *London: Institution of Engineering and Technology*. p. xxv;
Hind, B., 'How Dr. Who saved the day at Eastney', Portsmouth News, 19th January 2013[http://www.portsmouth.co.uk/nostalgia/how-dr-who-saved-the-day-at-eastney-1-4702566] Accessed 4/6/14;
Judd, E., 'Iconic Eastney building to be demolished', Portsmouth News, 10th July 2013[http://www.portsmouth.co.uk/news/politics/iconic-eastney-building-to-be-demolished-1-5267320] Accessed 4/6/14;
Marsh, B., Photograph depicting R.N. Gunnery School, Eastney, 1961[http://dovergrammar.co.uk/archives/Old-photos/1961Gunnery-school.html] Accessed 4/6/14

54 http://www.doctorwholocations.net/locations/frasergunneryrange

55 https://www.portsmouth.co.uk/news/politics/plans-for-130-homes-at-former-fraser-range-navy-site-are-unveiled-1-8368590 accessed 25 June 2020

57 http://derelictmisc.org.uk/fraser.html

58 https://www.portsmouth.co.uk/news/politics/plans-for-130-homes-at-former-fraser-range-navy-site-are-unveiled-1-8368590

59 https://www.portsmouth.co.uk/news/politics/plans-for-130-homes-at-former-fraser-range-navy-site-are-unveiled-1-8368590

60 Fraser Range Fort Cumberland Road Southsea Ref. No: 19/00420/FUL | Received: Thu 14 Mar 2019 | Validated: Wed 01 May 2019 | Status: Awaiting decision

61 Environment Agency representation to Portsmouth city council in response to the planning application; https://democracy.portsmouth.gov.uk/documents/s23698/Nitrate%20Mitigation%20%20measures%20to%20%enable%new%20housing%20development.pdf

62 https://www.endsreport.com/article/1668317;nitrates-crisis-portsmouth-ends-housing-freeze

63 Fiona Callingham 2020 *The News* 'Nitrates farmland could be "re-greened" in a bid to reduce pollution and allow housebuilding to resume'; 'Solent nitrates: "More than 10,000 new homes delayed' 8 January 2020 https://www.bbc.co.uk/news/uk-england-hampshire-501031464 accessed 26 February 2020

64 Rebecca Lambert Historic England Letter to Portsmouth City Council Planning 2 July 2019

65 CBA Email to Planning Case Officer, Portsmouth City Council 14 January 2020-06-25; Bob Sydes 2019 'Eastney Beach anti-tank defences" 2019 British Archaeology/November December p.65

66 SAVE Britain's Heritage Newsletter Winger 2019 'Fort Cumberland and Eastney'

67 Haslar Barracks Conservation Area Appraisal Mar 2018

68 The Planning Service Student Information Sheet – Tipner Redevelopment 2002 Portsmouth City Council;

'MoD land handed to Portsmouth Council for homes' The News 17 November 2014; https://www.portsmouth.gov.uk/ext/development-and-planning/regeneration/tipner-west-regeneration

69 'Portsmouth council unveils plans for car-free peninsula community' The News 19 September 2019; https://www.bbc.co.uk/news/uk-england-hampshire-49752640

70 Ella Jessel 2019 'Gensler reveals masterplan for 4,000 homes on Portsmouth 'super-peninsula' 20 SEPTEMBER (HTTPS://WWW.ARCHITECTSJOURNAL.CO.UK/NEWS/GENSLER-REVEALS-MASTERPLAN-FOR-4000-HOMES-ON-PORTSMOUTH-SUPER-PENINSULA/10044470.ARTICLE)

71 David Evans 2006 *Arming the Fleet. The Development of the Royal Ordnance Yards 1770-1945* Explosion Museum of Naval Firepower, English Heritage p.234

72 Wayne Cocroft *1999 Dangerous Energy: the archaeology of gunpowder and military explosives manufacture. Swindon: English Heritage. pp. 58–59;* HMS Excellent Training Establishment

73 Historic England "Details from listed building database (1387240)". National Heritage List for England. Retrieved 28 September 2015

74 https://www.derelictplaces.co.uk/main/misc-sites/5126-pounds-scrapyard-portsmouth.html#.Xjqy0y2cZEI

76 Emails from David Jenkins to Celia Clark 31 January, 3 February 2020; Celia Clark 2019 'Tipner – not a suitable place to live' The News 12 October 2019; 68 Steward Lock 2019 'Super peninsula? PCC's proposal to build in Tipner really is an act of irresponsible desperation'
The News October 7 2019 p. 17

77 Public Records Office 'Contamination of beaches trials at Tipner Point, Portsmouth' WO 189/1987 War Office, Ministry of Supply, Ministry of Defence: Chemical Defence Experimental Establishment, later Chemical and Biological Defence Establishment, Porton: Reports and Technical Papers 72 PRO WO 188/501'Persistency (including reports on other gases)
Includes 13 photographs depicting: Chemical Defence Research Department: chemical weapons - mustard gas: contaminated beaches; trials at Tipner Point, Portsmouth; blisters on skin; plate glass and raffia mat used in evaporation experiments; mustard gas under tropical conditions'. Dated 1932-1938. Date: 1930-40
78 Christopher Bellamy 4 June 1996 'Sixty secret mustard gas sites uncovered" Defence Correspondent Independent https://www.independent.co.uk/news/sixty-secret-mustard-gas-sites-uncovered-1335343.html

79 "New designs for seven-storey park and ride in Portsmouth revealed" Fiona Callingham 12 February 2020 The News

80 Naval Dockyards Society 2020 Objection to Tipner West planning application

81 https://rsis.ramsar.org/ris/720 accessed 26 September 2020

82 Ben Fishwick 2019 'Portsmouth's car-free £1bn Tipner West super peninsula with 4,000 homes branded a 'no go for nature' *The News* 26 September

83 Steward Lock 2019 'Super peninsula? PCC's proposal to build in Tipner really is an act of irresponsible desperation' The News October 7 2019 p. 17

84 Patrick Butler 2020 'Urban populations in south-east at greatest risk from air pollution' *The Guardian* 27 January p 13

85 Celia Clark 2019 'Tipner – not a suitable place to live' The News 12 October 2019

86 'Here's how to find out more about plans for a car-free community in Tipner West" Fiona Callingham The News 30 January 2020

87 Public Notices Planning Applications Town and Country Planning Acts The News May 29 2020 p.38

88 Celia Clark 2020 'Latest developments at Fraser Range, Portsmouth' Dockyards The Naval Dockyards Society May Volume 25 Number 1 pp.5-9; Ann Coats Naval *Dockyards* Society 2020

89 Portsmouth City Council Tipner West update 25 October 2020

Chapter 17
Field of fire to common, redoubt to rose garden; bastion and military lines to bosky walks

Introduction

If we're aware of their history, enjoying the summer scent of roses inside Lumps Fort on Southsea seafront, the wide expanses of Southsea Common, the Solent panorama from Portsdown Hill or climbing the steep steps on No. 1 Bastion in Gosport, there's a strange frisson and sense of privilege that we can now enter once forbidden military territory. Both Gosport and Portsmouth have gradually gained open space from releases of military land, once changes in weaponry and different enemies rendered it obsolete for defence. Much of Portsmouth and Gosport's open space is around their coastlines. They are vital to our enjoyment and health. Portsmouth has less greenspace per person than most other cities in the UK.

It's perhaps paradoxical that in a period when the privitisation of once publicly owned defence dedicated space such as Gunwharf is a current concern, that this chapter represents its opposite: the opening of once closed military spaces to unfettered public use. Stokes Bay Lines, now thickly planted with evergreen oaks is a popular open space. The Millennium walk linking former defence sites runs along Gosport's harbour shore, extends from Priddy's Hard, over the Millennium Bridge to No.1 Bastion facing Haslar. On the Portsmouth side it runs from the dockyard Victory Gate to Spur Redoubt.

Watching *Queen Elizabeth II* supercarrier approach harbour from Kings Bastion Old Portsmouth November 2019

Our second chapter described how Gosport Lines were mostly cleared in the 1930s and developed as leisure space, while Hampshire County Council restored No. 1 Bastion and its moat near Holy Trinity church in the 1980s. New initiatives, such as Gosport's Heritage Action Zones, where Historic England is working with Gosport Borough Council to address the challenge of regenerating significant built and natural heritage, led to the clearance of No. 1 Bastion of self- seeded sycamore trees, tons of rubbish and vegetation in six weeks in autumn 2019. The work was funded by a Coastal Revival fund grant. Further work is planned.[1] Chapter 12 described the open spaces of Stokes Bay Lines, while here we explore the open spaces gained from the military in Portsmouth.

In the late nineteenth century the elaborate glacis and moats of Portsmouth and Portsea were cleared as the population grew, leaving only the seaward towers, walls and moat, now much enjoyed in Old Portsmouth. Once the Portsmouth Garrison closed in 1960 the Polytechnic – later University - occupied the space kept clear for Portsmouth town's defences, which the terraces were built to face. Part of this area is still open land now called Ravelin Park. North of the railway line, Victoria Park, originally called the People's Park was laid out in 1878 on part of the site of the Portsea Lines.

Hilsea Lines with its triple defences: earth ramparts covering casemates with disappearing guns mounted on top, a brackish moat and the encircling tidal Portscreek is popular for walking and fishing. Southsea Common, once rough grazing land subject to flooding, was purchased from the

War Office in 1922 and laid out as a pleasure ground by Portsmouth Corporation. While the crest of Portsdown Hill is still strategically important to defence, we can also enjoy the incomparable panoramic views from its south and north facing slopes.

Rising sea levels

Rather than attacks by human enemies, the harbour communities' coastlines where so much of their scarce open space is concentrated are increasingly threatened by rising sea levels, a particularly acute threat to low-lying Portsmouth as one of the UK's only two island cities. The main area of Southsea Common is below sea level and therefore vulnerable to rising sea levels. In addition, the Common's shallow soil and thin grass cover are prone to flooding in heavy rain. Storms repeatedly smash and undermine the aging hundred-year-old concrete sea defences both east and west of Southsea Castle. In 2014 the sea punched a hole in the ancient stone wall of the moat in Old Portsmouth. Several times in recent years the beach has moved inland to flood the sunken Rock Gardens, killing many plants before the salt water was pumped out again. These were portents of the greater threat to come. As well as historic defence structures, housing near the seafront built on the Great Morass which is below sea level is vulnerable to flooding, including Celia and Deane Clark's house. Large areas of Portsmouth and Gosport will be submerged in the event of a 2c rise in temperatures. [2]

In 2010 the Institution of Civil Engineers and the Royal Institute of British Architects warned that cities including London, Bristol, Liverpool and the two island cities of Hull and Portsmouth were at serious risk of flooding.[3] The government first recognised Portsmouth's unique vulnerability by offering a grant of £62m to protect Portsea Island.[4] In 2018-20 the east and west sides of Portsea Island had their edges raised by rockfill to the west and a concrete wall around Tipner and Hilsea; in 2019/2020 work on Portscreek was under way.

There is no option not to build new sea defences, but there was a long-running dispute about their design – broadly either hard or 'soft' engineering, as well as the lack of opportunity for local people to contribute their expertise to its development. University research in the department of geography on public participation into coastal management concluded that in order to reduce conflict and future rejection of design options, community participation should not be viewed simply as a 'tick box' exercise, but a process in which to develop mutual understanding and agreed compromise.[5] Relevant local academic research like this did not influence the Portsmouth case, but another research and practice-based exercise which involved the University school of architecture did, as we explain below.

A local authority consortium, the East Solent Coastal Partnership was given the remit to manage the region's coastline. To defend Southsea seafront in 2014 they adopted 'Holding the Line', a 'hard' engineering approach – high terraced concrete walls – up to 3.8m tall under replaced shingle, from Old Portsmouth to Eastney. This design was based partly on the defences of Cleveley north of Blackpool, although that area faces a much higher tidal range and the concrete there had already failed in places.[6] In 2016 there was alarm about the damage a high wall blocking off the seafront would do to a city that derives a significant 12% of its income from tourism. An alternative design was developed by Dutch and British experts in architecture, landscape and engineering (including the East Solent Coastal Partnership) and Portsmouth university staff worked together to research international experience and to produce an alternative broadly 'soft' engineering design. Its strategic approach aimed to enhance the value, amenity, environment and ecology of Southsea Common, as well as addressing shortfalls in the existing proposals.[7] Based on Dutch experience it offered three lines of defence: the beach left untouched, mounding covering carparking and other facilities and a landward dyke to tackle rainwater flooding. The two alternative designs – broadly hard and soft engineering - were displayed and discussed in Portsmouth Cathedral in July 2016.

A petition asking that residents should be offered alternative designs as a genuine choice to reflect the fierce debates over these alternatives was launched on change.org on 21 August 2017.[8] By 30 August it attracted 2,985 supporters, but a councillor said it was a waste of time and not the appropriate place for public consultation, since the council did not recognise petitions - unless they are on its model, which is very difficult to use … and they had already commissioned the hard engineering designers. Requests for participatory design workshops drawing on local expertise were rejected because "the city did not wish to "passport the management of the community engagement process of the planning process for one of the largest civil engineering projects the city has undertaken."[9] In February 2018 Southsea Seafront Campaign's petition, with 3,984 signatures, 1424 of them local, was presented to the city council, but this was rejected by council lawyers because only postcodes rather than full postal addresses were included.

New designs and consultation were planned for February/March 2018, but the Chief Executive cancelled these. Following the change of ruling party in May, increased public involvement was promised. Several public exhibitions followed, evoking considerable public response, but despite requests for active public design workshops, what we were shown was lacking in detail and any local distinctiveness. The hard engineering design prevailed, although in response to public pressure the height of the concrete wall and terraces was reduced; it was proposed to increase the width of the beach by dredging to absorb wave energy instead.

However, in autumn 2019 there was still a shortfall of £17m in the total estimated cost of £131m.[10] The council voted to use its own funds to fill the gap. The key historic military areas including Long Curtain Moat and Southsea Castle were excluded from government funding - because nobody lives there! This ruling also excludes sea defences for Fort Cumberland and Fraser Battery. There was also opposition to the poor quality of the landscape design. Planning permission was granted in November 2019 in order to meet the government's funding deadline. In June 2020 the Treasury approved £97.8m for the Southsea Coastal Scheme to transform Southsea seafront with higher defences to protect it against rising sea levels and storm surges.[11] This entailed considerable change to coastal historic defence structures and sites: Long Curtain Moat, Southsea Common, the setting of the war memorials along the promenade, Southsea Castle, the Rock Gardens, South Parade Pier, Canoe Lake and Fort Cumberland. The first stage was to protect Long Curtain Moat in Old Portsmouth by a new seawall in front of the historic stone structure began that September.

We now examine the evolution of each once military and now public open space in Portsmouth in more detail, starting with the city's north and south edges: Southsea seafront, Southsea Marina, Victoria Park and Hilsea Lines. The development of the new Horsea Island Country Park follows. In 2019 the potential opening to the public of the historic landscape of Southwick House and Priory over the brow of Portsdown Hill was the subject of negotiations between the former owner and the Ministry of Defence.

Southsea Common

With very few parks or open spaces in its densely built up urban fabric, Portsmouth is unusual among coastal towns and cities such as Brighton or Blackpool in enjoying a large 62-hectare unbuilt-on open lung of green space on the southwestern part of the seafront - thanks to its long military occupation. The seaward view from Stokes Bay and from Southsea Common is spectacular.

Anthem of the Sea, Fort Gilkicker and hovercraft 2015

At any time of day we enjoy watching passing shipping in the Solent – cruise liners, ferries, naval ships, tankers, cargoe ships, hovercraft, yachts, fishing and rowing boats - and ships coming in and out of the harbour. Some evenings three huge cruise ships including *QEII* and *QMII* pass seaward from Southampton in stately procession. Naval families crowd the Round Tower to say goodbye to crewmembers and to welcome them home again.

USS United States passes Horse Sands Fort 2015

Huge American aircraft carriers such as the *USS Nimitz*, decks crowded with aircraft, have to anchor at Spithead because they are too large to enter the harbour. Portsmouth's role as crucible for national events is emotionally demonstrated in times of war. Crowding Spur Redoubt and Long Curtain people cheered as the Falklands fleet set sail in 1982 – and watched in horror as broken ships with missing crew returned; thousands watched the departure of the fleet to the Gulf in 2003, mostly in silence.

How was it that Southsea Common was never built on? In medieval times it was the former waste of the manor of Fratton or Froddington: uneven, ill-drained and marshy land, covered with scrub and gorse. At the time of the Dissolution of the Monasteries it was the property of the Hospital of St Nicholas, the Domus Dei. Only the Garrison Church survives as part of this medieval hospice. The hospital surrendered the land to the Crown in 1540, and as part of a wider scheme to protect the south coast Southsea Castle was built in 1544 to defend the harbour approaches. King Henry VIII personally witnessed the disastrous sinking of his flagship, the *Mary Rose* from the castle walls in 1545.

Development of the area that became Southsea Common was prohibited in order to keep open the field of fire from the town's elaborate ramparts. Meanwhile local people grazed their stock on the waste. Desultory efforts were made over the years to drain it. Surviving engravings give some idea of what it looked like. The *South East View of Portsmouth* of 1765 and other similar views show scrub, gorse, dirt tracks and pools of water criss-crossing the uneven land between Southsea Castle and Portsmouth's ramparts. The land was enclosed in 1785 but development was confined to a line marking its northern boundary, thus preserving a remarkable open space for recreational use today.

Both the Great Morass which stretched northwards from Southsea Castle far into present-day Southsea and the Little Morass which lay to the east of the ramparts are still described as 'imperfectly' drained on a sketch map in 1823. The land occupied by the Little Morass – to the east of Pier Road - is still susceptible to flooding. A spring reappears near the junction with Duisburg Way during heavy rain.

King's Birthday parade on Southsea Common 1907   JPI Media

Southsea began to develop from the 1810s as a residential suburb of Portsmouth and a seaside resort. The Common was levelled in 1831-43 and gradually laid out as a pleasure ground; Clarence Esplanade, defining the boundary was constructed in 1848. Housing development to the north and east of the Common continued in the 1840s and 1850s by Thomas Ellis Owen, a local developer who created the early garden suburb of Southsea. Its centrepiece is his Gothic design for St. Jude's Church built in 1851. The spire, grant-aided by the Admiralty was a seamark for the Swashway channel into the harbour until Homeheights, the yellow brick block of flats partially obscured it.

By 1870 avenues of holm oaks had been planted along some of the roads. After the council took a lease of the Common in 1884, a public promenade, the Ladies' Mile was laid out and used for parades. To the east the brackish Canoe Lake was created in 1886 to drain Craneswater, and in the late C19 and early C20 more avenues were planted. The housing development to the north was largely complete by 1900. The Common was still a waste, polluted by open drains, and frequently swept over by the sea. It was levelled, mainly by convict labour.[12] After 1922 the Common was purchased from the War Office by Portsmouth Corporation for use as a public park. More trees and gardens were planted. David Niven in his entertaining autobiography *The Moon's a Balloon* said "I shall always remember Southsea Common: flat greasy, wet and windswept, with a dejected flock of dirty sheep morosely munching its balding surface." He had just been expelled from school aged ten and was forced to go to a Southsea school for 'difficult' boys run by a brutal Commander Bollard and his wife.[13]

Further sports and recreational facilities including tennis courts and the clubhouse, now a popular café, were added between the wars and during the second half of the twentieth century. To the east of Serpentine Road, the land was parcelled out into four main parts, divided by roads. Southsea Recreation Ground was laid out in the 1920s and 30s to the northwest (between Serpentine Road, Clarence Parade, Avenue de Caen, and Clarence Esplanade) with tennis and softball courts, putting and bowling greens with pavilions. East of the Avenue de Caen there is a large skateboarding rink opened in 1978, which encloses an early twentieth century bandstand. Spectacular murals are painted inside the deep concrete bowls and on the exterior wall. There's quite a tradition of mural and graffiti painting in Portsmouth – especially on hoardings around empty building sites.

Ladies Mile in snow 2015

Running from northwest to southeast across these two areas, to the north of the sports development, is the late nineteenth century Ladies' Mile Walk with a double avenue of Huntingdon elms. Many magnificent trees were devastated by the great storms of 1987 and 1991 and by Dutch elm disease. Replacement disease-resistant elms have been planted at the eastern end of the Ladies Mile. In 2002 Southsea Common was listed as a historic landscape and in 2003 the seafront was designated Conservation Area No. 10.[14]

The area to the west of Serpentine Road in the western part of the Common is largely open, with a few scattered trees on the grass. Rows of holm oaks line some of the edges and the roads, which cross the Common. There are good views across the Common and to the sea, dominated by the monuments around the seafront. At the northwest end is the 1961 Clarence Pier in 'Googie Style", developed in California. Architectural historian Tom Dyckhoff told Celia Clark that it's one of his favourite buildings. In 2012 the Portsmouth Society applied to English Heritage to spotlist the pier when the council's Seafront Strategy proposed its demolition. Disappointingly, English Heritage did not consider it worth listing.[15]

Next to Clarence Pier is the Hovercraft Terminal – which is one of the very few passenger hovercraft services in the world still operating. The service to Ryde Isle of Wight began in 1959. Hovercraft were invented by Sir Christopher Cockerell in the 1950s and built at Saunders Roe, the flying boat company at Cowes in the Isle of Wight. Our aviation chapter gives more detail about this key Solent history.

The seafront is home to a series of Victorian monuments along Clarence Parade: the Peel or Shannon Naval Brigade Monument 1860; the Trafalgar Monument; the Chesapeake Monument designed by T J Willis and S J Nichol in 1862; the Trident Memorial by Macdonald Field and Co. of about 1860 re-erected in 1877; the Aboukir Memorial by Baker of Southsea of about 1875, restored in 1984; the Crimean Monument of 1856, and a Portland stone obelisk by H J Andrews, erected by the Portsmouth Debating Society. Clarence Esplanade has nineteen early C20 cast-iron lamp columns on the south side which are together listed Grade II.

War Memorial

The largest and most impressive monument on the seafront is the Royal Naval War Memorial of 1920/4 by Sir Robert Lorimer, who also designed the similar ones in Plymouth and Chatham – though their settings are very different. A slightly tapered square column sits on a stepped base. The Second World War memorial was added in 1955 to the north: a low walled enclosure terminating in winged pavilions listing many names flanked by moving sculptures of sailors in heavy weather gear. Photos, flowers and personal messages show how much those lost are still remembered. All these monuments will be affected by the building of the sea defences.

In the Common's northwest corner between Pembroke Road and Pier Road is a monument to Lieutenant-General Fitzclarence of 1852 by J Truefitt, and on the north side of the Common, along Southsea Terrace, there is a drinking fountain in memory of Charles McCheane of 1889, restored in 1977. Along Castle Esplanade is a row of early twentieth century lamp columns and three cast-iron and timber shelters of about 1900.

The eastern seafront is dominated by Southsea Castle. On either side, high nineteenth century batteries, later 'municipalized' with steps and random pieces of rock flank the castle.

Celebrations and contested space

Kite Festival on Southsea Common and Spinnaker Tower 2015

The Common still belongs to the city. Until the late twentieth century this military legacy on open un-built on grass expanse was sacrosanct. But that rule was abandoned and it began to be exploited more and more for a great variety of temporary events, whether they are historic vehicle rallies, funfairs, circuses, military ceremonies, kite festivals, the Southsea Show, the Americas Cup or the Victorious pop music festivals.

Its wide open space is ideal for ceremonial occasions. The solemn Drumhead Service at the start of the 50th Anniversary commemorations of the D-Day landings on June 5th 1994 was a National Service of Thanksgiving and Re-dedication. According to the leaflet, a drumhead service historically took place "to bless the colours and uplift the hearts and minds of the troops… Today's ceremony signifies the forces committed to the 'Great Adventure' that was to liberate Europe from the Occupation, to give thanks to those who took part in this momentous event and to remember those who did not return." It was attended by the archbishops of Canterbury and Westminster and other bishops and church heads, the Chief Rabbi, chaplains of the armed forces, the Queen and royal family, President Clinton and presidents of the Czech and Slovak republic, minsters of defence, kings of Norway and the Netherlands, prime ministers and veterans of the D-Day landing. The Queen sailed on *HMY Britannia* with historic aircraft overhead and the heads of state to review a flotilla at Spithead, which then accompanied her to Normandy. As a D-Day veteran, Celia's father attended the ceremony, and she, Deane and her brother remember shaking President Clinton's hand as he and other guests were welcomed by the crowds – in those much less formal and less security conscious days. In extreme contrast, the public were rigorously excluded by heavy concrete and steel barriers from approaching the 75th Anniversary ceremony in 2019, broadcast across the world on tv. Those who opposed the presence of President Trump were demonstrating in Guildhall Square.

Locked down at home during the 2020 pandemic made us value our parks, harbourside and seafront near our homes all the more. We find a sense of community in public spaces that feel safe as well as being accessible to those who cannot travel far. The closure of Southsea seafront's roads resulted in perceptibly cleaner air and a respite from traffic noise as well as offering valued gains in safe places to walk and cycle. But even before the outbreak of Covid-19 parks budgets were facing brutal cuts – around 40% on average – with decimated staff numbers. Parks are not statutory services, which led to councils to hire out green spaces for commercial events as a funding stream. Every summer the large Victorious Festival has taken over larger and larger areas of Southsea Common. It began opposite *HMS Victory*, in 2011 attracting 35,000 people. In its new location on the Common it brings in useful revenue to the council's coffers, depleted by years of local government cuts. Now extending over three days it has continued to grow, attracting up to 60,000 people per day in 2015. The 2016 festival brought in £8.47m income to the city, an increase of £2.5m on the year before. Hotels and b & bs were fully booked. The leader of the council expected "around a £10m boost to the city."[17]

At the end of August 2017 "tens of thousands of revellers" descended on Southsea "as the Victorious Festival rocked into its sixth year".[18] Kiosks, bar, the children's arena, stalls inside an extensive green-boarded fence were constructed over two weeks by 3,600 people. Access to the D-Day Museum was included in the price. A crowd of 30,000 attended the opening performances on 25 August, some of them staying in a campsite in Farlington overnight. 60,000 were expected on the second day. In 2019 a total of 150,000 people attended over the weekend.

"It combines international music headliners with local acts, traders, community groups and Southsea's own seafront attractions. It has made visiting the area on Victorious weekend a firm calendar commitment for many in and around the south coast, re-defining the festival scene as an affordable, value-for-money event that caters for all ages. It was short listed for 'Best Major Festival' and 'Best Family Festival' amongst heavyweights such as Glastonbury and the Isle Of Wight Festival".[19] In May 2020 the pandemic lockdown led to cancellation of the tenth festival.[20]

Americas Cup grandstand under construction 2015

Southsea ramparts, Castle Field and the seafront walk were also hired out each June to paying spectators for the four day Americas Cup, sponsored like the Spinnaker Tower by Emirates.[21] The cup was first contested in 1851 as a race around the Isle of Wight, making it the oldest trophy in international sport, predating the modern Olympic games by 45 years. To accommodate the spectators watching the races in the Solent a large area of Southsea Common was fenced off and a tall bandstand was erected on the highest part of the ramparts. There was no public access to the seafront or Southsea Castle – used for hospitality - for several weeks while barriers were assembled then taken down afterwards. Only restricted views of the rapid races of the hightech yachts built in the Camber were available to residents. This event was cancelled in 2020.

Seafront facilities – new and old

We now explore seafront facilities on the common from west to east. In 1982 Portsmouth City Council changed its tourist promotion from "Come to Sunny Southsea" to "Flagship of Maritime England", focusing on the dockyard heritage area and other defence heritage to attract holidaymakers. But indoor facilities and landscaping on the seafront still needed to be upgraded. In response, in 1988 adjoining the West Battery a series of brick pavilions was built for the Sea Life Centre, now called the Blue Reef Aquarium. Its tanks house local fish, sharks, rays and tropical species in an ocean display, a dramatic underwater viewing tunnel - and a resident family of otters. The play space behind the stone wall along the seafront promenade which was developed on the site of the former boating lake has an area of natural regeneration by seaside plants. The miniature railway which ran along the seawall which Celia's children used to enjoy has vanished, leaving just the end of its route marked in brick paving.

Rock Gardens Pavilion

In 1986 the old Rock Gardens Pavilion to the east of the castle was replaced by the three dark glass pyramids of the Pyramids Leisure Centre. Designed by Charles Smith Architects from Boston Spa its shape, including the steep grassed slopes is reminiscent of the Palais de sport in suburban Paris. Inside is a swimming pool, a performance space where many famous bands play and a less used seaward pavilion – which, disappointingly was not developed as a winter garden where local residents hoped to enjoy tropical planting while gazing out to sea. There are plans to replace the centre with other facilities.

To the south of the West Battery the ramparts slope down to an open grassed area with wonderful views of Spithead. A Chinese-style bandstand topped with the city's star and crescent crest is the focus for popular Sunday concerts in the summer. It was designed and made by local artist blacksmith Peter Clutterbuck, whose workshop is behind the south side of Palmerston Road.

Further west on the landward side of the Castle's nineteenth century ramparts is a car and coach park and the D-Day Museum, designed in 1984 by Ken Norrish, Portsmouth City Architect. Until 2018 a bedding scheme between two paths ran north between Southsea Castle and the D-Day Museum with a five-sided pond at the seaward end and a floral clock aligned on the Avenue de Caen and Palmerston Road. The clock and symbolic lily pond were replaced in 2018 by paving. Children now run through the surface level fountains near the castle.

Rock Gardens

To the east, between Castle Esplanade to the south and Clarence Esplanade, are the ornamental gardens enclosed in concrete walls featuring the city's star and crescent crest, built by unskilled workmen as part of a government Back to Work scheme in 1918 when botanical gardens were all the rage. Southsea Rock Gardens sunk into a hollow to seaward followed in 1928. Sheltered from cold sea winds, its tranquil winding paths between Westmoreland limestone rocks, tropical planting and pools are still much enjoyed by locals and by visitors alike. Closed and neglected in WWII, restoration was tough and hampered by storms and flooding by the sea in 1949 – also, as already said, a hazard in the late twentieth and early twentyfirst centuries when the sea and beach again moved inland; salt water endangering the plants had to be swiftly pumped out. When Celia Clark's children were small there was a menagerie with rabbits at one end and an aviary at the other. These were removed in 1986. A dedicated band of volunteers enhance the gardens, with donations for tools, benches and plants from the Victorious Festival. Jackie Baynes' publication celebrates this special place.[21]

The narrow, rectangular section between South Parade to the north and Clarence Esplanade to the south was laid out in the 1920s as two sunken Italian gardens. Along the south side of South Parade Pier are three early C20 lamp columns together listed grade II. The easternmost part of these gardens is a low-lying section known as The Dell.

South Parade Pier

The two parts of the Common are joined by South Parade, which runs east through a strip of ground between buildings to the north and the sea to the south, with South Parade Pier to the south. The pier has had a checkered history – of fire and storm damage. The South Parade Pier Company Limited was incorporated in 1877 and in the following year a South Parade Pier Order authorised construction of the first pier. Demonstrating how much Portsmouth's development has been controlled by the government and military priorities, to build it, in 1879 and 1880 the company was granted a lease on land by the War Department and a lease on the foreshore by the Commissioner of Woods. The pier was officially opened in July 1879. Further leases of land were granted in 1902 and 1905 from the Commissioner of Woods for an extension; a further lease was granted by the War Department in 1905. Total rents payable under these leases, set to expire in 1977, was £46 p.a. Initially the pier was 1000ft long and is now 600ft (183m).

However in 1904 the pavilion was almost totally destroyed by fire. The pier was then sold to Portsmouth Corporation for £10,782.[23] The new pier in iron, timber and stucco was designed by GE Smith in 1908-9 and officially reopened on 12 August 1908. In 1914 in an attempt to improve the financial prospects of Seaview Chain Pier in the Isle of Wight the Seaview Steam Packet Company was formed and began running a service between Seaview Chain Pier and South Parade Pier. This came a halt in September 1914 and was formally prevented from further running by the Admiralty in 1915. To hinder any invasion the pier was partly dismantled during World War II, but it must have been repaired in time for the troops to march to embark onto their landing crafts for the D-Day landings. Under the shingle to the east of the pier are the remains of rail tracks to move material to load onto vessels embarking for Normandy. The beach was closed to the public for the long years of the war.

South Parade Pier on fire

The pier later caught fire again several times, most famously in 1974 during filming of Ken Russell's rock opera film *Tommy*, when an overheated film cable destroyed the three-storey theatre. Shots of it burning are in the film. Deane Clark photographed Ken Russell watching the fire. A new much lower pavilion replaced the theatre. In the 1980s the pier's ballrooms were used several times a week for discos organised by Portsmouth Polytechnic students. The pier appeared in an episode of Mr. Bean: "Mind the Baby, Mr. Bean."

In 2012 public access to the pier, now listed Grade II, was closed as its condition deteriorated. Timbers from the seaward landing stage were torn away in storms and ended up on the beach. The South Parade Pier Trust, a social enterprise, was formed to own and operate the pier, but despite the precedent in Hastings Pier the council did not use its compulsory purchase powers to acquire it for handing on to the trust. Tommy Ware Senior and Tommy Ware Junior now own it. They commissioned the architect Chris Flint & Associates Waterlooville to repair and restore it. The cast iron pile structure, cast steel frame and concrete deck structure were seriously corroded and exposed to storms, rising sea levels – and fire. Their engineers salvaged as much as possible and identified the extent of decay. The work for the first two to three years was to stabilise the substructure. Tie rods were inserted; steel fabricators were on site during the four years of closure. The perimeter seating was reinstated. The spend was £3.5m over four years. This work was the winner of the Pier Society Award in 2017 and the Portsmouth Society Design Award Restoration Commendation in 2018. The judges, visitors and local residents were delighted to see the pier come back to life again, with many activities including a funfair and the large fish and chip restaurant Deep Blue with fine sea views. In 2019 the landing stage at the end was reconstructed; boat trips were planned.

D-Day Memorial to Canoe Lake

To the east of the Pier the road divides, with St Helen's Parade running north-east and the Esplanade continuing to the east. In the triangular plot of land between these two roads is the D-Day Memorial, centred on a replica concrete block modelled on the 'dragons' teeth': anti-tank defences on the road to the Hayling ferry and on the beach in front of Fraser Battery. Solemn parades commemorate the fallen here each year. In honour of Neil Gaiman, born in 1960 in Portchester, the very short road where the buses wait was renamed after his story: 'The Ocean at the End of the Lane'; in December 2019 this was also a play at the National Theatre.

From 1884-6 the last remnant of the Great Morass was transformed into the popular Canoe Lake, opened on 17 June 1886. Initially, like its equivalent in Gosport, it was used to sail model yachts – a popular pastime in Victorian and Edwardian times, until recently brought up to date with powered miniature naval vessels. The water is topped up from the sea by opening a sluice at high tide. Crabs, fish and moon jellyfish find their way in, and are delightedly fished for by children. We enjoy rides on popular swan and duck-shaped pedalos around the lake, which is also home to a flock of mute swans. During World War II the lake was used for experiments into countermeasures against magnetic mines.[2]

Lawns with scattered trees and a row of holm oaks along St Helen's Parade surround the lake. The angel-guarded Emmanuel Emmanuel Memorial Drinking Fountain (c 1870, listed grade II) commemorates the first of the large Jewish community to be elected to the council in 1844.[25] He recognised the potential for the development of Southsea as a watering place and was instrumental in the construction of Clarence Esplanade, Portsea Railway and Clarence Pier. He also helped to obtain the land that became Victoria Park from the Board of Ordnance. In the 1990s the fountain was restored by Richard Walker, a graduate of the university Restoration and Decorative studies course, who regilded it at his own expense.

At the eastern end of Canoe Lake is a C20 single-storey cafe and two children's playgrounds. Cumberland House (Grade II) inside its walled garden, a two-storey villa built in c 1830/40 is Portsmouth's Natural History Museum. Within the walled garden to the southeast is a 2017 butterfly house which replaced an earlier wooden structure, with butterflies cut into the steel supporting beams.

Lumps Fort into Southsea Rose Garden

To the east of Canoe Lake and playground is the mid-nineteenth century Lumps Fort which was purchased by the Council in 1932. We describe its military history in chapter 12. In the C20 it was partly dismantled and laid out as a formal rose garden. Jackie Baynes' book *Guns and Roses The Story of Lumps Fort and Canoe Lake Park. Their History & Development over Five Centuries* published by the Portsmouth Society in 2017 paints a vivid picture of the fort's transformation. Deane Clark remembers the tarmac of the parade ground piled up with random piles of bricks. He and his friends had a great time exploring the tunnels inside the fort's perimeter. After World War II the main area of the fort was cleared and refurbished. Lawns, paths and flowerbeds were constructed. The barrack block to the north was to be converted for kitchens and service quarters for a proposed open-air café, along with toilets and a new bandstand. Unfortunately loan applications to central government to make Lumps Fort 'a place of quiet and beauty' were turned down, so an area east of the fort was let for a year to install 'Playland and Ice-Time', based on one at the recent *Daily Mail* Ideal Homes Exhibition. An ice stage was erected under canvas, and the Southsea Shakespeare actors put on a two-week season in an open-air theatre. The £6,701 cost of laying out the Rose Garden was partially offset by the rent. The *Evening News* reported the Midget Town and Circus, popular singers Dorothy Squires and Billy Reid, and battle dioramas including 'the Allies Landing on D-Day'. In about 1950 castellated walls and entrance pillars were added – with the additional purpose of preventing visitors from falling into the gun emplacements!

Rose Garden

But by 1952 the council's efforts to commercialise Lumps Fort and make it pay for itself had failed, and Mr. EW Studley, the Parks Superintendent drew up plans to transform it into a garden with roses climbing over rustic pergolas and symmetrical beds of hybrid tea roses, a central sundial and circular seating in the gun emplacements. In summer it's a blaze of colour and perfume. It's a sheltered tranquil retreat in any time of year, just as the council intended when they first purchased it from the MOD. There's a view out to the sea in the main entrance through the original central musketry caponier. A carved stone cockleshell marks the beginning of the Shipwrights Way, in honour of the 12 marines who took part in the raid in December 1942, only two of whom returned. The slopes were terraced with steps to the top of the fort, which offers a marvellous panorama of Spithead, the Isle of Wight and the open sea. On the way up you pass through the Japanese garden, celebrating Portsmouth's twinning with Maizuru in Japan, opened in 2000. Maizuru also has a naval base and brickworks – just has Portsmouth once did. The garden was designed by Takashi Sawano, and includes a traditional Torrii arch, Japanese trees and shrubs and a 'historical stone' from the 16th century Tanabe Castle in Maizuru.

Dan Bernard

The stumps of the submarine barrier stretching towards the Isle of Wight emerge at low tide, and in the early evening a stately succession of cruise ships pass east and southwards - on their way from Southampton to France and the States.

The Rose Garden is often used for performances and wedding photographs. According to Myrtle Clark-Bremer, a film, *Roman Wedding* was made in 1967, probably on Super 8 cine film, in conjunction with a Workers' Educational Association course led by Barbara Guernier about the Romans. It featured Len Russell, a leading light in the Southsea Shakespeare Actors and several other actors in that group. Myrtle wrote the script in collaboration with Barbara.

The Southsea Rose Garden with its geometric paths and rose blossoms was the ideal setting for the wedding ceremony's long procession. The toga-clad celebrants chanted a Greek wedding hymn: 'Hymen Hymenaeus' as they made their way to the stone pedestal in the centre. In 2020 the Rose Gardens were a sanctuary and solace to many people. Celia and Deane Clark celebrated them in paintings, poetry and photographs in their booklet *Lockdown Spring*.

The red brick military engineering structure of the fort is highly visible – both inside and out, but there are still large covered areas that are underused except for storing gardening equipment. The city council's procurement process failed to raise potential occupants until 2019 when the Portsmouth Film Society prepared a bid under the city council's Business Concession Opportunity, led by Ayşegül Epengin of the university's school of film, media and communication. Plans for a film theatre and café were drawn up by their architect Deniz Beck to restore and convert the barrack block into a cinema and cafe. Disappointingly, Portsmouth City Council rejected the proposal, because they had doubts about the funding. These evocative spaces were empty and decaying; they might have come alive with film enthusiasts enjoying the cinema's unusual setting. Instead, in October 2019 there was a proposal to develop the rooms as a centre for the wellbeing of veterans. There are 24,000 ex-service men and women, their families and serving soldiers and sailors in Portsmouth – 10% of the city's population. The Solent NHS Trust supported the bid, emphasising its appropriateness in the year of the D-Day 75 commemorations. Another proposal is to install an art gallery in the space.

To the north is the beautiful new Southsea Tennis Pavilion, which won the Portsmouth Society's Best New Building award in 2018, as well as many other prizes. Its delicate gull wing roof and pale walls in Danish brick link a first floor cafe/bar to a generous outside terrace - the place to watch the tennis action. On the ground floor a popular wood and glass lined dividable hall is for community events. Wendy Perring of PAD Studio Architects was the designer and construction was managed by Rice Projects. A philanthropist and tennis enthusiast who lives nearby funded this special addition to our seafront.

The western skyline of Lumps Fort is dominated by the miniature tower of Portchester Castle at the highest point. This is part of Southsea Model Village developed from 1962 with winding paths around miniature buildings encircled by a model railway, still a popular tourist attraction. A London architect Mr. R Frost hired a former shipbuilder John Simmons to construct the buildings and layout according Mr. Frost's design to a 1:12 scale. Projecting to the west is a 125 foot long caponnier or musketry tunnel, one of Portsmouth's secret spaces. At the end inside is an impressive model of Portsmouth Guildhall and another of the Spinnaker Tower. Model ships including *HMS Victory* are in the adjacent expense shell and cartridge store. To the northwest of the fort and east of Cumberland House are bowling greens, with grass and hard tennis courts to the east.

Southsea Model Village

The shingle beach continually remakes itself at every tide, and it migrates: the West Winner at the end of the beach that we used to walk out on nearly a mile into the sea has disappeared, while the East winner on the Hayling Island side of the Langstone Harbour entrance, as well as Eastney beach, continues to grow. Pebbles bored with tracking sensors are monitoring this movement. Plants are colonizing Eastney Beach: vegetated shingle is a nationally rare habitat. Kite surfers soar over the waves off Hayling and Southsea, and in front of the Fraser Range is a popular nudist beach. Plans to convert Fraser Gunnery Range into flats and houses are discussed in our chapter on residential conversion.

The Glory Hole/Southsea Marina

Even the most intractable sites may be transformed: the Glory Hole on the Eastney peninsular, now Southsea Marina, is a case in point. As a remote area away from where people lived it was used to dump waste materials for many years. In 1912 permission was given by the Admiralty to infill areas on the Tipner Lake foreshore using ash, clinker and other 'innocuous' materials. The first area to be filled was the inlet to the east of the pumping station (now Henderson Road Caravan Site). Halcrow's 2008 Portsea Island Coastal Strategy Study said that for many years dockyard waste was deposited in the I,287 ha.pit which lists it as 'Former Naval landfill 1918-1960, uncontrolled tipping'. No records of what was dumped there were kept. By 1932 according to the OS map the area to the west of the caravan site was filled and a tramway constructed to transport materials from a refuse destructor producing the fill. According to the 2012 Eastern Solent Coastal Partnership report on contaminated land the Glory Hole was a much larger area which underlies the adjacent 1956-1985 naval housing estate. In 1991 120 naval families in the adjoining estate had to be evacuated and rehoused while the estate was sealed off for the topsoil to be replaced.

In 1984 Mike Hancock MP asked the Secretary of State for Defence what discussions had taken place about the dumping of material excavated from the Glory Hole, what materials were dumped there and whether there were records to indicate whether the waste was toxic or dangerous. He was told that of the tip was sold by the MOD to Portsmouth Corporation in 1971. [26] This was the eastern section where an agreement was reached in 1985 to the development, disposal and leasing of land for a proposed marina. So from 1980 to 1985 the tip's contents were investigated. Asbestos, cadmium, lead and mercury were found, while borun, chromium, cobalt, molydbenum and cynanide were said to 'exist at the site in backgroound levels'. The site was excavated and the waste was eventually removed in covered lorries to Ringwood (non-hazardous) and Purton (hazardous by covered lorry.[27] Southsea Marina was constructed in 1985-1987 in the excavation and a marina village built alongside it in 1986-1992. Now owned and operated by Premier Marinas Ltd the marina is connected to the harbour via a halfmile long channel and accessed via a tidal gate or sill which opens at half-tide.[28] The bar, café and first floor Bombay Bay restaurant are open to the public. Enjoying a curry while watching the sun go down is lovely.

Victoria Park

Postcard published by J. Welsh & Sons Portsmouth

As already mentioned, Alderman Emmanuel Emmanuel initiated the purchase of 15 acres of land by the Corporation from the Board of Ordnance when the fortifications of Portsea were levelled. It was designed as the People's Park by Alexander Mackenzie just to the north of Portsmouth Guildhall, to the northwest of Portsmouth and Southsea railway station. The first public park in Portsmouth, it was officially opened on 25 May 1878. A total area of around 15 acres (61,000m^2), it is planted with trees, shrubs and flowers. At the centre is an enclosed aviary with peacocks, parrots and other exotic birds and another area with rabbits and guinea pigs which children enjoy feeding. In 2017 there were plans to remove it, but local people wanted it to stay. An adventure playground offers children a more challenging experience than the usual swings and climbing frames.

The eastern perimeter of the park, once bounded to the south by the curving extension of the railway to the dockyard is now is hemmed in by tall blocks of student housing and a hotel in Stanhope Road. There is a new pedestrian entrance to the park opposite Portsmouth and Southsea Station. St. John's Roman Catholic Cathedral is to the north, and to the west the wall to HMS Nelson's wardroom is constructed in stone from the demolished ramparts.

Towards the eastern entrance of the park and Queen Street there are many memorials and monuments connected to the Royal Navy, including an ornamental cast iron fountain.

Chinese bell Victoria Park

The most exotic monument is the Chinese bell, enclosed inside a miniature pagoda. The Dagu bell guarded a fort to the Pearl River entrance ; it was rung to warn people of approaching enemy forces. It was taken as a war trophy by sailors from *HMS Orlando* who had been fighting in the Boxer Rebellion in 1900. It was then hung in a temple shaped memorial to their dead comrades in Victoria Park, where it remained for over a hundred years. In WWII a quick-witted gardener stopped it from being melted down to make armaments. Mark Lewis of Art and Soul Traders who occupied the Lodge in the park researched the bell's history over five years. Together with Kim Ip of Portsmouth Chinese Association he contacted the Chinese government, which had been campaigning for several years for the return of ancient artefacts. A hundred and five years after it was taken from its original site as a war trophy, Mark Lewis arranged for it to travel back to China in 2005, where it sits in a specially made museum in Tian Jin. In its place a cast iron replica made in the original Chinese foundry now hangs in the pagoda.[29]

Another memorial honours the officers and men of *HMS Powerful* who lost their lives in the Boer War. There are other obelisks and memorials situated in this area of the park. Occasionally events are held in Victoria Park such as the Green Fair and Proms in the Park.

In December 2019 the city council was awarded development funding of £251,000 by the National Lottery Heritage Fund to restore, improve, uncover and celebrate its long history and create a more inclusive space which already attracts half a million visitors a year, with a wider health and wellbeing programme for schools. The grant helped the city to progress their plans for a full grant of £2,050,800 to fund restoration of the park's historic design, improvements to facilities including toilets, paths and seating areas, a new educational programme and classroom space, support for volunteering, 'an illuminated fete' and community heritage and ecology open days.[30] In July 2020 the city council advertised for a design team to restore the 3.5ha park and its original Grade II features, create more inviting entrances, reinstate a large open space in the centre of the park, plant new trees, create a new performance space on the site of the bandstand and crate a new hub with education spaces and toilets. The design was the basis of a second round application to the National Heritage Lottery Fund.[31] The work was planned to be completed in 2023.

Hilsea Lines

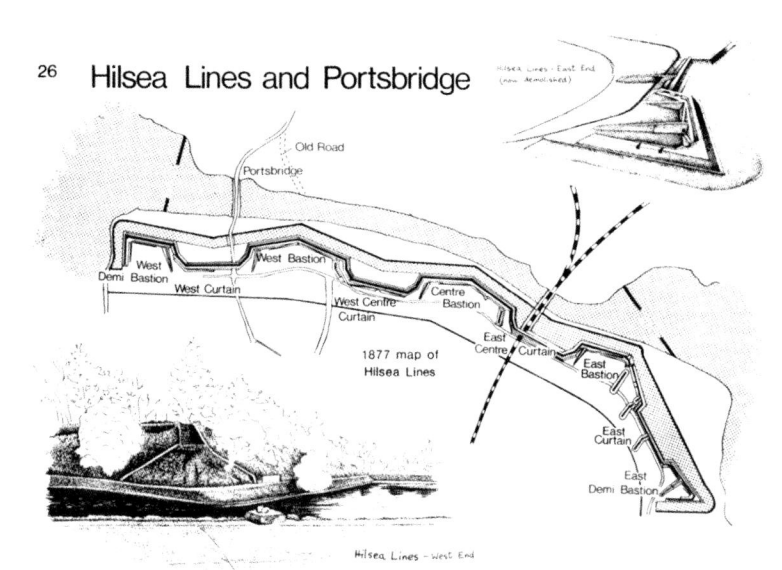

Hilsea Lines. *Portsmouth City and Fortress Unit 2*. Portsmouth Teachers' Local Studies Group 1980

From 1544 the first defences of Portsea Island protected Portsmouth's northern shore from inland attack for 300 years. A small star-shaped fort and earth bulwark was also constructed on the north side of the creek. In the mid eighteenth century there was a fear that Portsmouth's strong seaward defences would tempt the enemy to land along the coast and attack the city from the north, so between 1744 and 1763 extensive defensive lines were constructed on the Portsea Island side of the creek, under the supervision of John Peter Desmaretz. The Lines consisted of a 15–20-foot-wide (4.6–6.1 m) and 6-foot-deep (1.8 m) ditch backed by a 7–8-foot (2.1–2.4 m) rampart. Water was allowed to flow into the ditch from sluices at either end. The lines were modernized in 1815 at the end of the Napoleonic wars.

By the 1850s these were obsolete. In response to the renewed threat of a French invasion these lines were replaced at a cost of £260,000 from 1858 to 1871 by lines designed by Captain William Crossman, as part of the Palmerston Fortification Strategy. The red brick casements with bombproof vaulted ceilings were "designed to provide gun emplacements which afforded protection to the guns and their crews. The gun positions were arranged along the front of the casemate in a long room known as the gun gallery. Behind the gun positions were rooms, separated by curved corrugated iron blast shields, each of which provided living accommodation for five men. The guns in the casemates provided cross-fire to prevent the enemy crossing Ports Creek and to provide low profile fire along the front of the lines. A typical layout consisted of 2 gun galleries, various armaments stores, 9 accommodation rooms, a cook house and a wash room."[32]

They are largely built from clay and chalk thickly covered with trees. When completed they were 30 feet (9.1 m) high, 20m wide and 3000 yards (2.15km.) long. The Lines were divided into nine sections with five 'bastions', each roughly 300m long all set forward, close to the moat, and four 'curtains' roughly 200m long, set back, but linked to each bastion. They included special fortified bridges for road and rail access. It was originally planned to equip the lines with smoothbore guns, but these guns may never have been fitted. In 1886 the lines were equipped with a mix of RML 7inch guns and Rifled Breach Loading (RBL) 7 inch Armstrong guns on Moncrieff mountings fitted in newly constructed concrete emplacements on top of the lines. Further RBL 7 inch Armstrong guns were installed in the original casemates.

A model of the Lines featured in the 1862 International Exhibition. However, even before their completion they had been rendered obsolete by the 1859 Royal Commission on strengthening national defences and also by advances in artillery technology. The development of rifled artillery meant that it was now possible for an enemy to occupy the high ridge of Portsdown Hill north of Portsmouth and shell the naval base from several miles away without having to attack the Hilsea Lines. To counter the new threat a 'ring fortress' was constructed to protect the dockyard by land and sea - as described in Chapter 8 - and the Hilsea Lines were scaled back. A pair of forts planned behind the lines were not built. The guns were removed in 1903. A QF 6 pounder Hotchkiss was mounted on the lines during World War 1. A small number of guns were mounted on the lines during World War II.

The London, Brighton and South Coast Railway built a wooden bridge across the creek in 1847, allowing railway access to Portsmouth for the first time. An Admiralty order required the bridge to open between 2 and 3am on the first Sunday of every February. The swing bridge was supplanted by a drawbridge in 1909, which has been fixed in place since 1920.

Until the early C20 the other key constraint was that the only road access into and out of Portsmouth was via the London Road which entered Portsea Island through two narrow arches in the Lines: a pair of 15-foot-wide (4.6 m), 18- foot-high (5.5 m) tunnels and alongside, an 8-foot- wide (2.4 m), 11-foot-high (3.4 m) tunnel for pedestrians. In 1867 a retractable bridge crossed the creek to allow the passage of gunboats.[32]

Having only one way into and out of Portsmouth was no longer tenable. A large gap in the lines was opened up: the section containing the tunnels was demolished in 1922, and a wider bridge across the creek was built in 1927. Two large bus depots on either side of the London Road were built in the early 1930s and the Bastion café was built on the east side to offer refreshments. This was replaced in the late twentieth century, first by a garage, and later by a block of flats. The original high brick walls of the Lines are still visible behind it. The lines were crossed by a pedestrian bridge at Peronne Road in WWII, replaced in the late 1960s when the A27 was built on part of the channel along the north side of Ports Creek.

In 1932 the eastern lines were dynamited to make room for Portsmouth airport. A new road link, now the Eastern Road, offered a much needed second access to the mainland and to Portsmouth Airport. The housing development of Anchorage Park now occupies the site. In 1963 there was a suggestion that the creek should be redeveloped as the Dunkirk Memorial Channel. In the late 1960s the A27 was built on part of the channel, while the remainder was dredged. The first stage of sea defence work from Ports Creek Railway Bridge to Eastern Road bridge: earth embankments and a sloping rock revetment to provide the homes behind with coastal flood and erosion risk management were completed in 2015.[35]

Portsmouth Grammar School Playing Fields

Hilsea Playing Fields   The Portsmouth Grammar School

The longest sporting use of the land around Hilsea Lines is by Portsmouth Grammar School. The 17.5 acre playing fields now have 4 rugby pitches, an astroturf pitch for hockey and tennis, 5 netball courts, 7 tennis courts, 2 first class cricket squares and numerous rounders pitches. In the 1880s the Lieutenant-Governor of Portsmouth, Sir George Wills arranged for the school to provide playing fields part-time on the West Demi-Bastion four miles away from the school in Old Portsmouth. The boys travelled there by horse-drawn tram and teams changed in a shed. The first cricket match at Hilsea took place on 25 April 1885. During WWI chestnut, scots pine and fir trees were planted on the ramparts overlooking the playing fields to disguise the gun positions and fortifications from German aerial reconnaissance. A single cold water pipe was available for washing. Because there were no perimeter railings, boys could walk over the ramparts to the moat, where they skated unofficially when it froze in winter. In March 1931 the city sold an additional eight acres to the school, paid for by Sir Heath Harrison, a wealthy shipowner, alderman and county councillor. The casemates and ramparts were leased to the school for 999 years so that the bastion could be converted. Fives-courts, a tuck shop and changing rooms and bathrooms for 200 boys in the casemates were constructed. The pavilion was opened in 1931. At the outbreak of WWII the navy requisitioned the playing fields as outstations of HMS Excellent, the Naval Gunnery School at Whale Island and HMS Phoenix, the Firefighting School. The casemates housed civilian fire engines for the Fire Service based there. The playing fields were used to organise firefighters and bomb disposal teams sent into the blitzed city.

Royal Marines were trained there in preparation for an anticipated invasion. The RAF tethered a barrage balloon there. Heavy use took its toll and the ground condition deteriorated.

Hilsea casemates c.1980
The Portsmouth Grammar School

Once restored the school could use the playing fields again. Trees were planted to provide a wind-break at the far end of the green. Archery and the school scout group began to use the grounds. Flights by a manually catapulted glider by the Air Section of the Combined Cadet Force took place over the field. In 1972 on the centenary of the first FA Cup match the Royal Navy Field Gun Crew re-enacted the match.[36] The 17.5 acre playing fields now have 4 rugby pitches, an astroturf pitch for hockey and tennis, 5 netball courts, 7 tennis courts, 2 first class cricket squares and numerous rounders pitches. Those who did not wish to play rugby went on long runs – 'over the hill'. A PGS pupil gave a first hand account:

"There was at one end in the 50s/60s a vaulted ground floor room (the road access end ie south) with two huge baths at the rear with super sized brass taps. They were white tiled and each could hold about 10/12 pupils. Often they were quite muddy, but after freezing on the rugby pitch for two hours they were a welcome sight, and there were also some showers to finish off with. The other ranks changed in the barrel vaulted spaces with no daylight up the stairs. The raised gun emplacements (still with the brass rubbing strips in curves) were a good spot to stick one's clothes as there was too little room for all the kids. You had to climb up into them. No lockers obvs…The masters and the 1st Five-team had the ground floor rooms (casemates?) with windows - ha ha!"

"As the room was a brick barrel vault the acoustics were great: so much singing went on. Some songs sacred: 'Bread of Heaven' was popular though never ever sung at morning prayers, possibly other hymns but can't be sure. Other were WWI and 2 wartime rude songs such as… (to the tune of Colonel Bogey). Also sung straight...and in French too! La Marseillaise: 'Allons enfants de la Patrie...etc'. One Nazi memorabilia-minded 5th year (I shan't mention his name) used to sing the 'Horst Wessel Song'. Also 'The Red Flag: 'The peoples flag is scarlet red...'" [37]

In 2010 Portsmouth Grammar School celebrated 125 years of sport in the western-most end of Hilsea Lines with rugby and hockey matches on 11 December and a sports quiz between a team of professional sportsmen and women led by former pupil Roger Black MBE and Portsmouth Grammar School sports department staff and pupils on 1 October.[38]

Civilian purchase and development of Hilsea Lines

When the lines were built, large amounts of land were purchased by the Crown to the rear. Part of this, behind the west bastion was sold to the city of Portsmouth in 1926. A housing estate, a school and a recreation ground were developed on it. Between 1929 and 1930 the city purchased the east bastion, the curtain wall and the land behind them. The western end of the moat became known as the Hilsea Lagoon; in the mid-1930s work was carried out on the banks and it was developed as a boating lake. From 1929 to 1933 Portsmouth City Council, as owners of the land, created gardens, a tennis court and a hard levelled area intended for roller skating and dancing on the northwestern side of the Lines.

The land between the boating lake and the lines was converted into part of the Hilsea Bastion gardens. In 1938 a bridge was built across the boating lake section of the moat. It was pulled down in 1999 and replaced by the current lower arched structure. Sadly most of the gardens were destroyed by road widening in 1968-70 at Portscreek. In 1986 the city bought the parts of the lines that were not already under its control. The terraces that formed part of the gardens were demolished in 2000. Together with a level of neglect, there is little evidence of the grandeur of the original design.

Hilsea Lido

Hilsea Lido Pool for the People Trust

Alongside Portsceeek, Hilsea Lido, one of only two deep outdoor pools in the UK was officially opened on 24 July 1935. The City Engineer Joseph Parkin designed the swimming pool, cafe and splashpool. The total cost for the complex, built in the Modernist International Style was £36,000. Constructed over 7 months in 1935 as an employment initiative, it was to provide leisure facilities for the new housing developments in Hilsea. The Hisea Lido pool is 67m long, 18 metres wide and 4.6 metres deep. The main swimming pool could accommodate up to 900 swimmers and more than 1000 spectators. It was designed to provide drama and spectacle with three diving boards – the highest being a spectacular 10 metres – and water slides and water polo. In the 1930s the site hosted national diving championships and water spectaculars, including formation swimming and displays by the League of Health and Beauty.

Opened from 7am until 10pm from the end of May until the beginning of October, the complex included night illuminations and a PA system to entertain the crowds. Original admission charges into the main pool were 6d for adults (approx. 2.5p today) and 3d for children. The splashpool cost 1d for entry, car parking was 3d and parking your bicycle cost 1d – these facilities are free today! During WWII it was used for training by nearby military units.

Originally the water was pumped from Portscreek, filtrated and then aerated through the water fountains before entering the pool. This system operated every 8 hours. It's thought that because the tidal water of Portscreek was never of any great depth the water would be warmed by the sun prior to entering the pool (or perhaps people were just a bit hardier then!). Water now comes directly from the mains.

In 1975 Hilsea Lido – as well as South Parade Pier and a house with a green tiled roof on the seafront featured in the cult film and rock opera classic *Tommy*, with music by The Who and directed by Ken Russell.

When it became apparent that the city council planned to close Hilsea Lido at the end of the 2008 season a pressure group began investigating alternative ways of saving this much loved facility. Hilsea Lido Pool for the People Trust was formed and negotiations began with the aim of taking over Hilsea Lido, the Blue Lagoon and Hilsea Splashpool as a community asset transfer, as also happened at Southsea Skatepark.

While negotiations were going on, the HLPP Trust negotiated access to Hilsea Lido under license in order to complete low level maintenance, host profile-raising and fundraising events and provide a visible presence on site. Their first event was to take part in National Heritage's Heritage Weekend in 2009 and host a Family Fun Day. In preparation for the weekend the Probation Service were approached to for help from the Community Payback Team, which made a great contribution to the project and was the beginning of several very successful partnership schemes that have helped support the HLPP Trust. Hilsea Lido Pool for the People Trust is a community run organisation and registered charity (No. 1133714). Their aim is to provide the lido for public use, offering unique facilities to everybody with an entrance price we can all afford, regardless of income. In order to achieve this each year they must raise money in other ways.[39] In June 2010 the Trust signed a 99-year lease on Hilsea Lido and the Blue Lagoon. Unfortunately during the period of negotiation PCC decided that Hilsea Splashpool was "life expired" and therefore unfit for handover. In September 2010 the Trust took over both the Blue Lagoon and Hilsea Lido.

After years of rundown followed by closure in 2010 Hilsea Lido Pool for the People acquired the lido and the adjoining Blue Lagoon building on a 99-year lease from Portsmouth City Council. It was reopened in July 2014.[40]

Scheduled monument and conservation area – urban forest

In 1994 most of the Lines were scheduled as a Scheduled Ancient Monument protected by the Ancient Monuments and Archaeological Areas Act 1979. In 1967 their 2km length and 36.9 hectares were designated by the city council as Hilsea Lines Conservation Area. The ramparts are thickly covered with trees including walnuts, grown to supply gunstocks because of their hard wood. The lines were landscaped as a public park in 2000-2. The site is managed as a nature reserve, with interpretation boards about its history and wildlife.

"The Lines are the largest area not only of public open space within the city, but also tree planting. They in effect constitute an 'urban forest' which screens and separates the northern edge of Portsea Island from both Portscreek and the M27/A27 Motorway. It is the tree cover of the Lines which in many ways defines their character, and is also their most memorable feature as a piece of public space. In many instances the planting is dense, thick and overgrown. This 'wildness' is particularly notable on the ramparts and land to the immediate north of the Lines at the Centre Bastion, East Centre Curtain and East Bastions. Elsewhere, trees mingle with bushes, hedges and areas of managed and unmanaged grassland to create a patchwork effect to the north of the Lines. This area of planting and more open land is dissected by a number of woodland paths of varying widths, and in varying conditions — from open and free of impediments to quite heavily overgrown with trunks, branches and large root systems obstructing the path. Tree growth on the paths surrounding, and on the upper ramparts of East Centre (railway) Curtain is sufficiently thick to create the impression of being within a forest of much greater size."[41]

Other activities and ingenious new uses have developed inside some of the six angled rows of earth-covered bastions that flank the shore: a rifle range, band practice rooms with a specially designed sewage installation on recycling principles, and a boxing gym. A factory making luminous dials dumped radioactive materials on the adjacent Lines.

Bastion 3 is used by Volunteer Groups and the Portsmouth City Council funded the Hilsea Lines Ranger. Bastion 4 has a commercial use as a recording studio "Casemates Rehearsal Studios". Bastion 5, owned by Portsmouth City Council was in a derelict state, due to vandalism and destruction by local youths. It was listed on the Heritage At Risk Register maintained by Historic England. The bastion closest to the railway line continues to be badly damaged, while Bastion 6, which was manned by the Royal Artillery from the 1860s and was last used in 1946 to remove and recondition spare engine parts from Spitfires has a much brighter future – as the WWI Reminiscence Centre described in our Make art not war chapter.

The east and west sides of Portsea Island have recently had their edges raised by rockfill to the west and a concrete wall around Tipner and Hilsea; in 2019/2020 work on Portscreek was under way. Today the lines are accessible for most of their length, but overgrown and derelict in places. There is a walk around the Hilsea Lines by the Mountbatten Centre to the A27 Roundabout and back. The moat is used for fishing. In May 2006 it experienced some flooding. A raised sea defence wall to match those already built on the western sides of Portsea Island encloses the Lines. A preferable less damaging option might have been sluices at either end to control the tides. In 2010 the Lines were given a Green Flag Award. We now consider Portsmouth's newest park.

Horsea Island Country Park

The former Paulsgrove Amenity Tip 8 August 2006 Ian Hargreaves JPIMedia Limited

Sandwiched between Port Solent and Horsea Island - still an active MOD site - the former Paulsgrove landfill was developed from 2012 into an informal country park to help overcome the city's deficit of public open space and to meet Natural England's Accessible Natural Greenspace Standard. The main part of the park was created on top of older landfills over filled-in channels and previous infilling closer to the MOD's torpedo lake, which is still used to train naval and civilian divers. The 'Pyramids', adjacent to the M275 were created more recently. The tip closed in 2004 and was covered with soil which Onyx planted with wild flower seeds. More recently sapling trees were planted to create a new forest.

The park has 52 hectares (128 acres) of public open space. It was implemented through the PUSH Green Infrastructure Implementation Plan and funded by the EU Horizon 2020 research and innovation programme. Veolia, the company that operated the landfill site, was responsible for implementing the landscape proposals. It was designed to provide a new habitat for trees and greenery, to be known as the Queen's Diamond Jubilee Wood to celebrate her jubilee in 2012, one of 60 being created across the country. The aim was to plant about 50,000 trees and create grass areas, and wildflower meadows, footpaths, picnic areas and offer spectacular views across Portsmouth Harbour.

The community, including Portsmouth and Southsea Tree Wardens started the Diamond Wood off with 60 trees planted in March 2013, marked by a commemorative plaque. On 16 November 2019 a circle of 75 native trees was planted on the Port Solent far field, to commemorate the 75th Anniversary of the D-Day landings. The event was organised and funded by The Tree Council, supported by Portsmouth and Southsea Tree Wardens and Portsmouth City Council. D-Day veterans and their families took part together with dignitaries, tree wardens and other members of the community and community groups. Drought in 2020 affected the survival of some of them; the intention was to replace the casualties in the next planting season.[41] In August 2020 many of the remaining trees were damaged and uprooted and also needed to be replaced.

View of Goliath crane and city from Horsea Island Country Park 2020

In summer 2020 there were woods and informal footpaths and a diamond plaque adjacent to a planned entrance into the adjacent peaks of tipped material nearer the M275, nicknamed the 'Paulsgrove Alps". The highest is 41m high, but not yet open to the public. It appears that there is a plan to do so eventually once the tip is handed over by Veolia which operates the 'Portsmouth Materials Recovery Facility'. The peripheral footpath offers glimpses of the MOD activity at Horsea

Island, of the harbour, Gosport shore and the Goliath crane, but the view from the top of the Alps, once it's accessible will be much more impressive. We now move back in time to another park 'over the hill'.

Southwick Park

Aerial view of Southwick Park 2009   Bob Hunt Portsdown Tunnels

The historic landscape of Southwick Park has great potential as a public open space. In November 2016 the Ministry of Defence announced that the tri-service Defence College of Policing and Guarding was to close by 2025. As we explored in chapter 3 the descendent of its former owner, Mark Thistlethwayte negotiated with the Ministry of Defence for its return to Southwick Estates. The Crichel Down rules state that land "acquired under threat of compulsion... will normally be offered back to the former freeholder" apply - but the current value of the site as well as the terms on which a former owner may buy back a property requisitioned by the Crown for national defence are crucial to the outcome.

The story of Southwick Priory is intertwined with Portchester Castle two miles to the south. In 1135 King Henry I established an Augustinian priory at St. Mary's church inside the encircling walls of the castle. Within twenty years the canons, disturbed by the noise of the soldiers billeted in the castle, moved to Southwick. The lands the priory owned were probably enclosed as a deer park. The priory was suppressed in 1538 on the orders of Henry VIII. Today one wall of the priory, the monastic fishponds in the wooded area known as the Wilderness and earthworks survive.

After the Dissolution the priory lands came into lay hands. Sir Daniel Norton built a Jacobean house where the priory had stood, incorporating some of the ruins into it. To the east of the house a formal terraced garden was laid out with an orangery and parterres in the Dutch style. Kip's engraving gives a picture of the house and gardens, though these depictions are not always accurate. During the seventeenth century, gardens became larger and more formal, with terraces, parterres, avenues, canals and ornamental woodland known as wildernesses, reflecting the style of the Italian Renaissance and the influence of French and Dutch ideas. There is some archaeological evidence of the formal gardens created in the seventeenth century. As tastes changed the formal gardens were not maintained, but they survive as earthwork terraces named the Slopes, now part of a golf course. At the beginning of the 19th century the house was rebuilt on a new, higher site. An artificial lake 400 metres long was created by damming the waters of the River Wallington and the park was landscaped. The Park is now a significant historic landscape of great interest in the history of parks and gardens because changes that have taken place there are a reflection of differing styles in landscape gardening over the centuries.[42]

As we describe in Chapter 3, Southwick House of 1800 became a key location to the development of WWII. In 2004 the functions of HMS Dryad were transferred to HMS Collingwood in Fareham and the site reverted to its original name of Southwick Park. The Defence College of Policing and Guarding occupied the site from 2005. In 2016 the MOD put the site up for sale, saying that the land would be sold off to free up space to build thousands of homes, one of 13 sites which the government expected to provide 17,000 new places to live.[43] In 2019 after much negotiation Mr. Thistlethwayte was able to purchase 172 acres of parkland, the lake, priory and golf course surrounding the main base, to which as the inheritor of the former owner of Southwick Park he would like to add the rest of the site. His plans for it are outlined in Chapter 3.[44]

Although Portsmouth and Gosport did not gain from the philanthropic endowments of public parks and gardens seen in other towns dominated by civilian industry, as government defence priorities have evolved, people in the harbour communities have benefited from the release of military and naval land no longer needed for national defence into public open space.

Sources

N Pevsner and D Lloyd 1967 *The Buildings of England: Hampshire* pp. 459-63

The Growth of Southsea as a Naval Satellite and Victorian Resort Portsmouth Paper No 16, July 1972

W Curtis 1978 *Southsea: Its Story*

Portsmouth WEA Local History Group 2000 *Memories of Southsea*

A Triggs 2001 *Sunny Southsea: Memories of a Golden Age*

Portsmouth City Council Planning Service 2001 *The Seafront, Southsea Conservation Area, An Appraisal, (Consultation Draft)* June 2001

I T and C Lewis *Island of Portsea Maps 1833* Portsmouth Records Service; Bacon *Map of Portsmouth, 1883* Portsmouth Records Service: E Stanford *Map of Portsmouth, 1866* Portsmouth Records Service listed in Historic England 2002 Southsea Common https://historicengland.org.uk/listing/thelist/lisBacon%20Map%20of%20Portsmouth,%201883%20Portsmouth%20Records%20Service%20t-entry/1001624

Southsea Common Sunday June 5th 1994. A National Service of Commemoration, Thanksgiving and Re-Dedication"

Historic England Southsea Common registered under the Historic Buildings and Ancient Monuments Act 1953 within the Register of Historic Parks and Gardens for its special historic interest. Name: SOUTHSEA COMMON List entry Number: 1 0 0 1 6 2 4

https://historicengland.org.uk/listing/the-list/list-entry/1001624

https://www.ticketmaster.co.uk/americas-cup

Portsmouth City Council 2008 Portsmouth PPG17 Assessment Final Report: January 2008

Southsea Sea Defences: http://www.portsmouthisland.uk/southsea-common-s-sea-defences.html For the Cathedral

Presentation: http://www.portsmouthisland.uk/southsea-report-launch-presentation.html For the Questions and Answers:

https://drive.google.com/open?id=0B9rRhLkxDuh0WHNhaHR2ZVFJeUk For the Facebook campaign: https://www.facebook.com/SouthseaSeafrontAtRisk/

"Victoria Park, Portsmouth history". Portsmouth Now & Then. Retrieved 2010- 03-02. "History of Portsmouth". Portsmouth City Council. Retrieved 2010-03-02. "Victoria Park,

Portsmouth". Welcome to Portsmouth. Retrieved 2010-03-02. "Conservation Area 18 - History" (PDF). Portsmouth City Council. Retrieved 2010-03-02.

"Conservation Area 18". Portsmouth City Council. Retrieved 2010-03-02. Headley, Gwyn; Meulenkamp, Win (1986). *Follies a National Trust Guide.*

Jonathan Cape. pp. 84–85. ISBN 0-224-02105-2.28

"Historic bell returned to Chinese home" The News November 14 2005 South China Morning Post A18 News/Features Saturday November 14 2005

PA Vine 1990 *Hampshire Waterways. Middleton Press* p. 104, pp. 119- 121 ISBN 0-906520-84-3.

Garry Mitchell 1988 *Hilsea Lines and Portsbridge pp. 26–30. ISBN 0-947605-06- 1.*

The History of Hilsea Lines Military History Trail A self guided tour. Hilsea Lines Ranger Service n.d.
"Historic Campaigns - Ports Creek". Inland Waterways Association. 2006. Archived from the original on 26 February 2008. Retrieved 13 September 2009. Portsmouth City Council planning service Student Information Sheet

http://www.escp.org.uk/coastal-schemes/portsmouth/anchorage-park.

Northharbour Land reclamation" (PDF). Portsmouth City Council. March 1999.

Garry Mitchell 1988 p. 3. Archived from the original (PDF) on 28 September 2011. Retrieved 6 October 2009 *Hilsea Lines and Portsbridge.* pp. 1, 2-3, 7, 10, 12-13, 14-15, 23-26. ISBN 0-947605-06-1

John Webb 1977 *The Siege of Portsmouth in the Civil War.* Portsmouth City Council. pp. 14–15. ISBN 0-901559-33-4.

William G Gates 1946 *The Portsmouth That Has Passed*; Nigel Peake ed. 1987 *The Portsmouth that has Passed: With a Glimpse of Gosport.* Milestone Publications. p. 33. ISBN 1-85265-111-3.

Robert Hunt 1862 *Handbook to the industrial department of the international Exhibition* Stanford. p. 355.

Mike Osborne 2011 *Defending Hampshire The Military Landscape from Prehistory to the Present*. The History Press. p. 127. ISBN 9780752459868.

Michael Bateman & Raymond Riley eds. 1987 Institute of British Geographers. Conference *The Geography of defence*. Routledge. p. 69. ISBN 0-7099-3933-7. Bateman, Michael; Riley, Raymond Charles; Institute of British Geographers Conference *The Geography of defence*. Routledge. p. 72. ISBN 0-7099-3933-7.

Anthony Triggs 2002 *Portsmouth Airport*. bookcraft. p. 5. ISBN 1-84114-153-4.

Jane Smith 2002 *The book of Hilsea Gateway to Portsmouth*. Halsgrove. pp. 34, 53, 70, 72-74 ISBN 1-84114-131-3.

John Slater July 2006 "*Hilsea lido area action plan – preferred options*" (PDF). Portsmouth city council. p. 6. Retrieved 2010-09-18.

John Webb, Sarah Quail, Philip Haskell, Ray Riley 1997 *The Spirit of Portsmouth: A History*. Phillimore & Co. p. 66. ISBN 0-85033-617-1

"Mystery of rising water in moat". *The News*. Johnston Press Digital Publishing. 13 May 2006. Retrieved 13 September 2009.

"2.0 The character of conservation areas" (PDF). Portsmouth city council. 25 October 2006. p. Retrieved 20 March 2010.

"Hilsea Lines named as a top green leisure spot". *The News*. 28 July 2010. http://www.portsmouth.co.uk/news/defence/world-war-one-remembrance-centre-to- open-on-portsdown- hill-1-4939367 Fort Widley WW1RC Launch https://historicengland.org.uk/advice/heritage-at- risk/search-register/list-entry/1691179 Heritage England At Risk Register Entry

Conservation Area No.27 Hilsea Lines (Scheduled Ancient Monument) Conservation Area Appraisal & Management Guidelines www.portsmouth.gov.uk

https://www.naturvation.eu/nbs/portsmouth/horsea-island-country-park accessed 15 June 2020

https://www.pennymordaunt.com/single-post/2018/08/14/New-country-park-could- be-a- new-jewel-in-the-crown-for-Portsmouth accessed 15 June 2020 https://uknip.co.uk/2016/03/29/

horsea-island-country-park-plans-move-a-step- closer/ accessed 15 June 2020

John Sadden 2010 *Hilsea The winds of change* The Portsmouth Grammar School Monograph series No. 21 Nigel Watson 2008 *Independent Vision A History of The Portsmouth Grammar School* James H James (Publishers) Ltd.

References

1. https://www.gosport.gov.uk/article/1721/Heritage-Action-Zone-HAZ- accessed 5 May 2020

2. *The News* 21 December 2018 ' An interactive map highlighting the potentially devastating effects of global warming to the world's coastlines has provided a scary insight into the potential future of Portsmouth

3. Rupert Booth 2020 'British coastal cities threatened by rising sea 'must transform themselves. Hull and Portsmouth could be dramatically remodelled, suggests report' https://www.theguardian.com/environment/2010/jan/15/hull-venice-north-coastal-erosion accessed 7 July 2020

4. Environment Agency *FCERM Investment Programme* 2017 Row 626; tps://southseacoastalscheme.org.uk/southseas-new-coastal-defences-gets-green-light-with-nearly-100m-of-government-funding/ accessed 19 June 2020

5. Taye Olukayode Famuditi 'Developing local community participation within shoreline management in England: the role of Coastal Action Groups' University of Portsmouth PhD thesis 2016

6. https://www.theguardian.com/uk-news/2018/aug/06/new-27m-blackpool-sea-defences-are-failing-councillor-claims

7. http://projectcompass.co.uk/index.php/2017/06/11/portsmouth-elephant-cage-reports-southsea-common-/; W Menteth Abstract + PAPER Living and Sustainability pp. 342-359.pdf

8. https://www.change.org/p/portsmouth-city-council-give-residents-a-better-choice-of-flood-defence-proposals-for-southsea-seafront.

9 Cllr. Donna Jones quoting Claire Upton Brown city planning officer Email to Celia Clark 4 January 2018

10 Fiona Callingham 'Sea defences cash backlash. Local Enterprise Partnership says money was not available for project' *The News* 5 October 2019

11 https://southseacoastalscheme.org.uk/southseas-new-coastal-defences-gets-green-light-with-nearly-100m-of-government-funding/ accessed 25 June 2020

12 Sarah Quail 2000 *Southsea Past* Philimore Publishing p.113

13 David Niven 1971 *The Moon's a Balloon* Coronet Books p.24

14 https://historicengland.org.uk/listing/the-list/list-entry/1001624;

15 "The golden age of the English seaside pleasure pier was c1860-c1910, and the majority of surviving (and listed) examples date from that period. Clarence is one of only two piers dating from the post-war period; the other is the entrance building to Boscombe Pier, Dorset (1958-60) which is listed at Grade II. Clarence Pier, alongside Boscombe, is thus of some interest in that it illustrates the 1950s revitalisation of the British seaside resort before it was eclipsed by the lure of the Mediterranean; the spirit of that time is captured in its jaunty roofs, futuristic tower and bright colours, all of which make it a striking feature of the Southsea shoreline. These features certainly lend it something of the American Googie style (a kind of futuristic, space age inspired style named after a mid-century Los Angeles restaurant) which has attracted something of a cult following internationally. British examples are rare however, and thus deserving of appraisal. Despite its rarity value, Clarence Pier does not meet the criteria for listing. The design as a whole is disjointed, lacking the wit and seamless panache of US examples. Two English building comparators falling broadly within the Googie genre are cited in support of listing: Boscombe Pier and the former filling station at Markham Moor, Nottinghamshire (1960-1), both listed at Grade II for their dramatic concrete plate roofs; the former in the shape of a cantilevered boomerang, the latter a hyperbolic paraboloid. Clarence Pier does not possess the interest of either building however; here the hyperbolic paraboloid roof is perfunctory, an embellishment rather than the building's defining feature. The folded concrete roof of the former restaurant, a form broadly associated with Festival of Britain architecture, which gained popularity in the 1950s for covering large spans, is striking but lacks the bold sculptural profile which distinguishes some of the most impressive examples of this roof type… Intactness is a major consideration in designating post-war architecture, and the level of alteration to Clarence Pier takes it decisively below the threshold for listing. The integrity of the original design, particularly the frontage, has been significantly undermined by poor-quality ad hoc alterations and extensive recladding, while few interior features survive. While of local interest, Clarence Pier does not have the special interest in a national context required for listing." English Heritage Advice Report 2 April 2013

16 https://www.portsmouth.gov.uk/ext/documents-external/dev-cons-area-10-guidlines-theseafront.pdf

17 *The News* 26 August 2016 p.6

18 *The News* 28 August 2017 p.6

19 http://shapingportsmouth.co.uk/conferences/shaping-awards;*The News* 21 December 2018

20 https://www.expressfm.com/news/local-news/victorious-festival-2020-is-cancelled/ accessed 7 July 2020

21 https://www.ineosteamuk.com/en/articles/204_America-s-Cup-World-Series-returns-to- Portsmouth.html

22 *An Earthly Fairyland. The Story of Southsea Rock Gardens and the Rock Garden Pavilion* published by the Portsmouth Society in 2013

23 Martin Easdown, Linda Sage 2011 *Piers Of Hampshire & The Isle Of Wight* Amberley pp. 17– ISBN 9781445603551

24 David Moore 2013 *Portsmouth Lines and Southsea Defences* Solent Papers Number 12 Gosport. ISBN 09570302-3-1

25 A Tale of Two Port Jewish Communities by T Kushner 2002 City of Portsmouth Records of the Corporation ed. Gates, Singleton-Gates, Barnett, Blanchard, Windle, Riley; *The Evening News Portsmouth* A Reprint of the Jubillee Supplement Issued on April 27m 1927 pp. 25-27

26 https://hansard.parliament.uk/commons/1984-12-19/debates/5e85aadc-5191-470d-b197-90126b9745d5/GloryHolePortsmouth(Dumping);

27 1,287 ha. pit. 70% Non Hazardous fill and 10% Harzardous, to a depth of 3.20m.https://www.escp.org.uk/sites/default/files/coastal_schemes/Contaminated_Land.pdf accessed 12 July 2020; https://www.escp.org.uk/sites/default/files/documents/6.1_Appendices_PICSS-Appropriate-Assessment_0.pdf; https://www.escp.org.uk/sites/default/files/coastal_schemes/TR005_Annex_4.pdfp. 21;

The nearest non-hazardous landfill at Ringwood was 40 miles and to the nearest hazardous tip at Purton Swindon was 90 miles. The cost of moving the waste was estimated in 2008 at £2,782,372. https://www.escp.org.uk/sites/default/files/documents/6.1_Appendices_PICSS-Appropriate-Assessment_0.pdf; the deepest water in the marina is 2.4m.

28 https://eoceanic.com/sailing/harbours/399/southsea_marina; https://www.premiermarinas.com/UK-Marina-locations/Southsea-Marina

29 'Historic bell returns to Chinese home' Matt Dickinson *The* News 14 November 2005 p.14; Art & Soul Traders Chinese Dagu Bell 2005 mark@arftul-lodgers.org; 'It's Labell Epic Chinese joy as relic found in Portsmouth park finally goes home after a century away' The Portsmouth and District Post – January 2006 Issue – Page 76-77; 'Hand-Over: A century on and Chinese artefact returns to where it was taken from. Bell missed meltdown and is on its way home' Neil Evans www. portsmouth.co.uk 15. 2005

30 'National Lottery Support Will Breathe New Life Into Victoria Park' Shaping Portsmouth Newsletter December 2019 portsmouthcc.gov.uk>-blog/2530-victoria-park-national-lottery-fund

31 https://www.architectsjournal.co.uk/competitions/victoria-park-portsmouth/10047497.article?blocktitle=competitions&contentID=25242 accessed 8 July 2020

32 *The History of Hilsea Lines Military History Trail A self guided tour.* Hilsea Lines Ranger Service n.d.

33 Garry Mitchell 1988 *Hilsea Lines and Portsbridge*.pp. 26–29 ISBN 0-947605-06-1

34 "Historic Campaigns - Ports Creek". Inland Waterways Association. 2006.

35 https://southseacoastalscheme.org.uk/southseas-new-coastal-defences-gets-green-light-with-nearly-100m-of-government-funding/ accessed 25 June 2020

36 John Sadden 2010 *Hilsea The winds of change* The Portsmouth Grammar School Monograph series No. 21; Nigel Watson 2008 *Independent Vision A History of The Portsmouth Grammar School* James H James (Publishers) Ltd.

37 Mick Morris 2019 email to Celia Clark 3 March

38 *Hilsea 125. 2010 – a year of events in celebration of 125 years of PGS Sport at Hilsea*

39 http://hilsea-lido.org.uk

40 Jane *Smith 2002 The book of Hilsea Gateway to Portsmouth Halsgrove p.60 ISBN 1-84114-131-3*

41 Conservation Area No.27 Hilsea Lines (Scheduled Ancient Monument) Conservation Area Appraisal & Management Guidelines www.portsmouth.gov.uk

41 https://www.naturvation.eu/nbs/portsmouth/horsea-island-country-park accessed 15 June 2020
https://www.pennymordaunt.com/single-post/2018/08/14/New-country-park-could-be-a-new-jewel-in-the-crown-for-Portsmouth accessed 15 June 2020
https://uknip.co.uk/2016/03/29/horsea-island-country-park-plans-move-a-step-closer/ accessed 15 June 2020
Pauline Powell Email to Celia Clark 18 June 2020

42 Hampshire's Historic Parks and Gardens 2007

43 https://www.portsmouth.co.uk/our-region/fareham/former-southwick-park-landowner-opposes-plans-for-thousands-of-homes-on-site-1-7574639

44 Mark Thistlethwayte 2020 Interview with Celia Clark, Martin Marks and Deane Clark Southwick Park Estate Offices 18 February

Chapter 18
The bigger picture; local lessons

As we said at the beginning, this book is intended as a contribution to the debate on how to achieve sustainable regeneration of these often complex and challenging sites, focusing on a single small area in southeast England that contains in microcosm both all the challenges as well as many ingenious new uses of ex- military land. First we consider the wider international picture, and then lessons from the local experience documented in this book.

In the UK, unresolved is the irreconcilable three-way split between incompatible objectives:

> • the Treasury and National Audit Offices insistence on sales achieving the highest price in the shortest time, as stipulated in by H M Treasury in 1992;
> • government objectives on sustainable development, regeneration of former defence-dependent communities, environmental protection, decontamination and conservation of natural resources and historic buildings, as set out by HM Treasury in 1996; and
> • local communities' need for urban and rural regeneration and economic development, bringing back into use these once wealth generating sites.

For once defence-dependent communities, base conversion is a profoundly important and symbolic land use exchange. But research into reconstruction – and in particular its implications for the historic defence estate – locally around Portsmouth harbour or nationally, is rare in the UK, though not in other countries such as the United States. Although UK base conversion may be intermittently highly visible in land use change and development policy debates, especially in view of their potential to meet national housing targets, administrative and academic interest is not perhaps sustained because of the long timescales involved. This lack of research into the conversion of airfields, depots, barracks, dockyards, training grounds and fortifications into civilian uses, especially successful examples, leaves significant gaps in useful knowledge about the effect of base closures and the prospects for civilian reconstruction. There is very little systematic analysis of what follows base closures or of the search for sustainable futures for property formerly dedicated to national defence - leaving communities, governments, developers and planners with untested land use configurations, partnership structures, and financing strategies.[1]

Mechanisms for wider dissemination of experience and good practice are rare, except for the US Association of Defense Communities, a national lobby to which active and closing site managements and local authorities belong. Members share information and experience of post-defence planning via a frequent newsletter and annual conferences as sites are closed and redeveloped - at which point they leave as they move on to civilian futures.[2]

Would a UK based good practice guide based on successful examples of sustainable regeneration, be useful? Wayne Cocroft, Historic England's liaison officer with the Defence Infrastructure Organisation believes that UK defence sites are too diverse for useful communication between parties involved in their reuse. "The sites are so mixed in character and the outcomes so different, from total demolition to full conservation. Uses also vary from the very high value to sites returned to agriculture."[3] However, this is also true of the immense landholdings of the US Department of Defense.

The EU has linked particular ex-defence-dominated places to share experience, yielding examples of good practice, although the Portsmouth harbour communities have not gained from this opportunity. Examples are *ASCEND: Achieving the Socio-Economic Re-use of Former Military Land and Heritage. Model Management Framework*. Another was *Regeneration through Heritage. Understanding the Development Potential of Historic European Arsenals* in which English Heritage was a partner. The EU MAPS (Military Areas as Public Spaces) Handbook /A journey on the re-use of the former military assets based on a two year project linking eight European locations keen to reuse their forts, barracks and military lines is available at http://mapsnetwork. wordpress.com. Celia Clark was a keynote speaker at the final workshop in 2018 in Serres, Greece.

International conference series have documented and examined accumulated experience and good practice, but they are only one-offs or short-lived events. Examples are the biennial Wessex Institute of Technology's three *Defence Sites: Heritage and Future* events in Venice, Portsmouth, and Alicante from 2012-6 and the larger *Military Landscapes* conference organised and published by the University of Cagliari which attracted participants from South Korea and India in 2017.

Government state land disposal procedures vary from country to country and from free transfer to local interests – the US – to sale to the highest bidder – the UK and Germany. This variation directly affects the land use outcomes, and also the extent to which the local community gains from new jobs, cleaned land, reused buildings including those with heritage protection, new open space, housing, new educational and cultural facilities. What might be done to redress these gaps in accessible research needs to be addressed.

In the US surplus service homes are gifted free to the homeless and military hospitals and schools are transferred to local communities. Where there is sufficient local capacity, the whole process is controlled by a locally appointed Base Reuse Committee, whose remit is economic reconstruction. This is financed by the Department of Defense's Office of Economic Adjustment, which as far as is known has no equivalent in other countries – rather than being sold at maximum financial value. Celia Clark's study in 2000: *Vintage Ports or Deserted Dockyards: differing futures for naval heritage across Europe* found that in Sweden local authorities could acquire sites at current use value, consult widely so that the new uses matched local needs and aspirations, and sell, retaining the profit from increased value. The French had an independent agency to handle the sale of public land, and also gave it away to local groups. Perhaps we had something to learn from these examples.

Cross cultural research examining different countries' practice in disposing of redundant defence property in relation to other government objectives and evaluation of how useful exchanges of experience between different formerly defence-dominated communities have been would do much to inform and improve this on-going process towards sustainable regeneration.

The Portsmouth Harbour experience

Spithead as a sheltered anchorage for national fleets based in the dockyard in Portsmouth harbour and the many defensive structures developed to supply and protect it together made up a complex system, much of which is now redundant. As we have explored, in the twentieth and twentyfirst centuries many of these specialised sites and structures became obsolete and in need of new civilian life. We have explored why military sites close, where the local ones are, the challenges they present, as well as who has the power to influence what happens following closure.

As a hymn to Portsmouth: our home harbour and its communities we have drawn on many local people's accounts of how they have participated and shaped places in our area over the last half century - and how they continue to find new life for its diverse defence legacy. As a particular category of brownfields, once the many challenges they present are overcome, their potential contribution to local reconstruction is recognised. Among the many achievements we celebrate are examples with a wider significance, relevant to other once defence-dominated areas facing similar challenges of regeneration. Here we draw out lessons learnt from this local experience – and also offer more general guidance for this widespread but under reported process.

The sheer density, variety and complexity of defence sites around Portsmouth Harbour pose significant challenges to developers, local planning authorities communities, non-profit trusts, special interest groups and individuals who play a part in determining their future civilian roles. Secret facilities unknown to their host communities, missing infrastructure connecting them to their civilian surroundings, significant military contamination, aging listed or scheduled fabric and intangible heritage: local attachment to the armed forces - all these factors affect this complex transition. From the late nineteenth century to the present day the process of land release around the harbour to civilian reuse has gradually accelerated as the nation's armed force capability changes and reduces over time.

While Portsmouth remains the country's premier naval base, obsolete sites' density poses a challenge to local communities, planning authorities and developers' ingenuity and vision. There is clearly a public expectation that the new land uses will be of benefit to formerly defence-dependent communities – but whether this happens, how it is defined and how it is measured is still unclear. We suggest that in these cases sustainable regeneration brings beneficial local outcomes: land cleaned to the appropriate level, opportunities to develop new employment at similar or higher skill levels than those lost, new housing to meet social needs, new cultural facilities, accessible open space, reuse of buildings and sites for social purposes such as education, research, specialised industry....

This local context offers examples of vision and innovation in reuse by individuals and groups; pressure tor positive change by specialist interests; conservation plans by local authorities in anticipation of disposals; intractable and contaminated sites reused as marinas or new parks; civilians living in barracks or students learning in them; research into both the past and the future: new weapons systems, land and sea archaeology, conservation policy, historic objects and sites - and land use change achieved by considerable investment by developers.

The book was written inside a moving picture. We have documented changes as they've happened. The Defence Infrastructure Organisation's disposal list, which has on it many important local sites such as HMS Sultan and Southwick Park is subject to change and revision, which makes their inclusion into local plans and local responses about what new activities might be developed difficult. In 2020 an integrated defence and foreign policy review is likely to involve significant defence cuts was on the cards. Further land releases were expected as a result.

Decay, maintenance building and nature conservation

Decay from lack of maintenance of historic buildings leads to escalation in reuse costs, as exemplified in key buildings in the dockyard, Daedalus and Forts Rowner, Elson, Widley and Purbrook. Some forts have been abandoned to "controlled ruination". As time passes, memories of their military and naval past are lost and their historic significance fades. Use of authentic materials at Southsea Castle and Hot Walls Studios reflects the evolution of building conservation practice, while failure to use robust materials is apparent in the contrasting state in Gosport and Portsmouth of the Millennium Walk linking the defence sites around the harbour.

When unused sites are left undisturbed, conflicts may arise between natural and building conservation. In January 2020 Portsmouth and Gosport were identified by the National Diversity Network as some of the least biodiverse places in the UK. Local development results in sites being devoured by new roads and housing, as at Stokes Bay Lines and Fort Gomer. In Portsmouth's case the lack of diversity is because so much of it is developed, but as the Hampshire and Isle of Wight Wildlife Trust says, the built environment has gardens and public open spaces [including those gained from the MOD] which play a part in the network used by nature.[4] They were also essential in maintaining local people's physical and mental health during the months of lockdown.

The challenge of decontaminatiion

Tipner was proposed as a housing site to meet national targets because Portsmouth, as an island, has little developable land, but there are serious drawbacks including historic contamination to developing housing there. But another severely contaminated site: Eastney's Glory Hole had been transformed into the successful Southsea Marina, and the leadership and vision of the Portsmouth Naval Base Property Trust with developer Elite Homes continues to transform the contaminated former ordnance yard at Priddy's Hard in Gosport.

Local government, public consultation, civic societies and the DIO

Public consultation on land use change has evolved – in both methods and timing, across a spectrum from being genuinely responsive to local opinion to consult/ignore...as exemplified at Gunwharf, Haslar Hospital, Royal Clarence Yard, the interconnectors near Daedalus and Fort Cumberland for energy transfers between the UK and France, and the Southsea sea defences. In response to rising sea levels the proposals for Southsea's sea defences affected several scheduled military monuments, listed buildings, conservation areas and Southsea Common, a listed historic landscape kept clear of development for defensive reasons. Although Southsea Seafront Campaign's petition for a choice of alternative designs and their request to be directly involved in the design were not responded to, they successfully persuaded the engineer designers not to construct a high wall along the seafront. Instead the beaches were to be widened and rock armour added. Community Stakeholder meetings of interest groups were held over the long period of design, but were largely top-down and limited to group representatives, not open to participation by the wider community.

Gosport's conservation officer far-sightedly anticipated releases of defence facilities at Fort Blockhouse, Haslar Hospital, HMS Hornet and Royal Clarence Yard. Working with the Defence Estates/Defence Infrastructure Organisation, he produced conservation guidelines for these sites before their disposal, which facilitated their redevelopment. Gosport Borough Council as a small planning authority continues to be an active shaper of its historic defence legacy in close partnership with the MOD Defence Infrastucture Organisation, Historic England, Hampshire County Council and the Gosport Society in the first Heritage Action Zone, whose aim is to bring more defence heritage sites into active use. The community consultation integral to the process is a welcome development towards a more community-friendly direction. The local civic group the Gosport Society, with its unrivalled local knowledge and long experience of public participation in planning is a full partner. This partnership is a model, offering inspiration to other local initiatives to revalue their military heritage.

But while the DIO is working with locals in Gosport, its opacity about when Southwick Park (formerly HMS Dryad) would be released hindered new visions by Southwick Estates for the future of Southwick House and its historic landscape, despite the anticipated completion of the new £250m tri-service MOD College of Logistics, Policing and Administration at Worthy Down near Winchester in 2020.

More than 20,000 people marched through Gosport to protest about the closure of Haslar Hospital. The MOD paid for the Enquiry by Design on its post-defence future but no conditions to implement its findings were stipulated by Defence Estates when it was sold. The fully costed Veterans' Village proposed uses for all the buildings on the site, but the hospital still closed - and in the redevelopment only well off veterans are likely to be able to afford to live there.

Traditionally, local authorities have been crucial in shaping the planning and development of their local communities, developing plans which respond to local aspirations and needs for housing, employment, leisure facilities and open space, ensuring high environmental and design standards and well-restored historic assets to attract inward investment. As a consequence of austerity and longterm reductions in local government finance there has been a diminished capacity to manage change in the historic environment and a weakening of councils' ability to achieve benefit to their local communities in these unusual redevelopments. Cuts to councils' budgets were compounded during the lockdown phase of the 2020 pandemic by difficulties in setting up remote working. There were reductions and even a cessation in local planning activity in Portsmouth for several weeks as a result. In addition, expert in-house conservation advice was no longer available in Portsmouth from February 2020, when the conservation officer resigned. The Institute of Historic Building Conservation reported in its response to the Planning White Paper that since 2009 there had been a 48.7% fall in specialist conservation provision in local authorities, while 6% of councils had no access to advice. This is the critical frontline for protecting the historic environment and once lost it can never be recovered. Listed building owners and developers were no longer able to get detailed and proactive support and advice from local authorities to help them maintain and protect the heritage for which they are responsible.[5] The pandemic also resulted in considerable losses to local authorities which had invested in property – shopping centres, offices, airports - to bring in an income, as these were inactive during the lockdown.

Civic Voice, the national charity for civic societies with over 76,000 individual members expressed serious concerns about the 2020 Planning White Paper, in particular the removal of the opportunity to comment on planning applications, and the emphasis on growth over place-making. It accepted that the current planning system is not perfect, and that it had become over-complex and was not providing enough high-quality places. This frustrated local communities and led to a breakdown of trust in the system, with just 2% of the public trusting developers and only 7% trusting local authorities, when it comes to planning for large-scale development. It acknowledged that the current system was not working for communities, who felt that the balance of power had swung too far towards central government and developers. So Civic Voice welcomed the opportunity to consider needed reform. It emphasized how much planning matters. The system currently put important safeguards in place to prevent poor development outcomes that could have a lasting detrimental impact on people's lives and on our built, historic, and natural environment. It had a critical role in balancing social, economic, and environmental goals. Developers and local authorities acknowledged that positive community input through the planning process results in better outcomes. More traditional methods of communicating such as site notices and face-to-face discussions should continue, to complement more digital methods of engagement. (Civic Voice Response to the Planning White Paper October 2020).

Defence Heritage Revalued

Individuals' vision and patronage led to successful developments such as Defence of the Realm, the housing of the Overlord Embroidery in the rapidly designed and built D-Day Museum, the naval base Heritage Area's enhancement by the arrival, docking and restoration of the *Mary Rose*, *HMS Warrior* and *M33* as well as smaller vessels – all of which have enriched the harbour's defence heritage.

In the dockyard, heritage tourism accompanied by contemporary naval activity is a huge success. The Historic Dockyard has developed as a major tourist magnet; until the corona virus struck, it drew in thousands of visitors whose spending is a significant part of the local economy. The dynamic Portsmouth Naval Base Property Trust continues to transform the Heritage Area into a vibrant and enjoyable celebration of the navy's role in defending the country, with the help of large grants from the National Lottery Heritage Fund and other funds. More recently the trust took on the regeneration of the ordnance yard: Priddy's Hard across the harbour, adding a ferry pontoon, planning new uses and activities, financed by residential development. The Royal Naval Museum evolved into the muti-faceted National Museum of the Royal Navy, extending a wide reach over the harbour's defence museums and historic ships and boats. It is now responsible for the Submarine Museum and Explosion in Gosport as well as other naval heritage sites further afield.

Architects, surveyors, engineers and builders continue to transform defence sites around the harbour. Builders have had to tackle inaccessible sites such as the forts in the sea. Architects and surveyors play a critical role in several ways: in assessing the condition of historic fabric, in recommending viable new uses for it, in using their design skills – to restore the dockyard boathouses and storehouses, recreate the clock towers of the Vulcan and Storehouse 10, convert the sea forts and Portsmouth's Hot Walls, and provide community planning events whose findings were collated into collaborative masterplans which help to provide developers with planning certainty.

Positive factors that result in sustainable reuse include risk-taking/entrepreneurship. Developers taking on the challenge of investment in redevelopment of complex heritage-laden sites are essential to the regeneration of the harbour. The first developer at Eastney Barracks was a pioneer in barrack conversion, but several residential schemes for converting Gilkicker failed. Mike Clare's personal passion drove his investment in Spitbank and NoMansLand forts. Berkeley Homes took on and transformed Gunwharf and Royal Clarence Yard, predominantly for private housing, retail and leisure. Sunley Estates converted the two wings of St. George Barracks into houses and flats. National Regional Property Group proposed to convert the older buildings and add new houses at Fraser Battery.

English Heritage/Historic England's Portsmouth Hinterland project prepared the ground for Gosport's Heritage Action Zones, helping local groups to understand the significance of local historic defence sites and facilitating their plans to restore them and achieve their objectives for reuse. Their plans for reassembling Brunel's first surviving bridge, stored in Fort Cumberland are still awaited. As his birthplace, a Portsmouth location – for example adjacent to the Kings Bastion in Old Portsmouth - has been proposed.

Where innovation existed when a facility was first built is now matched by creativity in reuse this is particularly welcome. QinetiQ continues at the forefront of technological development in hull design at Haslar and weapons and communications on Portsdown Hill, while innovation in defence heritage reuse is demonstrated at Fort Cumberland. Historic England's vision is to restore the fort in partnership with local people. Still in public ownership, the aim is to develop it a vibrant community which rehabilitates veterans damaged by their service in the armed forces, alongside a mix of conservation policy generation and land- and sea-based archaeology. Its occupants are restoring the fort's decayed fabric: veterans, craftspeople, architects and distillers. This exemplar looks likely to be a beacon of how to meet the challenge of bringing a largely unrestored obsolete fortress to new life as an incubator for local enterprise.

As well as defence museums, connection with the armed forces continues at ShoreLeave at Haslar Hospital and Forgotten Veterans at Fort Cumberland. One original military use has continued as part of Councillor Peter Ashley's vision of using two of the Portsdown Hill forts for young people's activities. When the Activity Centres named after him took on the forts the stables which once housed soldiers' horses in Widley continued in their original use as an Equestrian Centre. Looking after the horses and ponies there offers local people work experience as well riding. Eastney swimming pool once used to train sailors to swim is now used by civilians, and Gosport's ramparts, Lumps Fort, Stokes Bay Lines and Hilsea Lines are much enjoyed public open space.

Volunteers are vital to preservation and interpretation of the harbour's defence legacy. The determination of the diving community to restore and convert No. 2 Battery in Gosport as a Diving Museum has deservedly won national museum accreditation. The Hovercraft Museum has restored and researched the pioneering development of civilian and military hovercraft in part of HMS Daedalus.

Education at all levels has taken over former defence buildings, from the nursery in St. George Barracks, Bay House, Portsmouth Grammar School and St. Vincent College to Portsmouth Polytechnic/University which has reoccupied the space of the cleared fortifications. It has also significantly replaced the contribution of defence activity to the local economy.

State of the national and local economy

The state of the local and national economy affects the timing and land uses achieved by redevelopment. Decisions taken by developers such as Land Securities, owners of Gunwharf are affected by national trends, for example, shopping, now being concentrated onto fewer sites, while shrinking and disappearing in traditional shopping areas. Covid-19 sharpened debate about the purpose of physical space, proving an accelerator to the debate about the future of physical retail, which had already led to the long decline of shopping in Gosport High Street and the closure of Southsea's two department stores.

Funding

Funding for change is critical. Philanthropy and the Heritage Lottery fund have played a large part. Sir John Smith paid for the rescue and first phase of the restoration of *HMS Warrior* and Sir Don Gosling contributed large funds to the restoration of *HMS Victory*. The Heritage Lottery Fund/National Lottery Heritage Fund has offered very large sums for reuse of the harbour's many historic buildings: for the Mary Rose museum, the D-Day Museum, Boathouse 4, the Diving Museum… But bids were not always successful, especially since fewer people were buying lottery tickets so less money was available. The failure of the initial NMRN bid for Royal Marines Museum transfer from Eastney was a case in point. The complexity of lottery applications relied on specialised expertise that must also be paid for.

Volunteers who take on military sites with plans to open them as museums find funding especially difficult. Bids to finance private museums are sometimes constricted by legal requirements for longterm occupation in addition to the oversubscription for Lottery funds. The contrasting experience of the Diving Museum and the Hovercraft Museum show how critical this is. The Diving Museum had agreed long term site occupancy and stability with the local council owners, whilst the latter's plans and ability to bid for grants were blocked by uncertainty over site development and the lack of a longer term lease. The Peter Ashley Activity Centres need a full repairing lease before they can apply for the substantial grants needed for restoration of their two forts.

Together all these strands – and many others – make Portsmouth Harbour the dynamic place it is to live in. Responses to sea level rise will be tested.

As a port Portsmouth has always looked over the horizon to the world. The supercarriers are stationed there to intervene in wars and protect trade routes, while freighters to the ferry port keep the country supplied with food and ferries carry goods and people across the Channel. Cultural heritage is now widely appreciated as an essential part of countries' society, environment and economy. Heritage is not just for elites, but essential for the life and survival of society.[10] Portsmouth Harbour's communities are fortunate in the rich physical legacy surviving from their long role in defence of the country and can be justifiably proud of what is being achieved in its civilian reinvention.

The case for changes to the system of disposal of public land

On the wider issue of reform of the state property disposal process in favour of more locally positive and sustainable outcomes, Brett Christophers' account of how much has been lost by disposals of property owned by the state is a salutary analysis of what is wrong with the current system.[6]

The detailed recommendations for a more community friendly procedure for defence disposals from the regional seminars which drew on local experience held in 1996, 2002 and 2017 are still valid. They relate to different stages in the process. As these sties are so complex, **a longer timescale for their regeneration** is critical to success.

Better care of historic defence buildings and structures

Historic England's Heritage At Risk Registers are intended as a spur to action – to renewed maintenance and to searches for new uses – disposal to the civilian property market or transfer to civilian bodies as community assets. Many defence sites particularly in the southeast, are on the Register. It is not acceptable for the MOD to state that 'austerity measures will continue to provide challenges for MOD heritage management. The effects are already being experienced with a decline in the condition of listed buildings and the scaling back of condition assessments as a result of budgetary constraints.' 7 With Crown immunity removed, the MOD must comply with planning statutes. Doing nothing for example with unused historic buildings should no longer be an option.

Neglected maintenance has allowed the condition of listed and scheduled structures to deteriorate, so eventual remedial costs will escalate. As a stitch in time saves nine, **funds allocated to maintenance of unused defence property are a worthwhile investment in their potential for future use – by the MOD or subsequent owners.**

Decay in key historic buildings in Portsmouth naval base are of particular concern. Vacant buildings are at greater risk of deterioration than occupied ones, where problems are more likely to be addressed before they become critical. Repairs to vacant buildings should be given due importance and allocating a risk category may aid this.8 The effect of deferring work, causing 'structural or weather tightness issues' and 'fabric deterioration', should be considered when deciding priority and urgency where buildings are vacant (as with all the structures on the At Risk Register). These structures may exhibit 'severe' vulnerability due to being in a 'coastal/high rain" area, which is clearly the case with Portsmouth dockyard. Although their state of disrepair may be discussed by the DIO and the local authority conservation officer, these negotiations were not in the public realm. Perhaps they could be reported to planning committees?

There needs to be a mechanism to **enable enforcement against neglect of listed buildings regardless of ownership.** The exclusion of active defence sites from local authorities' powers to issue Urgent Works and Repairs Notices requiring repairs to decayed historic buildings needs to be removed, especially where the MOD have no use for them, and they have long been empty and unused. The removal of the Crown Exemption was a precedent.

These buildings need maintenance – or disposal to new owners for reuse. Sites left empty for too long whilst MOD makes up its mind, lead to significant deterioration – for example, the Naval Academy. **The crucial route to sustainable regeneration for them and others on the register is for timely new and appropriate uses to be found, so that operational budgets also finance conservation.** This is the best way to secure their conservation and future.

Understanding a historic site is an essential stage in determining its sustainable future. Its setting, plan form and layout, condition, building materials and architectural features need to be taken into account. Priority should be given to retaining and enhancing its local character and distinctiveness and to enhancing its historic setting. The more significant a heritage asset, the greater the weight that should be given to its conservation and its capacity for change, and to the amount of detail in a planning application. Local planning authorities can assist developers' understanding these issues. This understanding, also enriched by **Archaeological Management Plans** and **Conservation Management Plans**, should be used to inform the constraints and opportunities available. Historic features should be retained where possible. **A condition report**

and **artefact survey** were carried out to identify the significance of Point Battery Portsmouth.

Specialist defence structures are particularly difficult to find sustainable new uses for. Gosport examples currently at risk are the listed Submarine Escape Training Tower and the Cavitation Tunnel. The Gosport Heritage Action Zones are a welcome development, funded by Historic England in partnership with Gosport Borough Council, Hampshire Cultural Trust and the Gosport Society

At **site level** the issues that arise are: urgency versus resource constraints, multi-designations (scheduled/listed) adding to their complexity, flood risk, coastal erosion, contamination, poor access, depressed land values, the unique and complex form of many structures, the difficulty of unravelling sites' significance, the many Buildings at Risk, the challenge of accessing funds and the failure to recognise the economic value of heritage. As a peninsula, Gosport has the added disadvantage of poor transport infrastructure.

Before release at local as well as national level – regular communication between the DIO and local authorities via MOD local community liaison representatives is essential to discuss proposed closures and disposals. This needs to be early, clear, transparent and timely - in order that local authorities, communities and developers have time in which to respond positively. As they once did, regular **meetings between local planning authorities and the MOD/Defence Infrastructure Organisation** need to take place - at sub-regional level, also involving local agencies such as the Partnership for Urban South Hampshire, to address disposals, the economic aspects of Heritage Assets and their sustainable reuse.

There should be a robust, transparent and accountable **developer selection procedure**. Treasury interests should not dominate issues of the national heritage. There should not necessarily be a presumption in favour of clearance and redevelopment. The MOD's preliminary valuation and shortlisting of bidders before the sale should be more open and transparent.

From the **developers'** point of view, historic military sites need a different/open approach in terms of planning and conservation requirements. According to Dave Craddock of Elite Homes which is a partner of Portsmouth Naval Base Property Trust at Priddy's Hard, a particular challenge is to overcome the protectionist approach that the very many consultees seem to have. These may include Historic England, Natural England, the Environment Agency, the Highways Agency, national amenity societies such as the Georgian and Victorian Societies and the Council for British Archaeology, County Ecologist, County Archaeologist, Local authority Conservation Department, Planning Officers, Council Planning Committees, Parish Council, local residents and organisations: a formidable list.

Local Planning Authorities need to be properly informed as to the significance of a site's heritage assets and landscaped setting and must stress the importance of 'front-loading' detailed site and building appraisals. **Frequent site meetings for major sites are necessary** and this needs resourcing. Excellent practice is the close supervision by Gosport's Conservation Officer of large sites such as Haslar Hospital via two weekly meetings on site which saved time and paper trails.

There is a need to look at 'Bigger Picture' benefits rather than specific losses to historic elements of buildings and landscape. The onus should be on creating **partnerships with owners of Historic Sites to facilitate best practice and sustainable design,** and to deliver appropriate density to create best value. Good Design and a creative approach are key to success. "We must breathe new life into unused Military Historic Sites by working in a collaborative and proactive way to deliver the best possible outcome for all parties" according to David Craddock. The **cost of enabling development is substantial** – and the developer can't pay all. The economic drivers need to be understood. Developers can't take on a site with indefinite costs, and **fixed site contracts are not possible on historic sites.**

It is also important to identify who is to pay for **maintenance and restoration of infrastructure** such as dock walls, culverts, basins, caissons, cranes, water, electricity and sewerage services. These may need to be separately funded via a sinking fund, which service charges to the new occupiers would not cover.

There should be **research and evaluation of heritage issues** by suitable independent bodies before the disposal process is implemented and an **independent assessment of development briefs** before the disposal process finalised. Conservation area status and listings should be open to public debate. **Section 106 Agreements** may be used by local authorities to secure funding from subsequent sales of parts of sites by the new owners to be set against the costs of conservation - as was done at Haslar Hospital. Any Section 106 Agreements involving historic sites should be approved by Historic England.

There should be **genuine involvement of local communities** which should begin as early as possible in the process, rather than waiting for the conventional planning process to begin. At the planning stage genuine community consultation – bottom-up as well as top-down – needs to be built into the regeneration process, especially where no public access was previously available. Methods include Heritage Open Days, site visits inviting feedback, Community Planning Events, Enquiry by Design, Planning for Real, public exhibitions... Local residents are resistant to change. Developers who gain public support for their proposals by genuine consultation benefit from faster and less contested process toward planning consents.

Developers who do not already acknowledge public involvement's importance need to learn the value of early local input and genuinely responsive consultation in easing their path to planning permission, and the MOD should consider working with expert bodies such as English Heritage with a view to drawing up **conservation management plans for historic sites** before disposal. Gosport's conservation officer's work with the MOD in preparation for releases stands out as a shining example of how important this is. Since early 2020 Portsmouth no longer has a conservation officer, which is of considerable concern when heritage tourism focused on the historic landscape is a major element of the local economy.

In the **design phase** the key to the success of the conversion of the Battery in Broad Street Old Portsmouth into artists' studios and café was striking a balance between the provision of modern fixtures and conveniences for a variety of contemporary uses, and the restoration and preservation of the historic fabric of the building. To this end, the design was driven by a focus on reversible and non-destructive work rather than material alterations. This essentially allowed (should the need arise) for the removal of any additions and the reversion of the structure to its current state with minimal visible changes.

New buildings should be sited so they are sensitive to the historic plan form of the site and its wider setting in the landscape. Enabling development should be considered in order to secure the future of historic buildings of high significance and sensitivity to change. Short-term solutions might include mothballing, temporary uses, carrying out urgent repairs, securing it and protecting the property from fire...

The way forward: in order to achieve greater collaboration and understanding in order to achieve the best out of every site, all parties must appreciate and consider other consultees' and stakeholders' positions as well as their own. A loss in one area can and should be a gain in another. To constantly expect the developer to bear the costs and to take all of the risks will mean fewer sites like these will get brought back to life. There needs to be appreciation that the costs of enabling these developments to go ahead arising from abnormal conditions is substantial and that something has to give to make it possible to pay for the long-term regeneration of the historic elements of the sites. **A positive and constructive approach to conservation is key.**

Master planning of large ex-defence sites responsive to their history and historic layout in accordance with local authority local plans and economic priorities - is a useful process in determining sustainable reuse. Other plans and supplementary planning guidance need to be taken into account. An example is the ARTches Project in Broad Street Old Portsmouth which was set out in the Seafront Masterplan Supplementary Planning Document of April 2013. Re-creation of lost employment including work using specialist high skills should be a priority in redevelopment as well as housing. CEMAST College, Innovation Centres, Solent Airport, business development are all positive examples of what can be done on ex-MOD sites.

Hybrid Outline Consents may be a useful tool. It is important to promote the intrinsic value of large military-heritage sites to the wider community, and their economic potential. For complicated, multi-phased redevelopments it is not reasonable to expect the developer to know precisely when each building will be tackled. Mass, form, layout, texture need to be considered before giving consent, then dealt with on a detailed basis, phasing the work on a critical path. Phasing also helps developers to secure and fund the reuse of large sites. Housing development should always be closely related to transport, education and social facility planning. Experiments in sustainable redevelopment such as Eco-towns may be appropriate to the redevelopment of ex-defence sites. Both national and local defence museums contribute substantially to the local economy.

In the **construction phase** contamination and pollution need to be addressed during redevelopment. Much of Daedalus soil is contaminated with oils, asbestos, aircraft solvents etc. The plan is to seal it in. Building dust/debris problems during redevelopment need much earlier, tighter control. The Dust Management Plan associated with the IFA2 Interconnector Converter was far too late with too little enforcement. It recognised Daedalus as a 'High Risk' site with 'Sensitive Receptors' (i.e. neighbours!). Limited water supplies in some parts of the site was not considered.

Funding: finding the financial resources for conversion is crucial. The Grade I listed building and scheduled monument and its setting at Point Battery in Old Portsmouth is owned by the city council, who acquired it following it decommissioning in 1958. Standing mostly empty for the past fifty years, maintaining its current condition became untenable without the means of it generating its own revenue. The council refurbished the building with the intention of creating a new artistic and cultural hub, thus securing a sustainable future for it.

Ways forward

Research about defence disposals and sustainable redevelopment in the UK and other countries is deposited in the Portsmouth History Centre of Portsmouth Central Library. This database, being developed with Portsmouth School of Architecture Conservation studies, is available for students and other researchers, to learn from experience and good practice.

Clawback where the developer pays the Treasury a proportion of the profit subsequently made over the first purchase price has operated several times at Gunwharf and elsewhere. These moneys, generated in the southeast region, might usefully fund a **MOD Conservation Group.**

At regional and national level the Local Government Association could help to **share experience between local authorities and local community groups** – perhaps via a dedicated website or e-publication. Publicity in national and local media is also important to publicise the effects of defence cuts on local communities as well as accounts of positive outcomes. A campaign is needed to win over hearts and minds to the potential of defence sites for local regeneration. As mentioned earlier in this chapter, consideration should also be given to setting up a **national group of all sectors concerned with defence sites**, on the model of the USA Association of Defense Communities, to share experience and to influence government. A national forum for all participants in the regeneration process would be useful to share experience and identify good practice. Methods might include a dedicated infranet, website, publications, regular seminars and conferences.

Policy recommendations based on findings by the speakers and seminar participants at the third seminar in October 2017 were intended to be presented to the **Defence Select Committee** - with a plea to the MOD Defence Infrastructure Organisation to institute a more orderly and locally responsive disposal process; to address neglected maintenance of historic defence buildings, and to encourage new sustainable uses for them, via the then chair of the Defence Select Committee: New Forest MP Julian Lewis. In 2019 the clerk to the Defence Select Committee responded to the chair of the Hampshire Buildings Preservation Trust that they were too busy to take the seminar's findings on board. Local MPs offered to raise the issue again.

The defence buildings this book is about were built of predominantly local materials with public money, mainly by generations of Portsmouth and Gosport craftsmen, so that Portsmouth Dockyard and its associated facilities could build, repair and supply ships. In the view of the Naval Dockyard Society they constitute public heritage, therefore the local authorities should be taking a leading role in their conservation. For the named structures to be restored to a useful condition, they should receive more of the operational naval base budget, with a higher level of annual maintenance than at present: quadrennial inspection reports on all listed buildings and quinquennial inspection reports for all scheduled monuments.

In tracking a moving picture, we are aware that there are gaps and mistakes in our account, which we acknowledge and would be happy to have corrected. We want to give particular thanks to the many people who have contributed to this book. Their memories bring alive the complex transition from military life to civilian reinvention in a small area of England, but, hopefully with lessons applicable to the country as a whole.

Accumulated local experience in reusing defence sites around Portsmouth Harbour has much to offer, but in order for it to be disseminated, there need to be greater opportunities to learn about it – not only locally but also with chances to draw on national and international experience.

As Emily Gee of Historic England said at the launch of Gosport's first Heritage Zone focused on historic defence buildings:

"We all have to work together to come up with interesting uses for these sites. There's something really quite special about seeing these lovely buildings come to life in unique and exciting ways, as well as spending time inside the buildings themselves. Every building had its own stories to tell, so it's crucial to give people more access to that... the more accessible they are, the more people will take an interest, and I think that lends itself to people caring more about where they live. It draws people in, without simply turning an area into one big museum; instead let's make our heritage a living, breathing space to explore and learn about."

References

1 Samer Bagaeen and Celia Clark 2016 *Sustainable Regeneration of Former Military Sites* Routledge p.1; Celia Clark 2000 *Vintage Ports or Deserted Dockyards: differing futures for naval heritage across Europe* Working Paper No. 57 115 pages Research Consultancy for University of the West of England Bristol ISBN I 86043 281 6)

2 https://www.defensecommunities.org

3 Email from Wayne Cocroft Historic England to Celia Clark 16 September 2016

4 Fiona Callingham 2019 'Calls for people to 'play their part' as Portsmouth revealed to be one of the least biodiverse places in the UK' *The News* 3 January 2020

5 Institute of Historic Building Conservation 2020 Local authority conservation capacity research for Planning White Paper Briefing to MPs and Peers
https://newsblogsnew.ihbc.org.uk/?p=28145 accessed 8 November 2020

6 Brett Christophers 2018 *The New Enclosure. The Appropriation of Public Land in Neoliberal Britain;* Hampshire Buildings Preservation Trust/Royal Town Planning Institute South East 2017 Sustainable Regeneration of Former Defence Sites. Policy Recommendations

7 *MOD Heritage Report* 2011–2013, 2014, para. 41

8 *Managing Heritage Assets* Historic England, 2009 pp.30, 31

Appendix

Futures for redundant defence sites: a research agenda
Dr. Celia Clark

www.celiaclark.co.uk

Abstract

The complex transformations of former military sites to new civilian life are rare subjects for research. There are many unanswered questions. Why do governments close them? How do different countries dispose of their surplus defence facilities? Who is responsible for cleaning contamination? Who influences the transition to civilian uses? What are the expected timescales between redundancy and regeneration? What new land uses result from the process? Who benefits and who loses? How can this major change from military to civilian activity be measured so that experience and good practice shared across different cultures and political systems?

There is little systematic analysis of the search for sustainable futures for property formerly dedicated to national defence - leaving communities, governments, developers and planners with untested land use configurations, partnership structures, and financing strategies. There is an expectation that the disposal of publicly owned land should result in public social, economic and environmental benefits, but these are not likely to happen if the land is valued in solely financial terms. Government state land disposal procedures also vary from free transfer to local interests to sale to the highest bidder. This variation directly affects the land use outcomes, and especially the extent to which the local community gains or does not do so. This paper explores what might be done to redress the lack of research into these complex transitions of defence sites, particularly those with significant built heritage, from public to private ownership in favour of community benefit to meet local needs.

The challenge of sustainable reuse

In response to geo-political, and economic change, defence cuts and developments in military technology, more and more military land and facilities are becoming surplus to countries' defence. Since 1979 the UK state has, for example, sold some 2 million hectares of defence, health and transport land to private owners – or about 10% of the entire British land mass.[1] In planning civilian futures for the significant subset: defence land, there are considerable difficulties. This task is perhaps particularly challenging in these politically turbulent times.

"Defence activities frequently have an impact upon the form and function of cities that long outlives the original military necessity that created them."[2] Examination of different countries' practice in disposing of redundant defence property in relation to other government objectives and local needs is needed. Evaluation of how useful exchanges of experience of post-defence reconstruction can be between different formerly defence-dominated communities also needs to be assessed, perhaps by developing transferable measures, although intangible values may be more difficult to measure. This would do much to inform and improve this on-going process towards sustainable regeneration.

Questions that arise are: why do military sites close? Where are they – and what's on the site? If there is contamination, who is responsible for clean up? Who has power to influence what happens next? How is the transition from naval, airforce or military to civilian uses managed? What influence do local authority planners have in determining landuse? How long a timescale is allowed: a 'quick fix' or a gradual process over a number of years? How do different successor owners – state, agencies, local authorities, and commercial developers - develop new activities? Is how they are financed a factor in what happens? What new land uses result from the process? Who gains and who loses? Can the systems and outcomes for local communities be improved?

If the reuse is to be sustainable, how is this defined? If local benefit rather than gain to the public purse is a priority, positive factors that result in sustainable reuse might include genuine community consultation with built in feedback and commitment by local authority planners and developers, long timescales, public investment in new infrastructure, vision and creativity by the new owners, and risk-taking/entrepreneurship by developers. A working definition of sustainable reuse might be the extent to which the local community gains from new jobs at the same or higher skill levels than those lost, cleaned land, reused buildings including those with heritage protection, accessible open space, new housing to meet local needs, new educational and cultural facilities. This paper explores what might be done to redress the current process in many countries including the UK in favour of local benefit

As brownfields, some of these sites may have a malign legacy of serious contamination to tackle, which may be unrecorded. They are also secret sites unknown to the surrounding community except to those who worked there, making local planning authorities' task more difficult. Best practice in reuse occurs, but in top-down systems there is a lack of local input or recompense for lost jobs.

A key area that needs extensive further investigation is how is state owned property disposed of in different countries? This varies process considerably from country to country – across a spectrum from free transfer to local interests: the US – to sale to the highest bidder – the UK and Germany. These very different processes: top-down procedures by central government or bottom up reconstruction, have a direct bearing on the new land uses that develop once the armed forces depart.

The United States, as a huge federation of states with millions of acres of defence land, has a locally friendly state property disposal process. Where there is sufficient local capacity, the whole redevelopment process is controlled by an appointed local Base Reuse Committee, whose remit is economic reconstruction according to local needs. This may be facilitated by finance from the Department of Defense's Office of Economic Adjustment, which as far as is known has no equivalent in other countries. Surplus service homes may be gifted free to the homeless and hospitals and schools and open space may be transferred without charge to local communities.

In contrast, in the UK where government power is centralised, redundant defence sites are usually first offered to other government departments and agencies; if not needed by them they are then sold to the highest bidder at maximum planning value within three years – with a slightly longer timescale and lower expected price for historic sites.[3]

There is a public expectation that the new uses will be of public benefit to formerly defence-dependent communities – but whether this happens, how it is defined and how it is measured is still unclear. A 2009 report by the British Ministry of Defence identified a lack of community benefit from disposal of defence land. Christophers [p.230] says that there have been no in-depth examinations of what public land disposal in Britain has meant specifically for those living in the vicinity of disposal sites, except for Julian Dobson's study 'In the Public Interest? Community Benefits from Ministry of Defence Land Disposals' funded by a Portsmouth charity. Dobson encountered 'a lack of overarching academic research and little to suggest the issue has been high on the national policy agenda' and 'minimal interest in the issue from central government'. He identified a perennial tension between short-term budgetary exigencies of the public bodies selling land – and the long-term needs of the local community. Community benefits tend not to correlate with sale price or 'value for money'.

Choices are made between the desire to maximise capital receipts from public land disposal and using the land for social benefit, and it was not surprising that benefit to local communities from a more considered but less lucrative approach is ignored by the Ministry of Defence in favour of maximum financial return to the defence budget [Christophers p. 278]. It is also important to establish whether the financial receipts go back to the country's Defence/War ministry; or to the Finance Ministry/Treasury, or to the army, navy or airforce.

Another variable is that some countries sell surplus sites at military use value rather than maximum planning value, allowing local governments to buy them, consult their communities about what's needed, and then gain from added value by the grant of planning permission and sale to new owners. In Celia Clark's earlier research in 2000 she found that in Sweden local authorities could acquire sites at current use value, consult widely so that the new uses matched local needs and aspirations, and sell, retaining the profit from increased value. The French had an independent agency to handle the sale of public land, and also gave it away to local groups.[4] Perhaps we had something to learn from these examples. These findings need to be updated, but funding for this research needs first to be identified.

Since many defence sites contain significant historic structures, a further question is who defines military heritage: Is it the Culture Ministry or the Ministry of Defence? In the UK Historic England recommends which defence sites should be legally protected to the Minister of Culture, though now the owner – in these cases, the Minister of Defence can comment before the decision whether to List properties for their Historic or Architectural interest is taken. Historic defence sites are particularly vulnerable to neglect and decay when they are owned by ministries or agencies who have no remit or funds to keep them in good repair. The big problem in the UK is that where the Ministry of Defence (MOD) has no use for a historic building - as they point out - they are not funded to keep it in good repair, ultimately leading to demolition, 'controlled ruination' or escalation in eventual reuse costs. Until 2006 the MOD was exempt from civil planning law, and was free to alter or demolish historic buildings without obtaining permission from the local authority. A further obstacle is that the MOD still cannot be prosecuted if the building is in an active base. It would be useful to establish whether this is a problem in other countries. There are lessons conservationists and historians focused on military landscapes could share - even between countries with widely different political systems.

Other differences are whether existing structures, including those with statutory protection are reused or demolished. While government agencies such as the governing body of Suomenlinna[6] and government appointed successor owners such as the Portsmouth Naval Base Property Trust[7] may be dedicated to sustainable reuse of the historic structures remaining on the site, if commercial developers paid a high price, they may clear the site to obtain high financial return from land uses such as leisure/retail and high end housing, as happened at Gunwharf in Portsmouth harbour.[8] These uses may not be what local communities need.

At present, local communities, planners, redevelopment agencies and developers find it hard to draw on similar communities' post-defence experience of economic and social reconstruction. More work is needed to redress these gaps in accessible research. Internationally transferable prototypes about how to achieve the successful regeneration of defence land, particularly historic sites, might be useful in this under-reported process – where there are so many stakeholders – even within very different political systems.

2. Directions for further research

The European Commission supported over 120 Europe-wide cultural heritage research projects between 1986 and 2007. In 2009 it published Preserving Our Heritage, Improving our Environment which included studies of restoration of historic buildings, monuments and works of art, often involving advanced technologies and non-destructive techniques. In 2008 it held a conference on Cultural Heritage in Ljubljana. But mechanisms for wider dissemination of experience and good practice about regenerating defence site are rare, except for the US Association of Defense Communities, a national lobby in Washington DC to which active and closing bases and local authorities belong. Members share information and experience of post-defence planning via a weekly newsletter and annual conferences as defence facilities are closed and redeveloped - at which point they leave as they move on to civilian futures.[9] Would countrywide associations modelled on the US example and/or international groups that identify successful examples of sustainable regeneration, be useful?

Cross-cultural research on the regeneration of former defence sites at yet hardly exists. Samer Bagaeen and Celia Clark's book *Sustainable Regeneration of Former Military Sites* published by Routledge launched on Governors Island New York in 2016 was the first to examine experience of post-defence reconstruction in very different countries: China, Taiwan, the Netherlands, the UK and the US, from the planning and social justice points of view. It analyses the transition from military to civilian life for these complex, contaminated, isolated, heritage laden and often contested sites in locations ranging from urban to remote and shows that the process is far from easy. The twelve case studies from different countries, many written by the participants, were chosen to answer some of the research questions identified above, reflecting different aspects and experience of the process. The book's purpose is to enable the diverse stakeholders in these projects to discover opportunities for reuse and learn from others' experiences of successful regeneration. The vexed issue of decontamination was not explored, nor were mechanisms of dissemination. To analyse such a widespread process in more detail will clearly require much more research and wider vehicles for dissemination.

Good practice guides exist, but are not perhaps widely enough known. For once defence-dependent communities, base conversion is a profoundly important and symbolic land use exchange, but research into reconstruction – and in particular its implications for the historic defence estate - is rare in the UK, though not in other countries such as the United States. Although UK base conversion may be intermittently highly visible in land use change and development policy debates, especially in view of defence sites' potential to meet national housing targets, administrative and academic interest is not perhaps sustained because of the long timescales involved. This lack of research in the UK and other countries into the conversion of airfields, depots, barracks, dockyards, training grounds and fortifications into civilian uses, especially the identification of successful examples, leaves significant gaps in useful knowledge about the effect of base closures and the prospects for civilian reconstruction. There is very little systematic analysis of what follows base closures or of the search for sustainable futures for property formerly dedicated to national defence - leaving communities, governments, developers and planners with untested land use configurations, partnership structures, and financing strategies. This paper proposes a new international network of scholars and practitioners to address themselves to researching and answering the questions outlined above. This would ideally include the existing groups mentioned below, with others already working in this field. This can only be achieved by investment from over-arching bodies with an international reach.

3. Existing international defence heritage groups; their roles and funding

Long established international organisations about military heritage exist, with different areas of interest. History is the focus of the International Fortress Council and Study Group. ICOFORT, the International Scientific Committee on Fortifications and Military Heritage established by ICOMOS in 2005 has regional sub-committees. It drafted a Charter on Fortifications and Related Heritage and guidelines for their Protection, Conservation and Interpretation adopted in 2019.[10] The Bonn International Centre for Conversion's research remit includes armament and arms control and supporting disarmament, the dynamics of violent conflict and understanding violence in social orders.[11]

Conversion might be defined as the maximum reuse of military resources, for example the replacement of military manufacturing with civilian jobs and rehabilitating ex-military personnel.[12] In contrast to these wider topics, this paper concentrates on the reuse of the physical sites abandoned by the military.

As well as these international organisations, there are also several European networks of military places, some of which have overlapping aims. The EU contributes to exchanges of experience, but mostly only on a time-limited site-by-site basis. Since the 1990s it has funded links between particular ex-defence-dominated places to share experience via INTEREG III, IVC, URBACT and other programmes that funded short-term projects linking places with similar challenges to find sustainable new uses for ex-defence sites.[13] ATFORT Atelier European Fortresses was set up to link significant European fortified heritage sites who share the belief that their preservation, coupled with their economic exploitation, can only be achieved by creative, mutual thinking, building on an exchange of experiences. Its website says that "Heritage is a visible expression of our common European culture and history, a tangible testimony of our roots without which our present would be impoverished and our future would become sterile. As such, it is an essential element of our local, regional, national and European identity…The historic environment of Europe is one of its greatest assets and an effective means of reconciling the needs of citizens, economy, community, environment, and society at large." Their overall objective is "to facilitate the adaptive re-use of fortified heritage sites by exploring solutions for successful approaches and methods to deal with enabling conditions and to create better frameworks for their exploitation."[14] They drafted a European Convention of Fortified Heritage in Helsinki in 2014 and identified examples of good practice. They established the International Research Center for the Valorisation of Military Architecture and Defence System in Forte Marghera in Venice.[15]

Another network, which appears to overlap with ATFORT, is EFFORT (European Federation of Fortified Sites) established in 2017 to share knowledge and practical expertise on military heritage such as walled towns, forts and defence lines. Its 2019 Declaration said "When they were built, military architecture was meant to be functional, modern and as durable as possible, using all technical and architectural know-how available. Today [they] often represent important challenges: spatial, socio-economic and energy-environmental…[and offer] great opportunities for a large range of local and regional stakeholders; … space for experimental approach in renovation, energy transition and spatial development… They demonstrate the intrinsic need to defend borders in the past, today they are symbols of our common history and challenges, uniting Europe instead of dividing it." 16 EFFORT's members signed a declaration in Venice in 2018, the European Year of Cultural Heritage. It claims to be the European professional representative of fortified sites, a member of the Europa Nostra managed European Heritage Alliance; it presents its views on coherent action to policymakers and stakeholders.

The limitation of such groups, apart from their overlapping aims – which are so far unexplained - is of course their geographical confinement to Europe, linked to EU funding, which excludes other countries outside the EU, including, sadly, in 2020, the UK. The scope of collaborative lobbying to influence national and EU policies is illustrated by KONVER, one of the earliest funded projects that focused on issues of defence conversion and the peace dividend. "Although the international events of 1989 triggered deep cuts in defence budgets and widespread job losses in the defence industry and military bases, the immediate response tended to be fragmented and muted. Significantly, the defence sector covered areas not traditionally supported by national or European regional policies… Consequently, the first responses emanated from within the EU with the introduction of the PERIFRA annual programme in 1991 (Arms Conversion Project 1995). This year was to prove a watershed, but the diversity and political sensitivities surrounding defence issues meant subsequent inter-local authority and pan-European collaboration occurred piecemeal, often addressing different agenda (Arms Conversion Project 1995; Association of County Councils 1992; European Defence Observatory 1994; Network Demilitarised 1994). This growth in local-based activity and the successful lobbying for KONVER II (with the UK securing 19% of the total allocation), this initiative has been implemented as a national programme led by the Department of Trade and Industry in Whitehall, with the regional offices, local authorities, Training and Enterprise Councils and other agencies perceived as 'bodies to be consulted.'[17]

But these initiatives took a long time to negotiate and the EU funding was seen as windfalls rather than funding for long-term reconstruction, even though it was the main source of public support for regeneration in the locality. Delays in implementation, the uncertainty of eligibility and the annual basis of early funding contributed to intervention often coming too late. Rather than being an integral part of any response to closure, rationalisation and job loss, these measures were only addressed to the need to ameliorate the worst effects of closures. Network Demilitarised and KONVER did open up the UK's southeast and eastern regions to EU funding, but this process could be described as money chasing problems rather than the resourcing of clear strategies for regional renewal and development.[18] In addition, as Connor Ryan said in his chapter 'Democracy, military bases, and marshmallows' in Celia Clark and Samer Bagaeen's book, elected politicians fight projected closures of bases. This response also inhibits post-defence planning.[19]

The European Commission reported in 1993 that the Community Initiatives were at their most effective when measures funded respond to locally generated ideas and translate into additional and tangible action on the ground via direct involvement of local and regional interests matching a top-down response from the Commission to a bottom-up approach to delivery. However the balance of partnerships – and hence, control of funding, was mostly held by the UK's central government.[20] This is an issue that is central to Celia Clark's PhD study of three former defence sites: which also explored different parties' ability to influence the land uses and public expectations of benefit to local communities and economies affected by the closure of military bases. [8]

The international Naval Dockyards Society inaugurated in 1996 aims to stimulate the production and exchange of information and research into naval dockyards and associated organisations. It runs annual conferences at the National Maritime Museum, publishes Transactions and a newsletter: Dockyards, runs tours and carries out campaigns to preserve dockyard sites.[21] One way forward to enhance exchanges of experience on their sustainable reuse is to deposit research about defence disposals and sustainable redevelopment to make it accessible to practitioners and scholars. The Portsmouth History Centre in the Central Library has a database of relevant material, which is being developed with Portsmouth School of Architecture Conservation studies. It is available for students and other researchers to learn from experience and good practice.

As a founder member of the Naval Dockyards Society, Celia Clark has a particular interest in architecture associated with the navy. She has examined EU projects that link naval bases, via visits to each other's sites and joint workshops. These include the EU RENAVAL project in 1994; the PACTE programme in 1995 which linked Den Helder, Portsmouth, Turku and Rostok; RENDOC in 1998 between Chatham, Karlskrona, Suomenlinna and Rochefort; and NAVARCH, a joint project between Karlskrona, Helsinki, Rochefort and Chatham to gain understanding of the common European architectural heritage associated with navies. This produced a four-language guidebook to typical naval structures: drydocks, storehouses, roperies, workshops, foundries and boathouses. In other linkages the focus was on military rather than naval architecture. ARCHWAY was led by the University of the West of England in Bristol. ASCEND (Achieving the Socio-Economic Re-use of Former Military and Heritage) between 2004 and 2006 produced a Model Management Framework for the transfer of military sites and heritage from the defence sector into civilian ownership and presented EU policy recommendations, analysing the contributions of Objectives 1 and 2 Operational Programmes. The

Sustainable Historic Arsenals Regeneration Partnership (SHARP) was formed to share lessons learned while seeking new futures for these culturally important but neglected former military sites. English Heritage (now Historic England) as a partner with Woolwich Arsenal published Regeneration through Heritage. Understanding the Development Potential of Historic European Assets in 2007. REPAIR in 2011 lined European sites while the ASIAURBS project's aim was to develop heritage tourism in China. From 2016-2018 eight European formerly defence-dominated places were linked in the MAPS project: Military Areas as Public Spaces. The MAPS handbook / 'A journey on the re-use of the former military assets' details the inclusive and dynamic process of transformation achieved, is available at http://mapsnetwork.wordpress.com. The effectiveness of these different model transformation processes or of the transferability of the EU's investment in these projects does not appear to have been recently evaluated. It would be productive to compare and analyse these different model processes.

These examples of exchange of experience are of course confined to Europe, but another mechanism that is intermittently available to discuss defence-civilian transformations are international conferences focused on brownfields reconstruction, urban regeneration and sustainability, planning, architecture, conservation and defence heritage where futures for ex-defence sites are intermittently visible. The Wessex Institute of Technology (WIT) in the UK has run several conferences on these topics. In 2002 they ran a new series: Defence Sites: Heritage and Future which took place in Portsmouth in 2012, Venice Arsenale in 2014 and Alicante in 2016. Speakers included many independent professionals and researchers rather than sponsored delegates from universities or governments. Although the papers were published by the WIT Press this may limit the extent to which their research is publicised and is likely to influence public policy.

The largest and most wide-ranging international conference on futures for defence sites was the four-day Military Landscapes A Future for Military Heritage held in 2017 in Sardinia to mark the 150tth anniversary of the decommissioning of Italian fortresses. Its focus was detailed study of conservation, monitoring, maintenance and restoration and strategies for economic, social and cultural enhancement. Speakers came from as far away as the Hebrides in Scotland, India and South Korea. It was curated by Giovanna Damiani and Donatella Rita Fiorino, also editors of the book.[22] Gratifyingly, Italian army and navy property experts charged with looking after and the use, reuse and redevelopment of military installations attended the conference. This link between current defence authorities and academic and practitioner researchers is particularly important if it is to result in changes in public policy practice. At this conference Celia Clark issued a questionnaire to delegates about the transfer and redevelopment process – asking what happens in the many different countries present, but without funding to proceed, little progress has so far been achieved.

4. Conclusion

There is still much research to be done, to distil and disseminate lessons from the wide-ranging experience of regeneration of military landscapes in different parts of the world, in order to assist the many participants in this as yet largely undocumented process. Knowledge-transfer mechanisms are needed to collate and transmit individual sites' experience. The US Association of Defence Communities offers a membership model exchange for property managers on active and closing bases. This paper is intended as a contribution to the debate on how to achieve locally beneficial sustainable regeneration of these often complex and challenging sites. Accumulated local experience in reusing defence sites has much to offer, but in order for it to be disseminated, there need to be greater opportunities to analyse and learn about it – not only locally but also drawing on national and international experience. Funding would have to come from governments, who have an interest in achieving objectives such as economic regeneration, from universities, from built environment professional institutes and from local communities.

References

1 Christophers, B. *The New Enclosure The Appropriation of Public Land in Neoliberal Britain* Verso London (2018) p.2

2 Ashworth, G. *War and the City* Routledge London (2012)

3 https://historicengland.org.uk/images-books/publications/disposal-heritage-assets/guidance-disposals-final-jun-10/, last accessed 2020/1/21

5 Clark, C. Vintage Ports or Deserted Dockyards: differing futures for naval heritage across Europe Research Paper University of the West of England Bristol (2000)

6 https://www.suomenlinna.fi/en/ accessed 2020/1/21

7 https://www.pnbpropertytrust.org accessed 2020/1/21

8 Clark, C. White Holes: Decision-making in Disposal of Ministry of Defence Heritage Sites PhD Thesis University of Portsmouth (2002)

9 Emily Gee *The News* Portsmouth July 11 2019

9 https://www.defensecommunities.org, last accessed 2020/1/21

10 https://www.icofort.org/copia-executive-committee, last accessed 2019/12/30

11 https://www.bicc.de/research-clusters/order-and-change/, last accessed 2017/5/21

12 Feldman, J.M. 'Industrial Conversion: A Linchpin for Disarmament and Development Chapter 10' in Geeraerts, G., Pauwels N. and Remacle, E. eds.: Dimensions of Peace and Security Peter Lang, Brussels (2006)

13 Camerin, F. Programmi e progetti europei di rigenerazione urbana e riuso delle aree militari in Italia. TRIA. , 17(1), 141-156. DOI: https://doi.org/10.6092/2281-4574/5314 (2017)

14 http://www.atfort.eu, last accessed 2019/12/31

15 http://www.atfort.eu, last accessed 2020/1/3

16 http://www.efforts-europe.eu/wpcontent/uploads/2018/09/DraftDeclarationVenice.pdf.

17 Sainsbury T., MP Hansard 21 October 1993 p. 304; *The Coherence of EU Regional Policy: Contrasting Perspectives on the its Structural Funds* Bachtier J., Turok I., Routledge London (2013) pp.115-117

18 Bachtier J., Turok, I.: *The Coherence of EU Regional Policy: Contrasting Perspectives on the its Structural Funds* I., Routledge London) pp.115-117 (2013)

19 Ryan C. 'Democracy, military bases, and marshmallows' in Bagaeen S. and Clark C.: Eds. Sustainable Regeneration of Former Military Sites Routledge London (2016) [p. 36]

20 Bachtier and Turok p.118

21 Davies, J. D., 'The Naval Dockyards Society: the first fifteen years' in *Defence Sites: Heritage and Future*. Clark C. and Brebbia C. A. eds. Transactions on the Built Environment. Wessex Institute of Technology Volume 123 pp. 3-14 www.witpress.com ISSN 1743-3509 Southampton (2012)

22 Damiani, G. and Rita Fiorino, D. Scenari per il Futuro del Patromonio Militare and Istituto Italiano dei Castelli/ *Military Landscapes A Future for Military Heritage* Università degli Studi di Cagliari, SKIRA editions Milano pp.286-269 (2017)

Index

A

Admiralty 9–337, 11, 16, 30, 32, 41, 51, 52, 56, 75, 121, 124, 150, 151, 222, 223, 224, 227, 270, 271, 285, 287, 298, 309, 315, 316, 324, 326, 348, 349, 350, 353, 354, 355, 356, 357, 358, 362, 366, 372, 374, 377, 384, 390, 402, 407, 411, 414

Admiralty Surface Weapons Establishment 309, 356, 357, 358

Americas Cup 251, 404, 405

B

Bay House 51, 349, 350

Beck Deniz 83, 183, 184, 185, 191, 327, 328, 330, 337, 385, 410, 447

Berkeley Homes 52, 63, 115, 206, 237, 238, 243, 244, 246, 254, 377, 378, 382, 383, 431

Board of Ordnance 38, 40, 42, 218, 220, 256, 257, 321, 374, 408

Bomb disposal 275

C

Charles II 30, 33, 34, 35

Civic Voice 63, 430

Civil War 30, 32, 34, 257, 285, 335, 422

Clarence Pier 41, 43, 403, 408, 424

Clawback 63, 64

Cocroft Wayne iv, 58, 72, 75, 83, 426, 438, 447, 452

Conservation Area 101, 102, 130, 225, 248, 273, 279, 280, 282, 308, 388, 396, 402, 417, 422, 423, 425

Crichel Down rules 55, 58, 66, 420

Crown Exemption 16, 16, 61, 73, 74, 74, 207, 225, 225

D

Daedalus 51, 56, 72, 90, 101, 268, 270, 273, 275, 276, 277, 278, 279, 280, 281, 282, 283, 298, 349

D-Day 12, 19, 20, 54, 55, 76, 159, 173, 174, 175, 176, 177, 178, 224, 253, 254, 271, 293, 295, 299, 302, 303, 309, 317, 371, 376, 404, 405, 406, 407, 408, 410, 419

D-Day Museum 173, 174, 175, 405, 406

Defence Estates/Defence Infrastructure Organisation 74

De Gomme Bernard 30, 33, 34, 35, 36, 447

E

Eastney iv, 12, 17, 24, 38, 91, 92, 100, 127, 128, 145, 154, 307, 323, 324, 325, 326, 328, 334, 338, 348, 352, 356, 358, 364, 366, 367, 369, 370, 371, 372, 383, 384, 387, 394, 396, 399, 410

East Solent Coastal Partnership 11, 63, 101, 399

Elite Homes 261, 262, 264, 267, 380, 429

Ellis Owen Thomas 11, 192, 402, 447, 452

Emery-Wallis Freddie 87, 88, 89, 103, 108, 145, 150, 153, 163, 168, 171, 190, 235, 313, 318, 447, 448

English Heritage/Historic England 62

EU 113, 115, 131, 136, 143, 392, 419, 426, 427, 443, 444, 445, 446

F

Fort Blockhouse 10, 10, 16, 16, 18, 18, 20, 20, 36, 36, 37, 45, 45, 51, 51, 75, 75, 91, 91, 101, 101, 102, 102, 192, 196, 198, 205, 284, 285, 286, 301, 301, 429

Fort Cumberland iv, 20, 21, 22, 23, 25, 38, 45, 80, 81, 91, 92, 101, 117, 188, 254, 284, 300, 306, 307, 313, 325, 326, 327, 334, 337, 352, 361, 366, 369, 383, 385, 386, 387, 394, 396, 400

Fort Fareham 52, 80, 81, 310, 311, 336

Fort Gomer 22, 52, 302, 303, 304

Fort Kilkicker 401

Fraser Battery Eastney 12

G

Gilkicker 21, 22, 34, 52, 86, 92, 188, 285, 294, 297, 298, 299, 300, 301, 303, 385, 401
Goodship Peter iv, 107, 108, 109, 110, 112, 116, 117, 129, 130, 131, 132, 133, 140, 158, 263, 264, 267, 448, 450
Gosport Borough Council iv, 15, 16, 20, 24, 43, 50, 51, 52, 56, 63, 85, 90, 91, 92, 101, 102, 116, 233, 234, 255, 260, 262, 263, 273, 275, 276, 279, 280, 281, 285, 295, 296, 298, 303, 308, 349, 350, 362, 377, 378, 379, 382, 388, 395, 398
Gosport Lines 35, 37, 39, 85, 101, 102, 294, 294, 294, 310, 376, 381, 398
Gosport Society iv, v, 28, 62, 83, 86, 92, 101, 103, 216, 239, 241, 242, 277, 283, 308, 350, 375, 380
Grange Airfield 268, 350
Great Morass 399, 401, 408
Greentree Hedley 150, 180, 232, 233, 237, 248, 250, 266, 448
Gunboat Sheds 248
Gunwharf ii, vii, 18, 21, 22, 24, 29, 38, 62, 64, 73, 74, 76, 87, 91, 92, 93, 100, 104, 115, 116, 180, 181, 205, 206, 207, 218, 219, 220, 221, 222, 223, 224, 225, 226, 227, 229, 230, 232, 233, 234, 235, 236, 237, 238, 239, 240, 241, 242, 243, 244, 245, 246, 247, 248, 249, 251, 252, 253, 254, 255, 264, 265, 266, 267, 309, 328, 331, 338, 365, 371, 374, 377, 379, 398, 429, 431, 432, 436, 441

H

Hampshire Buildings Preservation Trust v, 16, 26, 50, 53, 68, 81, 82, 89, 90, 96, 103, 111, 119, 124, 126, 133, 191, 234, 241, 242, 244, 246, 266, 393
Hampshire County Council iv, 16, 17, 23, 36, 51, 52, 56, 65, 87, 89, 91, 92, 101, 103, 108, 109, 114, 118, 134, 157, 233, 235, 257, 258, 299, 305, 313, 315, 339, 347, 349, 398
Harper Rob 91, 205, 285, 308, 448, 451
Haslar Barracks 15, 52, 90, 101, 102, 285, 308, 366, 388, 396
Haslar Gunboat Yard 52, 78, 81, 82, 101, 179, 285, 286, 288, 289, 290, 291, 292, 308, 353, 362
Haslar Hospital iv, 16, 20, 52, 63, 64, 71, 82, 91, 177, 209, 285, 379, 388
Henry VIII 30, 35, 85, 154, 175, 190, 285, 286, 294, 401, 420
Heritage Action Zone 10, 17, 36, 63, 101, 102, 277, 304
Hilsea Lines 17, 22, 29, 37, 38, 39, 40, 42, 44, 80, 81, 101, 163, 188, 309, 319, 324, 326, 365, 394, 398, 400, 413, 414, 416, 417, 418, 422, 423, 425
Historic England iv, v, 10, 16, 25, 57, 58, 62, 72, 73, 74, 77, 78, 80, 81, 82, 84, 85, 91, 100, 101, 102, 104, 123, 124, 133, 185, 188, 189, 190, 254, 263, 265, 283, 288, 301, 308, 311, 316, 327, 328, 334, 335, 336, 352, 361, 363, 370, 386, 387, 395, 396, 398, 418, 422, 426, 441, 445
HMS Daedalus 56, 270, 273, 283, 349
HMS Dolphin iv, 23, 24, 75, 179, 285, 286, 287
HMS Dryad 55, 319, 421
HMS Hornet 192, 205, 285, 288, 290, 292, 429
HMS Nelson 15, 56, 92, 98, 100, 110, 412
HMS Phoenix 43, 415
HMS Queen Elizabeth 24, 286
HMS Sultan 22, 52, 56, 67, 100, 101, 158, 158, 179, 179, 179, 180, 180, 248, 248, 269, 269, 282, 282, 303, 303, 304, 304, 304, 350, 350, 428
HMS Vernon v, 18, 62, 70, 87, 92, 100, 121, 152, 180, 222, 223, 224, 225, 227, 228, 232, 241, 244, 248, 252, 254, 265, 298, 340, 352, 392
HMS Vernon/Gunwharf 62, 254, 265
HMS Victory 16, 23, 89, 96, 107, 109, 113, 117, 118, 124, 125, 131, 134, 145, 146, 150, 151, 154, 155, 162, 404, 410
Horsea Island 12, 13, 17, 29, 43, 44, 45, 122, 222, 352, 389, 390, 400, 418, 419
Horsea Island Country Park 44, 400, 418, 419
Hovercraft v, 16, 163, 270, 272, 273, 274, 275, 277, 278, 279, 281, 282, 283, 302, 403
Hovercraft Museum v, 163, 270, 273, 274, 275, 277, 278, 279, 281, 282
Hull 35, 399, 423

I

Institution of Civil Engineers 399
IRA 23, 111
Isle of Wight 9, 17, 23, 24, 28, 29, 40, 42, 43, 45, 46, 54, 89, 90, 129, 132, 161, 162, 165, 168, 181, 186, 190, 251, 253, 267, 298, 309, 328, 329, 332, 359, 376, 379, 386, 387, 388, 394, 403, 405, 407, 409

K

King James's Gate 41, 97, 98, 182, 344

L

Langstone Harbour 12, 13, 38, 43, 284, 323, 325, 326, 383, 410
Little Morass 401
Long Curtain Moat 25, 63, 100, 400
Lumps Fort Southsea 321

M

Mary Rose 16, 18, 19, 23, 30, 85, 92, 98, 107, 108, 109, 110, 111, 113, 114, 116, 118, 120, 124, 126, 127, 129, 130, 131, 134, 136, 144, 145, 146, 149, 154, 156, 161, 163, 165, 174, 296, 401
Mary Rose Museum 19, 118, 146, 161
Millennium Tower 249, 250, 253
Ministry of Defence 16, 18, 19, 25, 29, 39, 47, 52, 53, 54, 55, 58, 59, 60, 62, 65, 66, 67, 68, 70, 71, 74, 76, 77, 80, 91, 107, 108, 113, 116, 125, 126, 129, 141, 145, 163, 167, 177, 233, 237, 245, 248, 251, 265, 267, 273, 284, 310, 315, 323, 329, 355, 356, 362, 377, 379, 395, 397, 400, 420, 440, 441, 446
MOD Defence Estates 83, 84
Mulberry Harbour 13

N

National Museum of the Royal Navy 16, 18, 25, 26, 105, 106, 113, 114, 128, 134, 144, 154, 157, 158, 160, 178, 179, 258, 263, 288, 325, 368, 369, 371
National Regional Property Group 188, 385
Naval Dockyards Society iv, v, 16, 26, 28, 72, 81, 82, 83, 123, 125, 130, 131, 132, 133, 161, 216, 335, 352, 387, 392, 393, 395, 397, 444, 446
Niven David 402, 424, 447, 450
No. 1 Bastion Gosport 22, 22

O

Old Portsmouth v, 11, 17, 22, 24, 31, 41, 44, 45, 46, 85, 87, 88, 91, 98, 100, 181, 182, 183, 190, 233, 246, 266, 313, 327, 328, 330, 336, 338, 365, 371, 385, 398, 399, 400, 415

P

Palmerston Forts Society v, 22, 24, 26, 45, 46, 91, 104, 165, 187, 190, 310, 313, 324, 335, 336, 337, 371
Planning White Paper 430
Point Barracks 38, 44, 87, 181, 338, 365
Point Battery 33, 34, 44, 101
Portsdown Hill 13, 17, 29, 39, 41, 43, 54, 78, 174, 186, 260, 284, 307, 309, 310, 311, 316, 320, 321, 323, 324, 335, 340, 352, 356, 357, 359, 360, 381, 393, 398, 399, 400, 414
Portsea 13, 14, 15, 17, 37, 38, 39, 40, 41, 42, 45, 56, 85, 93, 97, 98, 100, 103, 116, 120, 121, 132, 168, 181, 187, 190, 193, 219, 220, 240, 243, 245, 246, 247, 253, 254, 284, 309, 323, 325, 344, 345, 346, 347, 350, 385, 389, 398, 399, 408, 411, 413, 414, 417, 418, 422
Portsea Island 13, 17, 38, 39, 40, 41, 243, 284, 309, 323, 325, 389, 399, 411, 413, 414, 417, 418
Portsmouth Cathedral 399
Portsmouth City Council iv, v, 15, 17, 31, 40, 42, 45, 46, 50, 63, 64, 70, 82, 88, 89, 91, 92, 103, 107, 108, 114, 115, 129, 130, 132, 133, 143, 147, 164, 181, 182, 183, 189, 190, 225, 229, 230, 232, 233, 236, 237, 238, 239, 241, 242, 247, 248, 250, 251, 266, 310, 320, 321, 335, 360, 385, 389, 390, 393, 396, 405, 410, 416, 417, 418, 419, 422
Portsmouth Corporation 42, 52, 100, 145, 318, 399, 402, 407, 411
Portsmouth Naval Base Property Trust iv, 16, 18, 24, 28, 74, 77, 96, 101, 107, 108, 115, 116, 124, 131, 132, 138, 139, 146, 188, 218, 232, 262, 263, 267, 333, 336, 429, 430
Portsmouth Society 16, 26, 62, 82, 85, 86, 96, 97, 103, 104, 111, 120, 121, 124, 129, 136, 137, 142, 143, 148, 150, 156, 161, 162, 177, 180, 181, 188, 194, 230, 231, 234, 236, 239, 240, 241, 242, 243, 246, 247, 248, 250, 253, 265, 266, 330, 335, 345, 361, 370, 373, 393, 394, 403, 407, 408, 410, 424
Portsmouth University 127, 148, 149, 237
Port Solent 13, 232, 389, 393, 419
Priddy's Hard vii, 16, 17, 18, 20, 22, 24, 32, 37, 52, 63, 72, 83, 90, 91, 92, 101, 109, 114, 117, 119, 120, 127, 130, 157, 159, 218, 219, 232, 234, 241, 255, 256, 257, 258, 259, 260, 261, 262, 263, 265, 267, 290, 389, 398, 429, 430, 434
Prince Charles 16, 96, 154, 156, 159, 224, 241, 244

Pritchard Giles v, 102, 179, 304, 305, 308, 380, 448, 451

Q

QinetiQ 16, 16, 51, 51, 52, 52, 179, 179, 207, 285, 285, 286, 286, 286, 290, 291, 293, 298, 298, 298, 328, 352, 352, 353, 353, 354, 354, 355, 356, 358, 358, 359, 359, 359, 360, 362, 362, 362, 362, 384, 384, 385, 385, 385, 431
Quail Sarah v, 172–337, 191–337, 336–337, 423–446, 424–446
Queen Victoria's Railway Shelter 100

R

Ring Fortress 38
Rising sea levels 326, 399
Rock Gardens iv, 326, 399, 400, 406, 424
Round Tower 17, 31, 33, 34, 38, 41, 44, 92, 182, 184, 185, 284, 285, 401
Royal Clarence Yard v, v, 20, 20, 20, 36, 36, 42, 42, 42, 43, 43, 52, 52, 101, 101, 101, 115, 115, 215, 215, 234, 234, 256, 256, 262, 262, 264, 264, 267, 267, 331, 331, 353, 353, 366, 373, 375, 375, 375, 376, 376, 376, 377, 377, 378, 378, 378, 378, 379, 380, 395
Royal Institute of British Architects 399

S

Salvetti Geoffrey v, 28, 91, 104, 310, 313, 324, 325, 336, 337, 369, 370, 371, 394, 448, 451
Save Britain's Heritage 71
South Parade Pier 400, 406, 407, 417
Southsea Castle 20, 25, 30, 31, 34, 40, 42, 63, 73, 88, 163, 164, 165, 169, 170, 175, 177, 182, 190, 285, 321, 327, 399, 400, 401, 403, 405, 406
Southsea Coastal Scheme 400
Southsea Common 19, 19, 22, 22, 25, 25, 30, 30, 37, 37, 38, 38, 39, 39, 44, 44, 63, 63, 159, 159, 167, 167, 177, 177, 190, 190, 293, 293, 398, 398, 399, 400, 401, 402, 404, 405, 422, 429
Southsea Rose Garden 321, 322, 408, 409
Southwick House 54, 55, 174, 193, 194, 317, 400, 421, 429
Southwick Park 55, 207, 420, 421, 425, 428, 429
Spithead 9, 17, 29, 32, 38, 39, 43, 128, 129, 134, 154, 174, 181, 228, 253, 268, 284, 301, 303, 307, 323, 328, 359–363, 388, 401, 404, 406, 409
Square Tower 32, 92, 182, 218, 256, 257, 373, 374
St. George Barracks 36, 38, 90, 340, 366, 379, 381, 382, 383
Stokes Bay Lines 52, 293, 294, 295, 298, 301, 308, 398, 428, 431
St. Vincent Barracks 339–363
Submarine Museum 92, 106, 114, 159, 179, 191, 234, 288
Sunley Estates 52, 382, 431

T

Tate Ron 21, 53, 65, 70, 180, 451
Tipner 10, 17, 29, 38, 43, 44, 56, 63, 64, 100, 259, 316, 366, 388, 389, 390, 391, 392, 393, 396, 397, 399, 411, 418, 429
Trist Richard v, 107, 108, 132, 133, 162, 173, 174, 190, 191, 229, 266, 451, 452

U

Underwood Michael 249, 252, 253, 265, 452

V

Veterans' Village 63, 192, 205, 209, 210, 429
Victoria Park 42, 44, 100, 189, 246, 398, 400, 408, 411, 412, 422, 425
Victorious Festival 178, 178, 404, 404, 405, 405, 406, 406

W

Weeks Stephen v, 16, 85, 86, 103, 190, 451, 452
World Heritage Site 128
Wren Terry 142, 147, 150, 253, 452